Chevrolet Monte Carlo Automotive Repair Manual

by Curt Choate and John H Haynes

Member of the Guild of Motoring Writers

Models covered:

Chevrolet Monte Carlo with 267, 305, 350, 400, 402 & 454 cu in V8 engines and 200, 229, 231 and 262 cu in V6 engines, including turbocharged models 1970 through 1988

Does not include diesel engine information

Haynes UK   *(24055-5AA11)*
Sparkford Nr Yeovil
Somerset BA22 7JJ England

Haynes North America, Inc
859 Lawrence Drive
Newbury Park, CA 91320 USA

www.haynes.com

About this manual

Its purpose

The purpose of this manual is to provide comprehensive, useful and accessible automotive repair information, to help you get the best value from your vehicle. It can do so in several ways. It can help you decide what work must be done, even if you choose to have it done by a dealer service department or a repair shop; it provides information and procedures for routine maintenance and servicing; and it offers diagnostic and repair procedures to follow when trouble occurs.

We hope you use the manual to tackle the work yourself. For many simpler jobs, doing it yourself may be quicker than arranging an appointment to get the vehicle into a shop and making the trips to leave it and pick it up. More importantly, a lot of money can be saved by avoiding the expense the shop must pass on to you to cover its labor and overhead costs. An added benefit is the sense of satisfaction and accomplishment that you feel after doing the job yourself. However, this manual is not a substitute for a professional certified technician or mechanic. There are risks associated with automotive repairs. The ability to make repairs on a vehicle depends on individual skill, experience and proper tools. Individuals should act with due care and acknowledge and assume the risk of performing automotive repairs.

Using the manual

The manual is divided into Chapters. Each Chapter is divided into numbered Sections, which are headed in bold type between horizontal lines. Each Section consists of consecutively numbered paragraphs.

The reference numbers used in illustration captions pinpoint the pertinent Section and the Step within that Section. That is, illustration 3.2 means the illustration refers to Section 3 and Step (or paragraph) 2 within that Section.

Procedures, once described in the text, are not normally repeated. When it's necessary to refer to another Chapter, the reference will be given as Chapter and Section number. Cross references given without use of the word "Chapter" apply to Sections and/or paragraphs in the same Chapter. For example, "see Section 8" means in the same Chapter. References to the left or right side of the vehicle assume you are sitting in the driver's seat, facing forward.

This repair manual is produced by a third party and is not associated with an individual car manufacturer. If there is any doubt or discrepancy between this manual and the owner's manual or the factory service manual, please refer to factory service manual or seek assistance from a professional certified technician or mechanic. Even though we have prepared this manual with extreme care, neither the publisher nor the author can accept responsibility for any errors in, or omissions from, the information given.

NOTE

A **Note** provides information necessary to properly complete a procedure or information which will make the procedure easier to understand.

CAUTION

A **Caution** provides a special procedure or special steps which must be taken while completing the procedure where the Caution is found. Not heeding a Caution can result in damage to the assembly being worked on.

WARNING

A **Warning** provides a special procedure or special steps which must be taken while completing the procedure where the Warning is found. Not heeding a Warning can result in personal injury.

Acknowledgements

We are grateful for the help and cooperation of Tomco Industries, 1435 Woodson Road, St. Louis, Missouri 63132, for their assistance with technical information and certain illustrations. Wiring diagrams originated exclusively for Haynes North America, Inc. by Valley Forge Technical Information Services.

A book in the Haynes Automotive Repair Manual Series

ISBN-10: 1-85010-526-X
ISBN-13: 978-1-85010-526-8

Library of Congress Catalog Card Number 88-81714

Disclaimer

There are risks associated with automotive repairs. The ability to make repairs depends on individual skill, experience and proper tools. Individuals should act with due care and acknowledge and assume the risk of making automotive repairs. While every attempt is made to ensure that the information in this manual is correct, no liability can be accepted by the authors or publishers for loss, damage or injury caused by any errors in, or omissions from, the information given.

Contents

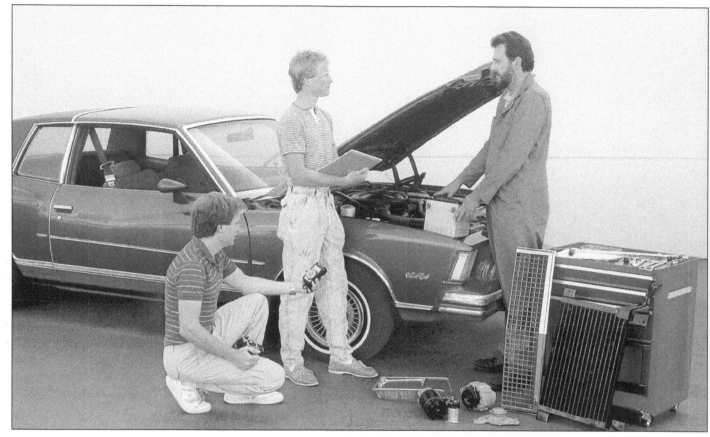

Haynes photographer, author and mechanic, with Chevrolet Monte Carlo

Introduction to the Chevrolet Monte Carlo

The Chevrolet Monte Carlo models covered in this manual are of conventional front engine/rear-wheel drive layout.

A variety of General Motors-built V6 and V8 engines were installed in these models over their long production run. The engine drives the rear wheels through either a manual or an automatic transmission via a drive-shaft and solid rear axle.

These vehicles are available in a two-door coupe body style only. Front suspension is independent, using upper and lower control arms, coil springs and shock absorbers with power assisted steering available on all models. The rear suspension consists of a solid rear axle with trailing arms, coil springs and shock absorbers.

Earlier models use drum brakes on all four wheels with front disc brakes being optional equipment, while later models feature standard disc-type brakes at the front and drums at the rear. Power assist was available on all models.

Vehicle identification numbers

The Vehicle Identification Number (VIN) is visible from outside the vehicle through the driver's side windshield

The body identification plate is usually found on the cowl or by the radiator support

Modifications are a continuing and unpublicized process in vehicle manufacturing. Since spare parts manuals and lists are compiled on a numerical basis, the individual vehicle numbers are essential to correctly identify the component required.

Vehicle Identification Number (VIN)

This very important identification number is stamped on a plate attached to the left side cowling just inside the windshield on the driver's side of the vehicle (see illustration). The VIN also appears on the Vehicle Certificate of Title and Registration. It contains information such as where and when the vehicle was manufactured, the model year and the body style.

Body identification plate

The body identification plate is located in the engine compartment on the upper surface of the radiator support/shroud or the cowl (see illustration).

Like the VIN it contains valuable information about the manufacture of the vehicle, as well as information on the options with which it is equipped. This plate is especially useful for matching the color and type of paint for repair work.

Engine identification number

Because of the wide variety of engines with which these models were equipped engine identification numbers can be found in a variety of locations (see illustrations).

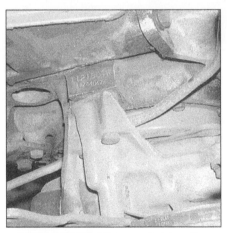

Engine identification number location
(all V8s and early V6s)

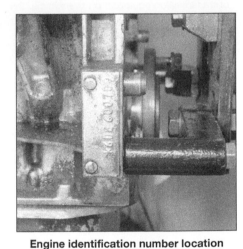

Engine identification number location
(later model V6s)

Engine identification

Up until 1977, most engines in GM automobiles were manufactured by the same division that marketed the body. However, since that time, GM automobiles come equipped with various engines, supplied by Oldsmobile, Chevrolet, Buick and Pontiac divisions of General Motors.

On 1972 and newer models, a check of the Vehicle Identification Number (VIN) will quickly determine from which GM plant your car's engine originated. The VIN number gives information concerning the engine, the year of the car, and so on. This metal tag is located on the dashboard in the left hand corner (driver's side), up against the windshield. On 1972 through 1980 models, the fifth digit of the VIN is the number that identifies the origin of the engine. On 1981 and newer models, the eighth digit of the VIN identifies the origin of the engine. The engine identification number on 1970 and 1971 models is not listed in the VIN code and is identified only by casting numbers located on the engine block. Refer to the accompanying chart for 1972 and later models to determine which engine you have.

There are some differences between the various makes of engines but overall, repair and maintenance procedures are nearly identical. Where differences occur, they will be

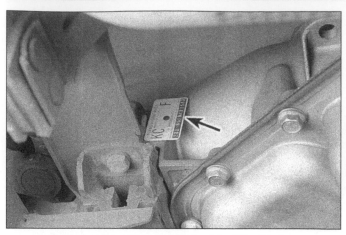

Typical automatic transmission identification tag

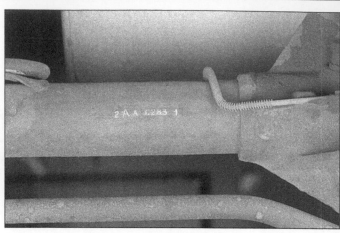

Typical rear axle identification number location

noted. A check of the Specifications Section at the beginning of each Chapter will alert the home mechanic to any differences in the various tolerances. Also check the Vehicle Emission Control Information label, located adjacent to the radiator, for additional engine information.

Transmission identification number

Depending on model year and the model of transmission equipped on the vehicle, the identification numbers or ID tag (see illustration) can be found in various locations:

Manual transmission identification number:

3-Speed Saginaw - lower right-hand side of the case, adjacent to the cover.

4-speed Borg Warner - rear vertical surface of the extension housing.

4-speed Muncie - rear right-hand side of the case flange.

4-speed Saginaw - lower right-hand side of case adjacent to the cover.

Automatic transmission identification number:

Powerglide - right rear surface of the oil pan.

Type 200 - lower right front forward surface of the transmission case.

Type 350 - right-hand vertical surface of the oil pan.

Type 400 - on blue tag right-hand side of the transmission.

YEAR	ENGINE	CUBIC TYPE	LITERS INCHES	VIN CODE	BUILT BY*
1972 through 1973					
	V8	350	5.7	H,J,K,L	C
	V8	400	6.6	R	C
	V8	402	6.6	S,U	C
	V8	454	7.4	V,W,	C
1974 through 1976					
	V8	305	5.7	Q	C
	V8	350	5.7	H,K,L,V	C
	V8	400	6.6	R,U	C
	V8	454	7.4	Y,Z,	C
1977 through 1982					
	V6	200	3.3	M	C
	V6	229	3.8	K	C
	V6	231	3.8	A,Z,3	B
	V8	267	4.4	J	C
	V8	305	5.7	H,U	C
	V8	350	5.7	L	C
1983 through 1988					
	V6	229	3.8	9	C
	V6	231	3.8	A	B
	V6	262	4.3	Z	C
	V8	305	5.7	H,G	C

B = Buick C = Chevrolet

Rear axle identification number

Information pertaining to the rear axle and differential can be found stamped on the front surface of the right side axle housing tube on most models (see illustration) or on a tag attached to one of the cover bolts. Some models also have final drive ratio and build date information stamped on the flange adjacent to the cover.

Buying parts

Replacement parts are available from many sources, which generally fall into one of two categories - authorized dealer parts departments and independent retail auto parts stores. Our advice concerning these parts is as follows:

Retail auto parts stores: Good auto parts stores will stock frequently needed components which wear out relatively fast, such as clutch components, exhaust systems, brake parts, tune-up parts, etc. These stores often supply new or reconditioned parts on an exchange basis, which can save a considerable amount of money. Discount auto parts stores are often very good places to buy materials and parts needed for general vehicle maintenance such as oil, grease, filters, spark plugs, belts, touch-up paint, bulbs, etc. They also usually sell tools and general accessories, have convenient hours, charge lower prices and can often be found not far from home.

Authorized dealer parts department: This is the best source for parts which are unique to the vehicle and not generally available elsewhere (such as major engine parts, transmission parts, trim pieces, etc.).

Warranty information: If the vehicle is still covered under warranty, be sure that any replacement parts purchased - regardless of the source - do not invalidate the warranty!

To be sure of obtaining the correct parts, have engine and chassis numbers available and, if possible, take the old parts along for positive identification.

Maintenance techniques, tools and working facilities

Maintenance techniques

There are a number of techniques involved in maintenance and repair that will be referred to throughout this manual. Application of these techniques will enable the home mechanic to be more efficient, better organized and capable of performing the various tasks properly, which will ensure that the repair job is thorough and complete.

Fasteners

Fasteners are nuts, bolts, studs and screws used to hold two or more parts together. There are a few things to keep in mind when working with fasteners. Almost all of them use a locking device of some type, either a lockwasher, locknut, locking tab or thread adhesive. All threaded fasteners should be clean and straight, with undamaged threads and undamaged corners on the hex head where the wrench fits. Develop the habit of replacing all damaged nuts and bolts with new ones. Special locknuts with nylon or fiber inserts can only be used once. If they are removed, they lose their locking ability and must be replaced with new ones.

Rusted nuts and bolts should be treated with a penetrating fluid to ease removal and prevent breakage. Some mechanics use turpentine in a spout-type oil can, which works quite well. After applying the rust penetrant, let it work for a few minutes before trying to loosen the nut or bolt. Badly rusted fasteners may have to be chiseled or sawed off or removed with a special nut breaker, available at tool stores.

If a bolt or stud breaks off in an assembly, it can be drilled and removed with a special tool commonly available for this purpose.

Most automotive machine shops can perform this task, as well as other repair procedures, such as the repair of threaded holes that have been stripped out.

Flat washers and lockwashers, when removed from an assembly, should always be replaced exactly as removed. Replace any damaged washers with new ones. Never use a lockwasher on any soft metal surface (such as aluminum), thin sheet metal or plastic.

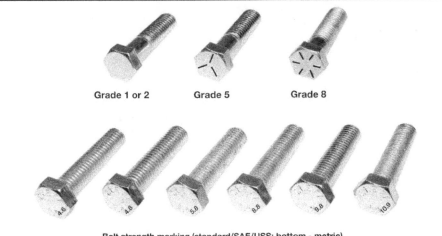

Grade 1 or 2   Grade 5   Grade 8

Bolt strength marking (standard/SAE/USS; bottom - metric)

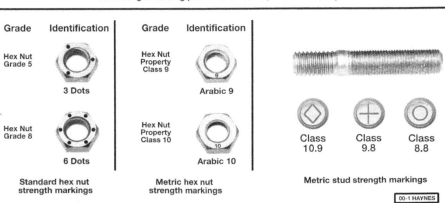

Grade	Identification
Hex Nut Grade 5	3 Dots
Hex Nut Grade 8	6 Dots

Standard hex nut strength markings

Grade	Identification
Hex Nut Property Class 9	Arabic 9
Hex Nut Property Class 10	Arabic 10

Metric hex nut strength markings

Class 10.9   Class 9.8   Class 8.8

Metric stud strength markings

00-1 HAYNES

Fastener sizes

For a number of reasons, automobile manufacturers are making wider and wider use of metric fasteners. Therefore, it is important to be able to tell the difference between standard (sometimes called U.S. or SAE) and metric hardware, since they cannot be interchanged.

All bolts, whether standard or metric, are sized according to diameter, thread pitch and length. For example, a standard 1/2 - 13 x 1 bolt is 1/2 inch in diameter, has 13 threads per inch and is 1 inch long. An M12 - 1.75 x 25 metric bolt is 12 mm in diameter, has a thread pitch of 1.75 mm (the distance between threads) and is 25 mm long. The two bolts are nearly identical, and easily confused, but they are not interchangeable.

In addition to the differences in diameter, thread pitch and length, metric and standard bolts can also be distinguished by examining the bolt heads. To begin with, the distance across the flats on a standard bolt head is measured in inches, while the same dimension on a metric bolt is sized in millimeters (the same is true for nuts). As a result, a standard wrench should not be used on a metric bolt and a metric wrench should not be used on a standard bolt. Also, most standard bolts have slashes radiating out from the center of the head to denote the grade or strength of the bolt, which is an indication of the amount of torque that can be applied to it. The greater the number of slashes, the greater the strength of the bolt. Grades 0 through 5 are commonly used on automobiles. Metric bolts have a property class (grade) number, rather than a slash, molded into their heads to indicate bolt strength. In this case, the higher the number, the stronger the bolt. Property class numbers 8.8, 9.8 and 10.9 are commonly used on automobiles.

Strength markings can also be used to distinguish standard hex nuts from metric hex nuts. Many standard nuts have dots stamped into one side, while metric nuts are marked with a number. The greater the number of dots, or the higher the number, the greater the strength of the nut.

Metric studs are also marked on their ends according to property class (grade). Larger studs are numbered (the same as metric bolts), while smaller studs carry a geometric code to denote grade.

It should be noted that many fasteners, especially Grades 0 through 2, have no distinguishing marks on them. When such is the case, the only way to determine whether it is standard or metric is to measure the thread pitch or compare it to a known fastener of the same size.

Standard fasteners are often referred to as SAE, as opposed to metric. However, it should be noted that SAE technically refers to a non-metric fine thread fastener only. Coarse thread non-metric fasteners are referred to as USS sizes.

Since fasteners of the same size (both standard and metric) may have different strength ratings, be sure to reinstall any bolts,

Metric thread sizes	Ft-lbs	Nm
M-6	6 to 9	9 to 12
M-8	14 to 21	19 to 28
M-10	28 to 40	38 to 54
M-12	50 to 71	68 to 96
M-14	80 to 140	109 to 154

Pipe thread sizes		
1/8	5 to 8	7 to 10
1/4	12 to 18	17 to 24
3/8	22 to 33	30 to 44
1/2	25 to 35	34 to 47

U.S. thread sizes		
1/4 - 20	6 to 9	9 to 12
5/16 - 18	12 to 18	17 to 24
5/16 - 24	14 to 20	19 to 27
3/8 - 16	22 to 32	30 to 43
3/8 - 24	27 to 38	37 to 51
7/16 - 14	40 to 55	55 to 74
7/16 - 20	40 to 60	55 to 81
1/2 - 13	55 to 80	75 to 108

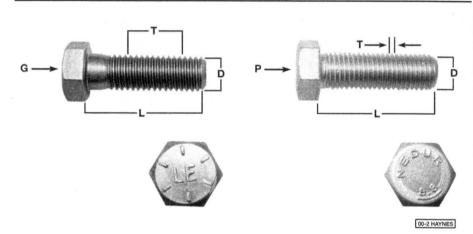

Standard (SAE and USS) bolt dimensions/grade marks

G   Grade marks (bolt strength)
L   Length (in inches)
T   Thread pitch (number of threads per inch)
D   Nominal diameter (in inches)

Metric bolt dimensions/grade marks

P   Property class (bolt strength)
L   Length (in millimeters)
T   Thread pitch (distance between threads in millimeters)
D   Diameter

studs or nuts removed from your vehicle in their original locations. Also, when replacing a fastener with a new one, make sure that the new one has a strength rating equal to or greater than the original.

Tightening sequences and procedures

Most threaded fasteners should be tightened to a specific torque value (torque is the twisting force applied to a threaded component such as a nut or bolt). Overtightening the fastener can weaken it and cause it to break, while undertightening can cause it to eventually come loose. Bolts, screws and studs, depending on the material they are made of and their thread diameters, have specific torque values, many of which are noted in the Specifications at the beginning of each Chapter. Be sure to follow the torque recommendations closely. For fasteners not assigned a specific torque, a general torque value chart is presented here as a guide. These torque values are for dry (unlubricated) fasteners threaded into steel or cast iron (not aluminum). As was previously mentioned, the size and grade of a fastener determine the amount of torque that can safely be applied to it. The figures listed here are approximate for Grade 2 and Grade 3 fasteners. Higher grades can tolerate higher torque values.

Fasteners laid out in a pattern, such as cylinder head bolts, oil pan bolts, differential cover bolts, etc., must be loosened or tightened in sequence to avoid warping the component. This sequence will normally be

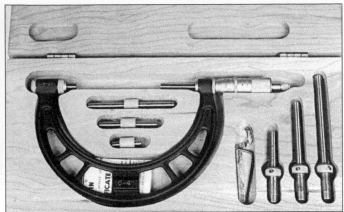

Micrometer set

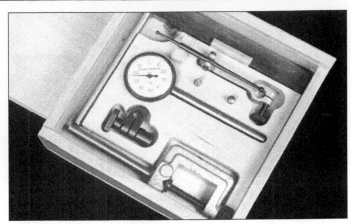

Dial indicator set

shown in the appropriate Chapter. If a specific pattern is not given, the following procedures can be used to prevent warping.

Initially, the bolts or nuts should be assembled finger-tight only. Next, they should be tightened one full turn each, in a criss-cross or diagonal pattern. After each one has been tightened one full turn, return to the first one and tighten them all one-half turn, following the same pattern. Finally, tighten each of them one-quarter turn at a time until each fastener has been tightened to the proper torque. To loosen and remove the fasteners, the procedure would be reversed.

Component disassembly

Component disassembly should be done with care and purpose to help ensure that the parts go back together properly. Always keep track of the sequence in which parts are removed. Make note of special characteristics or marks on parts that can be installed more than one way, such as a grooved thrust washer on a shaft. It is a good idea to lay the disassembled parts out on a clean surface in the order that they were removed. It may also be helpful to make sketches or take instant photos of components before removal.

When removing fasteners from a component, keep track of their locations. Sometimes threading a bolt back in a part, or putting the washers and nut back on a stud, can prevent mix-ups later. If nuts and bolts cannot be returned to their original locations, they should be kept in a compartmented box or a series of small boxes. A cupcake or muffin tin is ideal for this purpose, since each cavity can hold the bolts and nuts from a particular area (i.e. oil pan bolts, valve cover bolts, engine mount bolts, etc.). A pan of this type is especially helpful when working on assemblies with very small parts, such as the carburetor, alternator, valve train or interior dash and trim pieces. The cavities can be marked with paint or tape to identify the contents.

Whenever wiring looms, harnesses or connectors are separated, it is a good idea to identify the two halves with numbered pieces of masking tape so they can be easily reconnected.

Gasket sealing surfaces

Throughout any vehicle, gaskets are used to seal the mating surfaces between two parts and keep lubricants, fluids, vacuum or pressure contained in an assembly.

Many times these gaskets are coated with a liquid or paste-type gasket sealing compound before assembly. Age, heat and pressure can sometimes cause the two parts to stick together so tightly that they are very difficult to separate. Often, the assembly can be loosened by striking it with a soft-face hammer near the mating surfaces. A regular hammer can be used if a block of wood is placed between the hammer and the part. Do not hammer on cast parts or parts that could be easily damaged. With any particularly stubborn part, always recheck to make sure that every fastener has been removed.

Avoid using a screwdriver or bar to pry apart an assembly, as they can easily mar the gasket sealing surfaces of the parts, which must remain smooth. If prying is absolutely necessary, use an old broom handle, but keep in mind that extra clean up will be necessary if the wood splinters.

After the parts are separated, the old gasket must be carefully scraped off and the gasket surfaces cleaned. Stubborn gasket material can be soaked with rust penetrant or treated with a special chemical to soften it so it can be easily scraped off. A scraper can be fashioned from a piece of copper tubing by flattening and sharpening one end. Copper is recommended because it is usually softer than the surfaces to be scraped, which reduces the chance of gouging the part. Some gaskets can be removed with a wire brush, but regardless of the method used, the mating surfaces must be left clean and smooth. If for some reason the gasket surface is gouged, then a gasket sealer thick enough to fill scratches will have to be used during reassembly of the components. For most applications, a non-drying (or semi-drying) gasket sealer should be used.

Hose removal tips

Warning: *If the vehicle is equipped with air conditioning, do not disconnect any of the A/C hoses without first having the system depressurized by a dealer service department or a service station.*

Hose removal precautions closely parallel gasket removal precautions. Avoid scratching or gouging the surface that the hose mates against or the connection may leak. This is especially true for radiator hoses. Because of various chemical reactions, the rubber in hoses can bond itself to the metal spigot that the hose fits over. To remove a hose, first loosen the hose clamps that secure it to the spigot. Then, with slip-joint pliers, grab the hose at the clamp and rotate it around the spigot. Work it back and forth until it is completely free, then pull it off. Silicone or other lubricants will ease removal if they can be applied between the hose and the outside of the spigot. Apply the same lubricant to the inside of the hose and the outside of the spigot to simplify installation.

As a last resort (and if the hose is to be replaced with a new one anyway), the rubber can be slit with a knife and the hose peeled from the spigot. If this must be done, be careful that the metal connection is not damaged.

If a hose clamp is broken or damaged, do not reuse it. Wire-type clamps usually weaken with age, so it is a good idea to replace them with screw-type clamps whenever a hose is removed.

Tools

A selection of good tools is a basic requirement for anyone who plans to maintain and repair his or her own vehicle. For the owner who has few tools, the initial investment might seem high, but when compared to the spiraling costs of professional auto maintenance and repair, it is a wise one.

To help the owner decide which tools are needed to perform the tasks detailed in this manual, the following tool lists are offered: *Maintenance and minor repair, Repair/overhaul* and *Special.*

The newcomer to practical mechanics

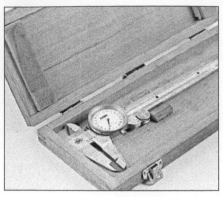

Dial caliper

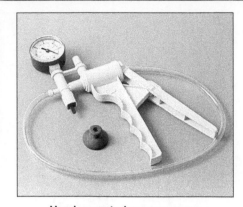

Hand-operated vacuum pump

Timing light

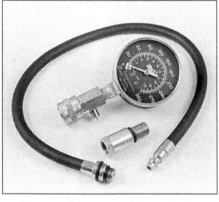

Compression gauge with spark plug
hole adapter

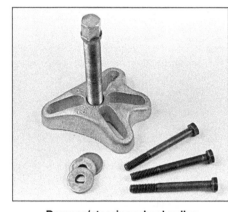

Damper/steering wheel puller

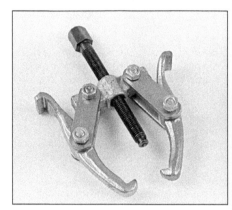

General purpose puller

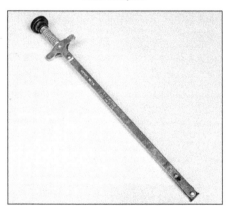

Hydraulic lifter removal tool

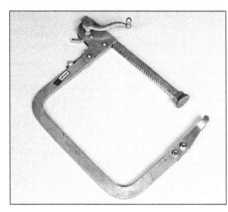

Valve spring compressor

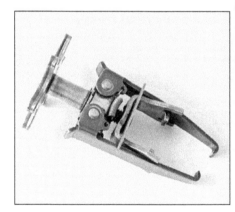

Valve spring compressor

Ridge reamer

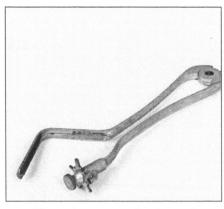

Piston ring groove cleaning tool

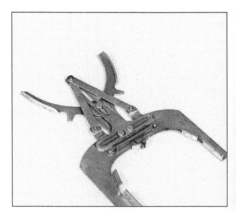

Ring removal/installation tool

Ring compressor

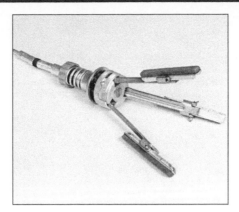

Cylinder hone

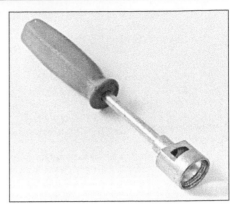

Brake hold-down spring tool

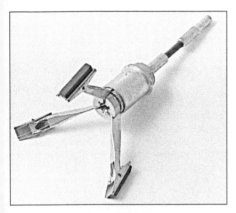

Brake cylinder hone

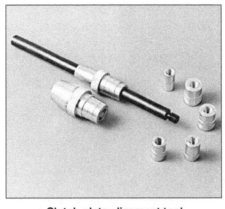

Clutch plate alignment tool

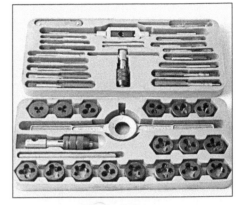

Tap and die set

should start off with the *maintenance and minor repair* tool kit, which is adequate for the simpler jobs performed on a vehicle. Then, as confidence and experience grow, the owner can tackle more difficult tasks, buying additional tools as they are needed. Eventually the basic kit will be expanded into the *repair and overhaul* tool set. Over a period of time, the experienced do-it-yourselfer will assemble a tool set complete enough for most repair and overhaul procedures and will add tools from the special category when it is felt that the expense is justified by the frequency of use.

Maintenance and minor repair tool kit

The tools in this list should be considered the minimum required for performance of routine maintenance, servicing and minor repair work. We recommend the purchase of combination wrenches (box-end and open-end combined in one wrench). While more expensive than open end wrenches, they offer the advantages of both types of wrench.

*Combination wrench set (1/4-inch to
    1 inch or 6 mm to 19 mm)*
Adjustable wrench, 8 inch
Spark plug wrench with rubber insert
Spark plug gap adjusting tool
Feeler gauge set
Brake bleeder wrench

*Standard screwdriver (5/16-inch x
    6 inch)*
Phillips screwdriver (No. 2 x 6 inch)
Combination pliers - 6 inch
Hacksaw and assortment of blades
Tire pressure gauge
Grease gun
Oil can
Fine emery cloth
Wire brush
Battery post and cable cleaning tool
Oil filter wrench
Funnel (medium size)
Safety goggles
Jackstands (2)
Drain pan

Note: *If basic tune-ups are going to be part of routine maintenance, it will be necessary to purchase a good quality stroboscopic timing light and combination tachometer/dwell meter. Although they are included in the list of special tools, it is mentioned here because they are absolutely necessary for tuning most vehicles properly.*

Repair and overhaul tool set

These tools are essential for anyone who plans to perform major repairs and are in addition to those in the maintenance and minor repair tool kit. Included is a comprehensive set of sockets which, though expensive, are invaluable because of their versatil-

ity, especially when various extensions and drives are available. We recommend the 1/2-inch drive over the 3/8-inch drive. Although the larger drive is bulky and more expensive, it has the capacity of accepting a very wide range of large sockets. Ideally, however, the mechanic should have a 3/8-inch drive set and a 1/2-inch drive set.

Socket set(s)
Reversible ratchet
Extension - 10 inch
Universal joint
*Torque wrench (same size drive as
    sockets)*
Ball peen hammer - 8 ounce
Soft-face hammer (plastic/rubber)
Standard screwdriver (1/4-inch x 6 inch)
*Standard screwdriver (stubby -
    5/16-inch)*
Phillips screwdriver (No. 3 x 8 inch)
Phillips screwdriver (stubby - No. 2)
Pliers - vise grip
Pliers - lineman's
Pliers - needle nose
Pliers - snap-ring (internal and external)
Cold chisel - 1/2-inch
Scribe
*Scraper (made from flattened copper
    tubing)*
Centerpunch
Pin punches (1/16, 1/8, 3/16-inch)
Steel rule/straightedge - 12 inch

*Allen wrench set (1/8 to 3/8-inch or
    4 mm to 10 mm)
A selection of files
Wire brush (large)
Jackstands (second set)
Jack (scissor or hydraulic type)*

Note: *Another tool which is often useful is an
electric drill with a chuck capacity of 3/8-inch
and a set of good quality drill bits.*

Special tools

The tools in this list include those which
are not used regularly, are expensive to buy,
or which need to be used in accordance with
their manufacturer's instructions. Unless
these tools will be used frequently, it is not
very economical to purchase many of them.
A consideration would be to split the cost
and use between yourself and a friend or
friends. In addition, most of these tools can
be obtained from a tool rental shop on a temporary basis.

This list primarily contains only those
tools and instruments widely available to the
public, and not those special tools produced
by the vehicle manufacturer for distribution to
dealer service departments. Occasionally,
references to the manufacturer's special
tools are included in the text of this manual.
Generally, an alternative method of doing the
job without the special tool is offered. However, sometimes there is no alternative to
their use. Where this is the case, and the tool
cannot be purchased or borrowed, the work
should be turned over to the dealer service
department or an automotive repair shop.

*Valve spring compressor
Piston ring groove cleaning tool
Piston ring compressor
Piston ring installation tool
Cylinder compression gauge
Cylinder ridge reamer
Cylinder surfacing hone
Cylinder bore gauge
Micrometers and/or dial calipers
Hydraulic lifter removal tool
Balljoint separator
Universal-type puller
Impact screwdriver
Dial indicator set
Stroboscopic timing light (inductive
    pick-up)
Hand operated vacuum/pressure pump
Tachometer/dwell meter
Universal electrical multimeter
Cable hoist
Brake spring removal and installation
    tools
Floor jack*

Buying tools

For the do-it-yourselfer who is just starting to get involved in vehicle maintenance
and repair, there are a number of options
available when purchasing tools. If maintenance and minor repair is the extent of the
work to be done, the purchase of individual
tools is satisfactory. If, on the other hand,
extensive work is planned, it would be a good
idea to purchase a modest tool set from one

of the large retail chain stores. A set can usually be bought at a substantial savings over
the individual tool prices, and they often
come with a tool box. As additional tools are
needed, add-on sets, individual tools and a
larger tool box can be purchased to expand
the tool selection. Building a tool set gradually allows the cost of the tools to be spread
over a longer period of time and gives the
mechanic the freedom to choose only those
tools that will actually be used.

Tool stores will often be the only source
of some of the special tools that are needed,
but regardless of where tools are bought, try
to avoid cheap ones, especially when buying
screwdrivers and sockets, because they
won't last very long. The expense involved in
replacing cheap tools will eventually be
greater than the initial cost of quality tools.

Care and maintenance of tools

Good tools are expensive, so it makes
sense to treat them with respect. Keep them
clean and in usable condition and store them
properly when not in use. Always wipe off any
dirt, grease or metal chips before putting
them away. Never leave tools lying around in
the work area. Upon completion of a job,
always check closely under the hood for
tools that may have been left there so they
won't get lost during a test drive.

Some tools, such as screwdrivers, pliers, wrenches and sockets, can be hung on a
panel mounted on the garage or workshop
wall, while others should be kept in a tool box
or tray. Measuring instruments, gauges,
meters, etc. must be carefully stored where
they cannot be damaged by weather or
impact from other tools.

When tools are used with care and
stored properly, they will last a very long
time. Even with the best of care, though,
tools will wear out if used frequently. When a
tool is damaged or worn out, replace it. Sub-
sequent jobs will be safer and more enjoy-
able if you do.

How to repair damaged threads

Sometimes, the internal threads of a nut
or bolt hole can become stripped, usually
from overtightening. Stripping threads is an
all-too-common occurrence, especially when
working with aluminum parts, because aluminum is so soft that it easily strips out.

Usually, external or internal threads are
only partially stripped. After they've been
cleaned up with a tap or die, they'll still work.
Sometimes, however, threads are badly damaged. When this happens, you've got three
choices:

1) *Drill and tap the hole to the next suitable
   oversize and install a larger diameter
   bolt, screw or stud.*
2) *Drill and tap the hole to accept a
   threaded plug, then drill and tap the plug
   to the original screw size. You can also
   buy a plug already threaded to the original size. Then you simply drill a hole to*

*the specified size, then run the threaded
plug into the hole with a bolt and jam
nut. Once the plug is fully seated,
remove the jam nut and bolt.*
3) *The third method uses a patented
   thread repair kit like Heli-Coil or Slimsert.
   These easy-to-use kits are designed to
   repair damaged threads in straight-
   through holes and blind holes. Both are
   available as kits which can handle a variety of sizes and thread patterns. Drill the
   hole, then tap it with the special
   included tap. Install the Heli-Coil and the
   hole is back to its original diameter and
   thread pitch.*

Regardless of which method you use,
be sure to proceed calmly and carefully. A little impatience or carelessness during one of
these relatively simple procedures can ruin
your whole day's work and cost you a bundle
if you wreck an expensive part.

Working facilities

Not to be overlooked when discussing
tools is the workshop. If anything more than
routine maintenance is to be carried out,
some sort of suitable work area is essential.

It is understood, and appreciated, that
many home mechanics do not have a good
workshop or garage available, and end up
removing an engine or doing major repairs
outside. It is recommended, however, that
the overhaul or repair be completed under
the cover of a roof.

A clean, flat workbench or table of com-
fortable working height is an absolute necessity. The workbench should be equipped with
a vise that has a jaw opening of at least four
inches.

As mentioned previously, some clean,
dry storage space is also required for tools,
as well as the lubricants, fluids, cleaning solvents, etc. which soon become necessary.

Sometimes waste oil and fluids, drained
from the engine or cooling system during
normal maintenance or repairs, present a disposal problem. To avoid pouring them on the
ground or into a sewage system, pour the
used fluids into large containers, seal them
with caps and take them to an authorized
disposal site or recycling center. Plastic jugs,
such as old antifreeze containers, are ideal
for this purpose.

Always keep a supply of old newspapers and clean rags available. Old towels are
excellent for mopping up spills. Many
mechanics use rolls of paper towels for most
work because they are readily available and
disposable. To help keep the area under the
vehicle clean, a large cardboard box can be
cut open and flattened to protect the garage
or shop floor.

Whenever working over a painted surface, such as when leaning over a fender to
service something under the hood, always
cover it with an old blanket or bedspread to
protect the finish. Vinyl covered pads, made
especially for this purpose, are available at
auto parts stores.

Jacking and towing

Jacking

The jack supplied with the vehicle should only be used for raising the vehicle when changing a tire or placing jackstands

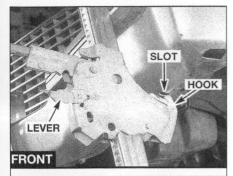

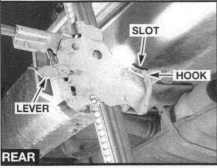

On earlier models, a bumper jack is used to raise the vehicle

under the frame. **Warning:** *Never work under the vehicle or start the engine while this jack is being used as the only means of support.*

The vehicle should be on level ground with the wheels blocked and the transmission in Park (automatic) or Reverse (manual). If the wheel is being replaced, loosen the wheel nuts one-half turn and leave them in place until the wheel is raised off the ground. Refer to Chapter 1 for information related to removing and installing the tire.

Place the jack in the slot in the bumper (early models) or under the vehicle jack locations (later models) in the indicated position **(see illustrations)**. Operate the jack with a slow, smooth motion until the wheel is raised off the ground.

Lower the vehicle, remove the jack and tighten the nuts (if loosened or removed) in a criss-cross sequence.

Towing

Vehicles can be towed with all four wheels on the ground, provided that speeds do not exceed 35 mph and the distance is not over 50 miles, otherwise transmission damage can result.

Towing equipment specifically designed for this purpose should be used and should be attached to the main structural members of the vehicle, not the bumper or brackets.

Safety is a major consideration when towing, and all applicable state and local laws must be obeyed. A safety chain system must be used for all towing.

While towing, the parking brake should be released and the transmission must be in Neutral. The steering must be unlocked (ignition switch in the Off position). Remember that power steering and power brakes will not work with the engine off.

On later models the jack is placed along the frame rails

Booster battery (jump) starting

Observe these precautions when using a booster battery to start a vehicle:

a) *Before connecting the booster battery, make sure the ignition switch is in the Off position.*
b) *Turn off the lights, heater and other electrical loads.*
c) *Your eyes should be shielded. Safety goggles are a good idea.*
d) *Make sure the booster battery is the same voltage as the dead one in the vehicle.*
e) *The two vehicles MUST NOT TOUCH each other!*
f) *Make sure the transaxle is in Neutral (manual) or Park (automatic).*
g) *If the booster battery is not a maintenance-free type, remove the vent caps and lay a cloth over the vent holes.*

Connect the red jumper cable to the positive (+) terminals of each battery **(see illustration)**.

Connect one end of the black jumper cable to the negative (-) terminal of the booster battery. The other end of this cable should be connected to a good ground on the vehicle to be started, such as a bolt or bracket on the body.

Start the engine using the booster battery, then, with the engine running at idle speed, disconnect the jumper cables in the reverse order of connection.

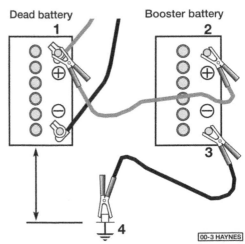

Make the booster battery cable connections in the numerical order shown (note that the negative cable of the booster battery is NOT attached to the negative terminal of the dead battery)

Automotive chemicals and lubricants

A number of automotive chemicals and lubricants are available for use during vehicle maintenance and repair. They include a wide variety of products ranging from cleaning solvents and degreasers to lubricants and protective sprays for rubber, plastic and vinyl.

Cleaners

Carburetor cleaner and choke cleaner is a strong solvent for gum, varnish and carbon. Most carburetor cleaners leave a dry-type lubricant film which will not harden or gum up. Because of this film it is not recommended for use on electrical components.

Brake system cleaner is used to remove grease and brake fluid from the brake system, where clean surfaces are absolutely necessary. It leaves no residue and often eliminates brake squeal caused by contaminants.

Electrical cleaner removes oxidation, corrosion and carbon deposits from electrical contacts, restoring full current flow. It can also be used to clean spark plugs, carburetor jets, voltage regulators and other parts where an oil-free surface is desired.

Demoisturants remove water and moisture from electrical components such as alternators, voltage regulators, electrical connectors and fuse blocks. They are non-conductive, non-corrosive and non-flammable.

Degreasers are heavy-duty solvents used to remove grease from the outside of the engine and from chassis components. They can be sprayed or brushed on and, depending on the type, are rinsed off either with water or solvent.

Lubricants

Motor oil is the lubricant formulated for use in engines. It normally contains a wide variety of additives to prevent corrosion and reduce foaming and wear. Motor oil comes in various weights (viscosity ratings) from 5 to 80. The recommended weight of the oil depends on the season, temperature and the demands on the engine. Light oil is used in cold climates and under light load conditions. Heavy oil is used in hot climates and where high loads are encountered. Multi-viscosity oils are designed to have characteristics of both light and heavy oils and are available in a number of weights from 5W-20 to 20W-50.

Gear oil is designed to be used in differentials, manual transmissions and other areas where high-temperature lubrication is required.

Chassis and wheel bearing grease is a heavy grease used where increased loads and friction are encountered, such as for wheel bearings, balljoints, tie-rod ends and universal joints.

High-temperature wheel bearing grease is designed to withstand the extreme temperatures encountered by wheel bearings in disc brake equipped vehicles. It usually contains molybdenum disulfide (moly), which is a dry-type lubricant.

White grease is a heavy grease for metal-to-metal applications where water is a problem. White grease stays soft under both low and high temperatures (usually from -100 to +190-degrees F), and will not wash off or dilute in the presence of water.

Assembly lube is a special extreme pressure lubricant, usually containing moly, used to lubricate high-load parts (such as main and rod bearings and cam lobes) for initial start-up of a new engine. The assembly lube lubricates the parts without being squeezed out or washed away until the engine oiling system begins to function.

Silicone lubricants are used to protect rubber, plastic, vinyl and nylon parts.

Graphite lubricants are used where oils cannot be used due to contamination problems, such as in locks. The dry graphite will lubricate metal parts while remaining uncontaminated by dirt, water, oil or acids. It is electrically conductive and will not foul electrical contacts in locks such as the ignition switch.

Moly penetrants loosen and lubricate frozen, rusted and corroded fasteners and prevent future rusting or freezing.

Heat-sink grease is a special electrically non-conductive grease that is used for mounting electronic ignition modules where it is essential that heat is transferred away from the module.

Sealants

RTV sealant is one of the most widely used gasket compounds. Made from silicone, RTV is air curing, it seals, bonds, waterproofs, fills surface irregularities, remains flexible, doesn't shrink, is relatively easy to remove, and is used as a supplementary sealer with almost all low and medium temperature gaskets.

Anaerobic sealant is much like RTV in that it can be used either to seal gaskets or to form gaskets by itself. It remains flexible, is solvent resistant and fills surface imperfections. The difference between an anaerobic sealant and an RTV-type sealant is in the curing. RTV cures when exposed to air, while an anaerobic sealant cures only in the absence of air. This means that an anaerobic sealant cures only after the assembly of parts, sealing them together.

Thread and pipe sealant is used for sealing hydraulic and pneumatic fittings and vacuum lines. It is usually made from a Teflon compound, and comes in a spray, a paint-on liquid and as a wrap-around tape.

Chemicals

Anti-seize compound prevents seizing, galling, cold welding, rust and corrosion in fasteners. High-temperature anti-seize, usually made with copper and graphite lubricants, is used for exhaust system and exhaust manifold bolts.

Anaerobic locking compounds are used to keep fasteners from vibrating or working loose and cure only after installation, in the absence of air. Medium strength locking compound is used for small nuts, bolts and screws that may be removed later. High-strength locking compound is for large nuts, bolts and studs which aren't removed on a regular basis.

Oil additives range from viscosity index improvers to chemical treatments that claim to reduce internal engine friction. It should be noted that most oil manufacturers caution against using additives with their oils.

Gas additives perform several functions, depending on their chemical makeup. They usually contain solvents that help dissolve gum and varnish that build up on carburetor, fuel injection and intake parts. They also serve to break down carbon deposits that form on the inside surfaces of the combustion chambers. Some additives contain upper cylinder lubricants for valves and piston rings, and others contain chemicals to remove condensation from the gas tank.

Miscellaneous

Brake fluid is specially formulated hydraulic fluid that can withstand the heat and pressure encountered in brake systems. Care must be taken so this fluid does not come in contact with painted surfaces or plastics. An opened container should always be resealed to prevent contamination by water or dirt.

Weatherstrip adhesive is used to bond weatherstripping around doors, windows and trunk lids. It is sometimes used to attach trim pieces.

Undercoating is a petroleum-based, tar-like substance that is designed to protect metal surfaces on the underside of the vehicle from corrosion. It also acts as a sound-deadening agent by insulating the bottom of the vehicle.

Waxes and polishes are used to help protect painted and plated surfaces from the weather. Different types of paint may require the use of different types of wax and polish. Some polishes utilize a chemical or abrasive cleaner to help remove the top layer of oxidized (dull) paint on older vehicles. In recent years many non-wax polishes that contain a wide variety of chemicals such as polymers and silicones have been introduced. These non-wax polishes are usually easier to apply and last longer than conventional waxes and polishes.

Conversion factors

Length (distance)
Inches (in)	X	25.4	= Millimetres (mm)	X 0.0394	= Inches (in)
Feet (ft)	X	0.305	= Metres (m)	X 3.281	= Feet (ft)
Miles	X	1.609	= Kilometres (km)	X 0.621	= Miles

Volume (capacity)
Cubic inches (cu in; in^3)	X	16.387	= Cubic centimetres (cc; cm^3)	X 0.061	= Cubic inches (cu in; in^3)
Imperial pints (Imp pt)	X	0.568	= Litres (l)	X 1.76	= Imperial pints (Imp pt)
Imperial quarts (Imp qt)	X	1.137	= Litres (l)	X 0.88	= Imperial quarts (Imp qt)
Imperial quarts (Imp qt)	X	1.201	= US quarts (US qt)	X 0.833	= Imperial quarts (Imp qt)
US quarts (US qt)	X	0.946	= Litres (l)	X 1.057	= US quarts (US qt)
Imperial gallons (Imp gal)	X	4.546	= Litres (l)	X 0.22	= Imperial gallons (Imp gal)
Imperial gallons (Imp gal)	X	1.201	= US gallons (US gal)	X 0.833	= Imperial gallons (Imp gal)
US gallons (US gal)	X	3.785	= Litres (l)	X 0.264	= US gallons (US gal)

Mass (weight)
Ounces (oz)	X	28.35	= Grams (g)	X 0.035	= Ounces (oz)
Pounds (lb)	X	0.454	= Kilograms (kg)	X 2.205	= Pounds (lb)

Force
Ounces-force (ozf; oz)	X	0.278	= Newtons (N)	X 3.6	= Ounces-force (ozf; oz)
Pounds-force (lbf; lb)	X	4.448	= Newtons (N)	X 0.225	= Pounds-force (lbf; lb)
Newtons (N)	X	0.1	= Kilograms-force (kgf; kg)	X 9.81	= Newtons (N)

Pressure
Pounds-force per square inch (psi; lbf/in^2; lb/in^2)	X	0.070	= Kilograms-force per square centimetre (kgf/cm^2; kg/cm^2)	X 14.223	= Pounds-force per square inch (psi; lbf/in^2; lb/in^2)
Pounds-force per square inch (psi; lbf/in^2; lb/in^2)	X	0.068	= Atmospheres (atm)	X 14.696	= Pounds-force per square inch (psi; lbf/in^2; lb/in^2)
Pounds-force per square inch (psi; lbf/in^2; lb/in^2)	X	0.069	= Bars	X 14.5	= Pounds-force per square inch (psi; lbf/in^2; lb/in^2)
Pounds-force per square inch (psi; lbf/in^2; lb/in^2)	X	6.895	= Kilopascals (kPa)	X 0.145	= Pounds-force per square inch (psi; lbf/in^2; lb/in^2)
Kilopascals (kPa)	X	0.01	= Kilograms-force per square centimetre (kgf/cm^2; kg/cm^2)	X 98.1	= Kilopascals (kPa)

Torque (moment of force)
Pounds-force inches (lbf in; lb in)	X	1.152	= Kilograms-force centimetre (kgf cm; kg cm)	X 0.868	= Pounds-force inches (lbf in; lb in)
Pounds-force inches (lbf in; lb in)	X	0.113	= Newton metres (Nm)	X 8.85	= Pounds-force inches (lbf in; lb in)
Pounds-force inches (lbf in; lb in)	X	0.083	= Pounds-force feet (lbf ft; lb ft)	X 12	= Pounds-force inches (lbf in; lb in)
Pounds-force feet (lbf ft; lb ft)	X	0.138	= Kilograms-force metres (kgf m; kg m)	X 7.233	= Pounds-force feet (lbf ft; lb ft)
Pounds-force feet (lbf ft; lb ft)	X	1.356	= Newton metres (Nm)	X 0.738	= Pounds-force feet (lbf ft; lb ft)
Newton metres (Nm)	X	0.102	= Kilograms-force metres (kgf m; kg m)	X 9.804	= Newton metres (Nm)

Vacuum
Inches mercury (in. Hg)	X	3.377	= Kilopascals (kPa)	X 0.2961	= Inches mercury
Inches mercury (in. Hg)	X	25.4	= Millimeters mercury (mm Hg)	X 0.0394	= Inches mercury

Power
Horsepower (hp)	X	745.7	= Watts (W)	X 0.0013	= Horsepower (hp)

Velocity (speed)
Miles per hour (miles/hr; mph)	X	1.609	= Kilometres per hour (km/hr; kph)	X 0.621	= Miles per hour (miles/hr; mph)

Fuel consumption*
Miles per gallon, Imperial (mpg)	X	0.354	= Kilometres per litre (km/l)	X 2.825	= Miles per gallon, Imperial (mpg)
Miles per gallon, US (mpg)	X	0.425	= Kilometres per litre (km/l)	X 2.352	= Miles per gallon, US (mpg)

Temperature
Degrees Fahrenheit = (°C x 1.8) + 32   Degrees Celsius (Degrees Centigrade; °C) = (°F - 32) x 0.56

*It is common practice to convert from miles per gallon (mpg) to litres/100 kilometres (l/100km),
where mpg (Imperial) x l/100 km = 282 and mpg (US) x l/100 km = 235

Safety first!

Regardless of how enthusiastic you may be about getting on with the job at hand, take the time to ensure that your safety is not jeopardized. A moment's lack of attention can result in an accident, as can failure to observe certain simple safety precautions. The possibility of an accident will always exist, and the following points should not be considered a comprehensive list of all dangers. Rather, they are intended to make you aware of the risks and to encourage a safety conscious approach to all work you carry out on your vehicle.

Essential DOs and DON'Ts

DON'T rely on a jack when working under the vehicle. Always use approved jackstands to support the weight of the vehicle and place them under the recommended lift or support points.

DON'T attempt to loosen extremely tight fasteners (i.e. wheel lug nuts) while the vehicle is on a jack - it may fall.

DON'T start the engine without first making sure that the transmission is in Neutral (or Park where applicable) and the parking brake is set.

DON'T remove the radiator cap from a hot cooling system - let it cool or cover it with a cloth and release the pressure gradually.

DON'T attempt to drain the engine oil until you are sure it has cooled to the point that it will not burn you.

DON'T touch any part of the engine or exhaust system until it has cooled sufficiently to avoid burns.

DON'T siphon toxic liquids such as gasoline, antifreeze and brake fluid by mouth, or allow them to remain on your skin.

DON'T inhale brake lining dust - it is potentially hazardous (see *Asbestos* below).

DON'T allow spilled oil or grease to remain on the floor - wipe it up before someone slips on it.

DON'T use loose fitting wrenches or other tools which may slip and cause injury.

DON'T push on wrenches when loosening or tightening nuts or bolts. Always try to pull the wrench toward you. If the situation calls for pushing the wrench away, push with an open hand to avoid scraped knuckles if the wrench should slip.

DON'T attempt to lift a heavy component alone - get someone to help you.

DON'T rush or take unsafe shortcuts to finish a job.

DON'T allow children or animals in or around the vehicle while you are working on it.

DO wear eye protection when using power tools such as a drill, sander, bench grinder, etc. and when working under a vehicle.

DO keep loose clothing and long hair well out of the way of moving parts.

DO make sure that any hoist used has a safe working load rating adequate for the job.

DO get someone to check on you periodically when working alone on a vehicle.

DO carry out work in a logical sequence and make sure that everything is correctly assembled and tightened.

DO keep chemicals and fluids tightly capped and out of the reach of children and pets.

DO remember that your vehicle's safety affects that of yourself and others. If in doubt on any point, get professional advice.

Asbestos

Certain friction, insulating, sealing, and other products - such as brake linings, brake bands, clutch linings, torque converters, gaskets, etc. - may contain asbestos. Extreme care must be taken to avoid inhalation of dust from such products, since it is hazardous to health. If in doubt, assume that they do contain asbestos.

Fire

Remember at all times that gasoline is highly flammable. Never smoke or have any kind of open flame around when working on a vehicle. But the risk does not end there. A spark caused by an electrical short circuit, by two metal surfaces contacting each other, or even by static electricity built up in your body under certain conditions, can ignite gasoline vapors, which in a confined space are highly explosive. Do not, under any circumstances, use gasoline for cleaning parts. Use an approved safety solvent.

Always disconnect the battery ground (-) cable at the battery before working on any part of the fuel system or electrical system. Never risk spilling fuel on a hot engine or exhaust component. It is strongly recommended that a fire extinguisher suitable for use on fuel and electrical fires be kept handy in the garage or workshop at all times. Never try to extinguish a fuel or electrical fire with water.

Fumes

Certain fumes are highly toxic and can quickly cause unconsciousness and even death if inhaled to any extent. Gasoline vapor falls into this category, as do the vapors from some cleaning solvents. Any draining or pouring of such volatile fluids should be done in a well ventilated area.

When using cleaning fluids and solvents, read the instructions on the container carefully. Never use materials from unmarked containers.

Never run the engine in an enclosed space, such as a garage. Exhaust fumes contain carbon monoxide, which is extremely poisonous. If you need to run the engine, always do so in the open air, or at least have the rear of the vehicle outside the work area.

If you are fortunate enough to have the use of an inspection pit, never drain or pour gasoline and never run the engine while the vehicle is over the pit. The fumes, being heavier than air, will concentrate in the pit with possibly lethal results.

The battery

Never create a spark or allow a bare light bulb near a battery. They normally give off a certain amount of hydrogen gas, which is highly explosive.

Always disconnect the battery ground (-) cable at the battery before working on the fuel or electrical systems.

If possible, loosen the filler caps or cover when charging the battery from an external source (this does not apply to sealed or maintenance-free batteries). Do not charge at an excessive rate or the battery may burst.

Take care when adding water to a non maintenance-free battery and when carrying a battery. The electrolyte, even when diluted, is very corrosive and should not be allowed to contact clothing or skin.

Always wear eye protection when cleaning the battery to prevent the caustic deposits from entering your eyes.

Household current

When using an electric power tool, inspection light, etc., which operates on household current, always make sure that the tool is correctly connected to its plug and that, where necessary, it is properly grounded. Do not use such items in damp conditions and, again, do not create a spark or apply excessive heat in the vicinity of fuel or fuel vapor.

Secondary ignition system voltage

A severe electric shock can result from touching certain parts of the ignition system (such as the spark plug wires) when the engine is running or being cranked, particularly if components are damp or the insulation is defective. In the case of an electronic ignition system, the secondary system voltage is much higher and could prove fatal.

Troubleshooting

Contents

Engine

1 Engine will not rotate when attempting to start

1   Battery terminal connections loose or corroded. Check the cable terminals at the battery: tighten or clean corrosion as necessary.

2   Battery discharged or faulty. If the cable connectors are clean and tight on the battery posts, turn the key to the 'On' position and switch on the headlights and/or windshield wipers. If these fail to function, the battery is discharged.

3   Automatic transmission not fully engaged in 'Park' or manual transmission clutch not fully depressed.

4   Broken, loose or disconnected wiring in the starting circuit. Inspect all wiring and connectors at the battery, starter solenoid (at lower right side of engine) and ignition switch (on steering column).

5   Starter motor pinion jammed on flywheel ring gear. If manual transmission, place gearshift in gear and rock the car to manually turn the engine. Remove starter (Chapter 5) and inspect pinion and flywheel (Chapter 2) at earliest convenience.

6   Starter solenoid faulty (Chapter 5).

7   Starter motor faulty (Chapter 5).

8   Ignition switch faulty (Chapter 12).

2   Engine rotates but will not start

1   Fuel tank empty.
2   Battery discharged (engine rotates slowly). Check the operation of electrical components as described in previous Section (see Chapter 1).
3   Battery terminal connections loose or corroded. See previous Section.
4   Carburetor flooded and/or fuel level in carburetor incorrect. This will usually be accompanied by a strong fuel odor from under the hood. Wait a few minutes, depress the accelerator pedal all the way to the floor and attempt to start the engine.
5   Choke control inoperative (Chapters 1 and 4).
6   Fuel not reaching carburetor. With ignition switch in 'Off' position, open hood, remove the top plate of air cleaner assembly and observe the top of the carburetor (manually move choke plate back if necessary). Have an assistant depress accelerator pedal fully and check that fuel spurts into carburetor. If not, check fuel filter (Chapters 1 and 4), fuel lines and fuel pump (Chapter 4).
7   Excessive moisture on, or damage to, ignition components (Chapter 1).
8   Worn, faulty or incorrectly adjusted spark plugs (Chapter 1).
9   Broken, loose or disconnected wiring in the starting circuit (see previous Section).
10 Distributor loose, thus changing ignition timing. Turn the distributor body as necessary to start the engine, then set ignition timing as soon as possible (Chapter 1).
11 Ignition condenser faulty (Chapter 5).
12 Broken, loose or disconnected wires at the ignition coil, or faulty coil (Chapter 5).

3   Starter motor operates without rotating engine

1   Starter pinion sticking. Remove the starter (Chapter 5) and inspect.
2   Starter pinion or engine flywheel teeth worn or broken. Remove the inspection cover at the rear of the engine and inspect.

4   Engine hard to start when cold

1   Battery discharged or low. Check as described in Section 1.
2   Choke control inoperative or out of adjustment (Chapters 1 and 4).
3   Carburetor flooded (see Section 2).
4   Fuel supply not reaching the carburetor (see Section 2).
5   Carburetor worn and in need of overhauling (Chapter 4).

5   Engine hard to start when hot

1   Choke sticking in the closed position (Chapter 1).
2   Carburetor flooded (see Section 2).

3   Air filter in need of replacement (Chapter 1).
4   Fuel not reaching the carburetor (see Section 2).
5   Thermac air cleaner faulty (Chapter 1).
6   EFE (heat riser) sticking in the closed position (Chapter 1).

6   Starter motor noisy or excessively rough in engagement

1   Pinion or flywheel gear teeth worn or broken. Remove the inspection cover at the rear of the engine and inspect.
2   Starter motor retaining bolts loose or missing.

7   Engine starts but stops immediately

1   Loose or faulty electrical connections at distributor, coil or alternator.
2   Insufficient fuel reaching the carburetor. Disconnect the fuel line at the carburetor and remove the filter (Chapter 1). Place a container under the disconnected fuel line. If equipped with HEI system (1975 - 1981), disconnect wiring connector marked 'BAT' from distributor cap. If conventional system (1970- 1974). disconnect the coil wire from the center of the distributor cap. These steps will prevent the engine from starting. Have an assistant crank the engine several revolutions by turning the ignition key. Observe the flow of fuel from the line. If little or none at all, check for blockage in the lines and/or replace the fuel pump (Chapter 4).
3   Vacuum leak at the gasket surfaces of the intake manifold and/or carburetor. Check that all mounting bolts (nuts) are tightened to specifications and all vacuum hoses connected to the carburetor and manifold are positioned properly and are in good condition.

8   Engine 'lopes' while idling or idles erratically

1   Vacuum leakage. Check mounting bolts (nuts) at the carburetor and intake manifold for tightness. Check that all vacuum hoses are connected and are in good condition. Use a doctor's stethoscope or a length of fuel line hose held against your ear to listen for vacuum leaks while the engine is running. A hissing sound will be heard. A soapy water solution will also detect leaks. Check the carburetor and intake manifold gasket surfaces.
2   Leaking EGR valve or plugged PCV valve (see Chapter 6).
3   Air cleaner clogged and in need of replacement (Chapter 1).
4   Fuel pump not delivering sufficient fuel to the carburetor (see Section 7).
5   Carburetor out of adjustment (Chapter 4).
6   Leaking head gasket. If this is suspected, take the car to a repair shop or GM dealer where this can be pressure checked without the need to remove the heads.

7   Timing chain or gears worn and in need of replacement (Chapter 2).
8   Camshaft lobes worn, necessitating the removal of the camshaft for inspection (Chapter 2).

9   Engine misses at idle speed

1   Spark plugs faulty or not gapped properly (Chapter 1).
2   Faulty spark plug wires (Chapter 1).
3   Faulty or incorrectly set contact breaker points (1970 - 1974 models only). Also check for excessive moisture on distributor components and/or damage (Chapter 1).
4   Carburetor choke not operating properly (Chapter 1).
5   Sticking or faulty emissions systems (see Troubleshooting in Chapter 6).
6   Clogged fuel filter and/or foreign matter in fuel. Remove the fuel filter (Chapter 1) and inspect.
7   Vacuum leaks at carburetor, intake manifold or at hose connections. Check as described in Section 8.
8   Incorrect idle speed (Chapter 1) or idle mixture (Chapter 4).
9   Incorrect ignition timing (Chapter 1).
10 Uneven or low cylinder compression. Remove plugs and use compression tester as per manufacturer's instructions.

10   Engine misses throughout driving speed range

1   Carburetor fuel filter clogged and/or impurities in the fuel system (Chapter 1). Also check fuel output at the carburetor (see Section 7).
2   Faulty or incorrectly gapped spark plugs (Chapter 1).
3   Incorrectly set ignition timing (Chapter 1).
4   Contact points faulty or incorrectly set (1970 - 1974 models only). At the same time check for a cracked distributor cap, disconnected distributor wires, or damage to the distributor components (Chapter 1).
5   Leaking spark plug wires (Chapter 1).
6   Emissions system components faulty (see Troubleshooting section, Chapter 6).
7   Low or uneven cylinder compression pressures. Remove spark plugs and test compression with gauge.
8   Weak or faulty ignition coil or condenser (1970 - 1974 models, see Chapter 5).
9   Weak or faulty HEI ignition system (1975 - 1981 models, see Chapter 5).
10 Vacuum leaks at carburetor, intake manifold or vacuum hoses (see Section 8).

11   Engine stalls

1   Carburetor idle speed incorrectly set (Chapter 1).
2   Carburetor fuel filter clogged and/or water and impurities in the fuel system (Chapter 1).
3   Choke improperly adjusted or sticking

(Chapter 1).
4   Distributor components damp, points out of adjustment or damage to distributor cap, rotor, etc. (Chapter 1).
5   Emission system components faulty (Troubleshooting section, Chapter 6).
6   Faulty or incorrectly gapped spark plugs. (Chapter 1). Also check spark plug wires (Chapter 1).
7   Vacuum leak at the carburetor, intake manifold or vacuum hoses. Check as described in Section 8.
8   Valve lash incorrectly set (Chapter 2).

12   Engine lacks power

1   Incorrect ignition timing (Chapter 1).
2   Excessive play in distributor shaft. At the same time check for worn or maladjusted contact points, faulty distributor cap, wires, etc. (Chapter 1).
3   Faulty or incorrectly gapped spark plugs (Chapter 1).
4   Carburetor not adjusted properly or excessively worn (Chapter 4).
5   Weak coil or condenser (Chapter 5).
6   Faulty HEI system coil (Chapter 5).
7   Brakes binding (Chapters 1 and 9).
8   Automatic transmission fluid level incorrect, causing slippage (Chapter 1).
9   Manual transmission clutch slipping (Chapter 1).
10   Fuel filter clogged and/or impurities in the fuel system (Chapter 1).
11   Emission control systems not functioning properly (see Troubleshooting, Chapter 6). ,
12   Use of sub-standard fuel. Fill tank with proper octane fuel.
13   Low or uneven cylinder compression pressures. Test with compression tester, which will also detect leaking valves and/or blown head gasket.

13   Engine backfires

1   Emissions systems not functioning properly (see Troubleshooting, Chapter 6).
2   Ignition timing incorrect (Section 1).
3   Carburetor in need of adjustment or worn excessively (Chapter 4).
4   Vacuum leak at carburetor, intake manifold or vacuum hoses. Check as described in Section 8.
5   Valve lash incorrectly set, and/or valves sticking (Chapter 2).
6   Valves sticking (Chapter 2).
7   Crossed spark plug wires (Chapter 1).

14   Pinging or knocking engine sounds on hard acceleration or uphill

1   Incorrect grade of fuel. Fill tank with fuel of the proper octane rating.
2   Ignition timing incorrect (Chapter 1).
3   Carburetor in need of adjustment (Chapter 4).
4   Improper spark plugs. Check plug type

with that specified on tune-up decal located inside engine compartment. Also check plugs and wires for damage (Chapter 1).
5   Worn or damaged distributor components (Chapter 1).
6   Faulty emission systems (see Troubleshooting, Chapter 6).
7   Vacuum leak. (Check as described in Section 8).

15   Engine 'diesels' (continues to run) after switching off

1   Idle speed too fast (Chapter 1).
2   Electrical solenoid at side of carburetor not functioning properly (not all models, see Chapter 4).
3   Ignition timing incorrectly adjusted (Chapter 1).
4   Thermac air cleaner valve not operating properly (see Troubleshooting, Chapter 6).
5   Excessive engine operating temperatures. Probable causes of this' are: malfunctioning thermostat, clogged radiator, faulty water pump. (See Chapter 3).

Engine electrical system

16   Battery will not hold a charge

1   Alternator drive belt defective or not adjusted properly (Chapter 1).
2   Electrolyte level too low or too weak (Chapter 1).
3   Battery terminals loose or corroded (Chapter 1).
4   Alternator not charging properly (Chapter 5).
5   Loose, broken or faulty wiring in the charging circuit (Chapter 5).
6   Short in vehicle circuitry causing a continual drain on battery.
7   Battery defective internally.

17   Ignition light fails to go out

1   Fault in alternator or charging circuit (Chapter 5).
2   Alternator drive belt defective or not properly adjusted (Chapter 1).

18   Ignition light fails to come on when key is turned

1   Ignition light bulb faulty (Chapter t0).
2   Alternator faulty (Chapter 5).
3   Fault in the printed circuit, dash wiring or bulb holder (Chapter 12).

Fuel system

19   Excessive fuel consumption

1   Dirty or choked air filter element (Chapter 1).

2   Incorrectly set ignition timing (Chapter 1).
3   Choke sticking or improperly adjusted (Chapter 1).
4   TCS emission system not functioning properly (not all cars, see Chapter 6).
5   Carburetor idle speed and/or mixture not adjusted properly (Chapters 1 and 4).
6   Carburetor internal parts excessively worn or damaged (Chapter 4).
7   Low tire pressure or incorrect tire size (Chapter 1).

20   Fuel leakage and/or fuel odor

1   Leak in a fuel feed or vent line (Chapter 6).
2   Tank overfilled. Fill only to automatic shut-off.
3   ECS emission system filter in need of replacement (Chapter 6).
4   Vapor leaks from ECS system lines (Chapter 6).
5   Carburetor internal parts excessively worn or out of adjustment (Chapter 4).

Engine cooling system

21   Overheating

1   Insufficient coolant in system (Chapter 1).
2   Fan belt defective or not adjusted properly (Chapter 1).
3   Radiator core blocked or radiator grille dirty and restricted (Chapter 3).
4   Thermostat faulty (Chapter 3).
5   Freewheeling clutch fan not functioning properly. Check for oil leakage at the rear of the cooling fan. indicating the need for replacement (Chapter 3).
6   Radiator cap not maintaining proper pressure. Have cap pressure tested by gas station or repair shop.
7   Ignition timing incorrect (Chapter 1).

22   Overcooling

1   Thermostat faulty (Chapter 3).
2   Inaccurate temperature gauge (Chapter 12).

23   External coolant leakage

1   Deteriorated or damaged hoses. Loose clamps at hose connections (Chapter 1).
2   Water pump seals defective. If this is the case, water will drip from the 'weep' hole in the water pump body (Chapter 3).
3   Leakage from radiator core or header tank. This will require the radiator to be professionally repaired (see Chapter 3 for removal procedures).
4   Engine drain plugs or water jacket freeze plugs leaking (see Chapters 2 and 3).

24   Internal coolant leakage

Note: *Internal coolant leaks can usually be detected by examining the oil. Check the dipstick and inside of valve cover for water deposits and an oil consistency like that of a milkshake.*

1   Faulty cylinder head gasket. Have the system pressure-tested professionally or remove the cylinder heads (Chapter 2) and inspect.
2   Cracked cylinder bore or cylinder head. Dismantle engine and inspect (Chapter 2).

25   Coolant loss

1   Overfilling system (Chapter 1).
2   Coolant boiling away due to overheating (see causes in Section 21).
3   Internal or external leakage (see Sections 22 and 23).
4   Faulty radiator cap. Have the cap pressure tested.

26   Poor coolant circulation

1   Inoperative water pump. A quick test is to pinch the top radiator hose closed with your hand while the engine is idling, then let loose. You should feel a surge of water if the pump is working properly (Chapter 3).
2   Restriction in cooling system. Drain, flush and refill the system (Chapter 1). If it appears necessary, remove the radiator (Chapter 3) and have it reverse-flushed or professionally cleaned.
3   Fan drive belt defective or not adjusted properly (Chapter 1).
4   Thermostat sticking (Chapter 3).

Clutch

27   Fails to release (pedal pressed to the floor - shift lever does not move freely in and out of reverse)

1   Improper linkage adjustment (Chapter 8).
2   Clutch fork off ball stud. Look under the car, on the left side of transmission.
3   Clutch disc warped, bent or excessively damaged (Chapter 8).

28   Clutch slips (engine speed increases with no increase in vehicle speed)

1   Linkage in need of adjustment (Chapter 8).
2   Clutch disc oil soaked or facing worn. Remove disc (Chapter 8) and inspect.
3   Clutch disc not seated in. It may take 30 or 40 normal starts for a new disc to seat.

29   Grabbing (shuddering) as clutch is engaged

1   Oil on clutch disc facings. Remove disc (Chapter 8) and inspect. Correct any leakage source.
2   Worn or loose engine or transmission mounts. These units may move slightly when clutch is released. Inspect mounts and bolts.
3   Worn splines on clutch gear. Remove clutch components (Chapter 8) and inspect.
4   Warped pressure plate or flywheel. Remove clutch components and inspect.

30   Squeal or rumble with clutch fully engaged (pedal released)

1   Improper adjustment; no lash (Chapter 8).
2   Release bearing binding on transmission bearing retainer. Remove clutch components (Chapter 8) and check bearing. Remove any burrs or nicks, clean and relubricate before reinstallation.
3   Weak linkage return spring. Replace the spring.

31   Squeal or rumble with clutch fully disengaged (pedal depressed)

1   Worn, faulty or broken release bearing (Chapter 8).
2   Worn or broken pressure plate springs (or diaphragm fingers) (Chapter 8).

32   Clutch pedal stays on floor when disengaged

1   Bind in linkage or release bearing. Inspect linkage or remove clutch components as necessary.
2   Linkage springs being over-traveled. Adjust linkage for proper lash. Make sure proper pedal stop (bumper) is installed.

Manual transmission

Note: *The following manual transmission Section references are to Chapter 7A, unless otherwise noted.*

33   Noisy in neutral with engine running

1   Input shaft bearing worn (Sections 10-12).
2   Damaged main drive gear bearing (Sections 10-12).
3   Worn countergear bearings (Sections 10-12).
4   Worn or damaged countergear anti-lash plate (Sections 10-12).

34   Noisy in all gears

1   Any of the above causes, and/or:
2   Insufficient lubricant (see checking procedures in Chapter 1).

35   Noisy in one particular gear

1   Worn, damaged or chipped gear teeth for that particular gear (Sections 10- 12).
2   Worn or damaged synchronizer for that particular gear (Sections 10- 12).

36   Slips out of high gear

1   Transmission loose on clutch housing (Section 3).
2   Shift rods interfering with engine mounts or clutch lever (Section 2).
3   Shift rods not working freely (Section 2).
4   Damaged mainshaft pilot bearing (Section 9).
5   Misalignment of transmission or clutch housing (Section 9).
6   Worn or improperly adjusted linkage (Section 2).

37   Difficulty in engaging gears

1   Clutch not releasing fully (see clutch adjustment, Chapter 8).
2   Loose, damaged or maladjusted shift linkage. Make a thorough inspection, replacing parts as necessary. Adjust as described in Section 2.

38   Oil leakage

1   Excessive amount of lubricant in transmission (see Chapter 1 for correct checking procedures. Drain lubricant as required).
2   Side cover loose or gasket damaged (Sections 7 - 8).
3   Rear oil seal or speedometer oil seal in need of replacement (Section 6).

Automatic transmission

Note: *Due to the complexity of the automatic transmission, it is difficult for the home mechanic to properly diagnose and service this component. For problems other than the following, the vehicle should be taken to a dealer service department or other qualified repair shop.*

39   Fluid leakage

1   Automatic transmission fluid is a deep red color. Fluid leaks should not be confused with engine oil, which can easily be blown by air flow to the transmission.

2   To pinpoint a leak, first remove all built-up dirt and grime from around the transmission. Degreasing agents and/or steam cleaning will achieve this. With the underside clean, drive the vehicle at low speeds so air flow will not blow the leak far from its source. Raise the vehicle and determine where the leak is coming from. Common areas of leakage are:

a) *Fluid Pan: Tighten mounting bolts and/or replace pan gasket as necessary (see Chapters 1 and 7).*
b) *Filler pipe: Replace the rubber seal where pipe enters transmission case.*
c) *Transmission oil lines: Tighten connectors where lines enter transmission case and/or replace lines.*
d) *Vent pipe: Transmission overfilled and/or water in fluid (see checking procedures, Chapter 1).*
e) *Speedometer connector: Replace the O-ring where speedometer cable enters transmission case (Chapter 7).*

40   General shift mechanism problems

1   Sections 4 and 5 in Chapter 7 Part B deal with checking and adjusting the shift linkage on automatic transmissions. Common problems which may be attributed to poorly adjusted linkage are:

a) *Engine starting in gears other than Park or Neutral*
b) *Indicator on shifter pointing to a gear other than the one actually being used*
c) *Vehicle moves when in Park*
2   Refer to Chapter 7 Part B to adjust the linkage.

41   Transmission will not downshift with accelerator pedal pressed to the floor

Sections 6 through 10 in Chapter 7 Part B deal with adjusting the downshift cable or downshift switch to enable the automatic transmission to downshift properly.

42   Engine will start in gears other than 'P' (Park) or 'N' (Neutral)

Sections 11 and 12 in Chapter 7 Part B deal with adjusting the neutral start switches used with automatic transmissions.

43   Transmission slips, shifts rough, is noisy or has no drive in forward or reverse

1   There are many probable causes for the above problems, but the home mechanic should concern himself only with one possibility; fluid level.
2   Before taking the vehicle to a specialist, check the level of the fluid and condition of the

fluid as described in Chapter 1. Correct fluid level as necessary or change the fluid and filter if needed. If problem persists, have a professional diagnose the probable cause.

Driveshaft

44   Leakage of fluid at front of driveshaft

Defective transmission rear oil seal. See Section 3 in Chapter 7 Part B for replacement procedures. While this is done, check the splined yoke for burrs or a rough condition which may be damaging the seal. If found, these can be dressed with crocus cloth or a fine dressing stone.

45   Knock or clunk when transmission is under initial load (just after transmission is put into gear)

1   Loose or disconnected rear suspension components. Check all mounting bolts and bushings (Chapter 1).
2   Loose driveshaft bolts. Inspect all bolts and nuts and tighten to torque specifications (Chapter 8).
3   Worn or damaged universal joint bearings. Test for wear (Chapter 8).

46   Metallic grating sound consistent with road speed

Pronounced wear in the universal joint bearings. Test for wear (Chapter 8).

47   Vibration

Note: *Before it can be assumed that the driveshaft is at fault, make sure the tires are perfectly balanced and perform the following test.*
1   Install a tachometer inside the car to monitor engine speed as the car is driven. Drive the car and note the engine speed at which the vibration (roughness) is most pronounced. Now shift the transmission to a different gear and bring the engine speed to the same point.
2   If the vibration occurs at the same engine speed (rpm) regardless of which gear the transmission is in, the driveshaft is NOT at fault since the driveshaft speed varies.
3   If the vibration decreases or is eliminated when the transmission is in a 1 different gear at the same engine speed, refer to the following probable causes.
4   Bent or dented driveshaft. Inspect and replace as necessary (Chapter 8).
5   Undercoating or built-up dirt, etc. on the driveshaft. Clean the shaft thoroughly and test.
6   Worn universal joint bearings. Remove and inspect (Chapter 8).
7   Driveshaft and/or companion flange out of balance. Check for missing weights on the

shaft. Remove driveshaft (Chapter 8) and reinstall 180° from original position. Retest. Have driveshaft professionally balanced if problem persists.

Rear axle

48   Noise - same when in drive as when vehicle is coasting

1   Road noise. No corrective procedures available.
2   Tire noise. Inspect tires and tire pressures (Chapter 1).
3   Front wheel bearings loose, worn or damaged (Chapter 1).

49   Vibration

1   See probable causes under 'Driveshaft'. Proceed under the guidelines listed for the driveshaft. If the problem persists. check the rear wheel bearings by raising the rear of the car and spinning the wheels by hand. Listen for evidence of rough (noisy) bearings. Remove and inspect (Chapter 8).

50   Oil leakage

1   Pinion oil seal damaged (Chapter 8).
2   Axle shaft oil seals damaged (Chapter 8).
3   Differential inspection cover leaking. Tighten mounting bolts or replace the gasket as required (Chapter 1).

Brakes

Note: *Before assuming a brake problem exists, check: that the tires are in good condition and are inflated properly (see Chapter 1); the front end alignment is correct; and that the vehicle is not loaded with weight in an unequal manner.*

51   Vehicle pulls to one side under braking

1   Defective, damaged or oil contaminated disc pad on one side. Inspect as described in Chapter 1. Refer to Chapter 9 if replacement is required.
2   Excessive wear of brake pad material or disc on one side. Inspect and correct as necessary.
3   Loose or disconnected front suspension components. Inspect and tighten all bolts to specifications (Chapter 1).
4   Defective caliper assembly. Remove caliper and inspect for stuck piston or damage (Chapter 9).

52   Noise (high pitched squeak without brake applied)

Front brake pads worn out. This noise

comes from the wear sensor rubbing against the disc. Replace pads with new ones immediately (Chapter 9).

53 Excessive brake pedal travel

1 Partial brake system failure. Inspect entire system (Chapter 1) and correct as required.
2 Insufficient fluid in master cylinder. Check (Chapter 1) and add fluid and bleed system if necessary.
3 Rear brakes not adjusting properly. Make a series of starts and stops while the vehicle is in 'R' (Reverse). If this does not correct the situation remove drums and inspect self-adjusters (Chapter 1).

54 Brake pedal appears spongy when depressed

1 Air in hydraulic lines. Bleed the brake system (Chapter 9).
2 Faulty flexible hoses. Inspect all system hoses and lines. Replace parts as necessary.
3 Master cylinder mountings insecure. Inspect master cylinder bolts (nuts) and torque-tighten to specifications.
4 Master cylinder faulty (Chapter 9).

55 Excessive effort required to stop vehicle

1 Power brake servo not operating properly (Chapter 9).
2 Excessively worn linings or pads. Inspect and replace if necessary (Chapter 1).
3 One or more caliper pistons (front wheels) or wheel cylinders (rear wheels) seized or sticking. Inspect and rebuild as required (Chapter 9).
4 Brake linings or pads contaminated with oil or grease. Inspect and replace as required (Chapter 1).
5 New pads or linings fitted and not yet 'bedded in'. It will take a while for the new material to seat against the drum (or rotor).

56 Pedal travels to floor with little resistance

Little or no fluid in the master cylinder reservoir caused by: leaking wheel cylinder(s); leaking caliper piston(s); loose, damaged or disconnected brake lines. Inspect entire system and correct as necessary.

57 Brake pedal pulsates during brake application

1 Wheel bearings not adjusted properly or in need of replacement (Chapter 1).
2 Caliper not sliding properly due to improper installation or obstructions. Remove and inspect (Chapter 9).
3 Rotor not within specifications. Remove the rotor (Chapter 9) and check for excessive lateral run-out and parallelism. Have the rotor professionally machined or replace it with a new one.

Suspension and steering

58 Vehicle pulls to one side

1 Tire pressures uneven (Chapter 1).
2 Defective tire (Chapter 1).
3 Excessive wear in suspension or steering components (Chapter 1).
4 Front end in need of alignment. Take car to a qualified specialist.
5 Front brakes dragging. Inspect braking system as described in Chapter 1.

59 Shimmy. shake or vibration

1 Tire or wheel out of balance or out of round. Have professionally balanced.
2 Loose, worn or out of adjustment wheel bearings (Chapter 1).
3 Shock absorbers and/or suspension components worn or damaged (Chapter 10).

60 Excessive pitching and/or rolling around corners or during braking

1 Defective shock absorbers. Replace as a set (Chapter 10).
2 Broken or weak coil springs and/or suspension components. Inspect as described in Chapter 11.

61 Excessively stiff steering

1 Lack of lubricant in steering box (manual) or power steering fluid reservoir (Chapter 1).
2 Incorrect tire pressures (Chapter 1).
3 Lack of lubrication at steering joints (Chapter 1).
4 Front end out of alignment.
5 See also Section 63 'Lack of power assistance'.

62 Excessive play in steering

1 Loose wheel bearings (Chapter 1).
2 Excessive wear in suspension or steering components (Chapter 1).
3 Steering gear out of adjustment (Chapter 10).

63 Lack of power assistance

1 Steering pump drive belt faulty or not adjusted properly (Chapter 1).
2 Fluid level low (Chapter 1).
3 Hoses or pipes restricting the flow. Inspect and replace parts as necessary.
4 Air in power steering system. Bleed system (Chapter 10).

64 Excessive tire wear (not specific to one area)

1 Incorrect tire pressures (Chapter 1).
2 Tires out of balance. Have professionally balanced.
3 Wheels damaged. Inspect and replace as necessary.
4 Suspension or steering components excessively worn (Chapter 1).

65 Excessive tire wear on outside edge

1 Inflation pressures not correct (Chapter 1).
2 Excessive speed on turns.
3 Front end alignment incorrect (excessive toe-in). Have professionally aligned.
4 Suspension arm bent or twisted.

66 Excessive tire wear on inside edge

1 Inflation pressures incorrect (Chapter 1).
2 Front end alignment incorrect (toe-out). Have professionally aligned.
3 Loose or damaged steering components (Chapter 1).

67 Tire tread worn in one place

1 Tires out of balance. Balance tires professionally.
2 Damaged or buckled wheel. Inspect and replace if necessary.
3 Defective tire.

Chapter 1
Tune-up and routine maintenance

Contents

Specifications

Recommended lubricants and fluids

Note: *Listed here are manufacturer recommendations at the time this manual was written. Manufacturers occasionally upgrade their fluid and lubricant specifications, so check with your local auto parts store for current recommendations.*

Engine oil type	API approved multigrade and fuel efficient engine oil
Viscosity	See accompanying chart
Automatic transmission fluid type	Dexron II automatic transmission fluid
Manual transmission fluid type	SAE 80W or 80W-90 GL-5 gear lubricant

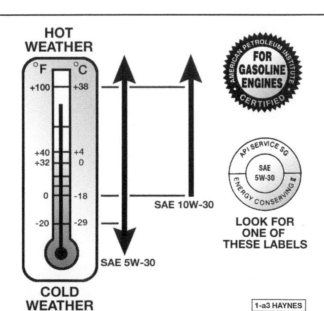

Engine oil viscosity chart - for best fuel economy and cold starting, select the lowest SAE viscosity grade for the expected temperature range

1-a3 HAYNES

Recommended lubricants and fluids (continued)

Differential lubricant type
    Standard differential .... SAE 80W or 80W-90 GL-5 gear lubricant
    Limited slip differential .... Add 4 fluid ounces of GM limited slip additive to the specified lubricant
Brake fluid type.. DOT 3 brake fluid
Power steering system fluid .... GM power steering fluid or equivalent
Steering gear fluid type.. GM power steering fluid or equivalent
Chassis grease .... SAE NLGI no. 2 chassis grease

Capacities*

Engine oil
    All V6 and V8 engines.. 4.0 quarts*
    *Add an additional 1 qt if filter is changed (V8 only)
Manual transmission
    All .... 3.0 pints*
    *Approx. to fill after draining
Automatic transmission
    Powerglide
        Drain and refill .... 3.0 quarts
        Dry .... 9.0 quarts
    Turbo Hydramatic 350
        Drain and refill .... 3.0 quarts
        Dry .... 10.0 quarts
    Turbo Hydramatic 200, 250 and 400
        Drain and refill .... 3.5 quarts
        Dry .... 11.0 quarts

Cooling system

1970 through 1974
    350 cid engine (all) .... 16 quarts
    400 cid engine
        1970 and 1971 only .... 23 quarts
        1972 only .... 24 quarts
        1974 only .... 16 to 17 quarts
    454 cid engine (all) .... 23 to 24 quarts
1975
    350/400 cid engine.. 18 quarts
    454 cid engine .... 23 quarts
1976 and 1977
    All .... 17 to 18 quarts
1978 through 1981
    200/229 cid engine (all) .... 18.8 quarts
    231 cid engine
        1978 only .... 14.79 quarts
        1979 through 1981... 15.4 quarts
    267 cid engine (all) .... 20.6 quarts
    305 cid engine (all) .... 19.2 quarts
1982 through 1988
    229 cu in V6 .... 15.0 quarts
    231 cu in V6 .... 12.0 quarts
    262 cu in V6 (1985 on)... 13.2 quarts
    267 cu in V8 .... 18.6 quarts
    305 cu in V8
        1982 thru 1984.. 16.3 quarts
        1985 and 1986 .... 15.6 quarts
        1987 and 1988 .... 16.6 quarts

All capacities are approximate. Add as necessary to bring to appropriate level.

Ignition system

Spark plug type and gap*

1970
    350 V8 .... AC R44S or equivalent @ 0.035 in
    400 V8 .... AC R44 or equivalent @ 0.035 in
    454 V8 .... AC R44T or equivalent @ 0.035 in
1971
    350 V8 .... AC R45TS or equivalent @ 0.035 in
    402 V8 .... AC R44TS or equivalent @ 0.035 in
    454 V8.. AC R43TS or equivalent @ 0.035 in

1972 through 1974
  350 V8 .... AC R44T or equivalent @ 0.035 in
  400 V8 .... AC R44T or equivalent @ 0.035 in
  402 V8 .... AC R44T or equivalent @ 0.035 in
  454 V8 .... AC R44T or equivalent @ 0.035 in
1975
  350 V8 .... AC R44TX or equivalent @ 0.060 in
  400 V8 .... AC R44TX or equivalent @ 0.060 in
  454 V8. .... AC R44TX or equivalent @ 0.060 in
1976 through 1977
  305 V8 .... AC R45TS or equivalent @ 0.045 in
  350 V8 .... AC R45TS or equivalent @ 0.045 in
  400 V8 .... AC R45TS or equivalent @ 0.045 in
1978 through 1980
  200 V6 (3.3L) .... AC R45TS or equivalent @ 0.060 in
  229 V6 (3.8L) .... AC R45TS or equivalent @ 0.045 in
  231 V6 (3.8L) .... AC R45TSX or equivalent @ 0.060 in
  231 V6 (3.8L turbo) .... AC R45TS or equivalent @ 0.040 in
  267 V8 (4.4L) .... AC R45TS or equivalent @ 0.045 in
  305 V8 (5.0L) .... AC R45TS or equivalent @ 0.045 in
1981 through 1983
  229 V6 (3.8L) .... AC R45TS or equivalent @ 0.045 in
  231 V6 (3.8L) .... AC R45TS8 or equivalent @ 0.080 in
  231 V6 (3.8L turbo) .... AC R45TS or equivalent @ 0.040 in
  267 V8 (4.4L) .... AC R45TS or equivalent @ 0.045 in
  305 V8 (5.0L) .... AC R45TS or equivalent @ 0.045 in
1984 through 1985
  229 V6 (3.8L) .... AC R45TS or equivalent @ 0.045 in
  231 V6 (3.8L) .... AC R45TSX or equivalent @ 0.060 in
  262 V6 (4.3L) .... AC R45TS or equivalent @ 0.035 in
  305 V8 (5.0L) VIN G .... AC R44TS or equivalent @ 0.045 in
  305 V8 (5.0L) VIN H .... AC R45TS or equivalent @ 0.045 in
1986
  231 V6 (3.8L) .... AC R45TSX or equivalent @ 0.060 in
  262 V6 (4.3L) .... AC R45TS or equivalent @ 0.035 in
  305 V8 (5.0L) VIN G .... AC R43TS or equivalent @ 0.035 in
  305 V8 (5.0L) VIN H .... AC R43TS or equivalent @ 0.035 in
1987 through 1988
  262 V6 (4.3L) .... AC R45TS or equivalent @ 0.035 in
  305 V8 (5.0L) VIN G AND H .... AC R45TS or equivalent @ 0.035 in

Refer to the Vehicle Emission Control Information label in the engine compartment; use the information there if it differs from that listed here.

Ignition point gap

1970 through 1974
  New .... 0.019 in
  Used .... 0.016 in
Dwell angle
  V8 engines .... 29 to 31-degrees

Engine firing order and distributor rotation

V6
  Firing order 1-6-5-4-3-2
  Distributor rotation .... clockwise

**200 and 229
cu in V6**

231 cu in V6

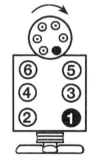

262 cu in V6

**Cylinder location and
distributor rotation**

*The blackened terminal shown
on the distributor cap indicates
the Number One spark plug
wire position*

24070-4 specs HAYNES

Engine firing order and distributor rotation (continued)

V8

Firing order 1-8-4-3-6-5-7-2
Distributor rotation .... clockwise

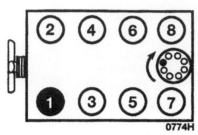

ALL V8 ENGINES WITH BREAKER POINT IGNITION

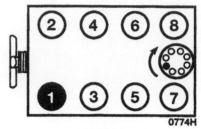

ALL V8 ENGINES WITH HEI IGNITION

Cylinder location and distributor rotation

The blackened terminal shown on the distributor cap indicates the Number One spark plug wire position

Ignition timing*

1970

350 V8 M/T	0-degrees BTDC @ 700 rpm
350 V8 A/T	4-degrees BTDC @ 600 rpm
400 V8 M/T	4-degrees BTDC @ 700 rpm
400 V8 A/T	8-degrees BTDC @ 600 rpm
454 V8 M/T	6-degrees BTDC @ 700 rpm
454 V8 A/T	6-degrees BTDC @ 600 rpm

1971

350 V8 2bbl M/T	2-degrees BTDC @ 600 rpm
350 V8 2bbl A/T	6-degrees BTDC @ 550 rpm
350 V8 4bbl M/T	4-degrees BTDC @ 600 rpm
350 V8 4bbl A/T	8-degrees BTDC @ 550 rpm
402 V8	8-degrees BTDC @ 600 rpm
454 V8 4bbl M/T (365 HP)	8-degrees BTDC @ 600 rpm
454 V8 4bbl M/T (425 HP)	8-degrees BTDC @ 700 rpm
454 V8 4bbl A/T (425 HP)	12-degrees BTDC @ 700 rpm

1972

350 V8 2bbl M/T	6-degrees BTDC @ 900 rpm
350 V8 2bbl A/T	6-degrees BTDC @ 600 rpm
350 V8 4bbl M/T	4-degrees BTDC @ 800 rpm
350 V8 4bbl A/T	8-degrees BTDC @ 600 rpm
400 V8 2bbl M/T	2-degrees BTDC @ 900 rpm
400 V8 2bbl A/T	6-degrees BTDC @ 600 rpm
454 V8 4bbl M/T (365 HP)	8-degrees BTDC @ 750 rpm
454 V8 4bbl A/T (365 HP)	8-degrees BTDC @ 600 rpm

1973

350 V8 2bbl M/T	8-degrees BTDC @ 900 rpm
350 V8 2bbl A/T	8-degrees BTDC @ 600 rpm
350 V8 4bbl M/T	8-degrees BTDC @ 900 rpm
350 V8 4bbl A/T	12-degrees BTDC @ 600 rpm
454 V8 4bbl M/T	10-degrees BTDC @ 900 rpm
454 V8 4bbl A/T	10-degrees BTDC @ 600 rpm

1974

350 V8 2bbl M/T	0-degrees BTDC @ 900 rpm
350 V8 2bbl A/T	8-degrees BTDC @ 600 rpm
350 V8 4bbl M/T	4-degrees BTDC @ 900 rpm
350 V8 4bbl A/T	8-degrees BTDC @ 600 rpm
400 V8	8-degrees BTDC @ 600 rpm
454 V8 4bbl M/T	10-degrees BTDC @ 800 rpm
454 V8 4bbl A/T	10-degrees BTDC @ 600 rpm

1975

350 V8 M/T	6-degrees BTDC @ 800 rpm
350 V8 A/T	6-degrees BTDC @ 600 rpm
400 V8	8-degrees BTDC @ 600 rpm
454 V8	16-degrees BTDC @ 600 rpm

Ignition timing*

1976
305 V8 California	0-degrees BTDC @ 600 rpm
305 V8 M/T	6-degrees BTDC @ 800 rpm
305 V8 A/T	8-degrees BTDC @ 600 rpm
350 V8 2bbl	6-degrees BTDC @ 600 rpm
350 V8 4bbl California	6-degrees BTDC @ 600 rpm
350 V8 4bbl M/T	8-degrees BTDC @ 800 rpm
350 V8 4bbl A/T	8-degrees BTDC @ 600 rpm
400 V8	8-degrees BTDC @ 600 rpm

1977
305 V8 California	6-degrees BTDC @ 500 rpm
305 V8 M/T	8-degrees BTDC @ 600 rpm
305 V8 A/T	8-degrees BTDC @ 500 rpm
350 V8 high altitude	8-degrees BTDC @ 600 rpm
350 V8 M/T	8-degrees BTDC @ 700 rpm
350 V8 A/T	8-degrees BTDC @ 500 rpm

1978
200 V6 M/T	8-degrees BTDC @ 700 rpm
200 V6 A/T	8-degrees BTDC @ 600 rpm
231 V6 M/T	15-degrees BTDC @ 600 rpm
231 V6 A/T	15-degrees BTDC @ 500 rpm
305 V8 California	6-degrees BTDC @ 500 rpm
305 V8 high altitude	8-degrees BTDC @ 500 rpm
305 V8 M/T	4-degrees BTDC @ 600 rpm
305 V8 A/T	4-degrees BTDC @ 500 rpm

1979
200 V6 M/T	8-degrees BTDC @ 700 rpm
200 V6 A/T	14-degrees BTDC @ 600 rpm
231 V6 M/T	15-degrees BTDC @ 600 rpm
231 V6 A/T	15-degrees BTDC @ 500 rpm
267 V8 M/T	4-degrees BTDC @ 600 rpm
267 V8 A/T	10-degrees BTDC @ 500 rpm
305 V8 M/T	4-degrees BTDC @ 600 rpm
305 V8 A/T	4-degrees BTDC @ 500 rpm
350 V8 A/T	8-degrees BTDC @ 500 rpm

1980
229 V6 M/T	8-degrees BTDC @ 700 rpm
229 V6 A/T	12-degrees BTDC @ 600 rpm
231 V6 A/T	15-degrees BTDC @ 600 rpm
231 V6 Turbo	15-degrees BTDC @ 650 rpm
267 V8 A/T	4-degrees BTDC @ 500 rpm
305 V8 M/T	4-degrees BTDC @ 700 rpm
305 V8 A/T	4-degrees BTDC @ 500 rpm

1981
229 V6	6-degrees BTDC @ idle
231 V6 A/T	15-degrees BTDC @ idle
231 V6 Turbo	15-degrees BTDC @ 650 rpm
267 V8 A/T	6-degrees BTDC @ 500 rpm
305 V8 M/T	6-degrees BTDC @ 700 rpm
305 V8 A/T	6-degrees BTDC @ 500 rpm

1982 through 1984
229 V6	0-degrees BTDC @ idle
231 V6 A/T	15-degrees BTDC @ idle
267 V8 A/T	2-degrees BTDC @ 600 rpm
305 V8 M/T	6-degrees BTDC @ 700 rpm
305 V8 A/T	6-degrees BTDC @ 500 rpm

1985 through 1987
231 V6	15-degrees BTDC @ Idle
262 V6	0-degrees BTDC @ idle
305 V8 VIN G	6-degrees BTDC @ 500 rpm
305 V8 VIN H	0-degrees BTDC @ 500 rpm

1988
All models	See emissions label

*Refer to the Vehicle Emission Control Information label in the engine compartment; use the information there if it differs from that listed here

Clutch

Clutch pedal freeplay
 1970 through 1972 .... 1 1/8 to 1 3/4 in
 1973 through 1981 .... 3/4 to 1 5/16 in

Brakes

Disc brake pad lining thickness (minimum) 1/8-in
Drum brake shoe lining thickness (minimum)........................... 1/16-in

Torque specifications   Ft-lb

Oil pan drain plug 20 (231 V6 30)
Spark plugs
 V8
   1970 through 1978 (5/8 in hex)................................... 15
   1979 and later (1 3/16 in hex) 22
 V6 (All) 20
Carburetor mounting nuts 12
TBI mounting bolts 11
Fuel inlet nut (fuel filter)... 18
Manual transmission fill plug .... 18 (Muncie 30)
Automatic transmission pan bolts 12
Rear differential /inspection plug.. 22
Rear axle cover bolts... 27
Brake caliper mounting bolts.. 35
Wheel lug nuts
 1970 65
 1971 through 1975 70
 1976 and later 80

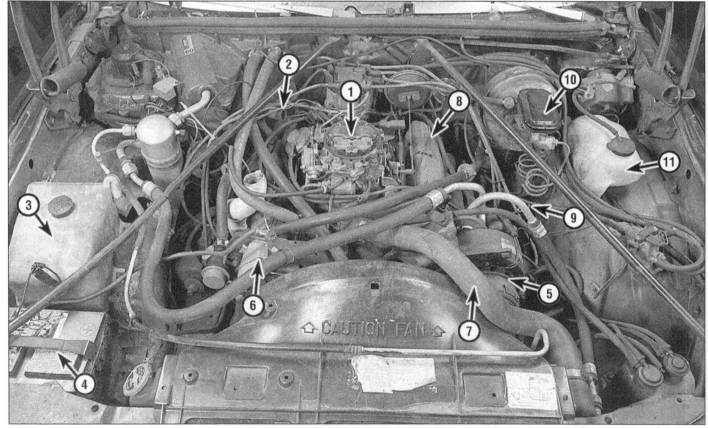

Typical engine compartment components

1	*Carburetor*	*4*	*Battery*	*8*	*Oil filler cap*
2	*Automatic transmission fluid*	*5*	*Drivebelt*	*9*	*Engine oil dipstick*
	dipstick	*6*	*Alternator*	*10*	*Brake fluid reservoir*
3	*Radiator coolant reservoir*	*7*	*Upper radiator hose*	*11*	*Windshield washer fluid reservoir*

1 Introduction

This Chapter was designed to help the home mechanic maintain his (or her) car for peak performance, economy, safety and longevity.

On the following pages you will find a maintenance schedule along with sections which deal specifically with each item on the schedule. Included are visual checks, adjustments and item replacements.

Servicing your car using the time/mileage maintenance schedule and the sequenced sections will give you a planned program of maintenance. Keep in mind that it is a full plan, and maintaining only a few items at the specified intervals will not give you the same results.

You will find as you service your car that many of the procedures can, and should, be grouped together, due to the nature of the job at hand. Examples of this are as follows:

If the car is fully raised for a chassis lubrication, for example, this is the ideal time for the following checks: manual transmission fluid, rear axle fluid, exhaust system, suspension, steering and the fuel system.

If the tires and wheels are removed, as during a routine tire rotation, go ahead and check the brakes and wheel bearings at the same time.

If you must borrow or rent a torque wrench, you will do best to service the spark plugs, repack (or replace) the wheel bearings and check the carburetor mounting torque all in the same day to save time and money.

The first step of this or any maintenance plan is to prepare yourself before the actual work begins. Read through the appropriate sections for all work that is to be performed, before you begin. Gather together all necessary parts and tools. If it appears you could have a problem during a particular job, don't hesitate to ask advice from your local parts man or dealer service department.

Typical engine compartment underside components

1	Radiator drain fitting	6	Lower radiator hose	11	Oil filter
2	Radiator	7	Balljoint	12	Exhaust pipe
3	Drivebelt and pulleys	8	Shock absorber	13	Brake caliper
4	Power steering hose	9	Shift linkage		
5	Steering gear	10	Oil drain plug	A	Lube points

Typical rear underside components

1	Muffler	4	Spring	7	Exhaust pipe
2	Driveshaft universal joint	5	Fuel tank	8	Parking brake cable
3	Suspension arm bushing	6	Rear axle housing differential cover	9	Shock absorber

Chevrolet Monte Carlo
Routine maintenance schedule

Every 250 miles or weekly - whichever comes first

Check the engine oil level (Section 2)
Check the engine coolant level (Section 2)
Check the windshield washer fluid level (Section 2)
Check the battery water level (if equipped with removable vent caps) (Section 2).
Check the tires and tire pressures (Section 3)

Every 3000 miles or 3 months - whichever comes first

All items listed above plus:
Check the automatic transmission fluid level (Section 2)
Check the power steering fluid level (Section 2)
Change engine oil and filter (Section 4)
Lubricate the chassis components (Section 5)
Check the cooling system (Section 6)
Check the brake master cylinder fluid level (Section 2)
Check the manual transmission fluid level (Section 2)
Check and service the battery (Section 33)
Inspect and replace if necessary, all underhood hoses (Section 35)
Inspect and replace if necessary, the windshield wiper blades (Section 36)

Every 7500 miles or 6 months - whichever comes first

All items listed above plus:
Check the rear axle fluid level (Section 2)*
Check the exhaust system (Section 7)*
Check the suspension and steering components (Section 8)*
Check and adjust (if necessary) the engine drive belts(Section 9)*
Check the fuel system components (Section 10)*
Check the clutch pedal free-play (manual transmission only Section 13)*
Rotate the tires (Section 14)
Check the Thermostatically Controlled air cleaner for proper operation (Section 15)
Check the EFE system (Section 17)*
Check the operation of the choke (Section 20)
Check the operation of the EGR valve (Section 21)
Check the braking system (Section 26)*
Check the starter safety switch (Section 34)
Check the seatbelts (Section 37)

Every 15,000 miles or 12 months - whichever comes first

All items listed above plus:
Replace the (PCV) valve (Section 11)
Replace the air filter and PCV filter (Section 12)
Replace the fuel filter (Section 18)
Replace the spark plugs (Section 24)
Check the distributor cap, rotor and spark plug wires (Section 28 and 32)
Replace the contact points and adjust dwell angle (1970 - 1974 only) (Section 31)
Check and adjust (if necessary) the engine idle speed (Section 16)
Check and adjust (if necessary) the ignition timing (Section 19)
Check the carburetor/throttle body mounting nut torque (Section 27)

Every 30,000 miles or 24 months - whichever comes first

All items listed above plus:
Change the rear axle fluid (Section 22)*
Check and repack the front wheel bearings (Section 24)
Change the automatic transmission fluid and filter (Section 25)**
Drain, flush and refill the cooling system (Section 29)
Check the ECS emissions system and replace the charcoal canister filter (Section 31)

* This item is affected by "severe" operating conditions as described below. If your vehicle is operated under severe conditions, perform all maintenance indicated with an asterisk (*) at 3000 mile/3 month intervals. Severe conditions are indicated if you mainly operate your vehicle under one or more of the following:
Operating in dusty areas
Towing a trailer
Idling for extended periods and/or low speed operation
Operating when outside temperatures remain below freezing and when most trips are less than four miles

** If operated under one or more of the following conditions, change the automatic transmission fluid every 12,000 miles:
In heavy city traffic where the outside temperature regularly reaches 90-degrees F (32-degrees C) or higher
In hilly or mountainous terrain
Frequent trailer pulling

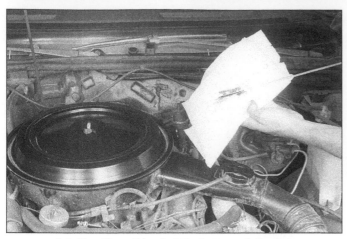

2.4 Checking oil level at the bottom of dipstick

2.6a Oil is added to the engine after removing the twist-off cap (arrow)

2 Fluid level checks

1   There are a number of components on a vehicle which rely on the use of fluids to perform their job. Through the normal operation of the car, these fluids are used up and must be replenished before damage occurs. See the *Recommended Lubricants* Section for the specific fluid to be used when adding is required. When checking fluid levels it is important that the car is on a level surface.

Engine oil

Refer to illustrations 2.4, 2.6a and 2.6b

2   The engine oil level is checked with a dipstick which is located at the side of the engine block. This dipstick travels through a tube and into the oil pan to the bottom of the engine.

3   The oil level should be checked preferably before the car has been driven, or about 15 minutes after the engine has been shut off. If the oil is checked immediately after driving the car, some of the oil will remain in the upper engine components, thus giving an inaccurate reading on the dipstick.

4   Pull the dipstick from its tube and wipe all the oil from the end with a clean rag. Insert the clean dipstick all the way back into the oil pan and pull it out again. Observe the oil at the end of the dipstick **(see illustration)**. At its highest point, the level should be between the 'Add' and 'Full' marks.

5   It takes approximately 1 quart of oil to raise the level from the 'Add' mark to the 'Full' mark on the dipstick. Do not allow the level to drop below the 'Add' mark as this may cause engine damage due to oil starvation. On the other hand, do not overfill the engine by adding oil above the 'Full' mark as this may result in oil-fouled spark plugs, oil leaks or oil seal failures.

6   Oil is added to the engine after removing a twist-off cap **(see illustrations)**. The cap should be duly marked 'Engine oil' or similar wording. An oil can spout or funnel will reduce spills as the oil is poured in.

7   Checking the oil level can also be a step towards preventative maintenance. If you find the oil level dropping abnormally, this is an indication of oil leakage or internal engine wear which should be corrected. If there are water droplets in the oil, or it is milky looking, this also indicates component failure and the engine should be checked immediately. The condition of the oil can also be checked along with the level. With the dipstick removed from the engine, take your thumb and index finger and wipe the oil up the dipstick, looking for small dirt particles or engine filings which will cling to the dipstick. This is an indication that the oil should be drained and fresh oil added (see Section 4).

Engine coolant

Warning: *Do not allow antifreeze to come in contact with your skin or painted surfaces of the vehicle. Flush contaminated areas immediately with plenty of water. Don't store new coolant or leave old coolant lying around where it's accessible to children or pets - they're attracted by its sweet taste. Ingestion of even a small amount of coolant can be fatal! Wipe up garage floor and drip pan coolant spills immediately. Keep antifreeze containers covered, and repair leaks in your cooling system immediately.*

8   Most vehicles are equipped with a pressurized coolant recovery system which makes coolant level checks very easy. A clear or white coolant reservoir attached to the inner fender panel is connected by a hose to the radiator cap. As the engine heats up during operation, coolant is forced from the radiator, through the connecting tube and into the reservoir. As the engine cools, this coolant is automatically drawn back into the radiator to keep the correct level.

9   The coolant level should be checked when the engine is cold. Merely observe the level of fluid in the reservoir, which should be at or near the 'Full cold' mark on the side of the reservoir. If the system is completely cooled, also check the level in the radiator by removing the cap. Some systems also have a

2.6b On some models, the oil filler cap is located in the rocker cover

'Full hot' mark to check the level when the engine is hot.

10   If your particular vehicle is not equipped with a coolant recovery system, the level should be checked by removing the radiator cap. However, the cap should not under any circumstances be removed while the system is hot, as escaping steam could cause serious injury. Wait until the engine has completely cooled, then wrap a thick cloth around the cap and turn it to its first stop. If any steam escapes from the cap, allow the engine to cool further. Then remove the cap and check the level in the radiator. It should be about 2 to 3 inches below the bottom of the filler neck.

11   If only a small amount of coolant is required to bring the system up to the proper level, regular water can be used. However, to maintain the proper antifreeze/water mixture in the system, both should be mixed together to replenish a low level. High-quality antifreeze offering protection to -20° should be mixed with water in the proportion specified on the container. Do not allow antifreeze to come in contact with your skin or painted surfaces of the car. Flush contacted areas

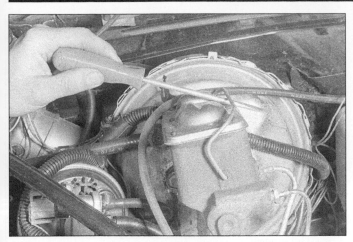

2.18a Early models have a cast iron brake master cylinder with a cover which must be removed to check the fluid level - use a screwdriver to unsnap the retainer

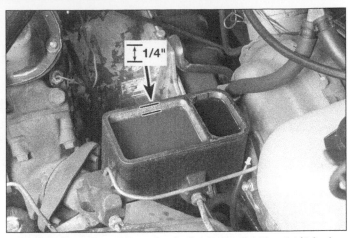

2.18b On early models, maintain the brake fluid level 1/4-inch below the top edge (arrow)

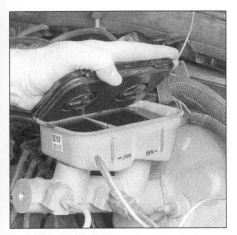

2.19 On late model vehicles, the fluid level inside the brake reservoirs is easily checked by observing the level from outside - fluid can be added to the reservoir after the cover is removed by prying up on the tabs

immediately with plenty of water.

12 On systems with a recovery tank. coolant should be added to the reservoir after removing the cap at the top of the reservoir. Coolant should be added directly into the radiator on systems without a coolant recovery tank.

13 As the coolant level is checked, observe the condition of the coolant. It should be relatively clear. If the fluid is brown or a rust color, this is an indication that the system should be drained, flushed and refilled (see Section 29).

14 If the cooling system requires repeated additions to keep the proper level, have the pressure radiator cap checked for proper sealing ability. Also check for leaks in the system (cracked hoses, loose hose connections, leaking gaskets, etc.).

Windshield washer

15 The fluid for the windshield washer system is located in a plastic reservoir. The level

inside the reservoir should be maintained at the Full' mark.

16 In milder climates, plain water can be used in the reservoir, but it should be kept no more than two-thirds full to allow for expansion if the water freezes. In colder climates, use windshield washer system antifreeze, available at any auto parts store, to lower the freezing point of the fluid. Mix the antifreeze with water in accordance with the manufacturer's directions on the container. **Caution:** *Don't use cooling system antifreeze - it will damage the vehicle's paint.*

Brake fluid

Refer to illustrations 2.18a, 2.18b and 2.19

17 The brake master cylinder is mounted on the front of the power booster unit in the engine compartment.

18 On earlier models it will be necessary to remove the reservoir cover to check the fluid level **(see illustrations).**

19 The fluid level on later models is readily visible through the translucent reservoir. The level should be above the MIN marks on the reservoirs **(see illustration).** If a low level is indicated, be sure to wipe the top of the reservoir cover with a clean rag to prevent contamination of the brake and/or clutch system before removing the cover.

20 When adding fluid, pour it carefully into the reservoir to avoid spilling it onto surrounding painted surfaces. Be sure the specified fluid is used, since mixing different types of brake fluid can cause damage to the system. See *Recommended lubricants and fluids* at the front of this Chapter or your owner's manual. **Warning:** *Brake fluid can harm your eyes and damage painted surfaces, so use extreme caution when handling or pouring it. Do not use brake fluid that has been standing open or is more than one year old. Brake fluid absorbs moisture from the air. Excess moisture can cause a dangerous loss of braking effectiveness.*

21 At this time the fluid and master cylinder can be inspected for contamination. The sys-

tem should be drained and refilled if deposits, dirt particles or water droplets are seen in the fluid.

22 After filling the reservoir to the proper level, make sure the cover is on tight to prevent fluid leakage.

23 The brake fluid level in the master cylinder will drop slightly as the pads and the brake shoes at each wheel wear down during normal operation.

24 If the master cylinder requires repeated additions to keep it at the proper level, it's an indication of leakage in the brake system, which should be corrected immediately. Check all brake lines and connections (see Section 26 for more information).

25 If, upon checking the master cylinder fluid level, you discover one or both reservoirs empty or nearly empty, the brake system should be bled (Chapter 9).

Manual transmission

26 Manual shift transmissions do not have a dipstick. The fluid level is checked by removing a plug in the side of the transmission case. Locate this plug and use a rag to clean the plug and the area around it.

27 With the vehicle components cold, remove the plug. If fluid immediately starts leaking out, thread the plug back into the transmission because the fluid level is alright. If there is no fluid leakage. completely remove the plug and place your little finger inside the hole. The fluid level should be just at the bottom of the plug hole.

28 If the transmission needs more fluid, use a syringe to squeeze the appropriate lubricant into the plug hole to bring the fluid up to the proper level.

29 Thread the plug back into the transmission and tighten it securely. Drive the car and check for leaks around the plug.

Automatic transmission

Refer to illustration 2.33

30 The fluid inside the transmission must be at normal operating temperature to get an accurate reading on the dipstick. This is done

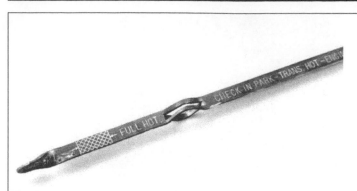

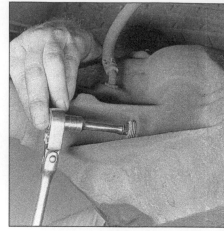

2.33 With the transmission at normal operating temperature the fluid level should be at the middle of the cross hatched area

2.38 Use a ratchet and extension to remove the differential fill plug

by driving the car for several miles, making frequent starts and stops to allow the transmission to shift through all gears.

31   Park the car on a level surface, place the selector lever in 'Park' and leave the engine running at an idle.

32   Remove the transmission dipstick (located on the right side, near the rear of the engine) and wipe all the fluid from the end of the dipstick with a clean rag.

33   Push the dipstick back into the transmission until the cap seats firmly on the dipstick tube. Now remove the dipstick again and observe the fluid on the end **(see illustration)**. The highest point of fluid should be between the 'Full' mark and 1/4 inch below the 'Full' mark.

34   If the fluid level is at or below the 'Add' marking the dipstick, add sufficient fluid to raise the level to the 'Full' mark. One pint of fluid will raise the level from 'Add' to 'Full'. Fluid should be added directly into the dipstick guide tube, using a funnel to prevent spills.

35   It is important that the transmission not be overfilled. Under no circumstances should the fluid level be above the 'Full' mark on the dipstick, as this could cause internal damage to the transmission. The best way to prevent overfilling is to add fluid a little at a time, driving the car and checking the level between additions.

36   Use only the specified transmission fluid. This information can be found in the *Recommended Lubricants* Section.

37   The condition of the fluid should also be checked along with the level. If the fluid at the end of the dipstick is a dark reddish-brown color, or if the fluid has a 'burnt' smell, the transmission fluid should be replaced with fresh. If you are in doubt about the condition of the fluid, purchase some new fluid and compare the two for color and smell.

Rear differential

Refer to illustrations 2.38 and 2.39

38   Like the manual transmission, the rear axle has an inspection and fill plug which must be removed to check the fluid level **(see illustration)**.

39   Remove the plug which is located either in the removable cover plate or on the side of the differential carrier **(see illustration)**. Use your little finger to reach inside the rear axle housing to feel the level of the fluid **(see illustration)**. It should be at the bottom of the plug hole.

40   If this is not the case, add the proper lubricant into the rear axle carrier through the plug hole. A syringe or a small funnel can be used for this.

41   Make certain the correct lubricant is used, as regular and Positraction rear axles require different lubricants. You can ascertain which type of axle you have by reading the stamped number on the axle tube *(See Vehicle Identification Numbers* at the front of this manual).

42   Tighten the plug securely and check for leaks after the first few miles of driving.

Power steering

Refer to illustration 2.47

43   Unlike manual steering, the power steering system relies on fluid which may, over a period of time, require replenishing.

44   The reservoir for the power steering pump will be located near the front of the engine, and can be mounted on either the left or right side.

45   The power steering fluid level should be checked only after the car has been driven, with the fluid at operating temperature. The front wheels should be pointed straight ahead.

46   With the engine shut off, use a rag to clean the reservoir cap and the areas around the cap. This will help to prevent foreign material from falling into the reservoir when the cap is removed.

47   Twist off the reservoir cap which has a built-in dipstick attached to it **(see illustration)**. Pull off the cap and clean the fluid at the bottom of the dipstick with a clean rag. Now reinstall the dipstick/cap assembly to get a fluid level reading. Remove the dip-

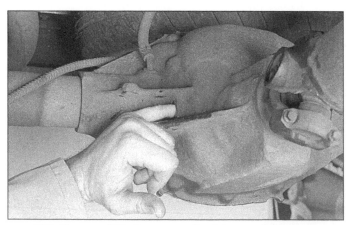

2.39 Use your little finger as a dipstick to make sure the differential oil level is even with the bottom of the opening

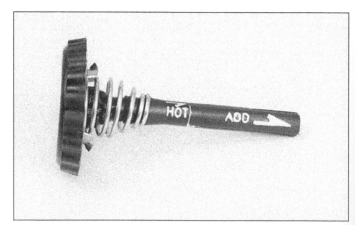

2.47 The markings on the power steering fluid dipstick indicate the safe range

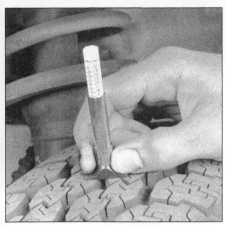

3.2 A tire tread depth indicator should be used to monitor tire wear - they are available at auto parts stores and service stations and cost very little

UNDERINFLATION

CUPPING

Cupping may be caused by:
- Underinflation and/or mechanical irregularities such as out-of-balance condition of wheel and/or tire, and bent or damaged wheel.
- Loose or worn steering tie-rod or steering idler arm.
- Loose, damaged or worn front suspension parts.

OVERINFLATION

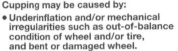
INCORRECT TOE-IN OR EXTREME CAMBER

FEATHERING DUE TO MISALIGNMENT

3.3 This chart will help you to determine the condition of the tires, the probable cause of the abnormal wear and the corrective action necessary

stick/cap and observe the fluid level. It should be at the 'Full hot' mark on the dipstick.

48   If additional fluid is required, pour the specified lubricant directly into the reservoir using a funnel to prevent spills.

49   If the reservoir requires frequent fluid additions, all power steering hoses, hose connections, the power steering pump and the steering box should be carefully checked for leaks.

3   Tire and tire pressure checks

Refer to illustrations 3.2, 3.3, 3.4a, 3.4b and 3.8

1   Periodic inspection of the tires may spare you the inconvenience of being stranded with a flat tire. It can also provide you with vital information regarding possible problems in the steering and suspension systems before major damage occurs.

2   The original tires on this vehicle are equipped with 1/2-inch wide bands that will appear when tread depth reaches 1/16-inch, at which time the tires can be considered worn out. Tread wear can be monitored with a simple, inexpensive device known as a tread depth indicator (see illustration).

3   Note any abnormal tread wear (see illustration). Tread pattern irregularities such as cupping, flat spots and more wear on one side than the other are indications of front end alignment and/or balance problems. If any of these conditions are noted, take the vehicle to a tire shop or service station to correct the problem.

4   Look closely for cuts, punctures and embedded nails or tacks. Sometimes a tire will hold air pressure for a short time or leak down very slowly after a nail has embedded itself in the tread. If a slow leak persists, check the valve stem core to make sure it's tight (see illustration). Examine the tread for an object that may have embedded itself in the tire or for a "plug" that may have begun to

leak (radial tire punctures are repaired with a plug that's installed in a puncture). If a puncture is suspected, it can be easily verified by spraying a solution of soapy water onto the puncture area (see illustration). The soapy solution will bubble if there's a leak. Unless the puncture is unusually large, a tire shop or service station can usually repair the tire.

5   Carefully inspect the inner sidewall of each tire for evidence of brake fluid leakage. If you see any, inspect the brakes immediately.

6   Correct air pressure adds miles to the

lifespan of the tires, improves mileage and enhances overall ride quality. Tire pressure cannot be accurately estimated by looking at a tire, especially if it's a radial. A tire pressure gauge is essential. Keep an accurate gauge in the vehicle. The pressure gauges attached to the nozzles of air hoses at gas stations are often inaccurate.

7   Always check tire pressure when the tires are cold. Cold, in this case, means the vehicle has not been driven over a mile in the three hours preceding a tire pressure check. A pressure rise of four to eight pounds is not

3.4a If a tire loses air on a steady basis, check the valve core first to make sure it's snug (special inexpensive wrenches are commonly available at auto parts stores)

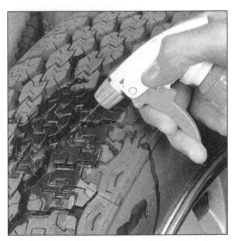

3.4b If the valve core is tight, raise the corner of the vehicle with the low tire and spray a soapy water solution onto the tread as the tire is turned slowly - slow leaks will cause small bubbles to appear

3.8 To extend the life of your tires, check the air pressure at least once a week with an accurate gauge (don't forget the spare!)

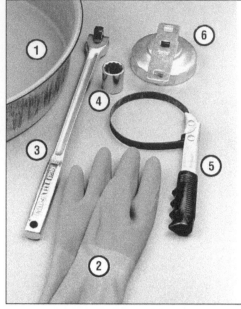

4.3 Tools and materials necessary for a typical oil and filter change

1   **Drain pan** - It should be fairly shallow in depth, but wide to prevent spills
2   **Rubber gloves** - When removing the drain plug and filter, you will get oil on your hands (the gloves will prevent burns)
3   **Breaker bar** - Sometimes the oil drain plug is tight, and a long breaker bar is needed to loosen it
4   **Socket** - To be used with the breaker bar or ratchet (must be the correct size to it the drain plug)
5   **Filter wrench** - This is a metal band-type wrench, which requires clearance around the filter to be effective
6   **Filter wrench** - This type fits on the bottom of the filter and can be turned with a ratchet or breaker bar (different-size wrenches are available for different types of filters)

uncommon once the tires are warm.

8   Unscrew the valve cap protruding from the wheel or hubcap and push the gauge firmly onto the valve stem **(see illustration)**. Note the reading on the gauge and compare the figure to the recommended tire pressure shown on the placard on the driver's side door pillar. Be sure to reinstall the valve cap to keep dirt and moisture out of the valve stem mechanism. Check all four tires and, if necessary, add enough air to bring them up to the recommended pressure.

9   Don't forget to keep the spare tire inflated to the specified pressure (refer to your owner's manual or the tire sidewall). Note that the pressure recommended for compact spare tires is higher than for the tires on the vehicle.

4   Engine oil and filter change

Refer to illustrations 4.3, 4.9, 4.14, 4.18, 4.19 and 4.20

1   Frequent oil changes may be the best form of preventative maintenance available for the home mechanic. When engine oil ages, it gets diluted and contaminated which ultimately leads to premature parts wear.

2   Although some sources recommend oil filter changes every other oil change, we feel that the minimal cost of an oil filter and the relative ease with which it is installed dictates that a new filter be used whenever the oil is changed.

3   Gather together all necessary tools and materials before beginning this procedure **(see illustration)**.

4   In addition, you should have plenty of clean rags and newspapers handy to mop up any spills. Access to the underside of the car is greatly improved if the car can be lifted on a hoist, driven onto ramps or supported by jack stands. Do not work under a car which is supported only by a bumper, hydraulic or scissors-type jack.

5   If this is your first oil change on the car,

it is a good idea to crawl underneath and familiarize yourself with the locations of the oil drain plug and the oil filter. Since the engine and exhaust components will be warm during the actual work, it is best to figure out any potential problems before the car and its accessories are hot.

6   Allow the car to warm up to normal operating temperature. If the new oil or any tools are needed, use this warm-up time to gather everything necessary for the job. The correct type of oil to buy for your application can be found in *Recommended Lubricants and fluids* at the beginning of this Chapter.

7   With the engine oil warm (warm engine oil will drain better and more built-up sludge will be removed with the oil), raise the vehicle for access beneath. Make sure the car is firmly supported. If jack stands are used they should be placed towards the front of the frame rails which run the length of the car.

8   Move all necessary tools, rags and newspaper under the car. Position the drain pan under the drain plug. Keep in mind that the oil will initially flow from the pan with some force, so place the pan accordingly.

9   Being careful not to touch any of the hot exhaust pipe components, use the wrench to remove the drain plug near the bottom of the oil pan **(see illustration)**. Depending on how hot the oil has become. you may want to wear gloves while unscrewing the plug the final few turns.

10   Allow the old oil to drain into the pan. It may be necessary to move the pan further under the engine as the oil flow reduces to a trickle.

11   After all the oil has drained, clean the drain plug thoroughly with a clean rag. Small metal filings may cling to this plug which could immediately contaminate your new oil.

12   Clean the area around the drain plug

4.9 The oil pan drain plug is located at the bottom of the pan and should be removed with a socket or box-end wrench - DO NOT use an open end wrench as the corners of the bolt can be easily rounded off

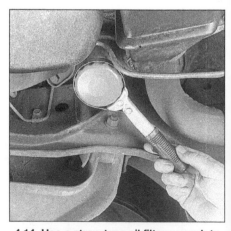

4.14 Use a strap-type oil filter wrench to loosen the filter. Note that the wrench is positioned at the bottom of the filter, where the filter has the most strength. If access makes removal difficult, other types of filter wrenches are available

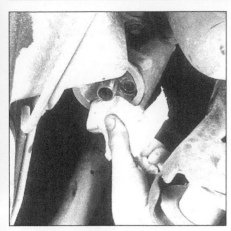

4.18 Clean the filter mount and check the old gasket

4.19 Lubricate the oil filter gasket with clean engine oil before installing the filter on the engine

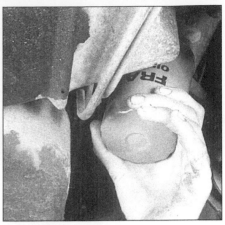

4.20 Tighten the new filter hand tight according to the instructions printed on the filter package (do not overtighten)

opening and reinstall the drain plug. Tighten the plug securely with your wrench.

13   Move the drain pan in position under the oil filter.

14   Now use the filter wrench to loosen the oil filter **(see illustration)**. Chain or metal band-type filter wrenches may distort the filter canister, but don't worry too much about this as the filter will be discarded anyway.

15   Sometimes the oil filter is on so tight it cannot be loosened, or it is positioned in an area which is inaccessible with a filter wrench. As a last resort, you can punch a metal bar or long screwdriver directly through the bottom of the canister (horizontally) and use this as a T-bar to turn the filter. If this must be done, be prepared for oil to spurt out of the canister as it is punctured.

16   Completely unscrew the old filter. Be careful, it is full of oil. Empty the old oil inside the filter into the drain pan.

17   Compare the old filter with the new one to make sure they are of the same type.

18   Use a clean rag to remove all oil, dirt and sludge from the area where the oil filter mounts to the engine **(see illustration)**. Check the old filter to make sure the rubber gasket is not stuck to the engine mounting surface. If this gasket is stuck to the engine (use a flashlight if necessary), remove it.

19   Open one of the cans of new oil and fill the new filter with fresh oil. Also smear a light coat of this fresh oil onto the rubber gasket of the new oil filter **(see illustration)**.

20   Screw the new filter to the engine following the tightening directions printed on the filter canister or packing box **(see illustration)**. Most filter manufacturers recommend against using a filter wrench due to possible overtightening or damage to the canister.

21   Remove all tools, rags, etc. from under the car, being careful not to spill the oil in the drain pan. Lower the car off its support devices.

22   Move to the engine compartment and locate the oil filler cap on the engine. In most cases there will be a screw-off cap on the rocker arm cover (at the side of the engine) or a cap at the end of a fill tube at the front of

the engine. In any case, the cap will most likely be labeled 'Engine Oil' or something similar.

23   If an oil can spout is used, push the spout into the top of the oil can and pour the fresh oil through the filler opening. A funnel placed into the opening may also be used.

24   Pour about 3 qts. of fresh oil into the engine. Wait a few minutes to allow the oil to drain to the pan, then check the level on the oil dipstick (see Section 2 if necessary).

Note: *When changing oil (or performing any operation which results in oil drainage or loss) on a turbocharged V6, perform the following steps BEFORE starting the engine:*

a) Disconnect the ignition switch connector /pink wire) from the HEI distributor.

b) Crank the engine over several times until the oil light goes out (do not exceed 30 seconds for each cranking interval).

c) Reconnect the pink wire to the distributor, then recheck the oil level. This procedure will ensure that the bearings in the turbocharger are not damaged due to lack of lubrication.

If the oil level is at or near the lower 'Add' mark, start the engine and allow the new oil to circulate.

25   Run the engine for only about a minute and then shut it off. Immediately look under the ear and check for leaks at the oil pan drain plug and around the oil filter: If either is leaking, tighten with a bit more force.

26   With the new oil circulated and the filter now completely full. recheck the level on the dipstick and add enough oil to bring the level to the 'Full' mark on the dipstick.

27   During the first few trips after an oil change, make a point to check for leaks and also the oil level.

28   The old oil drained from the engine cannot be reused in its present state and should be disposed of. Oil reclamation centers, auto repair shops and gas stations will normally accept the oil which can be refined and used again. After the oil has cooled, it can be drained into a suitable container (capped plastic jugs, topped bottles. milk cartons.

etc.) for transport to one of these disposal sites.

5   Chassis lubrication

Refer to illustrations 5.1 and 5.6

1   Refer to *Recommended lubricants and fluids* at the front of this Chapter to obtain the necessary grease, etc. You'll also need a grease gun **(see illustration)**. Occasionally plugs will be installed rather than grease fit-

5.1 Tools and materials necessary for chassis lubrication

1   *Engine oil - Light engine oil in a can like this can be used for door and hood hinges*

2   *Graphite spray - Used to lubricate lock cylinders*

3   *Grease - Grease, in a variety of types and weights, is available for use in a grease gun. Check the specifications for your requirements*

4   *Grease gun - A common grease gun, shown here with a detachable hose and nozzle, is needed for chassis lubrication. After use, clean it thoroughly!*

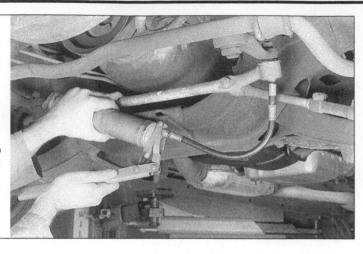

5.6 After wiping the grease fitting clean, push the nozzle firmly into place and pump the grease into the component. Usually about two pumps of the gun will be sufficient

Check for a chafed area that could fail prematurely.

Check for a soft area indicating the hose has deteriorated inside.

Overtightening the clamp on a hardened hose will damage the hose and cause a leak.

Check each hose for swelling and oil-soaked ends. Cracks and breaks can be located by squeezing the hose.

6.4 Hoses, like drivebelts, have a habit of failing at the worst possible time - to prevent the inconvenience of a blown radiator or heater hose, inspect them carefully as shown here

tings. If so, grease fittings will have to be purchased and installed.

2   Look under the vehicle and see if grease fittings or plugs are installed (*see the Typical engine compartment underside component illustration at the beginning of this Chapter*). If there are plugs, remove them and buy grease fittings, which will thread into the component. A dealer or auto parts store will be able to supply the correct fittings. Straight, as well as angled, fittings are available.

3   For easier access under the vehicle, raise it with a jack and place jackstands under the frame. Make sure it's safely supported by the stands. If the wheels are to be removed at this interval for tire rotation or brake inspection, loosen the lug nuts slightly while the vehicle is still on the ground.

4   Before beginning, force a little grease out of the nozzle to remove any dirt from the end of the gun. Wipe the nozzle clean with a rag.

5   With the grease gun and plenty of clean rags, crawl under the vehicle and begin lubricating the components.

6   Wipe the balljoint grease fitting nipple clean and push the nozzle firmly over it **(see illustration)**. Squeeze the trigger or handle on the grease gun to force grease into the component. The balljoints should be lubricated until the rubber seal is firm to the touch. Do not pump too much grease into the fittings as it could rupture the seal. For all other suspension and steering components, continue pumping grease into the fitting until it oozes out of the joint between the two components. If it escapes around the grease gun nozzle, the nipple is clogged or the nozzle is not completely seated on the fitting. Resecure the gun nozzle to the fitting and try again. If necessary, replace the fitting with a new one.

7   Wipe the excess grease from the components and the grease fitting. Repeat the procedure for the remaining fittings.

8   If equipped with a manual transmission, lubricate the shift linkage with a little multi-purpose grease.

9   On manual transmission equipped models, lubricate the clutch linkage pivot points with clean engine oil. Lubricate the pushrod-to-fork contact points with chassis grease.

10   While you are under the vehicle, clean and lubricate the parking brake cable, along with the cable guides and levers. This can be done by smearing some of the chassis grease onto the cable and its related parts with your fingers.

11   The steering gear seldom requires the addition of lubricant, but if there is obvious leakage of grease at the seals, remove the plug or cover and check the lubricant level. If the level is low, add the specified lubricant.

12   Open the hood and smear a little chassis grease on the hood latch mechanism. Have an assistant pull the hood release lever from inside the vehicle as you lubricate the cable at the latch.

13   Lubricate all the hinges (door, hood, etc.) with engine oil to keep them in proper working order.

14   The key lock cylinders can be lubricated with spray graphite or silicone lubricant, which is available at auto parts stores.

15   Lubricate the door weatherstripping with silicone spray. This will reduce chafing and retard wear.

6   Cooling system check

Refer to illustration 6.4

1   Many major engine failures can be attributed to a faulty cooling system. If the vehicle is equipped with an automatic transmission, the cooling system also cools the transmission fluid and thus plays an important role in prolonging transmission life.

2   The cooling system should be checked with the engine cold. Do this before the vehicle is driven for the day or after it has been shut off for at least three hours.

3   Remove the radiator cap by turning it to the left until it reaches a stop. If you hear a hissing sound (indicating there is still pressure in the system), wait until this stops. Now press down on the cap with the palm of your hand and continue turning to the left until the cap can be removed. Thoroughly clean the cap, inside and out, with clean water. Also clean the filler neck on the radiator. All traces

of corrosion should be removed. The coolant inside the radiator should be relatively transparent. If it is rust colored, the system should be drained and refilled (see Section 29). If the coolant level is not up to the top, add additional antifreeze/coolant mixture (see Section 2).

4   Carefully check the large upper and lower radiator hoses along with the smaller diameter heater hoses which run from the engine to the firewall. On some models the heater return hose runs directly to the radiator. Inspect each hose along its entire length, replacing any hose which is cracked, swollen or shows signs of deterioration. Cracks may become more apparent if the hose is squeezed **(see illustration)**. Regardless of condition, it's a good idea to replace hoses with new ones every two years.

5   Make sure that all hose connections are

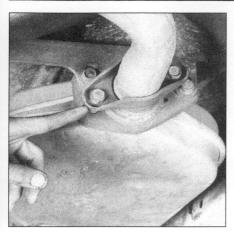

7.2 Checking exhaust system flanges and connections for signs of leaks

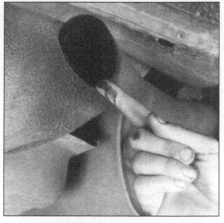

7.5 Check the tailpipe - black, sooty deposits at the end of the exhaust pipe may be an indication that the carburetor needs adjustment or the engine is in need of a tune-up

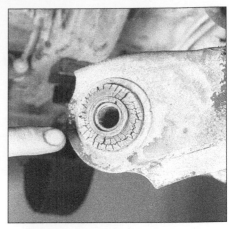

8.6a Cracked rubber bushings in the steering and suspension systems should be replaced

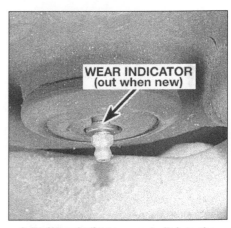

WEAR INDICATOR
(out when new)

8.6b Wear indicators are built into the lower balljoints to aid in their inspection

tight. A leak in the cooling system will usually show up as white or rust colored deposits on the areas adjoining the leak. If wire-type clamps are used at the ends of the hoses, it may be a good idea to replace them with more secure screw-type clamps.

6   Use compressed air or a soft brush to remove bugs, leaves, etc. from the front of the radiator or air conditioning condenser. Be careful not to damage the delicate cooling fins or cut yourself on them.

7   Every other inspection, or at the first indication of cooling system problems, have the cap and system pressure tested. If you don't have a pressure tester, most gas stations and repair shops will do this for a minimal charge.

---

7   Exhaust system check

Refer to illustrations 7.2 and 7.5

1   With the exhaust system cold (at least three hours after being driven), check the complete exhaust system from its starting point at the engine to the end of the tailpipe. This is best done on a hoist where full access is available.

2   Check the pipes and their connections for signs of leakage and/or corrosion indicating a potential failure. Check that all brackets and hangers are in good condition and are tight **(see illustration)**.

3   At the same time, inspect the underside of the body for holes, corrosion, open seams, etc. which may allow exhaust gases to enter the trunk or passenger compartment. Seal all body openings with silicone or body putty.

4   Rattles and other driving noises can often be traced to the exhaust system, especially the mounts and hangers. Try to move the pipes, muffler and catalytic converter (if equipped). If the components can come into contact with the body or driveline parts, secure the exhaust system with new mountings.

5   This is also an ideal time to check the running condition of the engine by inspecting

the very end of the tailpipe. The exhaust deposits here are an indication of engine tune. If the pipe is black and sooty, or bright white deposits are found here, the engine is in need of a tune-up including a thorough carburetor inspection and adjustment **(see illustration)**.

---

8   Suspension and steering check

Refer to illustrations 8.6a and 8.6b

1   Whenever the front of the car is raised for service it is a good idea to visually check the suspension and steering components for wear.

2   Indications of a fault in these systems are: excessive play in the steering wheel before the front wheels react; excessive sway around corners or body movement over rough roads; binding at some point as the steering wheel is turned.

3   Before the car is raised for inspection, test the shock absorbers by pushing downward to rock the car at each corner. If you push the car down and it does not come back to a level position within one or two bounces, the shocks are worn and need to be replaced. As this is done, check for squeaks and strange noises from the suspension components. Information on shock absorber and suspension components can be found in Chapter 10.

4   Now raise the front end of the car and support firmly by jack stands placed under the frame rails. Because of the work to be done, make sure the car cannot fall from the stands.

5   Grab the top and bottom of the front tire with your hands and rock the tire/wheel on its spindle. If there is movement of more than 0.005 in, the wheel bearings should be serviced (see Section 24).

6   Crawl under the car and check for loose bolts, broken or disconnected parts and deteriorated rubber bushings **(see illustra-**

tion) on all suspension and steering components. Look for grease or fluid leaking from around the steering box. Check the power steering hoses and their connections for leaks. Check the balljoints for wear **(see illustration)**.

7   Have an assistant turn the steering wheel from side to side and check the steering components for free movement, chafing or binding. If the steering does not react with the movement of the steering wheel, try to determine where the slack is located.

---

9   Drivebelt check, adjustment and replacement

Refer to illustrations 9.3, 9.4 and 9.8

1   The drivebelts, or V-belts as they are often called, are located at the front of the engine and play an important role in the overall operation of the engine and accessories. Due to their function and material makeup, the belts are prone to failure after a period of time and should be inspected and adjusted periodically to prevent major engine damage.

2   The number of belts used on a particular vehicle depends on the accessories installed.

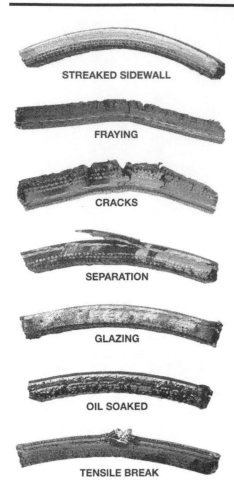

STREAKED SIDEWALL

FRAYING

CRACKS

SEPARATION

GLAZING

OIL SOAKED

TENSILE BREAK

9.3 Here are some of the more common problems associated with drivebelts (check the belts very carefully to prevent untimely breakdown)

Drivebelts are used to turn the alternator, power steering pump, water pump and air conditioning compressor. Depending on the pulley arrangement, more than one of these components may be driven by a single belt.

3   With the engine off, locate the drivebelts at the front of the engine. Using your fingers (and a flashlight, if necessary), move along the belts checking for cracks and separation of the belt plies. Also check for fraying and glazing, which gives the belt a shiny appearance **(see illustration)**. Both sides of each belt should be inspected, which means you will have to twist the belt to check the underside. Check the pulleys for nicks, cracks, distortion and corrosion.

4   The tension of each belt is checked by pushing on it at a distance halfway between the pulleys. Push firmly with your thumb and see how much the belt moves (deflects) **(see illustration)**. A rule of thumb is that if the distance from pulley center to pulley center is between seven and 11 inches, the belt should deflect 1/4-inch. If the belt travels between pulleys spaced 12 to 16 inches apart, the belt should deflect 1/2-inch.

5   If adjustment is needed, either to make the belt tighter or looser, it's done by moving the belt-driven accessory on the bracket.

6   For each component there will be an adjusting bolt and a pivot bolt. Both bolts must be loosened slightly to enable you to move the component.

7   After the two bolts have been loosened, move the component away from the engine to tighten the belt or toward the engine to loosen the belt.

8   Hold the accessory in position and check the belt tension. If it's correct, tighten the two bolts until just snug, then recheck the tension. If the tension is all right, tighten the bolts **(see illustration)**.

9   It will often be necessary to use some sort of pry bar to move the accessory while the belt is adjusted. If this must be done to gain the proper leverage, be very careful not to damage the component being moved or the part being pried against.

10   To replace a belt, follow the above procedures for drivebelt adjustment but slip the belt off the pulleys and remove it. Since belts tend to wear out more or less at the same time, it's a good idea to replace all of them at

the same time. Mark each belt and the corresponding pulley grooves so the replacement belts can be installed properly.

11   Take the old belts with you when purchasing new ones in order to make a direct comparison for length, width and design.

12   Adjust the belts as described earlier in this Section.

10   Fuel system check

1   There are certain precautions to take when inspecting or servicing the fuel system components. Work in a well ventilated area and do not allow open flames (cigarettes, appliance pilot lights, etc.) to get near the work area. Mop up spills immediately and do not store fuel-soaked rags where they could ignite.

2   The fuel system is under some amount of pressure, so if any fuel lines are disconnected for servicing, be prepared to catch the fuel as it spurts out. Plug all disconnected fuel lines immediately after disconnection to prevent the tank from emptying itself.

3   The fuel system is most easily checked with the car raised on a hoist where the components under the car are readily visible and accessible.

4   If the smell of gasoline is noticed while driving, or after the car has sat in the sun, the system should be thoroughly inspected immediately.

5   Remove the gas filler cap and check for damage, corrosion and a proper sealing imprint on the gasket. Replace the cap with a new one if necessary.

6   With the car raised, inspect the gas tank and filler neck for punctures, cracks or any damage. The connection between the filler neck and the tank is especially critical. Sometimes a rubber filler neck will leak due to loose clamps or deteriorated rubber; problems a home mechanic can usually rectify.

7   Do not under any circumstances try to repair a fuel tank yourself (except rubber

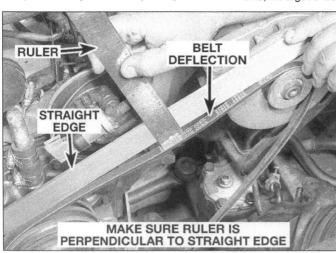

RULER

BELT DEFLECTION

STRAIGHT EDGE

MAKE SURE RULER IS PERPENDICULAR TO STRAIGHT EDGE

9.4 Drivebelt tension can be checked with a straightedge and a ruler

9.8 Adjust the belt tension by gently prying on the component as the adjustment bolt is tightened

11.4 Pliers are used to release the hose clamp and slide it away from the PCV valve

12.2 On most models, the first step in removing the air filter is unscrewing the wing nut

12.5 If light can be easily seen through the air filter element. the filter can be reused (if in doubt, replace the filter with a new one)

components) unless you have considerable experience. A welding torch or any open flame can easily cause the fuel vapors to explode if the proper precautions are not taken.

8   Carefully check all rubber hoses and metal lines leading away from the fuel tank. Check for loose connections, deteriorated hose, crimped lines or damage of any kind. Follow these lines up to the front of the car, carefully inspecting them all the way. Repair or replace damaged sections as necessary.

9   If a fuel odor is still evident after the inspection, refer to Section 31 on the evaporative emissions system and Section 16 for carburetor adjustment.

11   Positive Crankcase Ventilation (PCV) valve replacement

Refer to illustration 11.4

1   The PCV valve can usually be found pushed into one of the rocker arm covers at the side of the engine. There will be a hose connected to the valve which runs to either the carburetor or the intake manifold.

2   When purchasing a replacement PCV valve, make sure it is for your particular vehicle, model year and engine size.

3   Pull the valve (with the hose attached) from its rubber grommet in the rocker arm cover.

4   Using pliers or a screwdriver, depending on the type of clamp, loosen the clamp at the end of the hose and move the clamp upwards on the PCV hose **(see illustration)**.

5   Now pull the PCV valve from the end of the hose, noting its installed position and direction.

6   Compare the old valve with the new one to make sure they are the same.

7   Push the new valve into the end of the hose until it is fully seated.

8   Move the hose clamp down the hose and tighten the clamp securely around the end of the hose.

9   Inspect the rubber grommet in the cover

for damage and replace it with a new one if faulty.

10   Push the PCV valve and hose securely into the rocker arm cover.

11   More information on the PCV system can be found in Chapter 6.

12   Air filter and PCV filter replacement

Refer to illustrations 12.2, 12.5 and 12.12

1   At the specified intervals, the air filter and PCV filter should be replaced with new ones. A thorough program of preventative maintenance would call for the two filters to be inspected periodically between changes.

2   The air filter is located inside the air cleaner housing on the top of the engine. To remove the filter, unscrew the wing nut at the top of the air cleaner and lift off the top plate **(see illustration)**. If there are vacuum hoses connected to this plate, note their positions and disconnect them.

3   While the top plate is off, be careful not to drop anything down into the carburetor.

4   Lift the air filter out of the housing.

5   To check the filter, hold it up to strong sunlight, or place a flashlight or droplight on the inside of the ring-shaped filter **(see illustration)**. If you can see light coming through the paper element, the filter is alright. Check all the way around the filter.

6   Wipe the inside of the air cleaner clean with a rag.

7   Place the old filter (if in good condition) or the new filter (if specified interval has elapsed) back into the air cleaner housing. Make sure it seats properly in the bottom of the housing.

8   Connect any disconnected vacuum hoses to the top plate and reinstall the top plate with the wing nut.

9   On nearly all cars the PCV filter is also located inside the air cleaner housing. Remove the top plate as described previously and locate the filter on the side of the housing.

12.12 The PCV filter on most vehicles is located inside the air cleaner housing

10   Loosen the hose clamp at the end of the PCV hose leading to the filter. Disconnect the hose from the filter.

11   Remove the metal locking clip which secures the filter holder to the air cleaner housing. Pliers can be used for this.

12   Remove the filter and plastic holder from the inside of the air cleaner **(see illustration)**.

13   Compare the new filter with the old one to make sure they are the same.

14   Place the new filter assembly into position and install the metal locking clip on the outside of the air cleaner.

15   Connect the PCV hose and tighten the clamp around the end of the hose.

16   Reinstall the air cleaner top plate and any hoses which were disconnected.

17   A few engines will not have the PCV filter at the air cleaner, but rather the filter will be in the PCV hose at some point. To locate the filter, find the hose leading into the side of the air cleaner housing and follow this hose to the filter.

18   Replacing 'in-line' PCV filters is usually a simple matter of disconnecting the hose from the filter and then pushing a replacement filter into the hose.

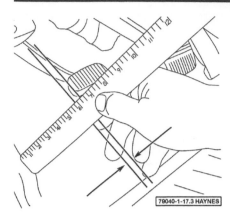

13.3 Clutch pedal free play is the distance the pedal moves before resistance is felt

13   Clutch pedal free play check and adjustment

Refer to illustration 13.3

Check

1   On manual transmission models, it is important to have the clutch free play at the proper point. Free play is the distance between the clutch pedal when it is all the way up and the point at which the clutch starts to disengage.
2   Slowly depress the clutch pedal until you can feel resistance. Do this a number of times until you can pinpoint exactly where the resistance is felt.
3   Now measure the distance the pedal travels before the resistance is felt **(see illustration)**.
4   If the distance is not as specified, the clutch pedal free play should be adjusted.

Adjustment

5   Disconnect the clutch fork return spring.
6   Hold the clutch pedal against the stop and loosen the jam nut so the clutch fork pushrod can be turned out of the swivel and

back against the clutch fork. The release bearing must contact the pressure plate fingers lightly.
7   Turn the clutch fork pushrod three and a half turns.
8   Tighten the jam nut, taking care not to change the pushrod length, and reconnect the return spring.
9   Recheck the free play.

14   Tire rotation

Refer to illustration 14.2

1   The tires should be rotated at the specified intervals and whenever uneven wear is noticed. Since the car will be raised and the tires removed anyway, this is a good time to check the brakes (see Section 26) and/or repack the wheel bearings (see Section 24). Read over these sections if this is to be done at the same time.
2   Refer to **the accompanying illustration** for the preferred tire rotation pattern.
3   See the information in *Jacking and Towing* at the front of this manual for the proper procedures to follow in raising the car and changing a tire: however, if the brakes are to be checked do not apply the parking brake as stated. Make sure the tires are blocked to prevent the car from rolling.
4   Preferably, the entire car should be raised at the same time. This can be done on a hoist or by jacking up each corner of the car and then lowering the car onto jack stands placed under the frame rails. Always use four jack stands and make sure the car is firmly supported all around.
5   After rotation, check and adjust the tire pressures as necessary and be sure to check wheel nut tightness.

15   Thermostatic air cleaner check

Refer to illustrations 15.3, 15.5 and 15.6

1   Some engines are equipped with a ther-

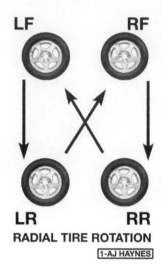

RADIAL TIRE ROTATION

14.2 Tire rotation diagram

mostatically controlled air cleaner which draws air to the carburetor from different locations, depending on engine temperature.
2   This is a visual check. If access is limited, a small mirror may have to be used.
3   Remove the engine cover and locate the damper door inside the air cleaner assembly. It's inside the long snorkel of the metal air cleaner housing **(see illustration)**.
4   If there is a flexible air duct attached to the end of the snorkel, leading to an area behind the grille, disconnect it at the snorkel. This will enable you to look through the end of the snorkel and see the damper inside.
5   The check should be done when the engine is cold. Start the engine and look through the snorkel at the damper, which should move to a closed position. With the damper closed, air cannot enter through the end of the snorkel, but instead enters the air cleaner through the flexible duct attached to the exhaust manifold and the heat stove passage **(see illustration)**.

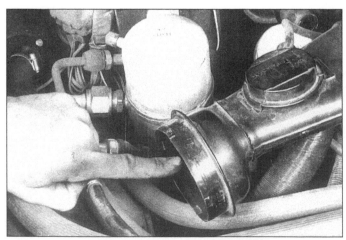

15.3 The operation of the damper door can be seen through the end of the air cleaner snorkel tube

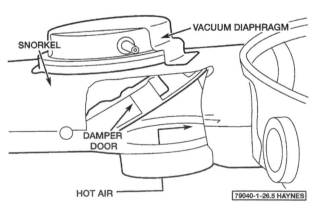

15.5 When the engine is cold, the damper door closes off the snorkel passage, allowing air warmed by the exhaust to enter the carburetor

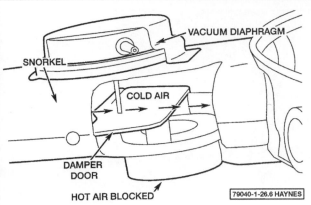

15.6 As the engine warms up, the damper door moves down to close off the heat stove passage and open the snorkel passage so outside air can enter the carburetor

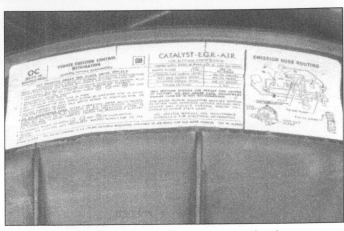

16.3 Engine compartment tune-up decal

6   As the engine warms up to operating temperature, the damper should open to allow air through the snorkel end **(see illustration)**. Depending on outside temperature, this may take 10 to 15 minutes. To speed up this check you can reconnect the snorkel air duct, drive the vehicle, then check to see if the damper is completely open.

7   If the thermo-controlled air cleaner isn't operating properly see Chapter 6 for more information.

16 Engine idle speed adjustment

Refer to illustrations 16.3 and 16.4

1   Engine idle speed is the speed at which the engine operates when no accelerator pedal pressure is applied. This speed is critical to the performance of the engine itself, as well as many engine sub-systems.

2   A hand-held tachometer must' be used when adjusting idle speed to get an accurate reading. The exact hook-up for these meters varies with the manufacturer, so follow the particular directions included.

3   Since the manufacturer used many different carburetors for their vehicles in the time period covered by this book, and each has its own peculiarities when setting idle

speed, it would be impractical to cover all types in this Section. Chapter 4 contains information on each individual carburetor used. The carburetor used on your particular engine can be found in the Specifications Section of Chapter 4. However, all vehicles covered in this manual should have a tune-up decal in the engine compartment, usually placed near the top of the radiator (see illustration). The printed instructions for setting idle speed can be found on this decal, and should be followed since they are for your particular engine.

4   Basically, for most applications, the idle speed is set by turning an adjustment screw located at the side of the carburetor (see illustration). This screw changes the linkage, in essence, depressing or letting up on your accelerator pedal. This screw may be on the linkage itself or may be part of the idle stop solenoid. Refer to the tune-up decal or Chapter 4.

5   Once you have found the idle screw, experiment with different length screwdrivers until the adjustments can be easily made, without coming into contact with hot or moving engine components.

6   Follow the instructions on the tune-up decal or in Chapter 4, which will probably include disconnecting certain vacuum or

electrical connections. To plug a vacuum hose after disconnecting it, insert a properly-sized metal rod into the opening, or thoroughly wrap the open end with tape to prevent any vacuum loss through the hose.

7   If the air cleaner is removed, the vacuum hose to the snorkel should be plugged.

8   Make sure the parking brake is firmly set and the wheels blocked to prevent the car from rolling. This is especially true if the transmission is to be in 'Drive'. An assistant inside the car pushing on the brake pedal is the safest method.

9   For all applications, the engine must be completely warmed-up to operating temperature, which will automatically render the choke fast idle inoperative.

17 EFE (heat riser) system check

Refer to illustrations 17.2 and 17.5

1   The heat riser valve with external counter weight (used on early models) and the Early Fuel Evaporation (EFE) system actuator and rod assembly (used on later models) both perform the same job, but function in a slightly different manner.

2   The heat riser is a valve located inside the right side exhaust pipe, near the junction

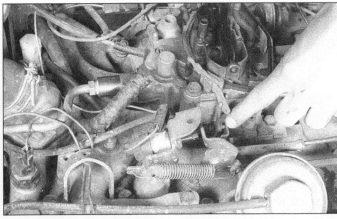

16.4 The carburetor idle speed screw located on the left side of the carburetor (arrow points to electrically-operated idle solenoid)

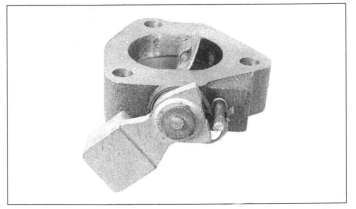

17.2 The EFE heat riser valve used is located at the connection between the exhaust manifold and the exhaust pipe (early model shown)

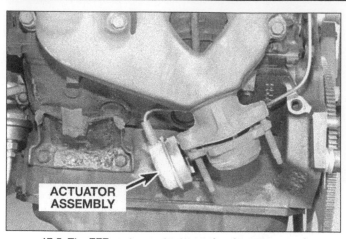

17.5 The EFE system actuator and rod system used on later models

18.6 Two wrenches are required to loosen carburetor fuel filter inlet nuts

between exhaust manifold and pipe. It can be identified by an external weight and spring **(see illustration)**.

3   With the engine and exhaust pipe cold, try moving the weight by hand. It should move freely.

4   Again with the engine cold, start the engine and observe the heat riser. Upon starting, the weight should move to the closed position. As the engine warms to normal operating temperature, the weight should move the valve to the open position, allowing a free flow of exhaust through the tailpipe. Since it could take several minutes for the system to heat up, you could mark the 'cold' weight position, drive the car, and then recheck the weight.

5   The EFE system also blocks off exhaust flow when the engine is cold. However, this system uses more precise temperature sensors and vacuum to open and close the exhaust pipe valve **(see illustration)**.

6   Locate the EFE actuator which is bolted to a bracket on the right side of the engine. It will have an actuating rod attached to it which will lead down to the valve inside the pipe. In some cases the entire mechanism, including actuator, will be located at the exhaust pipe-to-manifold junction.

7   With the engine cold, have an assistant start the engine as you observe the actuating rod. It should immediately move to close off the valve. Continue observing the rod, which should slowly open the valve as the engine warms. This process may take some time, so you might want to mark the position of the rod when the valve is closed, drive the car to reach normal operating temperature, then open the hood and check that the rod has moved to the open position.

8   Further information and testing procedures can be found in Chapter 6.

18   Fuel filter replacement

Warning: *Gasoline is extremely flammable, so take extra precautions when you work on any part of the fuel system. Don't smoke or*

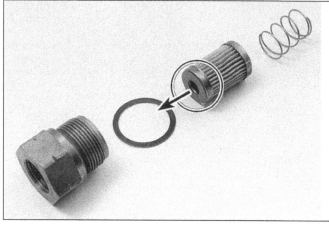

18.8 Fuel filter component layout (carbureted models)

allow open flames or bare light bulbs near the work area, and don't work in a garage where a natural gas-type appliance (such as a water heater or a clothes dryer) with a pilot light is present. Since gasoline is carcinogenic, wear latex gloves when there's a possibility of being exposed to fuel, and, if you spill any fuel on your skin, rinse it off immediately with soap and water. Mop up any spills immediately and do not store fuel-soaked rags where they could ignite. When you perform any kind of work on the fuel system, wear safety glasses and have a Class B type fire extinguisher on hand.

Carbureted models

Refer to illustrations 18.6 and 18.8

1   On carbureted models the fuel filter is located inside the fuel inlet nut at the carburetor. It's made of either pleated paper or porous bronze and cannot be cleaned or reused.

2   The job should be done with the engine cold (after sitting at least three hours). The necessary tools include open-end wrenches to fit the fuel line nuts. Flare-nut wrenches (which wrap around the nut) should be used if available. In addition, you have to obtain the replacement filter (make sure it's for your specific vehicle and engine) and some clean rags.

3   Remove the air cleaner assembly. If vac-

uum hoses must be disconnected, be sure to note their positions and/or tag them to ensure that they are reinstalled correctly.

4   Follow the fuel line from the fuel pump to the point where it enters the carburetor. In most cases the fuel line will be metal all the way from the fuel pump to the carburetor.

5   Place some rags under the fuel inlet fittings to catch spilled fuel as the fittings are disconnected.

6   With the proper size wrench, hold the fuel inlet nut immediately next to the carburetor body. Now loosen the fitting at the end of the metal fuel line. Make sure the fuel inlet nut next to the carburetor is held securely while the fuel line is disconnected **(see illustration)**.

7   After the fuel line is disconnected, move it aside for better access to the inlet nut. Don't crimp the fuel line.

8   Unscrew the fuel inlet nut, which was previously held steady. As this fitting is drawn away from the carburetor body, be careful not to lose the thin washer-type gasket on the nut or the spring, located behind the fuel filter. Also pay close attention to how the filter is installed **(see illustration)**.

9   Compare the old filter with the new one to make sure they're the same length and design.

10   Reinstall the spring in the carburetor body.

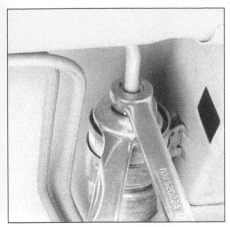

18.15 On fuel injected models, the fuel filter is located under the vehicle at the right frame rail - hold the filter body with an open-end wrench and loosen the fittings with a flare-nut wrench

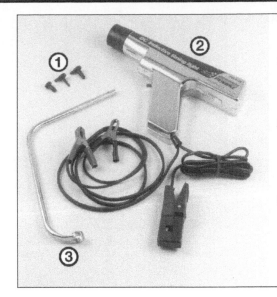

19.2 Tools needed to check and adjust the ignition timing

1 **Vacuum plugs** - Vacuum hoses will, in most cases, have to be disconnected and plugged. Molded plugs in various shapes and sizes are available for this
2 **Inductive pick-up timing light** - Flashes a bright concentrated beam of light when the number one spark plug fires. Connect the leads according to the instructions supplied with the light
3 **Distributor wrench** - On some models, the hold-down bolt for the distributor is difficult to reach and turn with a conventional wrench or socket. A special wrench like this must be used

11 Place the filter in position (a gasket is usually supplied with the new filter) and tighten the nut. Make sure it's not cross-threaded. Tighten it securely, but be careful not to overtighten it as the threads can strip easily, causing fuel leaks. Reconnect the fuel line to the fuel inlet nut, again using caution to avoid cross-threading the nut. Use a back-up wrench on the fuel inlet nut while tightening the fuel line fitting.
12 Start the engine and check carefully for leaks. If the fuel line fitting leaks, disconnect it and check for stripped or damaged threads. If the fuel line fitting has stripped threads, remove the entire line and have a repair shop install a new fitting. If the threads look all right, purchase some thread sealing tape and wrap the threads with it. Inlet nut repair kits are available at most auto parts stores to overcome leaking at the fuel inlet nut.

Fuel injected models

Refer to illustration 18.15
13 On fuel injected models, the fuel filter is mounted on the right frame rail. Raise the

vehicle and support it securely on jackstands.
14 Place a drain pan under the filter to catch fuel leakage.
15 Using a back-up wrench and a flare-nut wrench, disconnect the fuel lines from the filter **(see illustration)**.
16 Loosen the filter mounting bracket bolt and remove the filter.
17 Install the filter with the arrow on the filter facing the direction of fuel flow (towards the front of the vehicle).

19 Ignition timing adjustment

Refer to illustrations 19.2 and 19.6
Note: *It is imperative that the procedures included on the tune-up or Vehicle Emissions Control Information (VECI) label be followed when adjusting the ignition timing. The label will include all information concerning preliminary steps to be performed before adjusting the timing, as well as the timing specifications.*
1 At the specified intervals, whenever the ignition points have been replaced, the distributor removed or a change made in the fuel type, the ignition timing should be checked

and adjusted.
2 Locate the Tune-up or VECI label under the hood and read through and perform all preliminary instructions concerning ignition timing. Some special tools will be needed for this procedure **(see illustration)**.
3 Before attempting to check the timing, make sure the ignition point dwell angle is correct (see Section 30), and the idle speed is as specified (see Section 16).
4 If specified on the tune-up label, disconnect the vacuum hose from the distributor and plug the open end of the hose with a rubber plug, rod or bolt of the proper size.
5 Connect a timing light in accordance with the manufacturer's instructions. Generally, the light will be connected to power and ground sources and to the number one spark plug wire. The number one spark plug is the first spark plug on the right head (driver's side) as you are facing the engine from the front.
6 Locate the numbered timing tag on the front cover of the engine **(see illustration)**. It is located just behind the lower crankshaft pulley. Clean it off with solvent if necessary to read the printing and small grooves.

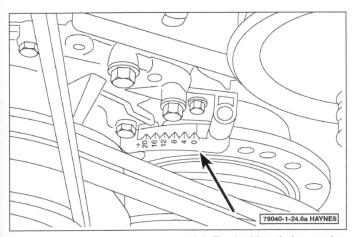

19.6 The ignition timing marks are located at the front of the engine

7   Use chalk or paint to mark the groove in the crankshaft pulley.

8   Put a mark on the timing tab in accordance with the number of degrees called for on the VECI label or the tune-up label in the engine compartment. Each peak or notch on the timing tab represents two degrees. The word Before or the letter A indicates advance and the letter 0 indicates Top Dead Center (TDC). As an example, if your vehicle specifications call for eight degrees BTDC (Before Top Dead Center), you will make a mark on the timing tab four notches before the 0.

9   Check that the wiring for the timing light is clear of all moving engine components, then start the engine and warm it up to normal operating temperature.

10   Aim the flashing timing light at the timing mark by the crankshaft pulley, again being careful not to come in contact with moving parts. The marks should appear to be stationary. If the marks are in alignment, the timing is correct.

11   If the notch is not lining up with the correct mark, loosen the distributor hold-down bolt and rotate the distributor until the notch is lined up with the correct timing mark.

12   Retighten the hold-down bolt and recheck the timing.

13   Turn off the engine and disconnect the timing light. Reconnect the vacuum advance hose, if removed, and any other components which were disconnected.

20   Carburetor choke check

Refer to illustration 20.3

1   The choke only operates when the engine is cold, and thus this check can only be performed before the car has been started for the day.

2   Open the hood and remove the top plate of the air cleaner assembly. It is held in place by a wing-nut at the center. If any vacuum hoses must be disconnected, make sure you tag the hoses for reinstallation in their original positions. Place the top plate and wing nut aside, out of the way of moving engine components.

3   Look at the top of the carburetor at the center of the air cleaner housing. You will notice a flat plate at the carburetor opening **(see illustration)**.

Mechanical choke

4   Have an assistant press the accelerator pedal to the floor. The plate should close fully. Start the engine while you observe the plate at the carburetor. Do not position your face directly over the carburetor, as the engine could backfire, causing serious burns. When the engine starts, the choke plate should open slightly.

5   Allow the engine to continue running at an idle speed. As the engine warms up to operating temperature, the plate should slowly open, allowing more cold air to enter through the top of the carburetor.

20.3 Carburetor choke plate is visible after removal of the air cleaner top plate

6   After a few minutes, the choke plate should be fully open to the vertical position.

7   You will notice that the engine speed corresponds with the plate opening. With the plate fully closed, the engine should run at a fast idle speed. As the plate opens, the engine speed will decrease.

Electric choke

8   This check should be performed while the outside air temperature is between 60 and 80°F. When the throttle is opened slightly, the choke valve should close completely.

9   Start the engine and check the time that elapses until the choke valve is fully open (3.2 minutes for 1978 and 1979 models; 3.5 minutes for 1980 and 1981 models).

10   If the choke fails to open fully in the allotted time, check the voltage at the choke heater connection (engine running). It should be 12 to 15 volts. If it is, the electric choke unit is faulty and must be replaced with a new one.

11   If the voltage is low, or zero, check all wires and connections. If the connections at the oil pressure switch are faulty, the oil warning light will be off when the key is "on" and the engine "off". If the fuse is blown, the radio or turn signal indicator will be inoperative.

12   If steps 10 and 11 do not correct the problem, replace the oil pressure switch with a new one.

21   Exhaust Gas Recirculation (EGR) valve check

Refer to illustration 21.2

1   the EGR valve is usually located on the intake manifold, adjacent to the carburetor. The majority of the time, when a fault develops in this emissions system it is due to a stuck or corroded EGR valve.

2   With the engine cold to prevent burns, reach under the EGR valve and manually push on the diaphragm **(see illustration)**.

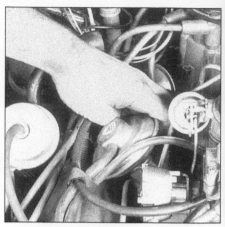

21.2 The diaphragm, located under the EGR valve, should move easily with finger pressure

Using moderate pressure, you should be able to press the diaphragm up and down within the housing.

3   If the diaphragm does not move or moves only with much effort, replace the EGR valve with a new one. If you are in doubt about the quality of the valve, go to your local parts store and compare the free movement of your EGR valve with a new valve.

4   Refer to Chapter 6 for further information on the EGR system.

22   Differential oil change

Refer to illustrations 22.6a, 22.6b, 22.6c and 22.8

1   Some differentials can be drained by removing the drain plug, while on others it's necessary to remove the cover plate on the differential housing. As an alternative, a hand suction pump can be used to remove the differential lubricant through the filler hole. If there is no drain plug and a suction pump isn't available, be sure to obtain a new gasket at the same time the gear lubricant is purchased.

2   Raise the vehicle and support it securely on jackstands. Move a drain pan, rags, newspapers and wrenches under the vehicle.

3   Remove the fill plug from the differential.

4   If equipped with a drain plug, remove the plug and allow the differential oil to drain completely. After the oil has drained, install the plug and tighten it securely.

5   If a suction pump is being used, insert the flexible hose. Work the hose down to the bottom of the differential housing and pump the oil out.

6   If the differential is being drained by removing the cover plate, remove the bolts on the lower half of the plate **(see illustration)**. Loosen the bolts on the upper half and use them to keep the cover loosely attached **(see illustration)**. Allow the oil to drain into the pan, then completely remove the cover **(see illustration)**.

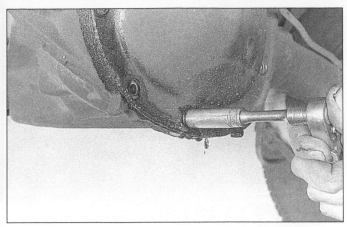

22.6a Remove the bolts from the lower edge of the cover . . .

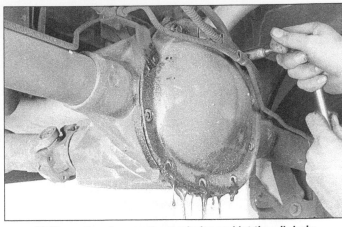

22.6b . . . then loosen the top bolts and let the oil drain

7   Using a lint-free rag, clean the inside of the cover and the accessible areas of the differential housing. As this is done, check for chipped gears and metal particles in the lubricant, indicating that the differential should be more thoroughly inspected and/or repaired.

8   Thoroughly clean the gasket mating surfaces of the differential housing and the cover plate. Use a gasket scraper or putty knife to remove all traces of the old gasket (see illustration).

9   Apply a thin layer of RTV sealant to the cover flange and then press a new gasket into position on the cover. Make sure the bolt holes align properly.

10   Place the cover on the differential housing and install the bolts. Tighten the bolts securely.

11   On all models, use a hand pump, syringe or funnel to fill the differential housing with the specified lubricant until it's level with the bottom of the plug hole.

12   Install the filler plug and tighten it securely.

23   Spark plug replacement

Refer to illustrations 23.2, 23.5a, 23.5b, 23.6 and 23.10

1   The spark plugs are located on each

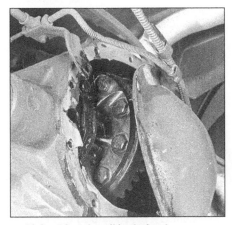

22.6c After the oil is drained, remove the cover

22.8 Carefully scrape the old gasket material off to ensure a leak-free seal with the new gasket

side of the engine and may or may not be easily accessible for removal. If the car is equipped with air conditioning or power steering, some of the plugs may be tricky to service in which case special extension or swivel tools will be necessary. Make a survey under the hood to ascertain if special tools will be needed.

2   In most cases, the tools necessary for spark plug replacement include a spark plug socket which fits onto a ratchet (spark plug sockets are padded inside to prevent damage to the porcelain insulators on the new

plugs), various extensions and a gap gauge to check and adjust the gaps on the new plugs (see illustration). A special plug wire removal tool is available for separating the wire boots from the spark plugs, but it isn't absolutely necessary. A torque wrench should be used to tighten the new plugs.

3   The best approach when replacing the spark plugs is to purchase the new ones in advance, adjust them to the proper gap and replace them one at a time. When buying the new spark plugs, be sure to obtain the correct

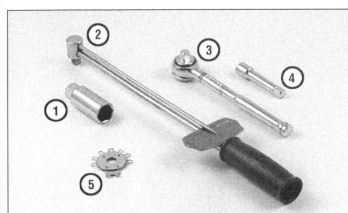

23.2 Tools necessary for changing spark plugs

1   **Spark plug socket** - This will have special padding inside to protect the spark plug's porcelain insulator

2   **Torque wrench** - Although not mandatory, using this tool is the best way to ensure the spark plugs are tightened properly

3   **Ratchet** - Standard hand tool to fit the spark plug socket

4   **Extension** - Depending on model and accessories, you may need special extensions and universal joints to reach one or more of the plugs

5   **Spark plug gap gauge** - This gauge for checking the gap comes in a variety of styles. Make sure the gap for your engine is included

23.5a Spark plug manufacturers recommend using a wire type gauge when checking the gap - if the wire does not slide between the electrodes with a slight drag, adjustment is required

23.5b To change the gap, bend the side electrode only, as indicated by the arrows, and be very careful not to crack or chip the porcelain insulator surrounding the center electrode

plug type for your particular engine. This information can be found on the Emission Control Information label located under the hood and in the factory owner's manual. If differences exist between the plug specified on the emissions label and in the owner's manual, assume that the emissions label is correct.

4   Allow the engine to cool completely before attempting to remove any of the plugs. While you're waiting for the engine to cool, check the new plugs for defects and adjust the gaps.

5   The gap is checked by inserting the proper thickness gauge between the electrodes at the tip of the plug **(see illustration)**. The gap between the electrodes should be the same as the one specified on the Emissions Control Information label. The wire should just slide between the electrodes with a slight amount of drag. If the gap is incorrect, use the adjuster on the gauge body to bend the curved side electrode slightly until the proper gap is obtained **(see illustration)**. If the side electrode is not exactly over the center electrode, bend it with the adjuster until it is. Check for cracks in the porcelain insulator (if any are found, the plug should not be used).

6   With the engine cool, remove the spark plug wire from one spark plug. Pull only on the boot at the end of the wire - do not pull on the wire. A plug wire removal tool should be used if available **(see illustration)**.

7   If compressed air is available, use it to blow any dirt or foreign material away from the spark plug hole. A common bicycle pump will also work. The idea here is to eliminate the possibility of debris falling into the cylinder as the spark plug is removed.

8   Place the spark plug socket over the plug and remove it from the engine by turning it in a counterclockwise direction.

9   Compare the spark plug with the chart shown on the inside back cover to get an indication of the general running condition of the engine.

10   Thread one of the new plugs into the hole until you can no longer turn it with your fingers, then tighten it with a torque wrench (if available) or the ratchet. A good idea is to slip a short length of rubber hose over the end of the plug to use as a tool to thread it into place **(see illustration)**. The hose will grip the plug enough to turn it, but will start to slip if the plug begins to cross-thread in the hole - this will prevent damaged threads and the accompanying repair costs.

11   Before pushing the spark plug wire onto

the end of the plug, inspect it following the procedures outlined in Section 28.

12   Attach the plug wire to the new spark plug, again using a twisting motion on the boot until it's seated on the spark plug.

13   Repeat the procedure for the remaining spark plugs, replacing them one at a time to prevent mixing up the spark plug wires.

24   Front wheel bearing check, repack and adjustment

Refer to illustrations 24.1, 24.6, 24.7, 24.8, 24.11 and 24.15

1   In most cases the front wheel bearings will not need servicing until the brake pads are changed. However, the bearings should be checked whenever the front of the vehicle is raised for any reason. Several items, including a torque wrench and special grease, are required for this procedure **(see illustration)**.

2   With the vehicle securely supported on jackstands, spin each wheel and check for noise, rolling resistance and free play.

3   Grasp the top of each tire with one hand and the bottom with the other. Move the wheel in and out on the spindle. If there's any

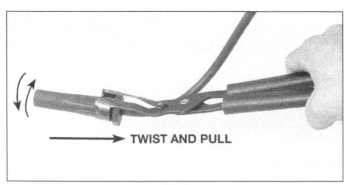

TWIST AND PULL

23.6 When removing a spark plug wire from a spark plug it is important to pull on the end of the boot and not on the wire itself. A slight twisting motion will also help

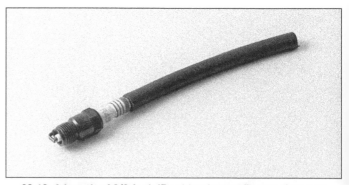

23.10 A length of 3/8-inch ID rubber hose will save time and prevent damaged threads when installing the spark plugs

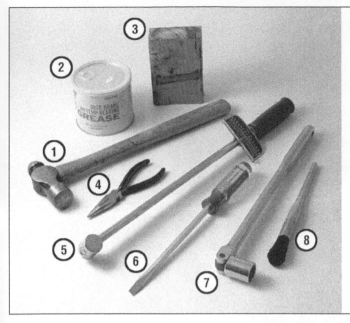

24.1 Tools and materials needed for front wheel bearing maintenance

1   **Hammer** - A common hammer will do just fine
2   **Grease** - High-temperature grease that is formulated for front wheel bearings should be used
3   **Wood block** - If you have a scrap piece of 2X4, it can be used to drive the new seal into the hub
4   **Needle-nose pliers** - Used to straighten and remove the cotter pin in the spindle
5   **Torque wrench** - This is very important in this procedure; if the bearing is too tight, the wheel won't turn freely - if it's loose, the wheel will "wobble" on the spindle. Either way, it could mean extensive damage
6   **Screwdriver** - Used to remove the seal from the hub (along screwdriver is preferred)
7   **Socket/breaker bar** - Needed to loosen the nut on the spindle if it's extremely tight
8   **Brush** - Together with some clean solvent, this will be used to remove old grease from the hub and spindle

noticeable movement, the bearings should be checked and then repacked with grease, or replaced if necessary.

4   Remove the wheel.
5   On disc brake equipped models, fabricate a wood block (1-1/16 inches by 1/2-inch by 2-inches long) which can be slid between the brake pads to keep them separated. Remove the brake caliper (Chapter 9) and hang it out of the way on a piece of wire.
6   Pry the dust cap out of the hub using a screwdriver or hammer and chisel **(see illustration)**.
7   Straighten the bent ends of the cotter pin, then pull the cotter pin out of the locking nut **(see illustration)**. Discard the cotter pin and use a new one during reassembly.
8   Remove the spindle nut and washer from the end of the spindle **(see illustration)**.
9   Pull the hub assembly out slightly, then

24.6 Dislodge the grease cap by working around the outer circumference with a hammer and chisel

24.7 Use needle-nose pliers to straighten the cotter pin and pull it out

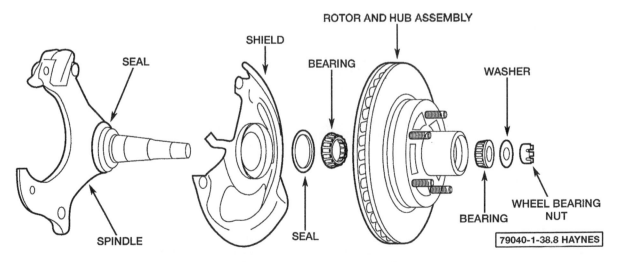

24.8 Front wheel bearing components

push it back into its original position. This should force the outer bearing off the spindle enough so it can be removed.

10   Pull the hub off the spindle.

11   Use a screwdriver to pry the seal out of the rear of the hub **(see illustration)**. As this is done, note how the seal is installed.

12   Remove the inner wheel bearing from the hub.

13   Use solvent to remove all traces of the old grease from the bearings, hub and spindle. A small brush may prove helpful; however make sure no bristles from the brush embed themselves inside the bearing rollers. Allow the parts to air dry.

14   Carefully inspect the bearings for cracks, heat discoloration, worn rollers, etc. Check the bearing races inside the hub for wear and damage. If the bearing races are defective, the hubs should be taken to a machine shop with the facilities to remove the old races and press new ones in. Note that the bearings and races come as matched sets and old bearings should never be installed on new races.

15   Use high-temperature front wheel bearing grease to pack the bearings. Work the grease completely into the bearings, forcing it between the rollers, cone and cage from the back side **(see illustration)**.

16   Apply a thin coat of grease to the spindle at the outer bearing seat, inner bearing seat, shoulder and seal seat.

17   Put a small quantity of grease inboard of each bearing race inside the hub. Using your finger, form a dam at these points to provide extra grease availability and to keep thinned grease from flowing out of the bearing.

18   Place the grease-packed inner bearing into the rear of the hub and put a little more grease outboard of the bearing.

19   Place a new seal over the inner bearing and tap the seal evenly into place with a hammer and block of wood until it's flush with the hub.

20   Carefully place the hub assembly onto the spindle and push the grease-packed outer bearing into position.

21   Install the washer and spindle nut. Tighten the nut only slightly (no more than 12

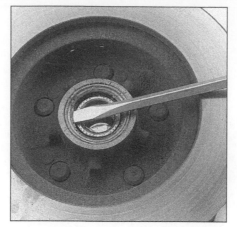

24.11 Use a screwdriver to pry the grease seal from the back side of the hub

ft-lbs of torque).

22   Spin the hub in a forward direction to seat the bearings and remove any grease or burrs which could cause excessive bearing play later.

23   Check to see that the tightness of the spindle nut is still approximately 12 ft-lbs.

24   Loosen the spindle nut until it's just loose, no more.

25   Using your hand (not a wrench of any kind), tighten the nut until it's snug. Install a new cotter pin through the hole in the spindle and spindle nut. If the nut slots don't line up, loosen the nut slightly until they do. From the hand-tight position, the nut should not be loosened more than one-half flat to install the cotter pin.

26   Bend the ends of the cotter pin until they're flat against the nut. Cut off any extra length which could interfere with the dust cap.

27   Install the dust cap, tapping it into place with a hammer.

28   Place the brake caliper near the rotor and carefully remove the wood spacer. Install the caliper (Chapter 9).

29   Install the tire/wheel assembly on the hub and tighten the lug nuts.

30   Grasp the top and bottom of the tire and

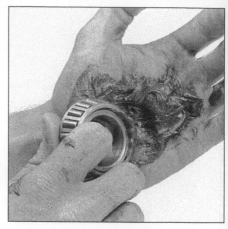

24.15 Work the grease completely into the rollers

check the bearings in the manner described earlier in this Section.

31   Lower the vehicle.

25   Automatic transmission fluid and filter change

Refer to illustrations 25.7, 25.9, 25.10 and 25.11

1   At the specified time intervals, the transmission fluid should be drained and replaced. Since the fluid will remain hot long after driving, perform this procedure only after the engine has cooled down completely.

2   Before beginning work, purchase the specified transmission fluid (see *Recommended lubricants and fluids* at the front of this Chapter) and a new filter.

3   Other tools necessary for this job include jackstands to support the vehicle in a raised position, a drain pan capable of holding at least eight pints, newspapers and clean rags.

4   Raise the vehicle and support it securely on jackstands.

5   With a drain pan in place, remove the front and side pan mounting bolts.

6   Loosen the rear pan bolts approximately four turns.

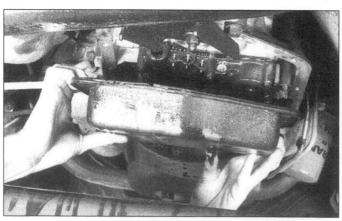

25.7 With the rear bolts in place but loose, pull the front of the pan down to let the fluid drain

25.9 After removing the pan it should be inspected for contamination and metal filings

25.10 After removing the bolts or screws, lower the filter from the transmission

25.11 Some models have an O-ring on the filter spout; if it does not come out with the filter, reach up into the opening with your finger to retrieve it

7   Carefully pry the transmission pan loose with a screwdriver, allowing the fluid to drain **(see illustration)**.
8   Remove the remaining bolts, pan and gasket. Carefully clean the gasket surface of the transmission to remove all traces of the old gasket and sealant.
9   Drain the fluid from the transmission pan and inspect it for metal filings or foreign materials **(see illustration)**. Thoroughly clean the pan with solvent and dry it with compressed air.
10   Remove the filter from the mount inside the transmission **(see illustration)**. Some models are retained with screws or a bolt, while others are pressed in.
11   Install a new filter and gasket or O-ring (if equipped) **(see illustration)**.
12   Make sure the gasket surface on the transmission pan is clean, then install a new gasket. Put the pan in place against the transmission and, working around the pan, tighten each bolt a little at a time until the final torque figure is reached.
13   Lower the vehicle and add the specified amount of automatic transmission fluid through the filler tube (see Section 6).
14   With the transmission in Park and the parking brake set, run the engine at a fast idle, but don't race it.
15   Move the gear selector through each range and back to Park. Check the fluid level.
16   Check under the vehicle for leaks during the first few trips.

26   Brake check

Note: *For detailed photographs of the brake system, refer to Chapter 9.*
Warning: *Brake system dust may contain asbestos, which is hazardous to your health. DO NOT blow it out with compressed air and DO NOT inhale it. DO NOT use gasoline or solvents to remove the dust. Use brake system cleaner or denatured alcohol only.*
1   In addition to the specified intervals, the brakes should be inspected every time the wheels are removed or whenever a defect is suspected.
2   To check the brakes, raise the vehicle and place it securely on jackstands. Remove the wheels (see *Jacking and towing* at the front of the manual, if necessary).

Disc brakes

Refer to illustrations 26.4 and 26.6
3   Disc brakes are used on the front wheels of later models. Extensive rotor damage can occur if the pads are not replaced when needed.
4   These vehicles are equipped with a wear sensor attached to the inner pad. This is a small, bent piece of metal which is visible from the inner side of the brake caliper. When the pad wears to the specified limit, the metal sensor rubs against the rotor and makes a squealing sound **(see illustration)**.
5   The disc brake calipers, which contain the pads, are visible with the wheels removed. There is an outer pad and an inner pad in each caliper. All pads should be inspected.
6   Each caliper has a "window" to inspect the pads. Check the thickness of the pad lining by looking into the caliper at each end and down through the inspection window at the top of the housing **(see illustration)**. If the wear sensor is very close to the rotor or the pad material has worn to about 1/8-inch or less, the pads should be replaced.
7   If you're unsure about the exact thickness of the remaining lining material, remove the pads for further inspection or replacement (refer to Chapter 9).
8   Before installing the wheels, check for leakage and/or damage (cracks, splitting, etc.) around the brake hose connections. Replace the hose or fittings as necessary, referring to Chapter 9.
9   Check the condition of the rotor. Look for score marks, deep scratches and burned spots. If these conditions exist, the hub/rotor assembly should be removed for servicing (see Section 24).

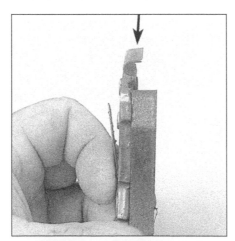

26.4 Most disc brake pads have built in wear indicators that contact the rotor and emit a squealing sound when the pads have worn to their limit

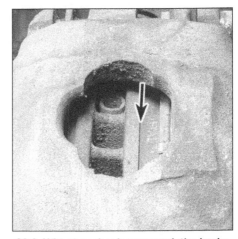

26.6 With the wheels removed, the brake pad lining (arrow) can be inspected through the caliper window and at each end of the caliper

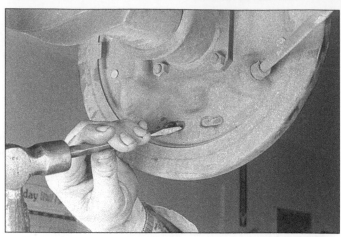

26.11 Use a hammer and chisel to remove the plug from the brake backing plate

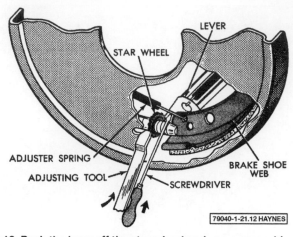

26.12 Push the lever off the star wheel and use a screwdriver or brake adjusting tool to turn the star wheel

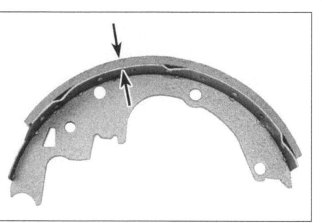

26.14 If the lining is bonded to the brake shoe, measure the lining thickness from the outer surface to the metal shoe, as shown here; if the lining is riveted to the shoe, measure from the lining outer surface to the rivet head

26.15 Check that all springs are in good condition

Drum brakes

Refer to illustrations 26.11, 26.12, 26.14, 26.15, 26.16 and 26.18

10   On front drum brakes, remove the hub/drum (see Section 24). On rear brakes, remove the drum by pulling it off the axle and brake assembly. If this proves difficult, make sure the parking brake is released, then squirt penetrating oil around the center hub areas. Allow the oil to soak in and try to pull the drum off again.

11   If the drum still cannot be pulled off, the brake shoes will have to be retracted. This is done by first removing the plug from the backing plate with a hammer and a chisel **(see illustration)**. If there is no plug stamped on the backing plate, drill a 1/2-inch hole into the round flat area on the backing plate (opposite side of the parking brake cable).

12   With the plug removed, push the lever off the star wheel and then use a small screwdriver to turn the star wheel, which will move the brake shoes away from the drum **(see illustration)**.

13   With the drum removed, do not touch any brake dust (see the Warning at the beginning of this Section.

14   Note the thickness of the lining material on both the front and rear brake shoes. If the material has worn away to within 1/16-inch of the recessed rivets or metal backing, the

shoes should be replaced (**see illustration** and Chapter 9). The shoes should also be replaced if they're cracked, glazed (shiny surface) or contaminated with brake fluid.

15   Make sure that all the brake assembly springs are connected and in good condition **(see illustration)**.

16   Check the brake components for any signs of fluid leakage. With your finger, carefully pry back the rubber cups on the wheel

cylinders located at the top of the brake shoes **(see illustration)**. Any leakage is an indication that the wheel cylinders should be overhauled immediately (Chapter 9). Also check brake hoses and connections for signs of leakage.

17   Wipe the inside of the drum with a clean

26.16 Leakage often occurs from the wheel cylinder located at the top of the brake shoes

26.18 The inside surface of the brake drum should be carefully inspected for cracks, scoring, hot-spots, etc

27.4 Evenly tighten the carburetor or throttle body mounting bolts (arrows) located at each corner

rag and brake cleaner or denatured alcohol. Again, be careful not to breath the dangerous asbestos dust.

18 Check the inside of the drum for cracks, score marks, deep scratches and hard spots, which will appear as small discolorations **(see illustration)** If these imperfections cannot be removed with fine emery cloth, the drum must be taken to a machine shop equipped to turn the drums.

19 If after the inspection process all parts are in good working condition, reinstall the brake drum (using a metal or rubber plug if the knockout was removed).

20 Install the wheels and lower the vehicle.

Parking brake

21 The parking brake operates from a foot pedal and locks the rear brake system. The easiest method of periodically checking the operation of the parking brake assembly is to park the vehicle on a steep hill with the parking brake set and the transmission in Neutral. If the parking brake cannot prevent the vehicle from rolling, it's in need of adjustment (see Chapter 9).

27 Carburetor/throttle body mounting nut/bolt torque check

Refer to illustration 27.4

1 The carburetor or throttle body unit is attached to the top of the intake manifold by several bolts or nuts. These fasteners can sometimes work loose from vibration and temperature changes during normal engine operation and cause a vacuum leak.

2 If you suspect that a vacuum leak exists at the bottom of the carburetor or throttle body, obtain a length of hose. Start the engine and place one end of the hose next to your ear as you probe around the base with the other end. You will hear a hissing sound if a leak exists (be careful of hot or moving engine components).

3 Remove the air cleaner assembly, tagging each hose to be disconnected with a piece of numbered tape to make reassembly easier.

4 Locate the mounting nuts or bolts at the base of the carburetor or throttle body **(see illustration)**. Decide what special tools or adapters will be necessary, if any, to tighten the fasteners.

5 Tighten the nuts or bolts to the specified torque. Don't overtighten them, as the threads could strip.

6 If, after the nuts or bolts are properly tightened, a vacuum leak still exists, the carburetor must be removed and a new gasket installed. See Chapter 4 for more information.

7 After tightening the fasteners, reinstall the air cleaner and return all hoses to their original positions.

28 Spark plug wire check

Refer to illustration 28.4

1 The spark plug wires should be checked at the recommended intervals and whenever new spark plugs are installed in the engine.

2 The wires should be inspected one at a time to prevent mixing up the order, which is essential for proper engine operation.

3 Disconnect the plug wire from one spark plug. To do this, grab the rubber boot, twist slightly and pull the wire free. Do not pull on the wire itself, only on the rubber boot **(see illustration 23.6)**.

4 Check inside the boot for corrosion, which will look like a white crusty powder **(see illustration)**. Push the wire and boot back onto the end of the spark plug. It should be a tight fit on the plug. If it isn't, remove the wire and use a pair of pliers to carefully crimp the metal connector inside the boot until it fits securely on the end of the spark plug.

5 Using a clean rag, wipe the entire length of the wire to remove any built-up dirt and grease. Once the wire is clean, check for holes, burned areas, cracks and other damage. Don't bend the wire excessively or the conductor inside might break.

6 Disconnect the wire from the distributor cap. A retaining ring at the top of the distribu-

tor may have to be removed to free the wires. Again, pull only on the rubber boot. Check for corrosion and a tight fit in the same manner as the spark plug end. Reattach the wire to the distributor cap.

7 Check the remaining spark plug wires one at a time, making sure they are securely fastened at the distributor and the spark plug when the check is complete.

8 If new spark plug wires are required, purchase a new set for your specific engine model. Wire sets are available pre-cut, with the rubber boots already installed. Remove and replace the wires one at a time to avoid mix-ups in the firing order. The wire routing is extremely important, so be sure to note exactly how each wire is situated before removing it.

29 Cooling system servicing (draining, flushing and refilling)

Refer to illustrations 29.5 and 29.7

1 Periodically, the cooling system should be drained, flushed and refilled. This is to replenish the antifreeze mixture and prevent rust and corrosion which can impair the performance of the cooling system and ultimately cause engine damage.

2 At the same time the cooling system is serviced. all hoses and the fill cap should be inspected and replaced if faulty (see Section 6).

3 As antifreeze is a poisonous solution. take care not to spill any of the cooling mixture on the vehicle's paint or your own skin. If this happens, rinse immediately with plenty of clear water. Also, it is advisable to consult your local authorities about the dumping of antifreeze before draining the cooling system. In many areas reclamation centers have been set up to collect automobile oil and drained antifreeze/water mixtures rather than allowing these liquids to be added to the sewage and water facilities.

4 With the engine cold. remove the radiator pressure fill cap.

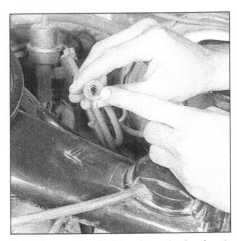

28.4 Inspecting inside of spark plug boot for corrosion

29.5 The radiator drain is located at the bottom (or lower side) of the radiator

29.7 Engine drain plugs are located on each side of the engine block

5   Move a large container under the radiator to catch the coolant as it is drained **(see illustration)**.
6   Drain the radiator. Most models are equipped with a drain plug at the bottom of the radiator which can be opened using a wrench to hold the fitting while the petcock is turned to the open position. If this drain has excessive corrosion and cannot be turned easily, or the radiator is not equipped with a drain, disconnect the lower radiator hose to allow the coolant to drain. Be careful that none of the solution is splashed on your skin or in your eyes.
7   If accessible, remove the two engine drain plugs **(see illustration)**. There is one plug on each side of the engine, about halfway back and on the lower edge near the oil pan rail. These will allow the coolant to drain from the engine itself.
8   On systems with an expansion reservoir,

disconnect the overflow pipe and remove the reservoir. Flush it out with clean water.
9   Place a cold water hose (a common garden hose is fine) in the radiator filler neck at the top of the radiator and flush the system until the water runs clean at all drain points.
10   In severe cases of contamination or clogging of the radiator, remove it (see Chapter 3) and reverse flush it. This involves simply inserting the cold pressure hose in the bottom radiator outlet to allow the clear water to run against the normal flow, draining through the top. A radiator repair shop should be consulted if further cleaning or repair is necessary.
11   Where the coolant is regularly drained and the system refilled with the correct antifreeze/inhibitor mixture there should be no need to employ chemical cleaners or descalers.
12   To refill the system. reconnect the radia-

tor hoses and install the drain plugs securely in the engine. Special thread sealing tape (available at auto parts stores) should be used on the drain plugs going into the engine block. Install the expansion reservoir and the overflow hose where applicable.
13   On vehicles without an expansion' reservoir, refill the system through the radiator filler cap until the water level is about three inches below the filler neck.
14   On vehicles with an expansion reservoir, fill the radiator to the base of the filler neck and then add more coolant to the expansion reservoir so that it reaches the 'FULL COLD' mark.
15   Run the engine until normal operating temperature is reached and with the engine idling, add coolant up to the correct level (see Section 2), then fit the radiator cap so that the arrows are in alignment with the overflow pipe. Install the reservoir cap.
16   Always refill the system with a mixture of high quality antifreeze and water in the proportion called for on the antifreeze container or in your owner's manual. Chapter 3 also contains information on antifreeze mixtures.
17   Keep a close watch on the coolant level and the various cooling hoses during the first few miles of driving. Tighten the hose clamps and/or add more coolant mixture as necessary.

30   Ignition point replacement and dwell angle adjustment

Ignition point replacement

Refer to illustrations 30.1, 30.2, 30.7, 30.9, 30.16 and 30.17
1   The ignition points must be replaced at regular intervals on vehicles not equipped

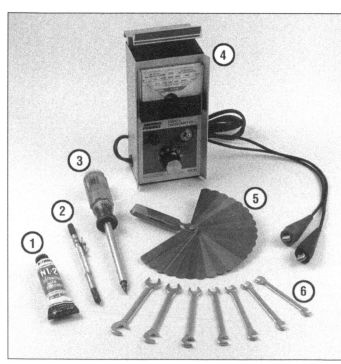

30.1 Tools and materials needed for contact point replacement and dwell angle adjustment

1   ***Distributor cam lube*** *- Sometimes this special lubricant comes with the new points; however, it's a good idea to buy a tube and have it on hand*
2   ***Screw starter*** *- This tool has special claws which hold the screw securely as it's started, which helps prevent accidental dropping of the screw*
3   ***Magnetic screwdriver*** *- Serves the same purpose as 2 above. If you don't have one of these special screwdrivers, you risk dropping the point mounting screws down into the distributor body*
4   ***Dwell meter*** *- A dwell meter is the only accurate way to determine the point setting (gap). Connect the meter according to the instructions supplied with it.*
5   ***Blade-type feeler gauges*** *- These are required to set the initial point gap (space between the points when they are open)*
6   ***Ignition wrenches*** *- These special wrenches are made to work within the tight confines of the distributor. Specifically, they are needed to loosen the nut/bolt which secures the leads to the points*

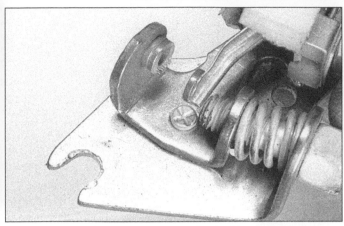

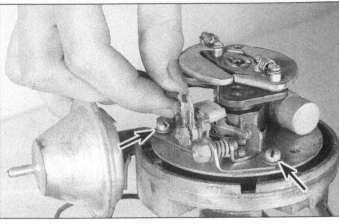

30.2 Although it is possible to restore ignition points that are pitted, burned and corroded (as shown here), they should be replaced instead

30.7 Loosen the nut and disconnect the primary and condenser wires from the points - note the ignition point mounting screws (arrows)

with electronic ignition. Occasionally the rubbing block will wear enough to require adjustment of the points. It's also possible to clean and dress them with a fine file, but replacement is recommended since they are relatively inaccessible and very inexpensive. Several special tools are required for this procedure (see illustration).

2 After removing the distributor cap and rotor (see Section 32), the ignition points are plainly visible. They can be examined by gently prying them open to reveal the condition of the contact surfaces (see illustration). If they re rough, pitted, covered with oil or burned, they should be replaced, along with the condenser. Caution: *This procedure requires the removal of small screws which can easily fall down into the distributor. To retrieve them, the distributor would have to be removed and disassembled. Use a magnetic or spring-loaded screwdriver and be extra careful.*

3 If not already done, remove the distributor cap by positioning a screwdriver in the slotted head of each latch. Press down on the latch and rotate it 1/2-turn to release the cap from the distributor body (see Section 32).

4 Position the cap (with the spark plug wires still attached) out of the way. Use a length of wire to hold it out of the way if necessary.

5 Remove the rotor (see Section 32 if necessary).

6 If equipped with a radio frequency interference shield (RFI), remove the mounting screws and the two-piece shield to gain access to the ignition points.

7 Note how they are routed, then disconnect the primary and condenser wire leads from the points (see illustration). The wires may be attached with a small nut (which should be loosened, but not removed) a small screw or by a spring loaded terminal. Note: *Some models are equipped with ignition points which include the condenser as an integral part of the point assembly. If your vehicle has this type, the condenser removal*

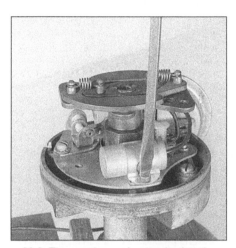

30.9 The condenser is attached to the breaker plate by a single screw

procedure below will not apply and there will be only one wire to detach from the points, rather than the two used with a separate condenser assembly.

8 Loosen the two screws which secure the ignition points to the breaker plate, but don't completely remove the screws (most ignition point sets have slots at these locations). Slide the points out of the distributor.

9 The condenser can now be removed from the breaker plate. Loosen the mounting strap screw and slide the condenser out or completely remove the condenser and strap (see illustration). If you remove both the condenser and strap, be careful not to drop the mounting screw down into the distributor body.

10 Before installing the new points and condenser, clean the breaker plate and the cam on the distributor shaft to remove all dirt, dust and oil.

11 Apply a small amount of distributor cam lube (usually supplied with the new points, but also available separately) to the cam lobes.

12 Position the new condenser and tighten the mounting strap screw securely.

30.16 Before adjusting the point gap, the rubbing block must be resting on one of the cam lobes (which will open the points)

13 Slide the new point set under the mounting screw heads and make sure the protrusions on the breaker plate fit into the holes in the point base (to properly position the point set), then tighten the screws securely .

14 Attach the primary and condenser wires to the new points. Make sure the wires are routed so they don't interfere with breaker plate or advance weight movement.

15 Although the gap between the contact points (dwell angle will be adjusted later, make the initial adjustment now, which will allow the engine to be started.

16 Make sure that the point rubbing block is resting on one of the high points of the cam (see illustration). If it isn't, turn the ignition switch to Start in short bursts to reposition the cam. You can also turn the crankshaft with a breaker bar and socket attached to the large bolt that holds the vibration damper in place.

17 With the rubbing block on a cam high point (points open), insert a 0.019-inch feeler gauge between the contact surfaces and use

30.17 With the points open, insert a 0.019-inch thick feeler gauge and turn the adjustment screw with an Allen wrench

30.29 With the door open, a 1/8-inch Allen wrench can be inserted into the adjustment screw socket and turned to adjust the point dwell

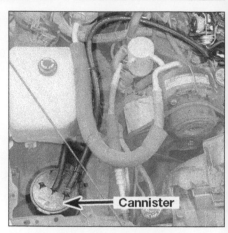

31.4 The charcoal canister is located near the front of the engine compartment on most vehicles and can be identified by the hoses attached to the top

an Allen wrench to turn the adjustment screw until the point gap is equal to the thickness of the feeler gauge **(see illustration)**. The gap is correct when a slight amount of drag is felt as the feeler gauge is withdrawn.

18   If equipped, install the RFI shield.

19   Before installing the rotor, check it as described in Section 32.

20   Install the rotor. The rotor is indexed with a square peg underneath on one side and a round peg on the other side. so it will fit on the advance mechanism only one way, If so, equipped, tighten the rotor mounting screws securely.

21   Before installing the distributor cap, inspect it as described in Section 32.

22   Install the distributor cap and lock the latches under the distributor body by depressing and turning them with a screwdriver.

23   Start the engine and check the dwell angle and ignition timing.

Dwell angle adjustment

Refer to illustration 30.29

24   Whenever new ignition points are installed or the original points are cleaned, the dwell angle must be checked and adjusted.

25   Precise adjustment of the dwell angle requires an instrument called a dwell meter. Combination tach/dwell meters are commonly available at reasonable cost from auto parts stores. An approximate setting can be obtained if a meter isn't available.

26   If a dwell meter is available, hook it up following the manufacturer's instructions.

27   Start the engine and allow it to run at idle until normal operating temperature is reached (the engine must be warm to obtain an accurate reading). Turn off the engine.

28   Raise the metal door in the distributor cap. Hold it in the open position with tape if necessary.

29   Just inside the opening is the ignition point adjustment screw. Insert a 1/8-inch Allen wrench into the adjustment screw socket **(see illustration)**.

30   Start the engine and turn the adjustment screw as required to obtain the specified dwell reading on the meter. Dwell angle specifications can be found in the Specifications section at the front of this Chapter or on the tune-up decal in the engine compartment. **Note:** *When adjusting the dwell, aim for the lower end of the dwell specification range. Then, as the points wear, the dwell will remain within the specified range over a longer period of time.*

31   Remove the Allen wrench and close the door on the distributor. Turn off the engine and disconnect the dwell meter, then check the ignition timing (see Section 19).

32   If a dwell meter isn't available, use the following procedure to obtain an approximate dwell setting.

33   Start the engine and allow it to idle until normal operating temperature is reached.

34   Raise the metal door in the distributor cap. Hold it in the open position with tape if necessary.

35   Just inside the opening is the ignition point adjustment screw. Insert a 1/8-inch Allen wrench into the adjustment screw socket **(see illustration 30.29)**.

36   Turn the Allen wrench clockwise until the engine begins to misfire, then turn the screw 1/2-turn counterclockwise.

37   Remove the Allen wrench and close the door. As soon as possible have the dwell angle checked and/or adjusted with a dwell meter to ensure optimum performance.

31   Evaporation Control System (ECS) filter replacement

Refer to illustrations 31.4, 31.5 and 31.6

1   The function of the ECS emissions system is to draw fuel vapors from the tank and carburetor, store them in a charcoal canister, and then burn these fumes during normal engine operation.

2   The filter at the bottom of the charcoal canister should be replaced at the specified intervals. If, however, a fuel odor is detected, the canister, filter and system hoses should immediately be inspected for fault.

3   To replace the filter, locate the canister at the front of the engine compartment. It will have between 3 and 6 hoses running out the top of it.

4   Remove the two bolts which secure the bottom of the canister to the body sheet metal **(see illustration)**.

5   Turn the canister upside-down and pull the old filter from the bottom of the canister **(see illustration)**. If you cannot turn the canister enough for this due to the short length of the hoses, the hoses must be duly marked with pieces of tape and then disconnected from the top.

6   Push the new filter into the bottom of the canister, making sure it is seated all the way around **(see illustration)**.

7   Place the canister back into position and tighten the two mounting bolts. Connect the various hoses if disconnected.

8   The ECS system is explained in more detail in Chapter 6.

31.5 Removing the filter from the bottom of the charcoal canister

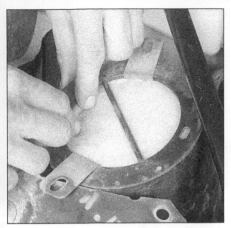

31.6 Make sure the new filter is fully seated all the way around the bottom of the canister

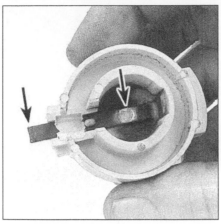

32.3a Check the distributor rotor contact (arrows) for wear and burn marks (later model remote coil-type distributor)

32 Distributor cap and rotor check and replacement

Note: *It's common practice to install a new distributor cap and rotor whenever new spark plug wires are installed. On models that have the ignition coil mounted in the cap, the coil will have to be transferred to the new cap.*

Check

Refer to illustrations 32.3a, 32.3b, 32.4 and 32.6

1   To gain access to the distributor cap, especially on a V6 engine, it may be necessary to remove the air cleaner assembly.

2   Loosen the distributor cap mounting screws (note that the screws have a shoulder

so they don't come completely out). On some models, the cap is held in place with latches that look like screws - to release them, push down with a screwdriver and turn them about 1/2-turn. Pull up on the cap, with the wires attached, to separate it from the distributor, then position it to one side.

3   The rotor is now visible on the end of the distributor shaft. Check it carefully for cracks and carbon tracks. Make sure the center terminal spring tension is adequate and look for corrosion and wear on the rotor tip **(see illustrations)**. If in doubt about its condition, replace it with a new one.

4   If replacement is required, detach the rotor from the shaft and install a new one. On some models, the rotor is press fit on the shaft and can be pried or pulled off **(see illustration)**. On other models, the rotor is attached to the distributor shaft with two screws.

5   The rotor is indexed to the shaft so it can only be installed one way. Press fit rotors have an internal key that must line up with a slot in the end of the shaft (or vice versa). Rotors held in place with screws have one square and one round peg on the underside that must fit into holes with the same shape.

6   Check the distributor cap for carbon tracks, cracks and other damage. Closely examine the terminals on the inside of the

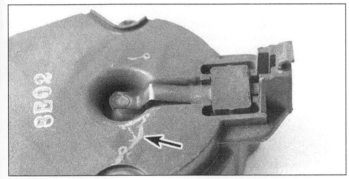

32.3b Carbon tracking on the rotor is caused by a leaking seal between the distributor cap and the ignition coil (always replace both the seal and rotor when this condition is present) (coil in cap-type distributor)

32.4 This type rotor can be pried off the distributor shaft with a screwdriver - other types have two screws which must be removed (be careful not to drop anything down into the distributor)

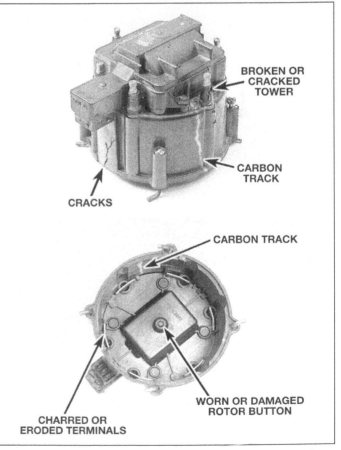

BROKEN OR CRACKED TOWER

CARBON TRACK

CRACKS

CARBON TRACK

WORN OR DAMAGED ROTOR BUTTON

CHARRED OR ERODED TERMINALS

32.6 Shown here are some of the common defects to look for when inspecting the distributor cap (if in doubt about its condition, install a new one)

32.12 Remove the two screws (arrows) that retain the plastic cover to the cap

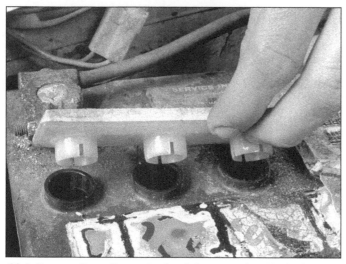

33.4   Remove the cell caps to check the water level in the battery - if the level is low, add distilled water only

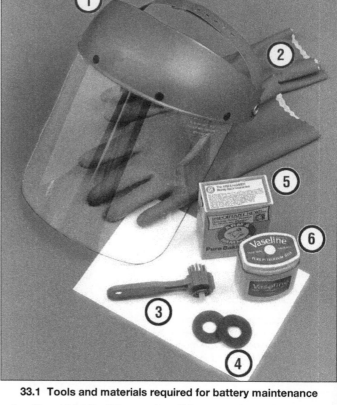

33.1 Tools and materials required for battery maintenance

1   *Face shield/safety goggles* - When removing corrosion with a brush, the acidic particles can easily fly up into your eyes
2   *Rubber gloves* - Another safety item to consider when servicing the battery - remember that's acid inside the battery!
3   *Battery terminal/cable cleaner* - This wire brush cleaning tool will remove all traces of corrosion from the battery cable
4   *Treated felt washers* - Placing one of these on each terminal, directly under the cable end, will help prevent corrosion (be sure to get the correct type for side-terminal batteries)
5   *Baking soda* - A solution of baking soda and water can be used to neutralize corrosion
6   *Petroleum jelly* - A layer of this on the battery terminal bolts will help prevent corrosion

cap for excessive corrosion and damage **(see illustration)**. Slight deposits are normal. Again, if in doubt about the condition of the cap, replace it with a new one. Be sure to apply a small dab of silicone lubricant to each terminal before installing the cap. Also, make sure the carbon brush (center terminal) is correctly installed in the cap - a wide gap between the brush and rotor will result in rotor burn-through and/or damage to the distributor cap.

Replacement
Conventional distributor

7   On models with a separately mounted ignition coil, simply separate the cap from the distributor and transfer the spark plug wires, one at a time, to the new cap. Be very careful not to mix up the wires!
8   Reattach the cap to the distributor, then tighten the screws or reposition the latches to hold it in place.

Coil-in-cap distributor
Refer to illustration 32.12
9   Use your thumbs to push the spark plug wire retainer latches away from the coil cover.
10   Lift the retainer ring away from the distributor cap with the spark plug wires attached to the ring. It may be necessary to work the wires off the distributor cap towers so they remain with the ring.
11   Disconnect the battery/tachometer/coil electrical connector from the distributor cap.
12   Remove the two coil cover screws and lift off the coil cover.
13   There are three small spade connectors on wires extending from the coil into the electrical connector hood at the side of the distributor cap. Note which terminals the wires are attached to, then use a small screwdriver to push them free.
14   Remove the four coil mounting screws and lift the coil out of the cap.

15   When installing the coil in the new cap, be sure to install a new rubber arc seal in the cap.
16   Install the coil screws, the wires in the connector hood, and the coil cover.
17   Install the cap on the distributor.
18   Plug in the coil electrical connector to the distributor cap.
19   Install the spark plug wires and retaining ring on the distributor cap.

33   Battery check and maintenance

Refer to illustrations 33.1, 33.4, 33.6a and 33.6b

Warning: *Certain precautions must be followed when checking and servicing the battery. Hydrogen gas, which is highly flammable, is always present in the battery cells, so keep lighted tobacco and all other open flames and sparks away from the battery. The*

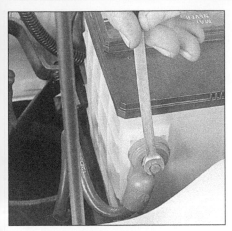

33.6a Make sure the battery terminal bolts are tight

Terminal end corrosion or damage.

Insulation cracks.

Chafed insulation or exposed wires.

Burned or melted insulation.

33.6b Typical battery cable problems

the regular brake pedal during the checks.

1   Manual transmission models are equipped with a clutch start switch, which prevents the engine from starting unless the clutch pedal is depressed. Automatic transmission models are equipped with a Park/Neutral safety switch, which prevents the engine from starting unless the shift lever is in Neutral or Park.

2   On automatic transmission vehicles, try to start the vehicle in each gear. The engine should crank only in Park or Neutral.

3   If equipped with a manual transmission, place the shift lever in Neutral. The engine should crank only with the clutch pedal depressed.

4   Make sure the steering column lock allows the key to go into the Lock position only when the shift lever is in Park (automatic transmission) or Reverse (manual transmission).

5   The ignition key should come out only in the Lock position.

35   Underhood hose check and replacement

General

Caution: *Replacement of air conditioning hoses must be left to a dealer service department or air conditioning shop that has the equipment to depressurize the system safely. Never remove air conditioning components or hoses until the system has been depressurized.*

1   High temperatures in the engine compartment can cause the deterioration of the rubber and plastic hoses used for engine, accessory and emission systems operation. Periodic inspection should be made for cracks, loose clamps, material hardening and leaks. Information specific to the cooling system hoses can be found in Section 9.

2   Some, but not all, hoses are secured to the fittings with clamps. Where clamps are used, check to be sure they haven't lost their tension, allowing the hose to leak. If clamps aren't used, make sure the hose has not expanded and/or hardened where it slips over the fitting, allowing it to leak.

Vacuum hoses

3   It's quite common for vacuum hoses, especially those in the emissions system, to be color coded or identified by colored stripes molded into them. Various systems require hoses with different wall thickness, collapse resistance and temperature resistance. When replacing hoses, be sure the new ones are made of the same material.

4   Often the only effective way to check a hose is to remove it completely from the vehicle. If more than one hose is removed, be sure to label the hoses and fittings to ensure correct installation.

5   When checking vacuum hoses, be sure to include any plastic T-fittings in the check.

electrolyte inside the battery is actually dilute sulfuric acid, which will cause injury if splashed on your skin or in your eyes. It will also ruin clothes and painted surfaces. When removing the battery cables, always detach the negative cable first and hook it up last!

1   Battery maintenance is an important procedure which will help ensure that you are not stranded because of a dead battery. Several tools are required for this procedure **(see illustration)**.

2   When checking/servicing the battery, always turn the engine and all accessories off.

3   A sealed (sometimes called maintenance-free), side-terminal battery is standard equipment on these vehicles. The cell caps cannot be removed, no electrolyte checks are required and water cannot be added to the cells. However, if a standard top-terminal aftermarket battery has been installed, the following maintenance procedure can be used.

4   Remove the caps and check the electrolyte level in each of the battery cells **(see illustration)**. It must be above the plates. There's usually a split-ring indicator in each cell to indicate the correct level. If the level is low, add distilled water only, then reinstall the cell caps. **Caution:** *Overfilling the cells may cause electrolyte to spill over during periods of heavy charging, causing corrosion and damage to nearby components.*

5   The external condition of the battery should be checked periodically. Look for damage such as a cracked case.

6   Check the tightness of the battery cable bolts **(see illustration)** to ensure good electrical connections. Inspect the entire length of each cable, looking for cracked or abraded insulation and frayed conductors **(see illustration)**.

7   If corrosion (visible as white, fluffy deposits) is evident, remove the cables from the terminals, clean them with a battery brush and reinstall them. Corrosion can be kept to a minimum by applying a layer of petroleum jelly or grease to the bolt threads.

8   Make sure the battery carrier is in good

condition and the hold-down clamp is tight. If the battery is removed (see Chapter 5 for the removal and installation procedure), make sure that no parts remain in the bottom of the carrier when it's reinstalled. When reinstalling the hold-down clamp, don't overtighten the bolt.

9   Corrosion on the carrier, battery case and surrounding areas can be removed with a solution of water and baking soda. Apply the mixture with a small brush, let it work, then rinse it off with plenty of clean water.

10   Any metal parts of the vehicle damaged by corrosion should be coated with a zinc-based primer, then painted.

11   Additional information on the battery, charging and jump starting can be found in the front of this manual and in Chapter 5.

34   Clutch start and Park/Neutral safety switch check

Warning: *During the following checks there is a chance that the vehicle could lunge forward, possibly causing damage or injuries. Allow plenty of room around the vehicle, apply the parking brake firmly and hold down*

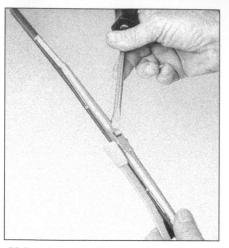

36.6 Using a small screwdriver pry up on the release pin and detach the blade assembly from the wiper arm pin

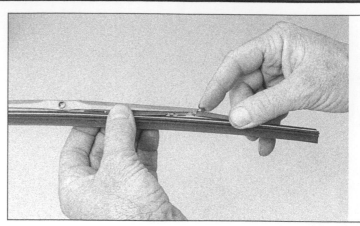

36.7 Depress the button and then slide the rubber element off of the wiper blade

Inspect the fittings for cracks and the hose where it fits over the fitting for distortion, which could cause leakage.

6   A small piece of vacuum hose (1/4-inch inside diameter) can be used as a stethoscope to detect vacuum leaks. Hold one end of the hose to your ear and probe around vacuum hoses and fittings, listening for the "hissing" sound characteristic of a vacuum leak. **Warning:** *When probing with the vacuum hose stethoscope, be very careful not to come into contact with moving engine components such as the drivebelt, cooling fan, etc.*

Fuel hose

Warning: *There are certain precautions which must be taken when inspecting or servicing fuel system components. Work in a well ventilated area and do not allow open flames (cigarettes, appliance pilot lights, etc.) or bare light bulbs near the work area. Mop up any spills immediately and do not store fuel soaked rags where they could ignite.*

7   Check all rubber fuel lines for deterioration and chafing. Check especially for cracks in areas where the hose bends and just before fittings, such as where a hose attaches to the fuel filter.

8   High quality fuel line, usually identified by the word *Fluoroelastomer* printed on the hose, should be used for fuel line replacement. Never, under any circumstances, use unreinforced vacuum line, clear plastic tubing or water hose for fuel lines.

9   Spring-type clamps are commonly used on fuel lines. These clamps often lose their

tension over a period of time, and can be "sprung" during removal. Replace all spring-type clamps with screw clamps whenever a hose is replaced.

Metal lines

10   Sections of metal line are often used for fuel line between the fuel pump and carburetor. Check carefully to be sure the line has not been bent or crimped and that cracks have not started in the line.

11   If a section of metal fuel line must be replaced, only seamless steel tubing should be used, since copper and aluminum tubing don't have the strength necessary to withstand normal engine vibration.

12   Check the metal brake lines where they enter the master cylinder and brake proportioning unit (if used) for cracks in the lines or loose fittings. Any sign of brake fluid leakage calls for an immediate thorough inspection of the brake system.

36   Wiper blade inspection and replacement

Refer to illustrations 36.6 and 36.7

1   The windshield wiper and blade assembly should be inspected periodically for damage, loose components and cracked or worn blade elements.

2   Road film can build up on the wiper blades and affect their efficiency, so they should be washed regularly with a mild detergent solution.

3   The action of the wiping mechanism can loosen the bolts, nuts and fasteners, so they should be checked and tightened, as necessary, at the same time the wiper blades are checked.

4   If the wiper blade elements (sometimes

called inserts) are cracked, worn or warped, they should be replaced with new ones.

5   Pull the wiper blade/arm assembly away from the glass.

6   Using a screwdriver, pry up on the release tab and pull the wiper blade assembly off of the wiper arm.

7   Depress the button and then slide the rubber element off of the wiper blade **(see illustration)**. To install a new element, press the button and slide the new piece into place. Once centered on the blade, it will lock into place.

8   The other method incorporates a spring-type retainer clip at the end of the removable element. When the retainer is pinched together, the element can slide out of the blade assembly. A small pair of pliers can be used to squeeze the retainer. When installing a new element, be certain that the metal insert passes through all of the retaining tabs of the blade assembly.

9   Reinstall the blade assembly on the arm, wet the windshield and check for proper operation.

37   Seatbelt check

1   Check the seatbelts, buckles, latch plates and guide loops for any obvious damage or signs of wear.

2   Make sure the seatbelt reminder light comes on when the key is turned on.

3   The seatbelts are designed to lock up during a sudden stop or impact, yet allow free movement during normal driving. The retractors should hold the belt against your chest while driving and rewind the belt when the buckle is unlatched.

4   If any of the above checks reveal problems with the seatbelt system, replace parts as necessary.

Chapter 2 Part A
Engines

Contents

Specifications

General

Displacement

V6 engines

3.3 liter (VIN M)	200 cubic inches
3.8 liter (VIN K and 9)	229 cubic inches
3.8 liter (VIN A, Z and 3)	231 cubic inches
4.3 liter (VIN Z)	262 cubic inches

V8 engines

4.4 liter (VIN J)	267 cubic inches
5.0 liter (VIN G, H, Q and U)	305 cubic inches
5.7 liter (VIN H, J, K, L and V)	350 cubic inches
6.6 liter (VIN R and U)	400 cubic inches
6.6 liter (VIN R and U)	402 cubic inches
7.4 liter (VIN V, W, Y and Z)	454 cubic inches

Bore and stroke

V6 engines

200	3.500 x 3.480 inches
229	3.736 x 3.480 inches
231	3.800 x 3.400 inches
262	4.000 x 3.480 inches

General (continued)

Bore and stroke
  V8 engines
    267 .... 3.500 x 3.480 inches
    305 .... 3.736 x 3.480 inches
    350 .... 4.000 x 3.480 inches
    400 small block .... 4.125 x 3.750 inches
    400 big block (1970 only) .... 4.125 x 3.760 inches
    402 .... 4.126 x 3.760 inches
    454 .... 4.250 x 4.000 inches
Cylinder numbers (front-to-rear)
  V6 engines
    Right bank (passenger's side) 2-4-6
    Left bank (driver's side) 1-3-5
  V8 engines
    Right bank (passenger's side) 2-4-6-8
    Left bank (driver's side) 1-3-5-7
Firing order
  V6 engines 1-6-5-4-3-2
  V8 engines 1-8-4-3-6-5-7-2
Compression pressure
  All V6 and V8 engines .... 100 to 150 psi
Maximum variation between cylinders .... 20 percent

200 and 229 cu in V6

231 cu in V6

262 cu in V6

24070-4 specs HAYNES

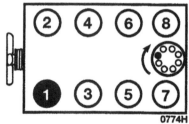

ALL V8 ENGINES WITH BREAKER POINT IGNITION

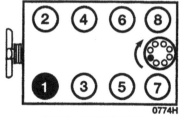

ALL V8 ENGINES WITH HEI IGNITION

Cylinder location and distributor rotation

The blackened terminal shown on the distributor cap indicates the Number One spark plug wire position

1970 models

Camshaft

Lobe lift (Intake)
  350 engine .... 0.2600 in
  400 (265 hp) engine .... 0.2235 in
  400 (330 hp) and 454 engines .... 0.2343 in
Lobe lift (Exhaust)
  350 engine .... 0.2733 in
  400 (265 hp) engine .... 0.2411 in
  400 (330 hp) engine .... 0.2343 in
  454 engine .... 0.2529 in
Camshaft journal diameter
  350 engine .... 1.8682 to 1.8692 inches
  454 engine .... 1.9487 to 1.9497 inches
Camshaft runout .... 0.015 in max

1971 and 1972 models
Camshaft
Lobe lift (Intake)
    350 (245 and 270 hp) engine.. 0.2600 in
    402 engine.. 0.2343 in
    454 engine.. 0.2714 in
Lobe lift (Exhaust)
    350 (245 and 270 hp) engine.. 0.2733 in
    402 engine.. 0.2343 in
    454 engine.. 0.2824 in
Camshaft journal diameter
    350 engine.. 1.8682 to 1.8692 inches
    All other engines.. 1.9482 to 1.9492 inches
Camshaft runout.. 0.0015 in max

1973 models
Camshaft
Lobe lift (Intake)
    350 engine.. 0.2600 in
    454 engine.. 0.2343 in
Lobe lift (Exhaust)
    350 engine.. 0.2733 in
    454 engine.. 0.2343 in
Camshaft journal diameter
    350 engine.. 1.8682 to 1.8692 inches
    454 engine.. 1.9482 to 1.9492 inches
Camshaft runout.. 0.0015 in max

1974 and 1975 models
Camshaft
Lobe lift (Intake)
    350 engine (except 245 hp)... 0.2600 in
    350 (245hp) engine.. 0.3000 in
    400 engine.. 0.2235 in
    454 engine.. 0.2590 in
Lobe lift (Exhaust)
    350 engine (except 245 hp)... 0.2733 in
    350 (245 hp) engine.. 0..3070 in
    400 engine.. 0.2411 in
    454 engine.. 0.2590 in
Camshaft journal diameter
    350 engine.. 1.8682 to 1.8692 inches
    400, 454 engines... 1.9482 to 1.9492 inches
Camshaft runout.. 0.0015 in max

1976 and 1977 models
Camshaft
Lobe lift (Intake)
    305 engine.. 0.2485 in
    350 and 400 engines... 0.2600 in
Lobe lift (Exhaust)
All engines .... 0.2733 in
Camshaft journal diameter
    305, 350 engines... 1.8682 to 1.8692 inches
    400 engine.. 1.9482 to 1.9492 inches
Camshaft runout.. 0.0015 in max

1978 and later models
Camshaft
V6 engines
    Lobe lift (Intake)
        200, 229 engines .... 0.2380 in
        231 engine .... N/A
        262 engine .... .2340 in

1978 and later models (Continued)

Camshaft
V6 engines
    Lobe lift (Exhaust)
        200, 229 engines .... 0.2600 in
        231 engine .... N/A
        262 engine .... 0.2570 in
    Camshaft journal diameter
        200, 229 and 262 engines .... 1.8682 to 1.8692 inches
        231 engine .... 1.785 to 1.786 inches
    Camshaft runout
        200, 229 and 262 engines .... 0.0015 in max
        231 engine .... N/A
V8 engines
    Lobe lift (Intake)
        267 engine .... 0.2380 in
        1978 though 1982
            305 engine VIN H.. 0.2484 in
            350 engine VIN L .... 0.2600 in
        1983 though 1988
            305 engine VIN H.. 0.2340 in
            305 engine VIN G.. 0.2690 in
    Lobe lift (Exhaust)
        267 engine .... 0.2600 in
        1978 though 1982
            305 engine VIN H.. 0.2667 in
            350 engine VIN L .... 0.2733 in
        1983 though 1988
            305 engine VIN H.. 0.2570 in
            305 engine VIN G.. 0.2760 in
    Camshaft journal diameter (all engines) .... 1.8682 to 1.8692 inches
    Camshaft runout (all engines)... 0.0015 in max
    Camshaft end play .... 0.004 to 0.012 in

Torque specifications   **ft-lbs** (unless otherwise indicated)

V6 engines

Cylinder head bolts
    200, 229 and 262 65
    231 80
Flywheel-to-crankshaft bolts... 60
Intake manifold
    200, 229 and 262 30
    231 45
Exhaust manifold
    200, 229 and 262 .... 20, 25 (inside)
    231 25
Spark plugs
    200, 229 and 262 22
    231 20
Camshaft sprocket
    200, 229 and 262 20
    231 22
Water pump
    200, 229 and 262 30
    231 7
Rocker arm shaft-to-cylinder head
    231 30
Rocker arm cover-to-cylinder head.. 4
Oil drain plug
    200, 229 and 262 20
    231 30
Oil pump-to-block or timing cover
    200, 229 and 262 65
    231 10
Torsional damper-to-crankshaft
    200, 229 and 262 60
    231 175

Rocker arm stud
    200, 229 and 262 50
Oil pan-to-cylinder block
    200, 229 and 262 .... large 30, small 10
    231 14
Special movable timing chain dampener bolt
    231 12
Bellhousing
    200, 229 and 262 30
    231 25
Flywheel housing cover
    200, 229 and 262 30
    231 10
Distributor hold-down clamp
    200, 229 and 262 20
    231 13
Oil pump regulator retainer.. 35
Oil pressure switch-to-cylinder block.. 23
Thermostat housing to intake manifold 13
Distributor hold-down clamp
    200, 229 and 262 25
    231 13
Fuel pump.. 20
Oil galley plugs 25

V8 engines

Crankcase front cover 7
Flywheel housing cover 7
Oil filter by-pass valve 7
Oil pan screws
    Small 7
    Large
        All engines (except big block engines) 6
        400, 402 and 454 big block engines.................................. 12
Oil pan to front cover screws
    400, 402 and 454 big block engines 7
Oil pump cover screws.. 7
Valve cover screws
    All engines (except big block engines)....................................... 4
    400, 402 and 454 big block engines 6
Camshaft sprocket bolt... 20
Clutch pressure plate bolts... 35
Distributor clamp bolts 20
Flywheel housing bolts 30
Exhaust manifold bolts 20
Inner bolts 350 engine 30
Intake manifold bolts 30
Water outlet bolts 30
Water pump bolts 30
Cylinder head bolts
    All engines (except big block engines)....................................... 65
    400, 402 and 454 big block engines 80
Oil pump bolts 65
Rocker arm stud 50
Flywheel bolts
    All engines (except big block engines)....................................... 60
    400, 402 and 454 big block engines 65
Torsional damper bolt
    All engines (except big block engines)....................................... 60
    400, 402 and 454 big block engines 85
Temperature sender unit 20
Oil pan drain plug 20
Spark plug
    5/8 hex 15
    13/16 hex 22

3.6 Turn the crankshaft until the line on the vibration damper is directly opposite the zero mark on the timing plate as shown here

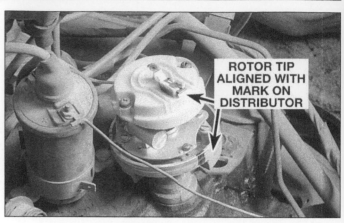

3.7 The rotor tip should point to the mark on the distributor housing

1   General information

This Part of Chapter 2 is devoted to in-vehicle repair procedures for the engine. All information concerning engine removal and installation and engine block and cylinder head overhaul can be found in Part B of this Chapter.

Since the repair procedures included in this Part are based on the assumption that the engine is still installed in the vehicle, if they are being used during a complete engine overhaul (with the engine already out of the vehicle and on a stand) many of the steps included here will not apply.

The specifications included in this Part of Chapter 2 apply only to the procedures found here. The specifications necessary for rebuilding the block and cylinder heads are included in Part B.

V6 and V8-engines used in the Monte Carlo varied in capacity according to the year of production. Engines of 267, 305, 350 and 400 cubic inch capacity are known as small block V8s while the 400 (1970 only), 402 and 454 cu in versions are referred to as Mark IV or big block V8 engines.

In 1978, a V6 engine of 231 cubic inch capacity was introduced, followed in 1979 by a 200 cu in, and in 1980 by a 229 cu in. In 1985 a 262 cu in. fuel injected V6 engine was introduced.

All engines are constructed of cast iron with overhead valves using balljoint-type or rail type rockers and hydraulic or roller lifters. A five-bearing crankshaft is used and full pressure lubrication is provided by a gear-type oil driven from the distributor shaft which in turn is geared to the camshaft.

2   Repair operations possible with the engine in the vehicle

Many major repair operations can be accomplished without removing the engine from the vehicle.

Clean the engine compartment and the exterior of the engine with some type of pres-sure washer before any work is done. A clean engine will make the job easier and will help keep dirt out of the internal areas of the engine.

Depending on the components involved, it may be a good idea to remove the hood to improve access to the engine as repairs are performed (refer to Chapter 11 if necessary).

If oil or coolant leaks develop, indicating a need for gasket or seal replacement, the repairs can generally be made with the engine in the vehicle. The oil pan gasket, the cylinder head gaskets, intake and exhaust manifold gaskets, timing cover gaskets and the crankshaft oil seals are accessible with the engine in place.

Exterior engine components, such as the water pump, the starter motor, the alternator, the distributor and the fuel system components, as well as the intake and exhaust manifolds, can be removed for repair with the engine in place.

Since the cylinder heads can be removed without pulling the engine, valve component servicing can also be accomplished with the engine in the vehicle.

Replacement of, repairs to, or inspection of the timing chain and sprockets, camshaft and the oil pump are all possible with the engine in place.

In extreme cases, caused by a lack of necessary equipment, repair or replacement of piston rings, pistons, connecting rods and rod bearings is possible with the engine in the vehicle. However, this practice is not recommended because of the cleaning and preparation work that must be performed on the components involved.

3   Top Dead Center (TDC) for number one piston - locating

Refer to illustrations 3.6 and 3.7

1   Top Dead Center (TDC) is the highest point in the cylinder that each piston reaches as it travels up and down when the crankshaft turns. Each piston reaches TDC on the compression stroke and again on the exhaust stroke, but TDC generally refers to piston position on the compression stroke. The timing marks on the vibration damper installed on the front of the crankshaft are referenced to the number one piston at TDC on the compression stroke.

2   Positioning the piston(s) at TDC is an essential part of many procedures such as rocker arm removal, valve adjustment, timing chain and sprocket replacement and distributor removal.

3   In order to bring any piston to TDC, the crankshaft must be turned using one of the methods outlined below. When looking at the front of the engine, normal crankshaft rotation is clockwise. **Warning:** *Before beginning this procedure, be sure to place the transmission in Neutral and unplug the wire connector at the distributor to disable the ignition system (electronic ignition) or remove the center wire from the distributor cap and ground it (conventional ignition).*

a) *The preferred method is to turn the crankshaft with a large socket and breaker bar attached to the vibration damper bolt threaded into the front of the crankshaft.*

b) *A remote starter switch, which may save some time, can also be used. Attach the switch leads to the S (switch) and B (battery) terminals on the starter motor. Once the piston is close to TDC, use a socket and breaker bar as described in the previous paragraph.*

c) *If an assistant is available to turn the ignition switch to the Start position in short bursts, you can get the piston close to TDC without a remote starter switch. Use a socket and breaker bar as described in paragraph a) to complete the procedure.*

4   Make a mark on the distributor housing directly below the number one spark plug wire terminal on the distributor cap. **Note:** *The terminal numbers may be marked on the spark plug wires near the distributor.*

5   Remove the distributor cap as described in Chapter 1.

6   Turn the crankshaft (see paragraph 3 above) until the line on the vibration damper is aligned with the zero mark on the timing plate

4.6a Remove the bolts (arrows) from the valve cover - 231 V6 shown

4.6b Early model small block and big block valve covers (small block shown) are held down by studs or bolts around the edges

4.6c The valve covers on later models are held down by three bolts which pass through the center of the cover

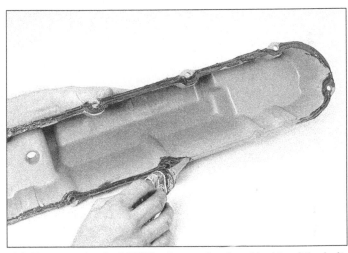

4.9 If your engine had RTV sealer, apply a bead inside of the bolt holes as shown

(see illustration). The timing plate and vibration damper are located low on the front of the engine, near the pulley that turns the drivebelt.

7   The rotor should now be pointing directly at the mark on the distributor housing **(see illustration)**. If it isn't, the piston may be at TDC on the exhaust stroke, so follow Step 8.

8   To get the piston to TDC on the compression stroke, turn the crankshaft one complete turn (360 degrees) clockwise. The rotor should now be pointing at the mark. When the rotor is pointing at the number one spark plug wire terminal in the distributor cap (which is indicated by the mark on the housing) and the ignition timing marks are aligned, the number one piston is at TDC on the compression stroke.

9   After the number one piston has been positioned at TDC on the compression stroke, TDC for any of the remaining cylinders can be located by turning the crankshaft 120 degrees at a time for V6 engines or 90 degrees for V8 engines and following the firing order (refer to the Specifications).

4   Valve covers - removal and installation

Refer to illustrations 4.6a, 4.6b, 4.6c and 4.9

1   Disconnect the negative cable from the battery.

2   Remove the air cleaner assembly.

3   Label and disconnect any emissions or vacuum lines and wires which are routed over the valve cover(s).

4   Remove any braces or brackets which block access to the cover(s), noting the way they are mounted to ease reassembly.

5   Detach any remaining accessories as necessary. If removal of the air conditioning compressor is required, unbolt the unit and carefully set it aside without disconnecting the refrigerant lines. **Warning:** *The air conditioning system is under high pressure. Do not disconnect any refrigerant line or fitting without first having the system discharged by an air conditioning technician.*

6   Remove the valve cover attaching bolts, along with any spark plug wire looms and hardware **(see illustrations)**.

7   Remove the valve cover. **Caution:** *Do not pry on the sealing flange. To do so may damage the surface, causing oil leaks. Tap on the sides of the cover with a rubber hammer until it pops loose.*

8   Thoroughly clean the mating surfaces, removing all traces of old gasket material. Use acetone or lacquer thinner and a clean rag to remove any traces of oil.

9   Prepare the cover for installation by applying a continuous bead of RTV sealer **(see illustration)** or a new gasket to the sealing flange. Use the same sealing method the vehicle had, otherwise new bolts of a different length may be required.

10   Install the cover(s) and bolts and tighten them securely. Do not overtighten them, as the covers may be deformed.

11   Reinstall the remaining parts in the reverse order of removal.

12   Run the engine and check for leaks.

5.2 On Buick 231 V6 engines unscrew the three rocker arm shaft bolts (arrows) then remove the rocker arms and shaft as an assembly

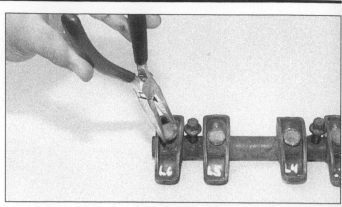

5.3a Remove the rocker arm retainers only if you have to inspect the shaft- the plastic retainers will probably break, so make sure you have new ones on hand - 231 V6 engines

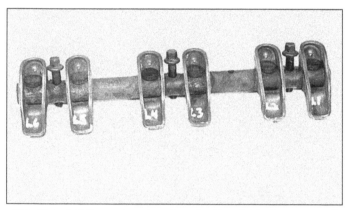

5.3b Reinstall the rockers (numbered with paint before removal) to the shaft in their original sequence - 231 V6 engines

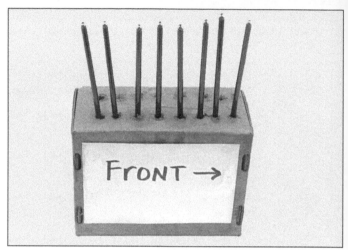

5.4 A perforated cardboard box can be used to store the pushrods to ensure that they are reinstalled in their original locations - note the label indicating the front of the engine

5   Rocker arms and pushrods - removal, inspection and installation

Removal

Refer to illustrations 5.2, 5.3a, 5.3b and 5.4

1   Refer to Section 4 and detach the valve cover(s) from the cylinder head(s).

2   Remove each of the rocker arm nuts. Place them at their correct location in a cardboard box or rack. The 231 V6 engine rocker arms and shafts can be removed as a unit once the retaining bolts are removed **(see illustration)**. **Note:** *On all engines except 231 V6, if the pushrods are the only items being removed, loosen each nut just enough to allow the rocker arms to be rotated to the side so the pushrods can be lifted out.*

3   Remove each rocker arm assembly, placing each component on the numbered box or rack. The 231 V6 engine rocker arms are held in place on the shaft with plastic retainers that can be pried out of place with pliers **(see illustrations)**. Small pieces of the retainer may break off inside the rocker arm

shafts; be sure to remove them during the cleaning process.

4   Remove the pushrods and store them separately to make sure they don't get mixed up during installation **(see illustration)**.

Inspection

5   Check each rocker arm for wear, cracks and other damage, especially where the pushrods and valve stems contact the rocker arm faces.

6   Make sure the hole at the pushrod end of each rocker arm is open.

7   Check each rocker arm pivot area for wear, cracks and galling. If the rocker arms are worn or damaged, replace them with new ones and use new pivot balls as well.

8   Inspect the pushrods for cracks and excessive wear at the ends. Roll each pushrod across a piece of plate glass to see if it's bent (if it wobbles, it's bent).

Installation

Refer to illustrations 5.10 and 5.11

9   Lubricate the lower end of each pushrod with clean engine oil or moly-base grease and install them in their original locations.

Make sure each pushrod seats completely in the lifter.

10   Apply moly-base grease to the ends of the valve stems and the upper ends of the pushrods before positioning the rocker arms

5.10 The ends of the pushrods and the valve stems should be lubricated with moly-based grease prior to installation of the rocker arms

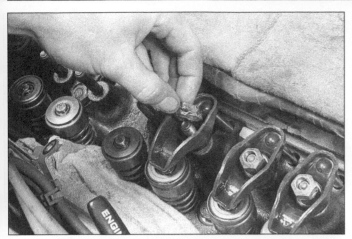

5.11 Moly-based grease applied to the pivot balls will ensure adequate lubrication until the oil pressure builds up when the engine is started

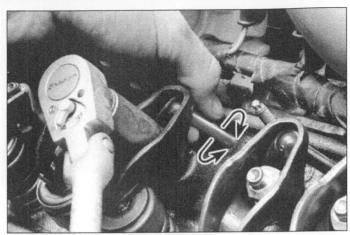

5.13 Rotate each pushrod as the rocker arm nut is tightened to determine the point at which all play is removed (zero lash), then tighten each nut an additional 3/4-turn

onto the cylinder head (see illustration).
11 Set the rocker arms in place, then install the pivot balls (except engines with rocker shafts) and nuts. Apply moly-base grease to the pivot balls or shafts to prevent damage to the mating surfaces before engine oil pressure builds up (see illustration). Be sure to install each nut with the flat side against the pivot ball.

Valve adjustment

Chevrolet-built V6 and V8 engines

Refer to illustration 5.13
12 Refer to Section 3 and bring the number one piston to top dead center on the compression stroke.
13 Tighten the rocker arm nuts (number one cylinder only) until all play is removed at the pushrods. This can be determined by rotating each pushrod between your thumb and index finger as the nut is tightened (see illustration). At the point where a slight drag is just felt as you spin the pushrod, all lash has been removed.
14 Tighten each nut an additional 3/4-turn (270-degrees) to center the lifters. Valve adjustment for cylinder number one is now complete.
15 Turn the crankshaft 120 degrees at a time for V6 engines or 90 degrees for V8 engines in the normal direction of rotation until the next piston in the firing order (number six for V6 engines or number eight for V8 engines) is at TDC on the compression stroke. The distributor rotor should be pointing in the direction of the next terminal on the cap (see Section 3 for additional information).
16 Repeat the procedure described in Steps 13 and 14 for the number six cylinder valves on V6 engines or number eight cylinder valves on V8 engines.
17 Follow the firing order sequence and adjust both valves for each cylinder before proceeding to the next cylinder. A cylinder number illustration and firing order sequence is also included - in the Specifications Section at the beginning of this Chapter.

18 Refer to Section 4 and install the valve covers. Start the engine, listen for unusual valvetrain noises and check for oil leaks at the valve cover joints.

Buick 231 V6 engine

19 These valves cannot be adjusted. If the clearance is excessive in the valve train, check that the pivot bolt is tightened to the proper torque Specification (or the nylon retainer is seated). Then inspect for worn components.

---

6 Valve spring, retainer and seals - replacement

Refer to illustrations 6.4, 6.8, 6.9, 6.16 and 6.17
Note: *Broken valve springs and defective valve stem seals can be replaced without removing the cylinder heads. Two special tools and a compressed air source are normally required to perform this operation, so read through this Section carefully and rent or buy the tools before beginning the job.*
1 Refer to Section 4 and remove the valve cover from the affected cylinder head. If all of the valve stem seals are being replaced, remove both valve covers.
2 Remove the spark plug from the cylinder which has the defective component. If all of the valve stem seals are being replaced, all of the spark plugs should be removed.
3 Turn the crankshaft until the piston in the affected cylinder is at top dead center on the compression stroke (refer to Section 3 for instructions). If you are replacing all of the valve stem seals, begin with cylinder number one and work on the valves for one cylinder at a time. Move from cylinder to cylinder following the firing order sequence (see specifications).
4 Thread an adapter into the spark plug hole and connect an air hose from a compressed air source to it (see illustration). Most auto parts stores can supply the air

6.4 Use compressed air and an air hose adapter (arrow) to hold the valves closed when the valve springs are removed - air hose adapters are commonly available from auto parts stores

hose adapter. **Note:** *Many cylinder compression gauges utilize a screw-in fitting that may work with your air hose quick-disconnect fitting.*
5 Remove the rocker arm(s) for the valve with the defective part and pull out the pushrod. If all of the valve stem seals are being replaced, all of the rocker arms and pushrods should be removed (refer to Section 5).
6 Apply compressed air to the cylinder. The valves should be held in place by the air pressure. If the valve faces or seats are in poor condition, leaks may prevent air pressure from retaining the valves - refer to the alternative procedure below.
7 If you don't have access to compressed air, an alternative method can be used. Position the piston at a point approximately 45-degrees before TDC on the compression stroke, then feed a long piece of nylon rope through the spark plug hole until it fills the combustion chamber. Be sure to leave the end of the rope hanging out of the engine so

6.8 Once the spring is depressed, the keepers can be removed with a small magnet or needle-nose pliers (a magnet is preferred to prevent dropping the keepers)

6.9 Some engines also have an umbrella type seal which fits over the valve guide boss

it can be removed easily. Use a large breaker bar and socket to rotate the crankshaft in the normal direction of rotation until slight resistance is felt.

8   Stuff shop rags into the cylinder head holes above and below the valves to prevent parts and tools from falling into the engine, then use a valve spring compressor to compress the spring/damper assembly. Remove the keepers with small needle-nose pliers or a magnet **(see illustration)**. **Note:** *A couple of different types of tools are available for compressing the valve springs with the head in place. One type, shown here, grips the lower spring coils and presses on the retainer as the knob is turned, while the other type utilizes the rocker arm stud and nut for leverage. Both types work very well, although the lever type is usually less expensive and won't work on all cylinders of engines with rocker shafts.*

9   Remove the spring retainer or rotator, oil shield and valve spring assembly, then remove the valve stem O-ring seal (the O-ring seal will most likely be hardened and will probably break when removed, so plan on installing a new one each time the original is removed). and the umbrella-type guide seal on later models engines **(see illustration)**. **Note:** *If air pressure fails to hold the valve in the closed position during this operation, the valve face and/or seat is probably damaged. If so, the cylinder head will have to be removed for additional repair operations.*

10   Wrap a rubber band or tape around the top of the valve stem so the valve won't fall into the combustion chamber, then release the air pressure. **Note:** *If a rope was used instead of air pressure, turn the crankshaft slightly in the direction opposite normal rotation.*

11   Inspect the valve stem for damage. Rotate the valve in the guide and check the end for eccentric movement, which would indicate that the valve is bent.

12   Move the valve up and down in the guide and make sure it doesn't bind. If the valve stem binds, either the valve is bent or

the guide is damaged. In either case, the head will have to be removed for repair.

13   Reapply air pressure to the cylinder to retain the valve in the closed position, then remove the tape or rubber band from the valve stem. If a rope was used instead of air pressure, rotate the crankshaft in the normal direction of rotation until slight resistance is felt.

14   Lubricate the valve stem with engine oil and install a new umbrella type guide seal (if equipped).

15   Install the spring/damper assembly and shield in position over the valve.

16   Install the valve spring retainer or rotator. Compress the valve spring assembly and carefully install the new O-ring seal in the lower groove of the valve stem. Make sure the seal isn't twisted - it must lie perfectly flat in the groove **(see illustration)**.

17   Position the keepers in the upper groove. Apply a small dab of grease to the inside of each keeper to hold it in place if necessary **(see illustration)**. Remove the pressure from the spring tool and make sure the keepers are seated.

18   Disconnect the air hose and remove the adapter from the spark plug hole. If a rope

was used in place of air pressure, pull it out of the cylinder.

19   Refer to Section 5 and install the rocker arm(s) and pushrod(s).

20   Install the spark plug(s) and hook up the wire(s).

21   Refer to Section 4 and install the valve cover(s).

22   Start and run the engine, then check for oil leaks and unusual sounds coming from the valve cover area.

7   Vibration damper - removal and installation

Refer to illustrations 7.5a, 7.5b, 7.6 and 7.7

1   Disconnect the negative cable from the battery.

2   Remove the fan shroud and fan assembly (see Chapter 3).

3   Remove the drivebelts (see Chapter 1).

4   Remove the pulley retaining bolts and detach the pulleys. **Caution:** *Before removing the pulleys, mark the location of each one in relation to the vibration damper in order to*

6.16 Make sure the O-ring seal under the retainer is seated in the groove and not twisted before installing the keepers

6.17 Apply a small dab of grease to each keeper as shown here before installation - it'll hold them in place on the valve stem as the spring is released

7.5a The vibration damper bolt (arrow) is usually very tight, so use a six-point socket and a breaker bar to loosen it (the three other bolts hold the pulley to the vibration damper)

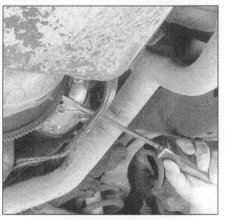

7.5b Have an assistant hold a large screwdriver or pry bar against the ring gear

7.6 Use the recommended puller to remove the vibration damper - if a puller that applies force to the outer edge is used, the damper will be damaged!

retain proper timing mark orientation during installation.

5   Raise the vehicle and support it securely on jackstands. Remove the lower bellhousing cover. Wedge a pry bar between the fly-wheel/driveplate teeth to prevent the crank-shaft from turning while loosening the large vibration damper bolt **(see illustrations)**.

6   Leave the damper bolt in place (with several threads still engaged), to provide the gear puller with something to push against. Use a puller to remove the damper **(see illustration)**. **Caution:** *Don't use a puller with jaws that grip the outer edge of the damper! The puller must be the type shown in the illustration that utilizes bolts to apply force to the damper hub only.*

7   To install the damper apply a dab of grease to the seal lips, position the damper on the nose of the crankshaft, align the key-way and start it onto the crankshaft with a special installation tool **(see illustration)**. Install the bolt. Tighten the bolt to the speci-fied torque.

8   The remaining installation steps are the reverse of the removal procedure.

8   Front crankshaft oil seal - replacement

Chevrolet V6 and V8 engines

Timing cover in place

Refer to illustrations 8.2 and 8.4

1   Remove the vibration damper as described in Section 7.

2   Carefully pry the seal out of the cover with a seal removal tool or a large screw-driver **(see illustration)**. Be careful not to dis-tort the cover or scratch the wall of the seal bore. If the engine has accumulated a lot of miles, apply penetrating oil to the seal-to-cover joint and allow it to soak in before attempting to pull the seal out.

3   Clean the bore to remove any old seal material and corrosion. Position the new seal in the bore with the open end of the seal fac-ing IN. A small amount of oil applied to the outer edge of the new seal will make installa-tion easier - don't overdo it!

4   Drive the seal into the bore with a large socket and hammer until it's completely seated **(see illustration)**. Select a socket that's the same outside diameter as the seal

7.7 Using a special tool to push the vibration damper onto the crankshaft

(a section of pipe can be used if a socket isn't available).

5   Reinstall the vibration damper.

Timing cover removed

Refer to illustrations 8.6 and 8.8

6   Use a punch or screwdriver and ham-mer to drive the seal out of the cover from the

8.2 Use a screwdriver or seal removal tool to pull out the front seal, being careful not to nick the crankshaft surface

8.4 Use the appropriate size socket to drive the new seal squarely into the front cover

8.6 While supporting the cover near the seal bore, drive the old seal out from the inside with a hammer and punch or screwdriver

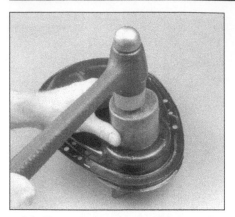

8.8 Clean the bore, then apply a small amount of oil to the outer edge of the seal and drive it squarely into the opening with a large socket or a piece of pipe and a hammer - do not damage the seal in the process!

8.9 Drive the oil shedder out of the cover using a punch from the front side

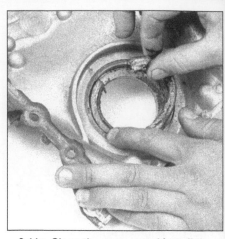

8.11a Clean the groove and install the new seal material with the ends butted together at the top of the cover

8.11b Install the shedder ring with a large socket and hammer, then stake the ring in place in three places with a small punch

8.12 Using a smooth hammer handle, press the seal into the groove completely around the opening

9.11 After covering the lifter valley, use a gasket scraper to remove all traces of sealant and old gasket material from the head and manifold mating surfaces

back side. Support the cover as close to the seal bore as possible (see illustration). Be careful not to distort the cover or scratch the wall of the seal bore. If the engine has accumulated a lot of miles, apply penetrating oil to the seal-to-cover joint on each side and allow it to soak in before attempting to drive the seal out.

7   Clean the bore to remove any old seal material and corrosion. Support the cover on blocks of wood and position the new seal in the bore with the open end of the seal facing IN. A small amount of oil applied to the outer edge of the new seal will make installation easier - don't overdo it!

8   Drive the seal into the bore with a large socket and hammer until it's completely seated **(see illustration)**. Select a socket that's the same outside diameter as the seal (a section of pipe can be used if a socket isn't available).

Buick 231 V6 engine

Refer to illustrations 8.9, 8.11a, 8.11b and 8.12

9   Remove the front timing chain cover

from the engine (see Section 12). Using a hammer and punch, drive the oil shedder from the backside of the cover **(see illustration)**.

10   Remove the old seal material from the groove in the front cover and clean the seal cavity thoroughly.

11   Press a new length of seal material into the groove. Position the ends, together, at the top of the groove. Drive the oil shedder into the cover and stake it in three places with a small punch **(see illustrations)**.

12   Use a wooden hammer handle to seat the seal material into the groove **(see illustration)**.

13   Lubricate the seal, reinstall the timing chain cover (see Section 12) and the vibration damper (see Section 7).

---

9   Intake manifold - removal and installation

Refer to illustrations 9.11, 9.12, 9.14a, 9.14b, 9.14c and 9.14d

1   Place protective covers on the fenders

and disconnect the negative cable from the battery.

2   Drain the cooling system (see Chapter 1).

3   Remove the air cleaner assembly.

4   Detach all coolant hoses from the intake manifold.

5   Label and then disconnect any wiring and vacuum, emission and/or fuel lines which attach to the intake manifold.

6   Remove the EGR valve, if necessary (see Chapter 6).

7   On engines with the distributor mounted through the manifold, remove the distributor (see Chapter 5).

8   Remove the carburetor or throttle body (see Chapter 4).

9   Remove the intake manifold bolts and any brackets, noting their locations, so they may be reinstalled in the same location.

10   Lift off the manifold. If it is difficult to break loose, carefully pry against a casting protrusion. **Caution:** *Do not pry against a sealing surface.* **Note:** *If your engine has a valley cover under the intake manifold, this*

9.12 Make sure the intake manifold gaskets are installed right side up or all the passages and bolt holes may not line up properly!

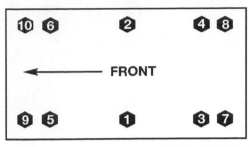

9.14a Intake manifold bolt tightening sequence for Buick 231 V6 engine

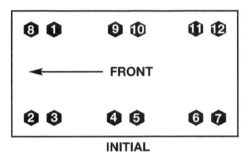
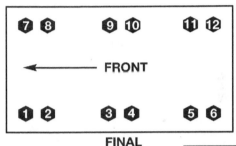

INITIAL   FINAL

9.14b Intake manifold bolt tightening sequence for 262 V6 engine

must be removed to access the valve lifters. Unbolt the cover and lift it out.

11   Stuff rags in the exposed ports and passages. Thoroughly clean all sealing surfaces, removing all traces of gasket material **(see illustration)**. Check the exhaust crossover passages below the carburetor for carbon blockage.

12   Position new gaskets and seals into place, using RTV sealer in the corners and to hold them in place **(see illustration)**. Use plastic gasket retainers, if provided. Follow the gasket manufacturer's instructions.

13   Set the manifold into position. Be sure all the bolt holes are lined up and start all the bolts by hand before tightening any.

14   Tighten the bolts in three stages, following the appropriate sequence **(see illustrations)** until the specified torque is reached.

15   The remainder of installation is the reverse of removal. Be sure to refill the cooling system and change the oil and filter (see Chapter 1), as coolant often gets into the oil during manifold removal.

16   Run the engine and check for fluid and vacuum leaks. Adjust the ignition timing, if the distributor was disturbed (see Chapter 1). Road test the vehicle, checking for proper operation of engine and accessories.

10   Exhaust manifolds - removal and installation

Refer to illustrations 10.11 and 10.14

1   Allow the engine to cool completely.

2   Place protective covers on the fenders.

3   Disconnect the negative cable from the battery.

4   Raise the vehicle and support it securely on jackstands.

5   Spray penetrating oil on the threads of fasteners holding the hot air shrouds and the exhaust pipe(s) to the manifold(s) you wish to remove. Go on to the next steps, allowing the penetrating oil to soak in.

6   Remove the air cleaner and hot air duct.

7   Detach the hot air shroud(s) or heat stove assembly from the manifold.

8   Label and then remove spark plug and oxygen sensor wires and any hoses such as EFE or AIR, as needed.

9   Unbolt any engine accessories which block access to the manifold(s) such as the alternator, air conditioning compressor or power steering pump. Refer to the appropri-

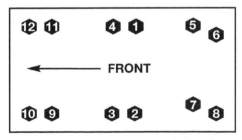

9.14c Intake manifold bolt tightening sequence for 200 and 229 cubic inch V6 engines and all small block V8 engines

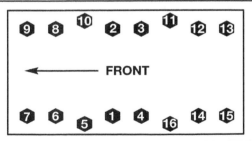

9.14d Intake manifold bolt tightening sequence for big block V8 engines

10.11 The spark plug heat shields are held in place by bolts located under the exhaust manifold

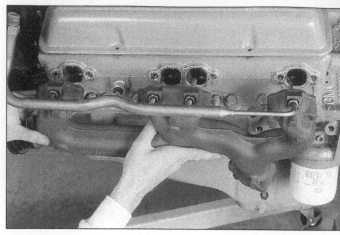

10.14 The exhaust manifold gaskets are held in place with sealant as the manifold is installed

ate Chapters (5, 3, and 10) for further information. When unbolting the air conditioning compressor, leave the refrigerant lines connected and tie the compressor aside without putting a strain on the fittings. **Warning:** *Do not disconnect any air conditioning line or fitting unless the system pressure has been discharged by a qualified A/C specialist.*

10 Remove the manifold-to-exhaust pipe fasteners. **Note:** *On some models it may be necessary to remove the starter (see Chapter 5), the lower bellhousing cover, and/or a front wheel for access to the manifolds.*

11 Bend back any lock tabs from the manifold bolt heads and remove the bolts and any spark plug heat shields **(see illustration)**, noting their types and locations for reassembly.

12 Remove the manifold(s) from the engine compartment.

13 Thoroughly clean and inspect the sealing surfaces, removing all traces of old gasket material. Look for cracks and damaged threads. If the gasket was leaking, take the manifold to an automotive machine shop to inspect/correct for warpage.

14 If replacing the manifold, transfer all parts to the new unit. Using new gaskets (when equipped), install the manifold **(see illustration)** and start the bolts by hand.

15 Working from the center outward, tighten the bolts to the specified torque. Bend the lock tabs (if equipped) against the bolt heads.

16 Reinstall the remaining components in the reverse order of removal.

17 Run the engine and check for exhaust leaks.

11 Cylinder heads - removal and installation

Refer to illustrations 11.7, 11.8a, 11.8b, 11.8c, 11.8d and 11.14

Caution: *The engine must be completely cool when the heads are removed. Failure to allow the engine to cool off could result in head warpage.*

Removal

1 Remove the intake manifold (see Section 9).

2 Remove the valve cover(s) as described in Section 4.

3 Remove the rocker arms and pushrods (see Section 5). **Note:** *If any pushrods or head bolts cannot be removed because of interference with components such as the vacuum brake booster, lift them up as far as possible and wrap a rubber band around them to keep them from dropping during head removal. Be sure to do the same during installation.*

4 Remove the exhaust manifolds (see Section 10).

5 Unbolt the engine accessories and brackets as necessary and set them aside. See Chapter 5 for alternator removal. When removing a power steering pump (see Chapter 10), leave the hoses attached and secure the unit in an upright position. Refrigerant lines should remain attached to air conditioning compressors (see Chapter 3). **Warning:** *The air conditioning system is under high pressure. Do not loosen any hose or fitting until the refrigerant gas has been discharged by an auto air conditioning technician.*

6 Recheck to be sure nothing is still connected to the head(s) you intend to remove. Remove anything that is still attached.

7 Using a new gasket, outline the cylinders and bolt pattern on a piece of cardboard **(see illustration)**. Be sure to indicate the front of the engine for reference. Punch holes at the bolt locations.

8 Loosen the head bolts in 1/4-turn increments until they can be removed by hand. Work from bolt to bolt in a pattern that's the reverse of the tightening sequence **(see illustrations)**. Store the bolts in the cardboard holder as they are removed; this will insure that the bolts are reinstalled in their original holes.

9 Lift the head(s) off the engine. If excessive resistance is felt, do not pry between the head and block as damage to the mating surfaces may result. To dislodge the head, place

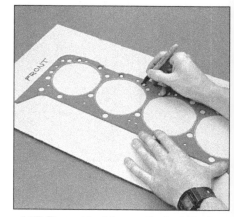

11.7 To avoid mixing up the head bolts, use a new gasket to transfer the bolt hole pattern to a piece of cardboard, then punch holes to accept the bolts

a block of wood against the end of it and strike the wood with a hammer. Store the head(s) on their sides to prevent damage to the gasket surface.

10 Cylinder head disassembly and inspection procedures are covered in detail in Chapter 2, Part B.

11 Use a gasket scraper to remove all traces of carbon and old gasket material, then clean the mating surfaces with lacquer thinner or acetone. If there's oil on the mating surfaces when the heads are installed, the gaskets may not seal properly and leaks may develop. When working on the block, cover the lifter valley with shop rags to keep out debris. Use a vacuum cleaner to remove any debris that falls into the cylinders.

12 Check the block and head mating surfaces for nicks, deep scratches and other damage. If damage is slight, it can be removed with a file. If the gasket seemed to be leaking, have the head(s) checked for cracks and warpage by an automotive machine shop. Surface damage or warpage can usually be corrected by machining. Cracks normally require replacement.

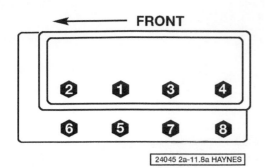

11.8a Cylinder head bolt tightening sequence for
231 V6 engine

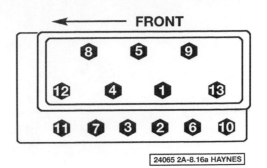

11.8b Cylinder head bolt tightening sequence for
200, 229, 262 V6 engines

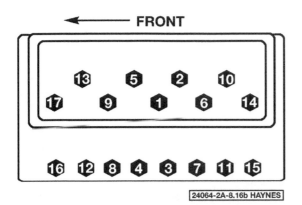

11.8c Cylinder head bolt tightening sequence for
small block V8 engines

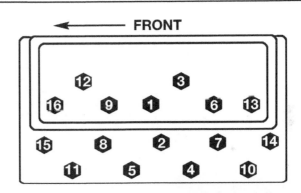

11.8d Cylinder head bolt tightening sequence for
big block V8 engines

Installation

13   Use a tap of the correct size to chase the threads in the head bolt holes. Mount each bolt in a vise and run a die down the threads to remove any corrosion and restore the threads. Dirt, corrosion, sealant and damaged threads will affect torque readings.

14   If an all-metal gasket is used, apply a thin, even coat of sealant such as K & W Copper Coat to both sides prior to installation. Metal gaskets must be installed with the raised bead UP **(see illustration)**. Most gaskets have "This side up" stamped into them. Composition gaskets must be installed dry -

don't use sealant. Follow the gasket manufacturer's instructions, if available. Position a new gasket over the dowel pins in the block.

15   Carefully position the head on the block without denting or disturbing the gasket.

16   Be sure the head bolt holes are clean. On all engines except 231 V6, apply nonhardening sealant such as Permatex no. 2 to the head bolt threads. Apply a light coating of engine oil to the head bolts of the 231 V6 engines.

17   Install the bolts in their original locations and tighten them finger tight. Follow the recommended sequence and tighten the bolts in several steps to the specified torque.

18   The remaining installation steps are the reverse order of removal. Check the index for the appropriate Sections for installation of the various components.

19   Add coolant, change the oil and filter (see Chapter 1), then start the engine and check carefully for leaks.

12   Timing cover, chain and sprockets

Refer to illustrations 12.7a, 12.7b, 12.11, 12.12a, 12.12b and 12.13

Removal

1   Disconnect the negative cable from the battery.

2   If equipped with a 231 V6 engine, remove the alternator and brackets as well as the distributor (see Chapter 5).

3   Drain the cooling system and disconnect the radiator hoses, heater hose (where applicable) and the small by-pass hose.

4   Remove the top radiator support, the fan shroud, the fan and pulley and the radiator (see Chapter 3).

5   Remove the crankshaft drive pulley and the vibration damper (see Section 7).

6   On all other engines, remove the water pump (see Chapter 3).

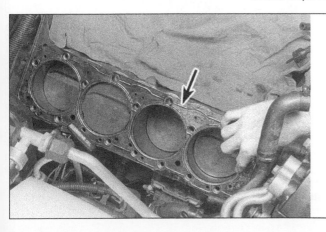

11.14 Steel gaskets must be installed with the raised bead (arrow) facing UP

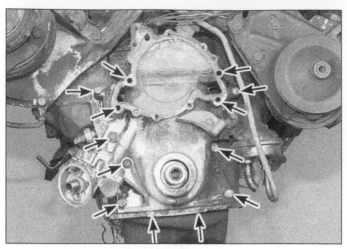

12.7a Timing chain cover bolt locations – Buick 231 V6 engine

12.7b A putty knife or screwdriver can be used to break the timing chain cover-to-block seal, but be careful when prying it off as damage to the cover may result - Chevrolet

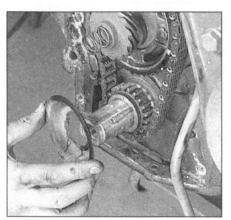

12.11 Oil slinger installation - typical

7   Remove the nuts and bolts that attach the cover to the engine, then pull the cover free **(see illustration)**. The 231 V6 engines have bolts attaching the oil pan to the cover.

8   Using vise-grip pliers, pull the dowel pins (if equipped) out of the engine block. Grind a chamfer on one end of each pin. Thoroughly clean all gasket mating surfaces (do not allow the old gasket material to fall into the oil pan), then wipe them with a cloth soaked in solvent.

9   Remove the fuel pump drive eccentric by unscrewing the bolt from the end of the camshaft (if equipped).

10   Some 231 V6 engines have a distributor drive gear mounted in front of the fuel pump eccentric (when removing them, note how they are aligned with each other and the camshaft).

11   On 231 V6 engines, slide the oil slinger off of the end of the crankshaft **(see illustration)**.

12   Turn the crankshaft until the marks on the camshaft and crankshaft sprockets are perfectly aligned **(see illustrations)**. DO NOT attempt to remove either sprocket or the chain until this is done. Also, do not turn the camshaft or the crankshaft after the sprockets are removed.

13   On all engines except 231 V6 engines, remove the bolts that attach the sprocket to the end of the camshaft **(see illustration)**.

14   Remove the timing chain and gears from the engine. Generally speaking, the camshaft sprocket, the crankshaft sprocket and the timing chain can be slipped off the shafts together. If resistance is encountered, it may be necessary to use two screwdrivers to carefully pry the sprockets off of the shafts. If extreme resistance is encountered (which may happen with the crankshaft sprocket), a gear puller will be required.

Installation

15   If the crankshaft and the camshaft are not disturbed while the timing chain and sprockets are out of place, then installation can begin with Step 18.

16   If the engine is completely dismantled, or if the crankshaft or camshaft are disturbed while the timing chain is off, then the No. 1 piston must be positioned at TDC before the timing chain and sprockets are installed. Align the hole in the camshaft sprocket with the dowel pin in the end of the camshaft, then slip the sprocket onto the end of the camshaft. On 231 V6 engines, turn the camshaft until the timing mark on the sprocket is pointed straight down. **Note:** *On all engines except the 231 V6, with the No. 1 piston at TDC, the timing mark on the camshaft sprocket must be pointed straight up* **(see illustrations 12.12a and 12.12b)**.

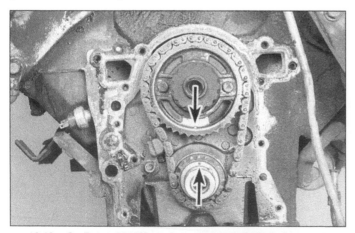

12.12a On Buick 231 V6 engines, the camshaft timing mark should be at the bottom (6 o'clock position) and the crankshaft mark at the top (12 o'clock position)

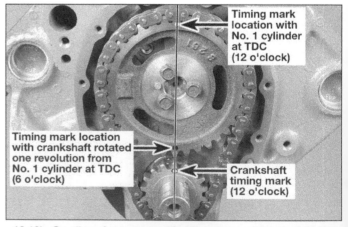

12.12b On all engines except 231 V6 engines, carefully note the positions of the timing marks on the camshaft sprocket and the crankshaft gear

12.13 Remove the three bolts from the end of the camshaft (arrows) – Chevrolet built engine shown

12.30 Before reinstalling the timing chain cover, insert the spring and button into the front of the camshaft

13.3 When checking the camshaft lobe lift, the dial indicator plunger must be positioned directly above the pushrod

17   Slip the crankshaft sprocket onto the end of the crankshaft (make sure the key and keyway are properly aligned), then turn the crankshaft until the timing mark on the sprocket is pointed straight up (all engines) and remove the camshaft sprocket from the camshaft

18   Lay the timing chain over the camshaft sprocket. Slip the other end of the chain over the crankshaft sprocket (keep the timing marks aligned as this is done) and reinstall the sprockets. When the sprockets are properly installed, the timing marks on the sprockets will be perfectly aligned opposite each other (all engines except 231 V6 engines; refer to **illustrations 12.12a and 12.12b** and Step 16).

19   Install the bolts that attach the sprocket to the camshaft, then recheck the alignment of the timing marks.

20   Be sure to tighten the camshaft sprocket retaining bolts to the specified torque (see Specifications).

21   The rest of the installation procedure is basically the reverse of removal with the following exceptions.

All engines except 231 V6

22   Ensure that all gasket surfaces are clean and free of excess gasket material.

23   Use a sharp knife to trim any protruding gasket material at the front of the oil pan.

24   Apply a 1/8-inch bead of RTV gasket sealer to the joint formed at the oil pan and engine block, as well as the front lip of the oil pan.

25   Coat the cover gasket with a non-setting sealant, position it on the cover, then loosely install the cover. First install the top four bolts loosely, then install two 1/4 - 20 x 1/2 inch long screws at the lower cover holes. Apply a bead of sealer on the bottom of the cover, then install the cover, tightening the screws alternately and evenly and at the same time aligning the dowel pins.

26   Remove the two 1/4 - 20 x 1/2 inch long screws and install the remaining cover bolts. Torque all cover bolts to the proper specifications.

27   Install the water pump using new gaskets (see Chapter 3).

28   Follow the removal steps in the reverse order for the remaining components.

Buick 231 V6 engine

Refer to illustration 12.30

29   Before re-installing the timing chain cover, remove the oil pump cover and pack petroleum jelly around the oil pump gears so that there is no air space left inside the pump. If this is not done, the pump may "lose its prime" and not begin pumping oil immediately when the engine is started.

30   Before reinstalling the timing chain cover, insert the spring and button into the front of the camshaft **(see illustration)**.

31   Install the pump cover, using a new gasket, and torque the bolts to specifications.

32   Make sure that the gasket surface of the block and timing chain cover are smooth and clean and install a new gasket on the cover.

33   Lubricate the vibration damper hub where it will go through the timing chain cover seal so that the seal will not be damaged when the engine is first started.

34   Using the dowel pins on the block, engage the dowel holes in the cover and position the cover against the block.

35   Apply sealer to the bolt threads and tighten the bolts to the Specifications.

36   Install the vibration damper, bolt and washer (see Section 7).

37   Follow the removal steps in the reverse order for the remaining components.

---

13   Camshaft, bearings and lifters - removal, inspection and installation

---

Refer to illustrations 13.3, 13.12 and 13.18

Camshaft lobe lift check

1   In order to determine the extent of cam lobe wear, the lobe lift should be checked prior to camshaft removal. Refer to Section 4

and remove the valve covers.

2   Position the number one piston at TDC on the compression stroke (see Section 3).

3   Beginning with the number one cylinder valves, mount a dial indicator on the engine and position the plunger against the top surface of the first rocker arm. The plunger should be directly above and in line with the pushrod **(see illustration)**.

4   Zero the dial indicator, then very slowly turn the crankshaft in the normal direction of rotation (clockwise) until the indicator needle stops and begins to move in the opposite direction. The point at which it stops indicates maximum cam lobe lift.

5   Record this figure for future reference, then reposition the piston at TDC on the compression stroke.

6   Move the dial indicator to the remaining number one cylinder rocker arm and repeat the check. Be sure to record the results for each valve.

7   Repeat the check for the remaining valves. Since each piston must be at TDC on the compression stroke for this procedure, work from cylinder to cylinder following the firing order sequence (refer to the Specifications).

8   After the check is complete, compare the results to the Specifications. If camshaft lobe lift is less than specified, cam lobe wear has occurred and a new camshaft should be installed.

Note: *On air-conditioned models, have the refrigerant discharged and remove the condenser prior to commencing the following procedure (see Chapter 3).*

Removal

9   Refer to the appropriate Sections and remove the intake manifold, the rocker arms, the pushrods and the timing chain and camshaft sprocket. The radiator, fuel pump and distributor should be removed as well (see Chapters 3, 4 and 5).

10   There are several ways to extract the lifters from the bores. A special tool designed to grip and remove lifters is manufactured by

many tool companies and is widely available, but it may not be required in every case. On newer engines without a lot of varnish buildup, the lifters can often be removed with a small magnet or even with your fingers. A machinists scribe with a bent end can be used to pull the lifters out by positioning the point under the retainer ring in the top of each lifter. **Caution:** *Do not use pliers to remove the lifters unless you intend to replace them with new ones (along with the camshaft). The pliers may damage the precision machined and hardened lifters, rendering them useless.*

11   Before removing the lifters, arrange to store them in a clearly labeled box to ensure that they are reinstalled in their original locations. Remove the lifters and store them where they will not get dirty. Do not attempt to withdraw the camshaft with the lifters in place.

12   Thread a long bolt of the proper thread into one of the camshaft sprocket bolt holes to use as a handle when removing the camshaft from the block **(see illustration)**.

13   Carefully pull the camshaft out. Support the cam near the block so the lobes do not nick or gouge the bearings as it is withdrawn.

Inspection

Camshaft and bearings

14   After the camshaft has been removed from the engine, cleaned with solvent and dried, inspect the bearing journals for uneven wear, pitting and evidence of seizure. If the journals are damaged, the bearing inserts in the block are probably damaged as well. Both the camshaft and bearings will have to be replaced.

15   Measure the bearing journals with a micrometer to determine if they are excessively worn or out-of-round.

16   Check the camshaft lobes for heat discoloration, score marks, chipped areas, pitting and uneven wear. If the lobes are in good condition and if the lobe lift measurements are as specified, the camshaft can be reused.

Conventional lifters

17   Clean the lifters with solvent and dry them thoroughly without mixing them up.

18   Check each lifter wall, pushrod seat and foot for scuffing, score marks and uneven wear. Each lifter foot (the surface that rides on the cam lobe) must be slightly convex, although this can be difficult to determine by eye **(see illustration)**. If the base of the lifter is concave, the lifters and camshaft must be replaced. If the lifter walls are damaged or worn (which is not very likely), inspect the lifter bores in the engine block as well. If the pushrod seats are worn, check the pushrod ends.

19   If new lifters are being installed, a new camshaft must also be installed. If a new camshaft is installed, use new lifters as well. Never install used lifters unless the original camshaft is used and the lifters can be installed in their original locations.

Roller lifters

20   Check the rollers carefully for wear and

13.12 A long bolt can be threaded into one of the camshaft bolt holes to provide a handle for removal and installation of the camshaft

damage and make sure they turn freely without excessive play. The inspection procedure for conventional lifters also applies to roller lifters.

21   Unlike conventional lifters, used roller lifters can be reinstalled with a new camshaft, and the original camshaft can be used if new lifters are installed.

Bearing replacement

22   Camshaft bearing replacement requires special tools and expertise that place it outside the scope of the home mechanic. Take the block to an automotive machine shop to ensure that the job is done correctly.

Installation

23   Lubricate the camshaft bearing journals and cam lobes with molybase grease or engine assembly lube.

24   Slide the camshaft into the engine. Support the cam near the block and be careful not to scrape or nick the bearings.

25   Lubricate and then install the thrust plate, if equipped.

26   Refer to Section 12 and install the timing chain and sprockets.

27   Lubricate the lifters with clean engine oil and install them in the block. If the original lifters are being reinstalled, be sure to return them to their original locations. If a new camshaft is being installed, be sure to install new lifters as well (except for engines with roller lifters).

28   The remaining installation steps are the reverse of removal.

29   Before starting and running the engine, change the oil and install a new oil filter (see Chapter 1).

---

14   Oil pan - removal and installation

Refer to illustrations 14.18, 14.21a and 14.21b

Warning: *Do not place any part of your body below the engine when it is supported solely*

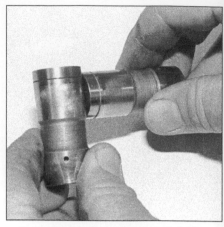

13.18 The foot of each lifter should be slightly convex - the side of another lifter can be used as a straightedge to check it; if it appears flat, it is worn and must not be reused

by a jack or hoist.

1   Disconnect the negative battery cable.

2   Remove the air cleaner assembly and set aside.

3   Remove the distributor cap to prevent breakage as the engine is raised (models with distributor at rear of engine only).

4   Unbolt the radiator shroud from the radiator support and hang the shroud over the cooling fan.

5   If necessary, remove the oil dipstick and dipstick tube.

6   Raise the car and support firmly on jack stands.

7   Drain the engine oil into a suitable container.

8   Disconnect the exhaust crossover pipe at the exhaust manifold flanges. Lower the exhaust pipes and suspend them from the frame with wire.

9   If equipped with an automatic transmission, remove the lower bellhousing cover.

10   If equipped with a manual transmission, remove the starter (see Chapter 5) and the flywheel cover.

11   Use the bolt at the center of the vibration damper to rotate the engine until the timing mark is straight down, at the 6 o'clock position. This will move the forward crankshaft throw upward, providing clearance at the front of the oil pan.

12   Remove the through bolt at each engine mount.

13   At this time the engine must be raised slightly to enable the oil pan to slide clear of the crossmember. The preferred method is to use an engine hoist or "cherry picker". Hook up the lifting chains as described in Part B, Section 5.

14   An alternative method can be used if extreme care is exercised. Use a floor jack and a block of wood placed under the oil pan. The wood block should spread the load across the oil pan, preventing damage or collapse of the oil pan metal. The oil pump pickup and screen is very close to the oil pan

14.18 Tilt the oil pan down (as shown) at the rear to clear the front crossmember

14.21a Sealant is applied at the area where the front gasket meets the side gasket (Chevrolet engine shown)

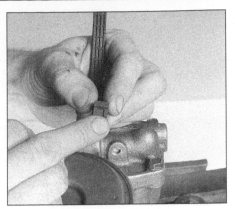

14.21b The front oil seal (Chevrolet engine) has an indentation which fits into the side gasket

bottom, so any collapsing of the pan may damage the pickup or prevent the oil pump from drawing oil properly.

15   With either method, raise the engine slowly until wood blocks can be placed between the frame front crossmember and the engine block. The blocks should be approximately three inches thick. Check clearances all around the engine as it is raised. Pay particular attention to the distributor and the cooling fan.

16   Lower the engine onto the wood blocks. Make sure it is firmly supported. If a hoist is being used, keep the lifting chains secured to the engine.

17   Remove the oil pan bolts. Note the different sizes used and their locations.

18   Remove the oil pan by tilting it downwards at the rear and then working the front clear of the crossmember **(see illustration)**. It may be necessary to use a rubber mallet to break the seal.

19   Before installing, thoroughly clean the gasket sealing surfaces on the engine block and on the oil pan. All sealer and gasket material must be removed.

20   Apply a thin film of sealer to the new side gaskets and fit them to the engine block. All bolt holes should line up properly. If equipped with retainers, mount the retainers over the gasket and secure the gasket to the oil pan.

21   Again using sealer, install the front and rear seals to the engine. Make sure the ends butt with the ends of the side gaskets **(see illustrations)**.

22   Lift the pan into position and install all bolts finger-tight. There is no specific order for tightening the bolts; however, it is a good policy to tighten the end bolts first.

23   Lower the engine onto its mounts and install the through bolts. Tighten the nuts/bolts securely.

24   Follow the removal steps in a reverse order. Fill the engine with the correct grade and quantity of oil, start the engine and check for leaks.

15   Oil pump - removal, inspection and installation

Buick 231 V6 engine

Refer to illustration 15.4

Removal

1   Unscrew and remove the oil filter (see Chapter 1).

2   Remove the screws attaching the oil pump cover assembly to the timing chain cover. Remove the cover assembly and slide out the oil pump gear.

Inspection

3   Wash off the gears with a proper solution and inspect for wear, scoring, etc. Replace any unserviceable gears with new ones.

4   Unscrew the oil pressure relief valve cap, spring and valve **(see illustration)**. Do not remove the oil filter by-pass valve and spring, as they are staked in place.

5   Wash the parts thoroughly in the proper solvent and inspect the relief valve for wear and scoring. Check to make sure that the relief valve spring is not collapsed or worn on its side. Any relief valve spring which is questionable should be replaced with a new one.

6   Check the relief valve in its bore in the cover. It should be an easy slip-fit only, and any side shake that can be felt is too much. The valve and/or cover should be replaced with a new one in this case.

7   The filter by-pass valve should be flat and free of nicks, cracks or warping and scratches.

8   Lubricate the pressure relief valve and spring, and install it in the bore of the oil pump case. Install the cap and gasket, and tighten the cap securely.

9   If the above inspections reveal wear or if high mileage/low oil pressure indicate a faulty pump, replace it with a new one.

10   If the condition of the pump is satisfactory at this point, remove the gears and pack the pocket full of petroleum jelly. Do not use chassis lube.

Installation

11   Re-install the gears, making sure that petroleum jelly is forced into every cavity of the gear pocket and between the teeth of the gears. The pump may not prime itself when the engine is started if the pump is not packed with the petroleum jelly.

12   Install the pump cover assembly screws and tighten them alternately and evenly. Torque-tighten to Specifications.

13   Install oil filter and check oil level with the dipstick. Pay close attention to the oil pressure gauge or warning light during the initial startup and driving period. Shut off the engine and inspect all work if a lack of pressure is indicated.

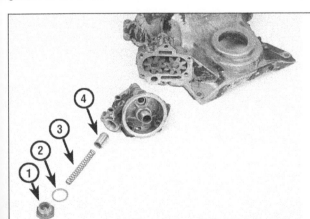

15.4 Oil pump cover and pressure relief components - typical 231 V6 engine

1   *Pressure relief cap*
2   *Copper sealing ring*
3   *Oil pressure relief spring*
4   *Oil pressure relief valve*

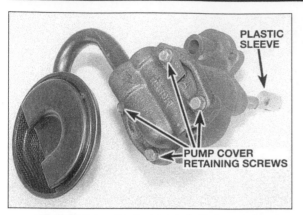

15.17 Remove the four cover bolts (arrows)

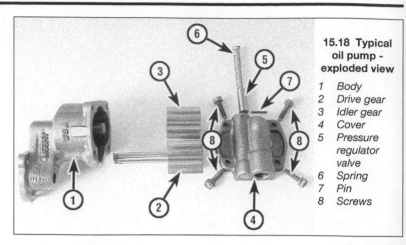

15.18 Typical oil pump - exploded view

1 Body
2 Drive gear
3 Idler gear
4 Cover
5 Pressure regulator valve
6 Spring
7 Pin
8 Screws

All engines except 231 V6

Refer to illustrations 15.17 and 15.18

Removal

14   Remove the oil pan as described in Section 14.
15   Remove the bolts securing the oil pump assembly to the main bearing cap. Remove the oil pump with its pickup tube and screen as an assembly from the engine block. Once the pump is removed, the oil pump driveshaft can be withdrawn from the block.

Inspection

16   In most cases it will be more practical and economical to replace a faulty oil pump with a new or rebuilt unit. If it is decided to overhaul the oil pump, check on internal parts availability before beginning.
17   Remove the pump cover retaining screws **(see illustration)** and the pump cover
18   Remove the idler gear, drive gear and shaft from the body **(see illustration)**.
19   Remove the pressure regulator valve retaining pin, the regulator valve and the related parts.
20   If necessary, the pick-up screen and pipe assembly can be extracted from the pump body.
21   Wash all the parts in solvent and thoroughly dry them. Inspect the body for cracks, wear or other damage. Similarly inspect the gears.
22   Check the drive gear shaft for looseness in the pump body, and the inside of the pump cover for wear that would permit oil leakage past the end of the gears.
23   Inspect the pick-up screen and pipe assembly for damage to the screen, pipe or relief grommet.
24   Apply a gasket sealant to the end of the pipe (pick-up screen and pipe assembly) and tap it into the pump body, taking care that no damage occurs. If the original press-fit cannot be obtained, a new assembly must be used to prevent air leaks and loss of pressure.
25   Install the pressure regulator valve and related parts.
26   Install the drive gear and shaft in the pump body, followed by the idler gear, with the smooth side towards the pump cover opening.

16.3 To insure proper balance, mark the flywheel's relation to the crankshaft

27   Install the cover and torque-tighten the screws to Specifications.
28   Turn the driveshaft to ensure that the pump operates freely.

Installation

29   To install, move the pump assembly into position and align the slot on the top end of the driveshaft with the drive tang on the lower end of the distributor. The distributor drives the oil pump, so it is essential that these two components mate properly.
30   Install the bolts and tighten them securely.
31   Make sure the oil pump screen is parallel with the oil rails. The screen must be in this position to fit into the oil pan properly.

16   Flywheel/driveplate - removal and installation

Refer to illustrations 16.3 and 16.4
1   Raise the vehicle and support it securely on jackstands, then refer to Chapter 7 and remove the transmission.
2   Remove the pressure plate assembly and clutch disc (see Chapter 8) (manual transmission equipped vehicles).
3   Use paint to draw a line from the flywheel to the end of the crankshaft for correct align-

16.4 A large screwdriver wedged in the starter ring gear teeth or one of the holes in the driveplate can be used to keep the flywheel/driveplate from turning as the mounting bolts are removed

ment during reinstallation **(see illustration)**.
4   Remove the bolts that secure the flywheel to the crankshaft rear flange. If difficulty is experienced in removing the bolts due to movement of the crankshaft, wedge a screwdriver to keep the flywheel from turning **(see illustration)**.
5   Remove the flywheel/driveplate from the crankshaft flange.
6   Clean any grease or oil from the flywheel. Inspect the surface of the flywheel for rivet grooves, burned areas and score marks. Light scoring can be removed with emery cloth. Check for cracked or broken teeth. Lay the flywheel on a flat surface and use a straightedge to check for warpage.
7   Clean the mating surfaces of the flywheel/driveplate and the crankshaft.
8   Position the flywheel/driveplate against the crankshaft, matching the alignment marks made during removal. Before installing the bolts, apply a non-hardening thread locking compound to the threads.
9   Wedge a screwdriver through the driveplate or into the ring gear teeth to keep the crankshaft from turning. Tighten the bolts to the specified torque in two or three steps, working in a criss-cross pattern.
10   The remainder of installation is the reverse of the removal procedure.

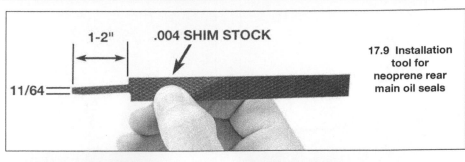

.004 SHIM STOCK

1-2"

11/64

17.9 Installation tool for neoprene rear main oil seals

17.5 Once the upper half of the seal has been adequately driven around the crankshaft, pull it out with a pair of pliers

17 Rear main oil seal - replacement (engine in car)

Two-piece neoprene type seal

Refer to illustrations 17.5, 17.9, 17.13 and 17.14

1   Always replace both halves of the rear main oil seal as a unit. While the replacement of this seal is much easier with the engine removed from the car, the job can be done with engine in place.

2   Remove the oil pan and oil pump as described previously in this Chapter.

3   Remove the rear main bearing cap from the engine.

4   Using a screwdriver, pry the lower half of the oil seal from the bearing cap.

5   To remove the upper half of the seal, use a small hammer and a brass pin punch to roll the seal around the crankshaft journal. Tap one end of the seal with the hammer and punch (be careful not to strike the crankshaft) until the other end of the seal protrudes enough to pull the seal out with a pair of pliers **(see illustration)**.

6   Clean all seal and and foreign material from the bearing cap and block. Do not use an abrasive cleaner for this.

7   Inspect components for nicks, scratches or burrs at all sealing surfaces.

8   Coat the seal lips of the new seal with light engine oil. Do not get oil on the seal mating ends.

9   Included in the purchase of the rear main oil seal should be a small plastic installation tool. If not included, make your own by cutting an old feeler gauge blade or shim

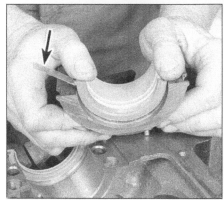

17.13 Use the protector tool (arrow) when pushing the seal into place

stock **(see illustration)**.

10   Position the narrow end of this installation tool between the crankshaft and the seal seat. The idea is to protect the new seal from being damaged by the sharp edge of the seal seat.

11   Raise the new upper half of the seal into position with the seal lips facing towards the front of the engine. Push the seal onto its seat, using the installation tool as a protector against the seal contacting the sharp edge.

12   Roll the seal around the crankshaft, all the time using the tool as a "shoehorn" for protection. When both ends of the seal are flush with the engine block, remove the installation tool, being careful not to withdraw the seal as well.

13   Install the lower half of the oil seal in the bearing cap, again using the installation tool to protect the seal against the sharp edge **(see illustration)**. Make sure the seal is firmly seated, then withdraw the installation tools.

14   Smear a bit of sealant on the bearing cap areas immediately adjacent to the seal ends **(see illustration)**.

15   Install the bearing cap (with seal) and torque the attaching bolts to about 10 to 12 ft-lbs only. Now tap the end of the crankshaft first rearward, then forward to line up the

17.14 Sealant should be used where the rear main cap touches the engine block

thrust surfaces. Retorque the bearing cap bolts to the proper Specification (see Chapter 2B).

Braided fabric type seal

Refer to illustrations 17.16a, 17.16b, 17.17 and 17.20

16   With the oil pan, oil pump and main bearing cap removed (see previous Sections), insert the special seal packing tool available at most auto parts stores **(see illustration)** against the seal. Drive the old seal into its groove until it is packed tight at each end **(see illustration)**.

17   Measure the amount which the seal was driven upwards, then add 1/16-inch. Cut two pieces that length from the old seal taken from the bearing cap. Use the bearing cap as

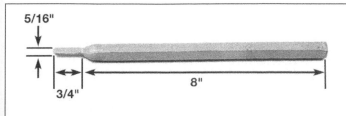

5/16"

3/4"

8"

17.16a Grind a piece of brass or aluminum rod to these dimensions as a rear seal driver

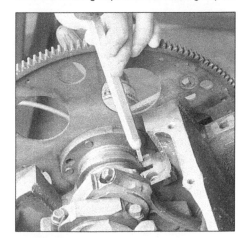

17.16b Drive each end of the seal into the groove until it feels tightly packed

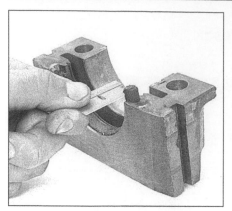

17.17 Use the bearing cap as a holding fixture when cutting short sections of the old seal

17.20 The new seal material used in the cap can be settled into the groove with a hammer handle or by taping it in with a hammer and large-diameter socket

17.23 To remove the seal from the housing, insert the tip of the screwdriver into each notch and pry the seal out

a guide when cutting **(see illustration)**.

18   Place a drop of sealant on each end of these seal pieces and then pack them into the upper groove to fill the gap made previously.

19   Trim the remaining material perfectly flush with the block. Be careful not to harm the bearing surface.

20   Install a new rope seal into the main bearing cap groove and push firmly all around using a hammer handle or special tool **(see illustration)**. Make sure the seal is firmly seated, then trim the ends flush with the bearing cap mating surface **(see illustration 17.17)**.

21   Install the cap and remaining components in reverse order, tightening all parts to Specifications (See Chapter 2B for main bearing cap torque specifications).

One-piece neoprene type seal

Refer to illustration 17.23

22   The one-piece rear main oil seal is installed in a bolt-on housing. Replacing this seal requires removal of the transmission, clutch assembly and flywheel (manual transmission) or the driveplate (automatic transmission). Refer to Chapter 7 for the transmission removal procedures and Section 16 of this Chapter for flywheel/driveplate removal.

23   Insert a screwdriver blade into the notches in the seal housing and pry out the old seal **(see illustration)**. Be sure to note how far it's recessed into the housing bore before removal so the new seal can be installed to the same depth. Although the seal can be removed this way, installation with the housing still mounted on the block requires the use of a special tool which attaches to the threaded holes in the crankshaft flange and then presses the new seal into place.

24   If the special installation tool is not available, remove the oil pan (see Section 14) and the bolts securing the housing to the block, then detach the housing and gasket. Whenever the housing is removed from the block a new seal and gasket must be installed. **Note:** *The oil pan must be removed because the two rearmost oil pan fasteners are studs that are part of the rear seal retainer, and the pan*

must come away enough to clear these studs to allow retainer removal and installation.

25   Clean the housing thoroughly, then apply a thin coat of engine oil to the lips of the new seal. Set the seal squarely into the recess in the housing, then, using two pieces of wood, one on each side of the housing, use a hammer or vise to press the seal into place.

26   Carefully slide the seal over the crankshaft and bolt the seal housing to the block. Be sure to use a new gasket, but don't use any gasket sealant.

27   The remainder of installation is the reverse of the removal procedure.

18 Engine mounts - check and replacement

Refer to illustration 18.7

1   Engine mounts seldom require attention, but broken or deteriorated mounts should be replaced immediately or the added strain placed on the driveline components may cause damage.

Check

2   During the check, the engine must be raised slightly to remove the weight from the mounts. Disconnect the negative battery cable from the battery.

3   Raise the vehicle and support it securely on jackstands, then position the jack under the engine oil pan. Place a large block of wood between the jack head and the oil pan, then carefully raise the engine just enough to take the weight off the mounts.

4   Check the mounts to see if the rubber is cracked, hardened or separated from the metal plates. Sometimes the rubber will split right down the center. Rubber preservative may be applied to the mounts to slow deterioration.

5   Check for relative movement between the mount plates and the engine or frame (use a large screwdriver or pry bar to attempt to move the mounts). If movement is noted, lower the engine and tighten the mount fasteners.

Replacement

6   Remove the fan shroud mounting screws and place the shroud over the fan.

7   Remove the engine mount through-bolts **(see illustration)**.

8   Disconnect the shift linkage where it connects the transmission to the body (see Chapter 7).

9   Raise the engine high enough to clear the clevis brackets. Do not force the engine up too high. If it touches anything before the mounts are free, remove the part for clearance. Place a block of wood between the oil pan and crossmember as a safety precaution.

10   Unbolt the mount from the engine block and remove it from the vehicle. **Note:** *On vehicles equipped with self-locking nuts and bolts, replace them with new ones whenever they are disassembled. Prior to assembly, remove hardened residual adhesive from the engine block holes with a proper-size bottoming tap.*

11   Attach the new mount to the engine block and install the fasteners in the appropriate locations. Tighten the fasteners securely.

12   Remove the wooden block and lower the engine into place. Install the through-bolts and tighten the nuts securely.

13   Complete the installation by reinstalling all parts removed to gain access to the mounts.

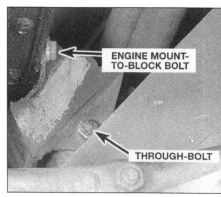

ENGINE MOUNT-TO-BLOCK BOLT

THROUGH-BOLT

18.7 Typical engine mounting

Chapter 2 Part B
General engine overhaul procedures

Contents

Specifications

General

Displacement
V6 engines
3.3 liter (VIN M)	200 cubic inches
3.8 liter (VIN K and 9)	229 cubic inches
3.8 liter (VIN A, Z and 3)	231 cubic inches
4.3 liter (VIN Z)	262 cubic inches

V8 engines
4.4 liter (VIN J)	267 cubic inches
5.0 liter (VIN G, H, Q and U)	305 cubic inches
5.7 liter (VIN H, J, K, L and V)	350 cubic inches
6.6 liter (VIN R and U)	400 cubic inches
6.6 liter (VIN R and U)	402 cubic inches
7.4 liter (VIN V, W, Y and Z)	454 cubic inches

Bore and stroke
V6 engines
200	3.500 x 3.480 inches
229	3.736 x 3.480 inches
231	3.800 x 3.400 inches
262	4.000 x 3.480 inches

V8 engines
267	3.500 x 3.480 inches
305	3.736 x 3.480 inches
350	4.000 x 3.480 inches
400 small block	4.125 x 3.750 inches
400 big block (1970 only)	4.125 x 3.760 inches
402	4.126 x 3.760 inches
454	4.250 x 4.000 inches

Cylinder numbers (front-to-rear)
V6 engines
Right bank (passenger's side)	2-4-6
Left bank (driver's side)	1-3-5

V8 engines
Right bank (passenger's side)	2-4-6-8
Left bank (driver's side)	1-3-5-7

General (continued)

Firing order	
V6 engines	1-6-5-4-3-2
V8 engines	1-8-4-3-6-5-7-2
Compression pressure	
All V6 and V8 engines	100 to 150 psi
Maximum variation between cylinders	20 percent

1970 models

Engine block

Cylinder bore diameter	
350 engine	3.9995 to 4.0025 inches
400 engine	4.1246 to 4.1274 inches
454 engine	4.2495 to 4.2525 inches
Out-of-round	0.002 in max
Taper	0.005 in max

Piston and piston rings

Piston clearance in bore	
350 engine	0.0027 in max
400 engine	0.0034 in max
454 engines	0.0049 in max
Piston ring clearance in groove	
400 (265 hp) engine	
Top	0.0012 to 0.027 in
2nd	0.0012 to 0.0032 in
Oil control	0.000 to 0.005 in
350 engine	
Top	0.0012 to 0.0032 in
2nd	0.0012 to 0.0027 in
Oil control	0.000 to 0.005 in
400 (330 hp) and 454 engines	
Top	0.0017 to 0.0032 in
2nd	0.0017 to 0.0032 in
Oil control	0.005 to 0.0065 in
Piston ring end gap	
350 engine	
Top	0.010 to 0.020 in
2nd	0.013 to 0.025 in
Oil control	0.015 to 0.058 in
400, 454 engines	
Top	0.010 to 0.020 in
2nd	0.010 to 0.020 in
Oil control	0.015 to 0.055 in
Piston pin diameter	
350 engine	0.9270 to 0.9273 in
400, 454 engines	0.9895 to 0.9898 in
Clearance in piston	0.001 in max
Interference fit in rod	0.0008 to 0.016 in

Crankshaft

Main journal diameters	
350 engine	
Journals 1-2-3-4	2.4484 to 2.4493 inches
Journal 5	2.4479 to 2.4488 inches
400, 454 engines	
Journal 1	2.7485 to 2.7494 inches
Journals 2-3-4	2.7481 to 2.7490 inches
Journal 5	2.7484 to 2.7488 inches
Journal taper	0.001 in max
Journal out of round	0.001 in max
Main bearing oil clearance	
No 1	0.002 in max
No 2	0.0035 in max
Crankshaft endplay	
350, 400 engines	0.002 to 0.006 in
454 engine	0.006 to 0.010 in
Crankpin diameter	
350, 400 (265 hp) engines	2.099 to 2.100 inches
400 (330 hp) and 454 engines	2.199 to 2.200 inches

Crankpin taper	0.001 in max
Crankpin out of round	0.001 in max
Rod bearing oil clearance	0.0035 in
Rod side clearance	
350 and 400 (265 hp) engines	0.008 to 0.014 in
454 engine	0.015 to 0.021 in
400 (330 hp) engine	0.015 to 0.023 in

Valve system

Lifter	Hydraulic
Valve lash	One turn down from zero
Valve face angle	45°
Valve seat angle	46°
Valve seat width	
Intake	1/32 to 1/16 inch
Exhaust	1/16 to 3/32 inch
Valve stem clearance	
Intake	0.0012 to 0.0029 in
Exhaust	0.0012 to 0.0029 in
Valve spring free length	
350, 400 (265 hp) engines	2.03 inches
400 (330 hp) engine	2.12 inches
454 engine	2.09 inches
Valve spring installed height	
350 engine	1 23/32 inches
400, 454 engines	1 7/8 inches
Damper free length	
All engines except 454	1.94 inches
454 engine	1.94 to 2.00 inches

1971 models

Engine block

Cylinder bore diameter	
350 engine	3.9995 to 4.0025 inches
402 engine	4.1246 to 4.1274 inches
454 engine	4.2495 to 4.2525 inches
Out-of-round	0.002 in max
Taper	0.005 in max

Pistons and piston rings

Piston clearance in bore	
350 engine	0.0027 in
402 engine	0.0035 in
454 engine	0.0049 in
Piston ring clearance in groove	
350 (245 hp) engine	
Top	0.0012 to 0.0032 in
2nd	0.0012 to 0.0032 in
Oil control	0.002 to 0.007 in
350 (270 hp) engine	
Top	0.0012 to 0.0032 in
2nd	0.0012 to 0.0027 in
Oil control	0.005 in
402, 454 engines	
Top	0.0017 to 0.0032 in
2nd	0.0017 to 0.0032 in
Oil control	0.0005 to 0.0065 in
Piston ring end gap	
All engines	
Top	0.010 to 0.0200 in
350 engine	
2nd	0.013 to 0.025 in
402, 454 engines	
2nd	0.010 to 0.020 in
All engines	
Oil control	0.015 to 0.055 in
Piston pin diameter	
350 engine	0.9270 to 0.9273 in
402, 454 engines	0.9895 to 0.9898 in
Clearance in piston	0.001 in max
Interference fit in rod	0.0008 to 0.0016 in

1971 models (continued)

Crankshaft

Main journal diameters
    350 engine
        Journals 1-2-3-4 .... 2.4484 to 2.4493 inches
        Journal 5 .... 2.4479 to 2.4488 inches
    402 engine
        Journals 1-2 .... 2.7487 to 2.7496 inches
        Journals 3-4 .... 2.7481 to 2.7490 inches
        Journal 5 .... 2.7473 to 2.7483 inches
    454 engine
        Journal 1 .... 2.7485 to 2.7494 inches
        Journals 2-3-4 .... 2.7481 to 2.7490 inches
        Journal 5 .... 2.7478 to 2.7488 inches
Journal taper .... 0.001 in max
Journal out of round .... 0.001 in max
Main bearing oil clearance
    No 1 .... 0.002 in max
    No 2-3-4-5 .... 0.0035 in max
Crankshaft end play
    350 engine .... 0.002 to 0.006 in
    402, 454 engines .... 0.006 to 0.010 in
Crankpin diameter
    350 engine .... 2.099 to 2.100 inches
    402, 454 engines .... 2.199 to 2.200 inches
Crankpin taper .... 0.001 in max
Crankpin out of round .... 0.001 in max
Rod bearing oil clearance .... 0.0035 in max
Rod side clearance
    350 engine .... 0.008 to 0.014 in
    402 engine .... 0.013 to 0.023 in
    454 engine .... 0.015 to 0.021 in

Valve system

Lifter .... Hydraulic
Rocker arm ratio
    350 engine .... 1.50 : 1
    All other engines .... 1.70 : 1
Valve lash .... One turn down from zero
Valve face angle .... 45°
Valve seat angle .... 46°
Valve seat width
    Intake .... 1/32 to 1/16 inch
    Exhaust .... 1/16 to 3/32 inch
Stem clearance
    Intake .... 0.0010 to 0.0027 in
    Exhaust .... 0.0012 to 0.0027 in
Valve spring (outer) free length
    350 engine .... 2.03 inches
    402, 454 engines .... 2.12 inches
Valve spring (outer) installed height
    350 engine .... 1 23/32 inches
    402, 454 engines .... 1 7/8 inches
Valve spring (inner) free length
    402, 454 engines only .... 2.06 inches
Valve spring (inner) installed height
    402, 454 engines only .... 1 23/32 inches
Damper free length
    350 engine only .... 1.94 inches

1972 models

Engine block

Cylinder bore diameter
    350 engine .... 3.9995 to 4.0025 inches
    402 engine .... 4.1246 to 4.1274 inches
    454 engine .... 4.2495 to 4.2525 inches
Out-of-round .... 0.002 in max
Taper .... 0.005 in max

Pistons and piston rings

Piston clearance in bore
    350 engine... 0.0027 in
    402 engine... 0.0035 in
    454 engine... 0.0049 in

Piston ring clearance in groove
    350 (165 hp) engine
        Top... 0.0012 to 0.0032 in
        2nd... 0.0012 to 0.0032 in
        Oil control .... 0.002 to 0.007 in

350 (175 hp) engine
        Top... 0.0012 to 0.0032 in
        2nd... 0.0012 to 0.0027 in
        Oil control .... 0.005 in max
    402, 454 engines
        Top... 0.0017 to 0.0032 in
        2nd... 0.0017 to 0.0032 in
        Oil control .... 0.005 to 0.0065 in

Piston ring end gap
    402, 454 engines
        Top... 0.010 to 0.020 in
        2nd... 0.010 to 0.020 in
        Oil control .... 0.015 to 0.055 in
    350 engine
        Top... 0.010 to 0.020 in
        2nd... 0.013 to 0.025 in
        Oil control .... 0.015 to 0.035 in

Piston pin diameter
    350 engine... 0.9270 to 0.9273 in
    402.454 engines... 0.9895 to 0.9898 in
Clearance in piston... 0.001 in max
Interference fit in rod .... 0.008 to 0.0016 in

Crankshaft

Main journal diameters
    350 engines
        Journal 1 .... 2.4484 to 2.4493 inches
        Journals 2-3-4............................. 2.4481 to 2.4490 inches
        Journal 5 .... 2.4479 to 2.4488 inches
    402 engine
        Journals 1-2................................. 2.7487 to 2.7496 inches
        Journals 3-4................................. 2.7481 to 2.7490 inches
        Journal 5 .... 2.7473 to 2.7483 inches
    454 engine
        Journal 1 .... 2.7485 to 2.7494 inches
        Journals 2-3-4............................. 2.7481 to 2.7490 inches
        Journal 5 .... 2.7478 to 2.7488 inches
Crankshaft journal taper .... 0.001 in max
Crankshaft journal out of round........................... 0.001 in max
Main bearing oil clearance
    No 1 .... 0.002 in max
    No 2-5 .... 0.0035 in max
Crankshaft end play
    350 engine... 0.002 to 0.006 in
    402, 454 engines... 0.006 to 0.010 in
Crankpin diameters
    350 engine... 2.099 to 2.100 inches
    402.454 engines... 2.199 to 2.200 inches
Crankpin taper .... 0.001 in max
Crankpin out of round... 0.001 in max
Rod bearing oil clearance .... 0.0035 in max
Rod side clearance
    350 engine... 0.008 to 0.014 in
    402 engine... 0.013 to 0.023 in
    454 engine... 0.015 to 0.021 in

Valve system

Lifter... Hydraulic
Valve lash... One turn down from zero
Valve face angle.. 45°

1972 models (continued)

Valve seat angle .... 46°
Valve seat width (Intake) .... 1/32 to 1/16 inch
Valve seat width (Exhaust) .... 1/16 to 3/32 inch
Valve stem clearance
    Intake .... 0.0010 to 0.0027 in
    Exhaust
        350 engine .... 0.0012 to 0.0029 in
        402. 454 engines .... 0.0012 to 0.0027 in
Valve spring (outer) free length
    350 engine .... 2.03 inches
    402, 454 engines .... 2.12 inches
Valve spring installed height
    350 engine .... 1 23/32 inches
    402. 454 engines .... 1 7/8 inches
Valve spring (inner) free length
    402. 454 engines only .... 2.06 inches
Valve spring (inner) installed height
    402, 454 engines only .... 1 25/32 inches

1973 models

Engine block

Cylinder bore diameter
    350 engine .... 3.9995 to 4.0025 inches
    454 engine .... 4.2495 to 4.2525 inches
Out-of-round .... 0.002 in max
Taper .... 0.005 in max

Pistons and piston rings

Piston clearance in bore (max)
    350 engine .... 0.0027 in
    454 engine .... 0.0035 in
Piston ring clearance in groove
    350 (145 hp) engine
        Top .... 0.0012 to 0.0032 in
        2nd .... 0.0012 to 0.0032 in
        Oil control .... 0.002 to 0.007 in
    350 (175 hp) engine
        Top .... 0.0012 to 0.0032 in
        2nd .... 0.0012 to 0.0027 in
        Oil control .... 0.005 in max
    454 engine
        Top .... 0.0017 to 0.0032 in
        2nd .... 0.0017 to 0.0032 in
        Oil control .... 0.0005 to 0.0065 in
Piston ring end gap
    454 engine
        Top .... 0.010 to 0.020 in
        2nd .... 0.010 to 0.020 in
        Oil control .... 0.015 to 0.055 in
    350 engine
        Top .... 0.010 to 0.020 in
        2nd .... 0.013 to 0.025 in
        Oil control .... 0.015 to 0.055 in
Piston pin diameter
    All engines except 454 .... 0.9270 to 0.9273 in
    454 engine .... 0.9895 to 0.9898 in
Clearance in piston .... 0.001 in max
Interference fit in rod .... 0.008 to 0.0016 in

Crankshaft

Main journal diameters
    350 engine
        Journal 1 .... 2.4484 to 2.4493 inches
        Journals 2-3-4 .... 2.4481 to 2.4490 inches
        Journal 5 .... 2.4479 to 2.4488 inches
    454 engine
        Journal 1 .... 2.7485 to 2.7494 inches
        Journals 2-3-4 .... 2.7481 to 2.7490 inches
        Journal 5 .... 2.7478 to 2.7488 inches

Journal taper...	0.001 in max
Journal out of round	0.001 in max
Main bearing oil clearance	
Journal 1..	0.002 in max
Journals 2-3-4-5..	0.0035 in max
Crankshaft end play	
350 engine...	0.002 to 0.006 in
454 engine...	0.006 to 0.010 in
Crankpin diameter	
All engines except 454	2.099 to 2.100 inches
454 engine...	2.199 to 2.200 inches
Crankpin tape	0.001 in max
Crankpin out of round..	0.001 in max
Rod bearing oil clearance	
All engines except 454	0.0013 to 0.0035 in
454 engine...	0.0009 to 0.0025 in
Rod side clearance	
All engines except 454	0.008 to 0.014 in
454 engine...	0.015 to 0.021 in

Valve system

Lifter...	Hydraulic
Valve lash...	One turn down from zero
Valve face angle...	45°
Valve seat angle...	46°
Valve seat width (Intake)	1/32 to 1/16 inch
Valve seat width (Exhaust)...	1/16 to 3/32 inch
Valve stem clearance	
All engines	
Intake	0.0010 to 0.0027 in
350 engine	
Exhaust	0.0012 to 0.0029 in
454 engine	
Exhaust	0.0012 to 0.0027 in
Valve spring free length	
350 engine...	2.03 inches
454 engine...	2.10 inches
Valve spring installed height	
350 engine...	1 23/32 inches
454 engine...	1 7/8 inches
Damper free length	
All engines except 454	1.94 inches
454 engine...	1.86 inches

1974 models

Engine block

Cylinder bore diameter	
350 engine...	3.9995 to 4.0025 inches
400 engine...	4.1246 to 4.1274 inches
454 engine...	4.2495 to 4.2525 inches
Out-of-round...	0.002 in max
Taper..	0.005 in max

Piston and piston rings

Piston clearance in bore (max)	
350 (145 hp) engine..	0.0025 in
350 (160 hp) engine..	0.0027 in
400 engine...	0.0034 in
454 engine...	0.0035 in
Piston ring clearance in groove	
350 (145 hp) engine	
Top...	0.0012 to 0.0027 in
2nd...	0.0012 to 0.0032 in
Oil control	0.005 in max
350 (160 hp) engine	
Top...	0.0012 to 0.0032 in
2nd...	0.012 to 0.0027 in
Oil control	0.002 to 0.07 in

1974 models (continued)

Piston ring clearance in groove
  400 engine
    Top .... 0.0012 to 0.0027 in
    2nd .... 0.0012 to 0.0032 in
    Oil control .... 0.005 in
  454 engine
    Top .... 0.0017 to 0.0032 in
    2nd .... 0.0017 to 0.0032 in
    Oil control .... 0.0005 to 0.0065 in
Piston ring end gap
  All engines
    Top .... 0.010 to 0.020 in
  All engines except 350
    2nd .... 0.010 to 0.020 in
  350 (145 hp) engine
    2nd .... 0.010 to 0.020 in
  350 (160 hp) engine
    2nd .... 0.013 to 0.025 in
  All engines
    Oil control .... 0.015 to 0.055 in
Piston pin diameter
  All engines except 454 .... 0.9270 to 0.9273 in
  454 engine .... 0.9895 to 0.9898 in
Clearance in piston .... 0.001 in max
Interference fit in rod .... 0.008 to 0.0016 in

Crankshaft

Main journal diameters
  350 engine
    Journal 1 .... 2.4484 to 2.4493 inches
    Journals 2-3-4 .... 2.4482 to 2.4490 inches
    Journal 5 .... 2.4479 to 2.4488 inches
  400 engine
    Journals 1-2-3-4 .... 2.6484 to 2.6493 inches
    Journal 5 .... 2.6497 to 2.6488 inches
  454 engine
    Journal 1 .... 2.7485 to 2.7494 inches
    Journals 2-3-4 .... 2.7481 to 2.7490 inches
    Journal 5 .... 2.7478 to 2.7488 inches
Journal taper .... 0.001 in max
Journal out of round .... 0.001 in max
Main bearing oil clearance
  No 1 journal .... 0.002 in max
  All others .... 0.0035 in
Crankshaft end play
  All engines except 454 .... 0.002 to 0.006 in
  454 engine .... 0.006 to 0.010 in
Crankpin diameter
  All engines except 454 .... 2.099 to 2.100 inches
  454 engine .... 2.199 to 2.200 inches
Crankpin taper .... 0.001 in max
Crankpin out of round .... 0.001 in max
Rod bearing oil clearance
  All engines except 454 .... 0.0013 to 0.0035 in
  454 engine .... 0.0009 to 0.0025 in
Rod side clearance
  All engines except 454 .... 0.0008 to 0.014 in
  454 engine .... 0.015 to 0.021 in

Valve system

Lifter .... Hydraulic
Valve lash .... One turn down from zero
Valve face angle .... 45°
Valve seat angle .... 46°
Valve seat width
  Intake .... 1/32 to 1/16 inch
  Exhaust .... 1/16 to 3/32 inch
Valve stem clearance
  Intake .... 0.0010 to 0.0027 in

Exhaust
    350 engine .... 0.0012 to 0.0029 in
    400, 454 engines .... 0.0012 to 0.0027 in
Valve spring free length
    350 engine
        Intake .... 2.03 inches
        Exhaust .... 1.91 inches
    400 engine
        (Intake and exhaust)... 2.03 inches
    454 engine
        (Intake and exhaust)... 2.10 inches
Valve spring installed height
    350 engine .... 1 5/8 inches
    400 engine .... 1 23/32 inches
    454 engine .... 1 7/8 inches
Damper free length
    All engines except 454 .... 1.94 inches
    454 engine.. 1.86 inches

1975 models

Engine block

Cylinder bore diameter
    350 engine... 3.9995 to 4.0025 inches
    400 engine... 4.1246 to 4.1274 inches
    454 engine... 4.2495 to 4.2525 inches
Out-of-round.. 0.002 in max
Taper .... 0.005 in max

Pistons and piston rings

Piston clearance in bore
    350 (2 barrel carburetor) engine....................................... 0.0025 in max
    350 (4 barrel carburetor) engine....................................... 0.0027 in max
    400 engine.. 0.0034 in max
    454 engine.. 0.0035 in max
Piston ring clearance in groove
    350 (2 barrel carburetor) and 400 engines
        Top.. 0.0012 to 0.0027 in
        2nd.. 0.0012 to 0.0032 in
        Oil control .... 0.005 in max
    350 (4 barrel carburetor) engine
        Top.. 0.0012 to 0.0032 in
        2nd.. 0.0012 to 0.0032 in
        Oil control .... 0.005 in max
    454 engine
        Top.. 0.0017 to 0.0032 in
        2nd.. 0.0017 to 0.0032 in
        Oil control .... 0.0005 to 0.0065 in
Piston ring end gap
    All engines
        Top.. 0.010 to 0.020 in
    350 (2 barrel carburetor) engine
        2nd.. 0.010 to 0.020 in
    350 (4 barrel carburetor) engine
        2nd.. 0.013 to 0.025 in
    400 and 454 engines
        2nd.. 0.010 to 0.020 in
    All engines
        Oil control .... 0.015 to 0.055 in
Piston pin diameter
    All engines except 454 .... 0.9895 to 0.9898 in
Clearance in piston.. 0.001 in max
Interference fit in rod .... 0.0008 to 0.0016 in

Crankshaft

Main journal diameters
    350 engine
        Journal 1 .... 2.4484 to 2.4493 inches
        Journals 2-3-4 .... 2.4481 to 2.4490 inches
        Journal 5 .... 2.4479 to 2.4488 inches

1975 models (continued)

Crankshaft

Main journal diameters
  400 engine
    Journals 1-2-3-4 .... 2.6484 to 2.6493 inches
    Journal 5 .... 2.6479 to 2.6488 inches
  454 engine
    Journal 1 .... 2.7485 to 2.7494 inches
    Journals 2-3-4 .... 2.748t to 2.7490 inches
    Journal 5 .... 2.7478 to 2.7488 inches
Journal taper .... 0.001 in max
Journal out of round .... 0.001 in max
Main bearing oil clearance
  No 1 Journal .... 0.002 in max
  All others .... 0.0035 in max
Crankshaft end play
  350 engine .... 0.002 to 0.006 in
  400, 454 engines .... 0.006 to 0.010 in
Crankpin diameter
  All engines except 454 .... 2.099 to 2.100 inches
  454 engine .... 2.199 to 2.200 inches
Crankpin taper .... 0.001 in max
Crankpin out of round .... 0.001 in max
Rod bearing oil clearance
  350, 400 engines .... 0.0013 to 0.0035 in
  454 engine .... 0.0009 to 0.0025 in
Rod side clearance
  350, 400 engines .... 0.008 to 0.014 in
  454 engine .... 0.015 to 0.021 in

Valve system

Lifter .... Hydraulic
Valve lash .... One turn down from zero
Valve face angle .... 45°
Valve seat angle .... 46°
Seat width
  Intake .... 1/32 to 1/16 inch
  Exhaust .... 1/16 to 3/32 inch
Valve stem clearance
  All engines
    Intake .... 0.0010 to 0.0027 in
  350 engine
    Exhaust .... 0.0012 to 0.0029 in
  400, 454 engines
    Exhaust .... 0.0012 to 0.0027 in
Valve spring free length
  350, 400 engines .... 2.03 inches
  454 engine .... 2.10 inches
Valve spring installed height
  350 400 engines .... 1 23/32 inches
  454 engine .... 1 7/8 inches
Damper free length
  350, 400 engines .... 1.94 inches
  454 engine .... 1.86 inches

1976 models

Engine block

Cylinder bore diameter
  305 engine .... 3.7350 to 3.7385 inches
  350 engine .... 3.9995 to 4.0025 inches
  400 engine .... 4.1246 to 4.1274 inches
Out-of-round .... 0.002 in max
Taper .... 0.005 in max

Pistons and piston rings

Piston clearance in bore
  305, 350 engines .... 0.0027 in max
  400 engine .... 0.0034 in max

Piston ring clearance in groove
    Top
        305, 350 engines .... 0.0012 to 0.0032 in
        400 engine .... 0.0012 to 0.0027 in
    2nd
        305, 350 engines .... 0.0012 to 0.0027 in
        400 engine .... 0.0012 to 0.0032 in
    Oil control
        All models .... 0.005 in max
Piston ring end gap
    Top
        All engines... 0.010 to 0.020 in
    2nd
        305 engine .... 0.010 to 0.025 in
        350 (2 barrel carb), 400 engines .... 0.010 to 0.020 in
        350 (4 barrel carburetor engine)................................. 0.013 to 0.025 in
    Oil control
        All engines... 0.015 to 0.055 in
Piston pin diameter
    All engines .... 0.9270 to 0.9273 in
Clearance in piston... 0.001 in
Interference fit in rod .... 0.008 to 0.0016 in

Crankshaft

Main journal diameters
    305, 350 engines
        Journal 1 2 .... 2.4484 to 2.4493 inches
        Journals 2-3-4.. 2.4481 to 2.4490 inches
        Journal 5 .... 2.4479 to 2.4488 inches
    400 engine
        Journals 1-2-3-4.. 2.6484 to 2.6493 inches
        Journal 5 .... 2.6479 to 2.6483 inches
Journal taper.. 0.001 in max
Journal out of round .... 0.001 in max
Main bearing oil clearance
    No 1 journal .... 0.002 in max
    All others .... 0.0035 in max
Crankshaft end play
    All engines.. 0.002 to 0.006 in
Crankpin diameter
    All engines.. 2.099 to 2.100 inches
Crankpin taper .... 0.001 in max
Crankpin out of round... 0.001 in max
Rod bearing oil clearance... 0.0035 in max
Rod side clearance
    All engines.. 0.008 to 0.014 in

Valve system

Lifter.. Hydraulic
Valve lash.. 3/4 turn down from zero
Valve face angle.. 45°
Valve seat angle.. 46°
Valve seat width
    Intake .... 1/32 to 1/16 inch
    Exhaust .... 1/16 to 3/32 inch
Valve stem clearance
    Intake .... 0.0010 to 0.0027 in
    Exhaust .... 0.0010 to 0.0027 in
Valve spring free length
    All engines.. 2.03 inches
Valve spring installed height
    305, 350 engines.. 1 23/32 inches
    400 engine .... 1 7/8 inches
Damper free length
    All engines.. 1.94 inches

1977 models

Engine block
Cylinder bore diameter
    305 engine.. 3.7350 to 3.7385 inches
    350 engine.. 3.9995 to 4.0025 inches
Out-of-round.. 0.002 in max
Taper .... 0.005 in max

Pistons and piston rings
Piston clearance in bore
    305, 350 engines... 0.0027 in max
Piston ring clearance in groove
    305, 350 engines Top.. 0.0012 to 0.0032 in
    305 and 350 engines 2nd.................................... 0.0012 to 0.0027 in
    All engines Oil control .... 0.005 in max
Piston ring end gap
    All engines Top.. 0.010 to 0.020 in
    305 engine 2nd.. 0.010 to 0.025 in
    350 (4 barrel carburetor engine)......................... 0.013 to 0.025 in
    All engines Oil control .... 0.015 to 0.055 in
Piston pin diameter
    All engines... 0.9270 to 0.9273 in
Clearance in piston... 0.001 in
Interference fit in rod .... 0.0008 to 0.0016 in

Crankshaft
Main journal diameters
    305, 350 engines
        Journal 1 .... 2.4484 to 2.4493 inches
        Journals 2-3-4.. 2.4481 to 2.4490 inches
        Journal 5 .... 2.4479 to 2.4488 inches
Journal taper.. 0.001 in max
Journal out of round .... 0.001 in max
Main bearing oil clearance
    No 1 Journal.. 0.002 in
    All others... 0.0035 in max
Crankshaft end play
    All engines... 0.002 to 0.006 in
Crankpin diameter
    305, 350 engines... 2.099 to 2.100 inches
Crankpin taper .... 0.001 in max
Crankpin out of round.. 0.001 in max
Rod bearing oil clearance.. 0.0035 in max
Rod side clearance
    All engines... 0.008 to 0.014 in

Valve system
Lifter .... Hydraulic
Valve lash... 3/4 turn down from zero
Valve face angle.. 45°
Valve seat angle.. 46°
Valve seat width
    Intake .... 1/32 to 1/16 inch
    Exhaust .... 1/16 to 3/32 inch
Valve stem clearance
    Intake .... 0.0010 to 0.0027 in
    Exhaust .... 0.0010 to 0.0027 in
Valve spring free length
    All engines... 2.03 inches
Valve spring installed height
    305, 350 engines... 1 23/32 inches
Damper free length
    All engines... 1.86 inches

1978 and later models

V6 engines

Engine block

Cylinder bore diameter
200 engine	3.4995 to 3.5025 inches
229 engine	3.7350 to 3.7385 inches
231 engine	3.7995 to 3.8025 inches
262 engine	3.9995 to 4.0025 inches
Out-of-round	0.002 in max
Taper	0.001 in max

Pistons and piston rings

Piston clearance in bore
200, 229and 262 engines	0.0007 to 0.0017 in
231 engine	0.0008 to 0.0020 in

Piston ring clearance in groove

200, 229 and 262 engines
Top	0,0012 to 0.0032 in
2nd	0.0012 to 0.0032 in
Oil control	0.002 to 0.007 in

231 engine
Top	0.003 to 0.005 in
2nd	0.010 to 0.020 in
Oil control	0,0035 in max

Piston ring end gap

200, 229 and 262 engines
Top	0.010 to 0.020 in
2nd	0.010 to 0.025 in
Oil control	0.015 to 0.055 in

231 engine
Top	0.010 to 0.020 in
2nd	0.010 to 0.020 in
Oil control	0.015 to 0.035 in

Piston pin diameter
200, 229 and 262 engines	0.9270 to 0.9273 in
231 engine	0.9391 to 0.9394 in

Clearance in piston
200, 229 and 262 engines	0.00025 to 0.00035 in
231 engine	0.0004 to 0.0007 in

Interference fit in rod
200, 229 and 262 engines	0.0008 to 0.0016 in
231 engine	0.0007 to 0.0017 in

Crankshaft

Main journal diameters

200, 229 and 262 engines
Front	2.4484 to 2.4493 in
Intermediate	2.4481 to 2.4490 in
Rear	2.4479 to 2.4488 in

231 engine
All	2.4995 inches

Journal taper
200, 229 and 262 engines	0.0002 in max
231 engine	0.0015 in max

Journal out of round
200, 229 and 262 engines	0.0002 in max
231 engine	0.0015 in max

Main bearing oil clearance

200, 229 and 262 engines
Front	0.0008 to 0.0020 in
Intermediate	0.0011 to 0.0023 in
Rear	0.0017 to 0.0032 in

231 engine
All	0.0004 to 0.0015 in

Crankshaft endplay
200, 229 and 262 engines	0.002 to 0.006 in
231 engine	0.004 to 0.008 in

1978 and later models

V6 engines

Crankpin diameter	
200, 229 and 262 engines	2.0986 to 2.0998 in
231 engine	2.2495 to 2.2487 in
Crankpin taper	
200, 229 and 262 engines	0.001 in max
231 engine	0.0015 in
Crankpin out of round	
200, 229 and 262 engines	0.001 in max
231 engine	0.0015 in max
Rod bearing oil clearance	
200, 229 and 262 engines	0.0013 to 0.0035 in
231 engine	0.0005 to 0.0026 in
Rod side clearance	
200, 229 and 262 engines	0.008 to 0.014 in
231 engine	0.006 to 0.027 in

Valve system

Lifter	Hydraulic
Valve lash	
200, 229 and 262 engines, hydraulic	One turn down from zero
231 engine	Non adjustable
Valve face angle	45°
Valve seat angle	
200, 229 and 262engines	46°
231 engine	45°
Valve seat width	
Intake	1/32 to 1/16 inch
Exhaust	1/16 to 3/32 inch
Valve stem clearance	
200, 229 and 262 engines	
Intake	0.0010 to 0.0027 in
Exhaust	0.0010 to 0.0027 in
231 engine	
Intake	0.0015 to 0.0032 in
Exhaust	0.0015 to 0.0032 in
Valve spring free length	
200, 229 and 262 engines	2.03 inches
231 engine	N/A

V8 engines

Engine block	
Cylinder bore diameter	
267 engine	3.4995 to 3.5025 inches
305 engine	3.7350 to 3.7385 inches
350 engine	3.9995 to 4.0025 inches
Out-of-round	0.002 in max
Taper	0.001 in max

Pistons and piston rings

Piston clearance in bore (all engines)	0.0027 in max
Piston ring clearance in groove (all engines)	
Top	0.0012 to 0.0032 in
2nd	0.0012 to 0.0032 in
Oil control	0.002 to 0.007 in
Piston ring end gap (all engines)	
Top	0.010 to 0.020 in
2nd	0.010 to 0.025 in
Oil control	0.015 to 0.055 in
Piston pin diameter (all engines)	0.9270 to 0.9273 in
Clearance in piston (all engines)	0.001 in max
Interference fit in rod (all engines)	0.0008 to 0.0016 in

Crankshaft

Main journal diameters (all engines)	
Journal 1	2.4484 to 2.4493 inches
Journals 2-3-4	2.4481 to 2.4490 inches
Journal 5	2.4479 to 2.4488 inches

Journal taper (all engines).. 0.001 in max
Journal out of round (all engines) .... 0.001 in max
Main bearing oil clearance (all engines)
    Journal 1.. 0.001 to 0.0015 in
    Journals 2-3-4.. 0.001 to 0.0025 in
    Journal 5.. 0.0025 to 0.0035 in
Crankshaft end play (all engines).. 0.002 to 0.006 in
Crankpin diameter (all engines) .... 2.0986 to 2.0998 inches
Crankpin taper (all engines) .... 0.001 in max
Crankpin out of round (all engines) .... 0.001 in max
Rod bearing oil clearance (all engines)...................................... 0.0035 in max
Rod side clearance (all engines) .... 0.008 to 0.014 in

Valve system

Lifter .... Hydraulic
Valve lash... 1 turn down from zero
Valve face angle (all engines)... 45°
Valve seat angle (all engines)... 46°
Valve seat width (all engines)
    Intake .... 1/32 to 1/16 inch
    Exhaust .... 1/16 to 3/32 inch
Valve stem clearance (all engines)... 0.0010 to 0.0027 in
Valve spring free length (all engines) .... 2.03 inches
Valve spring installed height (all engines) .... 1 23/32 inches +/- 1/32
Damper free length (all engines) .... 1.86 inches

Torque specifications*   ft-lbs (unless otherwise indicated)

V6 engines

Rod bearing bolts
    200, 229 and 262 engines... 45
    231 engine.. 40
Main bearing bolts
    200, 229 and 262 engines... 70
    231 engine.. 100

V8 engines

Connecting rod cap bolts
    All engines (except big block engines0............................ 45
    400, 402 and 454 big block engines 50
Main bearing cap bolts
    All engines (except big block engines)............................. 70
    400, 402 and 454 big block engines 110
Outer bolts with 4 bolt caps (All engines)................................. 65

* **Note:** *Refer to Chapter 2 Part A for additional torque specifications.*

1   General information

Included in this portion of Chapter 2 are the general overhaul procedures for the cylinder heads and internal engine components.

The information ranges from advice concerning preparation for an overhaul and the purchase of replacement parts to detailed, step-by-step procedures covering removal and installation of internal engine components and the inspection of parts.

The following Sections have been written based on the assumption that the engine has been removed from the vehicle. For information concerning in-vehicle engine repair, as well as removal and installation of the external components necessary for the overhaul, see Part A of this Chapter and Section 7 of this Part.

The Specifications included in this Part are only those necessary for the inspection and overhaul procedures which follow. Refer to Part A for additional Specifications.

2   Engine overhaul - general information

Refer to illustration 2.4

It's not always easy to determine when, or if, an engine should be completely overhauled, as a number of factors must be considered.

High mileage is not necessarily an indication that an overhaul is needed, while low mileage doesn't preclude the need for an overhaul. Frequency of servicing is probably the most important consideration. An engine that's had regular and frequent oil and filter changes, as well as other required maintenance, will most likely give many thousands of miles of reliable service. Conversely, a neglected engine may require an overhaul very early in its life.

Excessive oil consumption is an indication that piston rings, valve seals and/or valve guides are in need of attention. Make sure that oil leaks aren't responsible before decid-

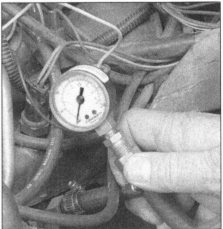

2.4 The sender is located near the base of the distributor on V8 engines

ing that the rings and/or guides are bad. Perform a cylinder compression check to determine the extent of the work required (see Section 3).

Check the oil pressure with a gauge installed in place of the oil pressure sending unit **(see illustration)** and compare it to the

3.5 If your engine has a coil-in-cap distributor, disconnect the wire from the BAT terminal on the distributor cap when checking the compression

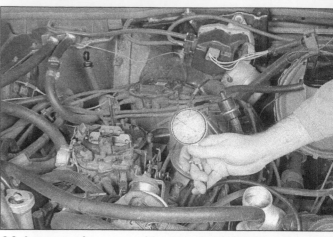

3.6 A compression gauge with a threaded fitting for the plug hole is preferred over the type that requires hand pressure to maintain the seal

Specifications. If it's extremely low, the bearings and/or oil pump are probably worn out.

Loss of power, rough running, knocking or metallic engine noises, excessive valve train noise and high fuel consumption rates may also point to the need for an overhaul, especially if they're all present at the same time. If a complete tune-up doesn't remedy the situation, major mechanical work is the only solution.

An engine overhaul involves restoring the internal parts to the specifications of a new engine. During an overhaul, the piston rings are replaced and the cylinder walls are reconditioned (rebored and/or honed). If a rebore is done by an automotive machine shop, new oversize pistons will also be installed. The main bearings, connecting rod bearings and camshaft bearings are generally replaced with new ones and, if necessary, the crankshaft may be reground to restore the journals. Generally, the valves are serviced as well, since they're usually in less-than-perfect condition at this point. While the engine is being overhauled, other components, such as the distributor, starter and alternator, can be rebuilt as well. The end result should be a like-new engine that will give many trouble-free miles. **Note:** *Critical cooling system components such as the hoses, drivebelts, thermostat and water pump MUST be replaced with new parts when an engine is overhauled. The radiator should be checked carefully to ensure that it isn't clogged or leaking (see Chapter 3). Also, we don't recommend overhauling the oil pump - always install a new one when an engine is rebuilt.*

Before beginning the engine overhaul, read through the entire procedure to familiarize yourself with the scope and requirements of the job. Overhauling an engine isn't difficult, but it is time consuming. Plan on the vehicle being tied up for a minimum of two weeks, especially if parts must be taken to an automotive machine shop for repair or reconditioning. Check on availability of parts and make sure that any necessary special tools

and equipment are obtained in advance. Most work can be done with typical hand tools, although a number of precision measuring tools are required for inspecting parts to determine if they must be replaced. Often an automotive machine shop will handle the inspection of parts and offer advice concerning reconditioning and replacement. **Note:** *Always wait until the engine has been completely disassembled and all components, especially the engine block, have been inspected before deciding what service and repair operations must be performed by an automotive machine shop.* Since the block's condition will be the major factor to consider when determining whether to overhaul the original engine or buy a rebuilt one, never purchase parts or have machine work done on other components until the block has been thoroughly inspected. As a general rule, time is the primary cost of an overhaul, so it doesn't pay to install worn or substandard parts.

As a final note, to ensure maximum life and minimum trouble from a rebuilt engine, everything must be assembled with care in a spotlessly clean environment.

3 Cylinder compression check

Refer to illustrations 3.5 and 3.6

1   A compression check will tell you what mechanical condition the upper end (pistons, rings, valves, head gaskets) of your engine is in. Specifically, it can tell you if the compression is down due to leakage caused by worn piston rings, defective valves and seats or a blown head gasket. **Note:** *The engine must be at normal operating temperature and the battery must be fully charged for this check. Also, the choke valve must be all the way open to get an accurate compression reading (if the engine's warm, the choke should be open).*

2   Begin by cleaning the area around the spark plugs before you remove them (com-

pressed air should be used, if available, otherwise a small brush or even a bicycle tire pump will work). The idea is to prevent dirt from getting into the cylinders as the compression check is being done.

3   Remove all of the spark plugs from the engine (see Chapter 1).

4   Block the throttle wide open.

5   On vehicles with "point-type" ignition, detach the coil wire from the center of the distributor cap and ground it on the engine block. Use a jumper wire with alligator clips on each end to ensure a good ground. On electronic ignition equipped vehicles, the ignition circuit should be disabled by unplugging the "BAT" wire going to the distributor **(see illustration)**.

6   Install the compression gauge in the number one spark plug hole **(see illustration)**.

7   Crank the engine over at least seven compression strokes and watch the gauge. The compression should build up quickly in a healthy engine. Low compression on the first stroke, followed by gradually increasing pressure on successive strokes, indicates worn piston rings. A low compression reading on the first stroke, which doesn't build up during successive strokes, indicates leaking valves or a blown head gasket (a cracked head could also be the cause). Deposits on the undersides of the valve heads can also cause low compression. Record the highest gauge reading obtained.

8   Repeat the procedure for the remaining cylinders and compare the results to the Specifications.

9   Add some engine oil (about three squirts from a plunger-type oil can) to each cylinder, through the spark plug hole, and repeat the test.

10   If the compression increases after the oil is added, the piston rings are definitely worn. If the compression doesn't increase significantly, the leakage is occurring at the valves or head gasket. Leakage past the valves may be caused by burned valve seats and/or

faces or warped, cracked or bent valves.

11 If two adjacent cylinders have equally low compression, there's a strong possibility that the head gasket between them is blown. The appearance of coolant in the combustion chambers or the crankcase would verify this condition.

12 If the compression is unusually high, the combustion chambers are probably coated with carbon deposits. If that's the case, the cylinder heads should be removed and decarbonized.

13 If compression is way down or varies greatly between cylinders, it would be a good idea to have a leak-down test performed by an automotive repair shop. This test will pinpoint exactly where the leakage is occurring and how severe it is.

4 Vacuum gauge diagnostic checks

Refer to illustration 4.5

1 A vacuum gauge provides valuable information about what is going on in the engine at a low cost. You can check for worn rings or cylinder walls, leaking head or intake manifold gaskets, vacuum leaks in the intake manifold, restricted exhaust, stuck or burned valves, weak valve springs, improper valve timing, and ignition problems. Vacuum gauge readings are easy to misinterpret, however, so they should be used in conjunction with other tests to confirm the diagnosis.

2 Both the absolute readings and the rate of needle movement are important for accurate interpretation. Most gauges measure vacuum in inches of mercury (in-Hg). The following references to vacuum assume the diagnosis is being performed at sea level. As elevation increases (or atmospheric pressure decreases), the reading will decrease. For every 1,000 foot increase in elevation above approximately 2000 feet, the gauge readings will decrease about one inch of mercury.

3 Connect the vacuum gauge directly to intake manifold vacuum, not to ported (throttle body) vacuum. Be sure no hoses are left disconnected during the test or false readings will result. **Note:** *Do not disconnect engine sensors or vacuum solenoids to connect the vacuum gauge. Disconnected engine control components can affect engine operation and produce abnormal vacuum gauge readings.*

4 Before you begin the test, warm the engine up completely. Block the wheels and set the parking brake. With the transmission in Park, start the engine and allow it to run at normal idle speed. **Warning:** *Carefully inspect the fan blades for cracks or damage before starting the engine. Keep your hands and the vacuum gauge clear of the fan and do not stand in front of the vehicle or in line with the fan when the engine is running.*

5 Read the vacuum gauge; an average, healthy engine should normally produce about 17 to 22 inches of vacuum with a fairly steady gauge needle at idle. Refer to the following vacuum gauge readings and what they indicate about the engine's condition **(see illustration)**:

6 A low steady reading usually indicates a leaking intake manifold gasket. this could be at one of the cylinder heads, between the upper and lower manifolds, or at the throttle body. Other possible causes are a leaky vacuum hose or incorrect camshaft timing.

7 If the reading is 3 to 8 inches below normal and it fluctuates at that low reading, suspect an intake manifold gasket leak at an intake port or a faulty fuel injector.

8 If the needle regularly drops about two to four inches at a steady rate, the valves are probably leaking. Perform a compression check or leakdown test to confirm this.

9 An irregular drop or downward flicker of the needle can be caused by a sticking valve or an ignition misfire. Perform a compression check or leakdown test and inspect the spark plugs to identify the faulty cylinder.

10 A rapid needle vibration of about four inches at idle combined with exhaust smoke indicates worn valve guides. Perform a leakdown test to confirm this. If the rapid vibration occurs with an increase in engine speed, check for a leaking intake manifold gasket or head gasket, weak valve springs, burned valves, or ignition misfire.

11 A slight fluctuation - one inch up and down - may mean ignition problems. Check all the usual tune-up items and, if necessary, run the engine on an ignition analyzer.

12 If there is a large fluctuation, perform a compression or leakdown test to look for a weak or dead cylinder or a blown head gasket.

13 If the needle moves slowly through a wide range, check for a clogged PCV system or intake manifold gasket leaks.

14 Check for a slow return of the gauge to a normal idle reading after revving the engine by quickly snapping the throttle open until the engine reaches about 2,500 rpm and let it shut. Normally the reading should drop to near zero, rise about 5 inches above normal idle reading, and then return to the previous idle reading. If the vacuum returns slowly and doesn't peak when the throttle is snapped shut, the rings may be worn. If there is a long delay, look for a restricted exhaust system (often the muffler or catalytic converter). One way to check this is to temporarily disconnect the exhaust ahead of the suspected part and repeat the test.

5 Engine removal - methods and precautions

If you've decided that an engine must be removed for overhaul or major repair work, several preliminary steps should be taken.

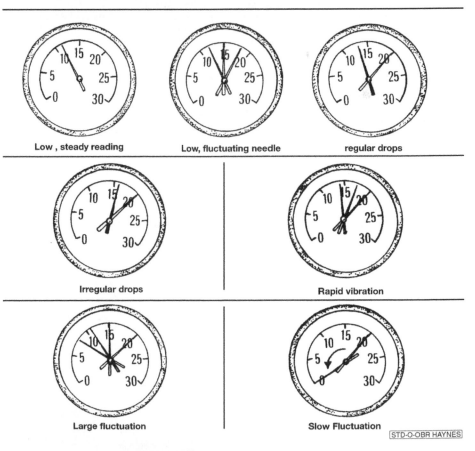

4.5 Typical vacuum gauge diagnostic readings

Low , steady reading   Low, fluctuating needle   regular drops

Irregular drops   Rapid vibration

Large fluctuation   Slow Fluctuation

STD-O-OBR HAYNES

Locating a suitable place to work is extremely important. Adequate work space, along with storage space for the vehicle, will be needed. If a shop or garage isn't available, at the very least a flat, level, clean work surface made of concrete or asphalt is required.

Cleaning the engine compartment and engine before beginning the removal procedure will help keep tools clean and organized.

An engine hoist or A-frame will also be necessary. Make sure the equipment is rated in excess of the combined weight of the engine and accessories. Safety is of primary importance, considering the potential hazards involved in lifting the engine out of the vehicle.

If the engine is being removed by a novice, a helper should be available. Advice and aid from someone more experienced would also be helpful. There are many instances when one person cannot simultaneously perform all of the operations required when lifting the engine out of the vehicle.

Plan the operation ahead of time. Arrange for or obtain all of the tools and equipment you'll need prior to beginning the job. Some of the equipment necessary to perform engine removal and installation safely and with relative ease are (in addition to an engine hoist) a heavy duty floor jack, complete sets of wrenches and sockets as described in the front of this manual, wooden blocks and plenty of rags and cleaning solvent for mopping up spilled oil, coolant and gasoline. If the hoist must be rented, make sure that you arrange for it in advance and perform all of the operations possible without it beforehand. This will save you money and time.

Plan for the vehicle to be out of use for quite a while. A machine shop will be required to perform some of the work which the do-it-yourselfer can't accomplish without special equipment. These shops often have a busy schedule, so it would be a good idea to consult them before removing the engine in order to accurately estimate the amount of time required to rebuild or repair components that may need work.

Always be extremely careful when removing and installing the engine. Serious injury can result from careless actions. Plan ahead, take your time and a job of this nature, although major, can be accomplished successfully.

---

6   Engine - removal and installation

Warning 1: *The air conditioning system is under high pressure! Have a dealer service department or service station discharge the system before disconnecting any A/C system hoses or fittings.*
Warning 2: *Gasoline is extremely flammable, so take extra precautions when you work on any part of the fuel system. Don't smoke or allow open flames or bare light bulbs near the work area, and don't work in a garage where*

a natural gas-type appliance (such as a water heater or a clothes dryer) with a pilot light is present. Since gasoline is carcinogenic, wear latex gloves when there's a possibility of being exposed to fuel, and, if you spill any fuel on your skin, rinse it off immediately with soap and water. Mop up any spills immediately and do not store fuel-soaked rags where they could ignite. When you perform any kind of work on the fuel system, wear safety glasses and have a Class B type fire extinguisher on hand.

Removal

Refer to illustrations 6.5 and 6.20

1   Disconnect the negative cable from the battery.
2   Cover the fenders and cowl and remove the hood (see Chapter 11).
Special pads are available to protect the fenders, but an old bedspread or blanket will also work.
3   Remove the air cleaner assembly.
4   Drain the cooling system (see Chapter 1).
5   Label the vacuum lines, emissions system hoses, wiring connectors, ground straps and fuel lines, to ensure correct reinstallation, then detach them. Pieces of masking tape with numbers or letters written on them work well **(see illustration)**. If there's any possibility of confusion, make a sketch of the engine compartment and clearly label the lines, hoses and wires.
6   Label and detach all coolant hoses from the engine.
7   Remove the cooling fan, shroud and radiator (see Chapter 3).
8   Remove the drivebelts (see Chapter 1).
9   **Warning:** *Gasoline is extremely flammable, so extra precautions must be taken when working on any part of the fuel system. DO NOT smoke or allow open flames or bare light bulbs near the vehicle. Also, don't work in a garage if a natural gas appliance with a pilot light is present. Disconnect the fuel lines running from the engine to the chassis (see Chapter 4). Plug or cap all open fittings/lines.*
10 Disconnect the throttle linkage (and TV linkage/speed control cable, if equipped) from the engine (see Chapter 4).

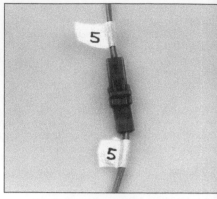

6.5 Flag each wire and hose to ease reassembly

11 On power steering equipped vehicles, unbolt the power steering pump (see Chapter 10). Leave the lines/hoses attached and make sure the pump is kept in an upright position in the engine compartment (use wire or rope to restrain it out of the way).
12 On A/C equipped vehicles, unbolt the compressor (see Chapter 3) and set it aside. Do not disconnect the hoses.
13 Drain the engine oil (see Chapter 1) and remove the filter.
14 Remove the starter motor (see Chapter 5).
15 Remove the alternator (see Chapter 5).
16 Unbolt the exhaust system from the engine (see Chapter 4).
17 If you're working on a vehicle with an automatic transmission, refer to Chapter 7 and remove the torque converter-to-driveplate fasteners.
18 Support the transmission with a jack. Position a block of wood between them to prevent damage to the transmission. Special transmission jacks with safety chains are available - use one if possible.
19 Attach an engine sling or a length of chain to the lifting brackets on the engine.
20 Roll the hoist into position and connect the sling to it **(see illustration)**. Take up the slack in the sling or chain, but don't lift the engine. **Warning:** *DO NOT place any part of your body under the engine when it's supported only by a hoist or other lifting device.*

6.20 Properly attached engine chain

21 Remove the transmission-to-engine block bolts.

22 Remove the engine mount-to-frame bolts.

23 Recheck to be sure nothing is still connecting the engine to the transmission or vehicle. Disconnect anything still remaining.

24 Raise the engine slightly. Carefully work it forward to separate it from the transmission. If you're working on a vehicle with an automatic transmission, be sure the torque converter stays in the transmission (clamp a pair of vise-grips to the housing to keep the converter from sliding out). If you're working on a vehicle with a manual transmission, the input shaft must be completely disengaged from the clutch. Slowly raise the engine out of the engine compartment. Check carefully to make sure nothing is hanging up.

25 Remove the flywheel/driveplate and mount the engine on an engine stand.

Installation

26 Check the engine and transmission mounts. If they're worn or damaged, replace them.

27 If you're working on a manual transmission equipped vehicle, install the clutch and pressure plate (see Chapter 7). Now is a good time to install a new clutch.

28 Carefully lower the engine into the engine compartment - make sure the engine mounts line up.

29 If you're working on an automatic transmission equipped vehicle, guide the torque converter into the crankshaft following the procedure outlined in Chapter 7.

30 If you're working on a manual transmission equipped vehicle, apply a dab of high-temperature grease to the input shaft and guide it into the crankshaft pilot bearing until the bellhousing is flush with the engine block.

31 Install the transmission-to-engine bolts and tighten them securely. **Caution:** *DO NOT use the bolts to force the transmission and engine together!*

32 Reinstall the remaining components in the reverse order of removal.

33 Add coolant, oil, power steering and transmission fluid as needed.

34 Run the engine and check for leaks and proper operation of all accessories, then install the hood and test drive the vehicle.

35 Have the A/C system recharged and leak tested.

---

7 Engine rebuilding alternatives

The do-it-yourselfer is faced with a number of options when performing an engine overhaul. The decision to replace the engine block, piston/connecting rod assemblies and crankshaft depends on a number of factors, with the number one consideration being the condition of the block. Other considerations are cost, access to machine shop facilities, parts availability, time required to complete the project and the extent of prior mechanical experience on the part of the do-it-yourselfer.

Some of the rebuilding alternatives include:

Individual parts - If the inspection procedures reveal that the engine block and most engine components are in reusable condition, purchasing individual parts may be the most economical alternative. The block, crankshaft and piston/connecting rod assemblies should all be inspected carefully. Even if the block shows little wear, the cylinder bores should be surface-honed.

Crankshaft kit - This rebuild package consists of a reground crankshaft and a matched set of pistons and connecting rods. Piston rings and the necessary bearings will be included in the kit. These kits are commonly available for standard cylinder bores, as well as for engine blocks which have been bored to a regular oversize.

Short block - A short block consists of an engine block with renewed crankshaft and piston/connecting rod assemblies already installed. All new bearings are incorporated and all clearances will be correct. The existing cylinder head(s), camshaft, valve train components and external parts can be bolted to the short block with little or no machine shop work necessary.

Long block - A long block consists of a short block plus an oil pump, oil pan, cylinder heads, valve covers, camshaft and valve train components, timing sprockets, timing chain and timing cover. All components are installed with new bearings, seals and gaskets incorporated throughout. The installation of manifolds and external parts is all that is necessary.

Used engine assembly - While overhaul provides the best assurance of a like-new engine, used engines available from wrecking yards and importers are often a very simple and economical solution. Many used engines come with warranties, but always give any engine a thorough diagnostic check-out before purchase. Check compression, vacuum and also for signs of oil leakage. If possible, have the seller run the engine, ether in the vehicle or on a test stand so you can be sure it runs smoothly with no knocking or other noises.

Give careful thought to which alternative is best for you and discuss the situation with local automotive machine shops, auto parts dealers or parts store countermen before ordering or purchasing replacement parts.

---

8 Engine overhaul - disassembly sequence

1 It's much easier to disassemble and work on the engine if it's mounted on a portable engine stand. A stand can often be rented quite cheaply from an equipment rental yard. Before the engine is mounted on a stand, the flywheel/driveplate should be removed from the engine.

2 If a stand isn't available, it's possible to disassemble the engine with it blocked up on the floor. Be extra careful not to tip or drop the engine when working without a stand.

3 If you're going to obtain a rebuilt engine, all external components must come off first, to be transferred to the replacement engine, just as they will if you're doing a complete engine overhaul yourself. These include:

Alternator and brackets
Emissions control components
Distributor, spark plug wires and spark plugs
Thermostat and housing cover
Water pump
Carburetor or throttle body
Intake/exhaust manifolds
Oil filter
Engine mounts
Clutch and flywheel/driveplate
Engine rear plate

Note: *When removing the external components from the engine, pay close attention to details that may be helpful or important during installation. Note the installed position of gaskets, seals, spacers, pins, brackets, washers, bolts and other small items.*

4 If you're obtaining a short block, which consists of the engine block, crankshaft, pistons and connecting rods all assembled, then the cylinder heads, oil pan and oil pump will have to be removed as well. See *Engine rebuilding alternatives* for additional information regarding the different possibilities to be considered.

5 If you're planning a complete overhaul, the engine must be disassembled and the internal components removed in the following order:

Valve covers
Intake and exhaust manifolds
Rocker arms and pushrods
Valve lifters
Cylinder heads
Timing cover
Timing chain and sprockets
Camshaft
Oil pan
Oil pump
Piston/connecting rod assemblies
Crankshaft and main bearings

6 Before beginning the disassembly and overhaul procedures, make sure the following items are available. Also, refer to *Engine overhaul - reassembly sequence* for a list of tools and materials needed for engine reassembly.

Common hand tools
Small cardboard boxes or plastic bags for storing parts
Gasket scraper
Ridge reamer
Vibration damper puller
Micrometers
Telescoping gauges
Dial indicator set
Valve spring compressor
Cylinder surfacing hone
Piston ring groove cleaning tool

9.2 A small plastic bag, with an appropriate label, can be used to store the valve train components so they can be kept together and reinstalled in the correct guide

9.3 Use a valve spring compressor to compress the spring, then remove the keepers from the valve stem

Electric drill motor
Tap and die set
Wire brushes
Oil gallery brushes
Cleaning solvent

9   Cylinder head - disassembly

Refer to illustrations 9.2, 9.3 and 9.4
Note: *New and rebuilt cylinder heads are commonly available for most engines at dealerships and auto parts stores. Due to the fact that some specialized tools are necessary for the disassembly and inspection procedures, and replacement parts may not be readily available, it may be more practical and economical for the home mechanic to purchase replacement heads rather than taking the time to disassemble, inspect and recondition the originals.*
1   Cylinder head disassembly involves removal of the intake and exhaust valves and related components. If they're still in place, remove the rocker arm nuts, pivot balls and rocker arms from the cylinder head studs. Label the parts or store them separately so they can be reinstalled in their original locations.
2   Before the valves are removed, arrange to label and store them, along with their related components, so they can be kept separate and reinstalled in the same valve guides they are removed from **(see illustration)**.
3   Compress the springs on the first valve with a spring compressor and remove the keepers **(see illustration)**. Carefully release the valve spring compressor and remove the retainer, the spring and the spring seat (if used).
4   Pull the valve out of the head, then remove the oil seal from the guide. If the valve binds in the guide (won't pull through), push it back into the head and deburr the area around the keeper groove with a fine file or whetstone **(see illustration)**.

5   Repeat the procedure for the remaining valves. Remember to keep all the parts for each valve together so they can be reinstalled in the same locations.
6   Once the valves and related components have been removed and stored in an organized manner, the heads should be thoroughly cleaned and inspected. If a complete engine overhaul is being done, finish the engine disassembly procedures before beginning the cylinder head cleaning and inspection process.

10   Cylinder head - cleaning and inspection

1   Thorough cleaning of the cylinder heads and related valve train components, followed by a detailed inspection, will enable you to decide how much valve service work must be done during the engine overhaul. **Note:** *If the engine was severely overheated, the cylinder head is probably warped (see Step 12).*

9.4 If the valve won't pull through the guide, deburr the edge of the stem end and the area around the top of the keeper groove with a file

Cleaning
2   Scrape all traces of old gasket material and sealing compound off the head gasket, intake manifold and exhaust manifold sealing surfaces. Be very careful not to gouge the cylinder head. Special gasket removal solvents that soften gaskets and make removal much easier are available at auto parts stores.
3   Remove all built up scale from the coolant passages.
4   Run a stiff wire brush through the various holes to remove deposits that may have formed in them.
5   Run an appropriate size tap into each of the threaded holes to remove corrosion and thread sealant that may be present. If compressed air is available, use it to clear the holes of debris produced by this operation. **Warning:** *Wear eye protection when using compressed air!*
6   Clean the rocker arm pivot stud threads, when equipped, with a wire brush.
7   Clean the cylinder head with solvent and dry it thoroughly. Compressed air will speed the drying process and ensure that all holes and recessed areas are clean. **Note:** *Decarbonizing chemicals are available and may prove very useful when cleaning cylinder heads and valve train components. They are very caustic and should be used with caution. Be sure to follow the instructions on the container.*
8   Clean the rocker arm components and pushrods with solvent and dry them thoroughly (don't mix them up during the cleaning process). Compressed air will speed the drying process and can be used to clean out the oil passages.
9   Clean all the valve springs, spring seats, keepers and retainers (or rotators) with solvent and dry them thoroughly. Do the components from one valve at a time to avoid mixing up the parts.
10   Scrape off any heavy deposits that may have formed on the valves, then use a motorized wire brush to remove deposits from the valve heads and stems. Again, make sure the valves don't get mixed up.

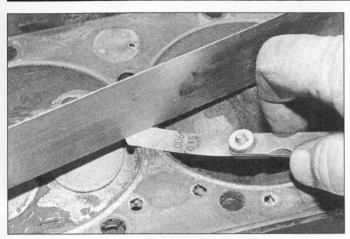

10.12 Check the cylinder head gasket surface for warpage by trying to slip a feeler gauge under the straightedge (see the Specifications for the maximum warpage allowed and use a feeler gauge of that thickness)

10.14a A dial indicator can be used to determine the valve stem-to-guide clearance (move the valve stem as indicated by the arrows)

Inspection

Note: *Be sure to perform all of the following inspection procedures before concluding that machine shop work is required. Make a list of the items that need attention. The inspection procedures for the lifters and rocker arms, as well as the camshafts, can be found in Chapter 2A.*

Cylinder head

Refer to illustrations 10.12, 10.14a and 10.14b

11   Inspect the head very carefully for cracks, evidence of coolant leakage and other damage. If cracks are found, check with an automotive machine shop concerning repair. If repair isn't possible, a new cylinder head should be obtained.

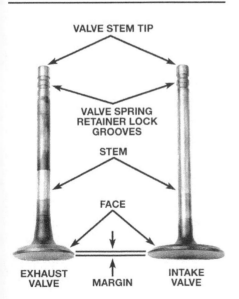

10.15 Check for valve wear at the points shown here - use a micrometer to measure the stem diameter at several points

12   Using a straightedge and feeler gauge, check the head gasket mating surface for warpage **(see illustration)**. If the warpage exceeds the specified limit, it can be resurfaced at an automotive machine shop. **Note:** *If the heads are resurfaced, the intake manifold flanges will also require machining.*

13   Examine the valve seats in each of the combustion chambers. If they're pitted, cracked or burned, the head will require valve service that's beyond the scope of the home mechanic.

14   Check the valve stem-to-guide clearance by measuring the lateral movement of the valve stem with a dial indicator attached securely to the head **(see illustration)**. The valve must be in the guide and approximately 1/16-inch off the seat. The total valve stem movement indicated by the gauge needle must be divided by two to obtain the actual clearance. After this is done, if there's still some doubt regarding the condition of the valve guides they should be checked by an automotive machine shop (the cost should be minimal). Oversize valves may be used to compensate for guide wear **(see illustration)**.

Valves

Refer to illustrations 10.15 and 10.16

15   Carefully inspect each valve face for uneven wear, deformation, cracks, pits and burned areas **(see illustration)**. Check the valve stem for scuffing and galling and the neck for cracks. Rotate the valve and check for any obvious indication that it's bent. Look for pits and excessive wear on the end of the stem. The presence of any of these conditions indicates the need for valve service by an automotive machine shop.

16   Measure the margin width on each valve **(see illustration)**. Any valve with a margin narrower than specified will have to be replaced with a new one.

Valve components

Refer to illustrations 10.18 and 10.19

17   Check each valve spring for wear (on

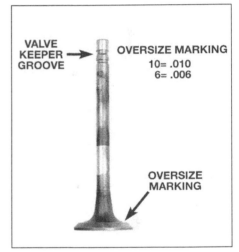

10.14b Some engines are equipped with slightly oversized valves from the factory. Valves with larger diameter stems are available to compensate for guide wear; however, a machine shop must resize the guides

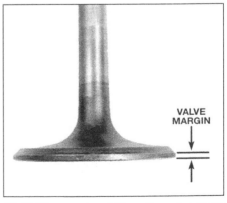

10.16 The margin width on the valve must be as specified (if no margin exists, the valve cannot be reused)

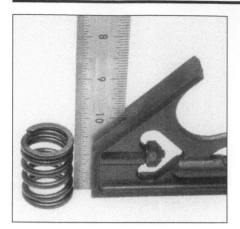

10.18 Check each valve spring for squareness

10.19 The exhaust valve rotators can be checked by turning the inner and outer sections in opposite directions - feel for smooth movement and excessive play

12.3 Positive type valve seals can be installed with a special installer or an appropriate-size deep socket - tap them down until they seat on the valve guide

the ends) and pits. Any springs that are shorter than specified have sagged and should not be reused. The tension of all springs should be checked with a special fixture before deciding that they're suitable for use in a rebuilt engine (take the springs to an automotive machine shop for this check).

18   Stand each spring on a flat surface and check it for squareness **(see illustration)**. If any of the springs are distorted or sagged, replace all of them with new parts.

19   Check the spring retainers (or rotators) and keepers for obvious wear and cracks. Any questionable parts should be replaced with new ones, as extensive damage will occur if they fail during engine operation. Make sure the rotators operate smoothly with no binding or excessive play **(see illustration)**.

Rocker arm components

20   Check the rocker arm faces (the areas that contact the pushrod ends and valve stems) for pits, wear, galling, score marks and rough spots. Check the rocker arm pivot contact areas and pivot as well. Look for cracks in each rocker arm and nut or bolt.

21   Inspect the pushrod ends for scuffing and excessive wear. Roll each pushrod on a flat surface, like a piece of plate glass, to determine if it's bent.

22   Check the rocker arm studs in the cylinder heads for damaged threads and secure installation.

23   Any damaged or excessively worn parts must be replaced with new ones.

24   If the inspection process indicates that the valve components are in generally poor condition and worn beyond the limits specified, which is usually the case in an engine that's being overhauled, reassemble the valves in the cylinder head and refer to Section 10 for valve servicing recommendations.

11   Valves - servicing

1   Because of the complex nature of the job and the special tools and equipment needed, servicing of the valves, the valve seats and the valve guides, commonly known

as a valve job, should be done by a professional.

2   The home mechanic can remove and disassemble the heads, do the initial cleaning and inspection, then reassemble and deliver them to a dealer service department or an automotive machine shop for the actual service work. Doing the inspection will enable you to see what condition the heads and valvetrain components are in and will ensure that you know what work and new parts are required when dealing with an automotive machine shop.

3   The dealer service department, or automotive machine shop, will remove the valves and springs, recondition or replace the valves and valve seats, recondition the valve guides, check and replace the valve springs, spring retainers or rotators and keepers (as necessary), replace the valve seals with new ones, reassemble the valve components and make sure the installed spring height is correct. The cylinder head gasket surfaces will also be resurfaced if they're warped.

4   After the valve job has been performed by a professional, the heads will be in like-

new condition. When the heads are returned, be sure to clean them again before installation on the engine to remove any metal particles and abrasive grit that may still be present from the valve service or head resurfacing operations. Use compressed air, if available, to blow out all the oil holes and passages.

12   Cylinder head - reassembly

Refer to illustrations 12.3, 12.5a, 12.5b, 12.6 and 12.9

1   Regardless of whether or not the heads were sent to an automotive repair shop for valve servicing, make sure they are clean before beginning reassembly.

2   If the heads were sent out for valve servicing, the valves and related components will already be in place. Begin the reassembly procedure with Step 8.

3   If originally equipped, install new positive type seals on each of the intake valve

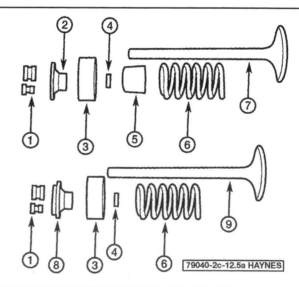

12.5a Typical engine valves and related components - exploded view

1   *Keepers*
2   *Retainer*
3   *Oil shield*
4   *O-ring stem seal*
5   *Umbrella or positive type seal*
6   *Spring and damper*
7   *Intake valve*
8   *Retainer/rotator*
9   *Exhaust valve*

79040-2c-12.5a HAYNES

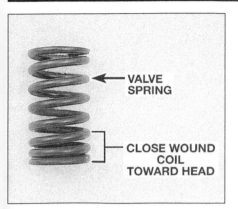

12.5b If the valve springs are wound closer at one end, install them as shown

12.6 Apply a small dab of grease to each keeper as shown here before installation - it will hold them in place on the valve stem as the spring is released

12.9 Be sure to check the valve spring installed height (the distance from the top of the seat/shims to the top of the shield)

guides. Using a hammer and a deep socket, gently tap each seal into place until it is completely seated on the guide **(see illustration)**. Do not twist or cock the seals during installation or they will not seal properly on the valve stems. Umbrella-type seals (if originally equipped) are installed over the valves after the valves are in place (see Step 4). **Note:** *Early model engines are typically equipped with O-ring stem seals only from the factory. Some later models engines are equipped with a positive type guide seal on the intake valve and O-ring stem seals on both the intake and exhaust valves. Other late model engines are equipped with a positive type guide seal on the intake valve, an umbrella type seal on the exhaust valve and O-ring stem seals on both the intake and exhaust valves.*

4   Beginning at one end of the head, lubricate and install the first valve. Apply moly-base grease or clean engine oil to the valve stem.

5   Drop the spring seat or shim(s) over the valve guide and set the valve springs, shield and retainer (or rotator) in place **(see illustrations)**.

6   Compress the springs with a valve spring compressor and carefully install the O-ring oil seal in the lower groove of the valve stem. Make sure the seal is not twisted - it must lie perfectly flat in the groove. Apply a

small dab of grease to each keeper to hold it in place if necessary **(see illustration)** and position the keepers in the upper groove, then slowly release the spring compressor.

7   Repeat the procedure for the remaining valves. Be sure to return the components to their original locations - do not mix them up!

8   Check to see that all the keepers are seated in the retainers and installed properly.

9   Check the installed valve spring height with a ruler graduated in 1/32-inch increments or a dial caliper. If the heads were sent out for service work, the installed height should be correct (but don't automatically assume that it is). The measurement is taken from the top of each spring seat or shim(s) to the top of the oil shield (or the bottom of the retainer/rotator, the two points are the same) **(see illustration)**. If the height is greater than specified, shims can be added under the springs to correct it. **Caution:** *Do not, under any circumstances, shim the springs to the point where the installed height is less than specified.*

10   On Chevrolet built engines, apply moly-base grease to the rocker arm faces and the ball pivot, then install the rocker arms and pivots on the cylinder head studs. Thread the nuts on three or four turns only at this time

(when the heads are installed, the nuts will be tightened following a specific procedure).

13   Pistons/connecting rod assembly - removal

Refer to illustrations 13.1, 13.3 and 13.6
Note: *Prior to removing the piston/connecting rod assemblies, remove the cylinder heads, the oil pan and the oil pump (except on Buick V6 engines) by referring to the appropriate Sections in Chapter 2A.*

1   Use your fingernail to feel if a ridge has formed at the upper limit of ring travel (about 1/4-inch down from the top of each cylinder). If carbon deposits or cylinder wear have produced ridges, they must be completely removed with a special tool **(see illustration)**. Follow the manufacturer's instructions provided with the tool. Failure to remove the ridges before attempting to remove the piston/connecting rod assemblies may result in piston breakage.

2   After the cylinder ridges have been removed, turn the engine upside-down so the crankshaft is facing up.

3   Before the connecting rods are re-

13.1 A ridge reamer is required to remove the ridge from the top of the cylinder - do this before removing the pistons!

13.3 Check the connecting rod side clearance with a feeler gauge as shown

13.6 To prevent damage to the crankshaft journals and cylinder walls, slip sections of hose over the rod bolts before removing the pistons

14.1 Checking crankshaft end play with a dial indicator

moved, check the end play with feeler gauges. Slide them between the first connecting rod and the crankshaft throw until the play is removed **(see illustration)**. The end play is equal to the thickness of the feeler gauge(s). If the end play exceeds the service limit, new connecting rods will be required. If new rods (or a new crankshaft) are installed, the end play may fall under the specified minimum (if it does, the rods will have to be machined to restore it - consult an automotive machine shop for advice if necessary). Repeat the procedure for the remaining connecting rods.

4   Check the connecting rods and caps for identification marks. If they aren't plainly marked, use a small center punch to make the appropriate number of indentations on each rod and cap (1, 2, 3, etc., depending on the engine type and cylinder they're associated with).

5   Loosen each of the connecting rod cap nuts 1/2-turn at a time until they can be removed by hand. Remove the number one connecting rod cap and bearing insert. Don't drop the bearing insert out of the cap.

6   Slip a short length of plastic or rubber hose over each connecting rod cap bolt to protect the crankshaft journal and cylinder wall as the piston is removed **(see illustration)**.

7   Remove the bearing insert and push the connecting rod/piston assembly out through the top of the engine. Use a wooden hammer handle to push on the upper bearing surface in the connecting rod. If resistance is felt, double-check to make sure that all of the ridge was removed from the cylinder.

8   Repeat the procedure for the remaining cylinders.

9   After removal, reassemble the connecting rod caps and bearing inserts in their respective connecting rods and install the cap nuts finger tight. Leaving the old bearing inserts in place until reassembly will help prevent the connecting rod bearing surfaces from being accidentally nicked or gouged.

14.3 Checking crankshaft end play with a feeler gauge

10   Don't separate the pistons from the connecting rods (see Section 18 for additional information).

14   Crankshaft - removal

Refer to illustrations 14.1, 14.3, 14.4a and 14.4b

Note: *The crankshaft can be removed only after the engine has been removed from the vehicle. It's assumed that the flywheel or driveplate, vibration damper, timing chain, oil pan, oil pump and piston/connecting rod assemblies have already been removed. If your engine is equipped with a one-piece rear main oil seal, the seal housing must be unbolted and separated from the block before proceeding with crankshaft removal.*

1   Before the crankshaft is removed, check the end play. Mount a dial indicator with the stem in line with the crankshaft and just touching one of the crank throws **(see illustration)**.

2   Push the crankshaft all the way to the rear and zero the dial indicator. Next, pry the

14.4a Use a center punch or number stamping dies to mark the main bearing caps to ensure that they are reinstalled in their original locations on the block (make the punch marks near one of the bolt heads)

crankshaft to the front as far as possible and check the reading on the dial indicator. The distance that it moves is the end play. If it's greater than specified, check the crankshaft thrust surfaces for wear. If no wear is evident, new main bearings should correct the end play.

3   If a dial indicator isn't available, feeler gauges can be used. Gently pry or push the crankshaft all the way to the front of the engine. Slip feeler gauges between the crankshaft and the front face of the thrust main bearing to determine the clearance **(see illustration)**.

4   Check the main bearing caps to see if they're marked to indicate their locations. They should be numbered consecutively from the front of the engine to the rear. If they aren't, mark them with number stamping dies or a center punch **(see illustration)**. Main bearing caps generally have a cast-in arrow, which points to the front of the engine **(see illustration)**. Loosen the main bearing cap

14.4b The arrow on the main bearing cap indicates the front of the engine

15.1 Use a hammer and a large punch to knock the core plugs sideways in their bores, then pull them out with pliers

15.10 A large socket on an extension can be used to drive the new core plugs into the bores

bolts 1/4-turn at a time each, until they can be removed by hand. Note if any stud bolts are used and make sure they're returned to their original locations when the crankshaft is reinstalled.

5   Gently tap the caps with a soft-face hammer, then separate them from the engine block. If necessary, use the bolts as levers to remove the caps. Try not to drop the bearing inserts if they come out with the caps.

6   Carefully lift the crankshaft out of the engine. It may be a good idea to have an assistant available, since the crankshaft is quite heavy. With the bearing inserts in place in the engine block and main bearing caps, return the caps to their respective locations on the engine block and tighten the bolts finger tight.

15 Engine block - cleaning

Refer to illustrations 15.1 and 15.10

1   Remove the core plugs from the engine block. To do this, knock one side of the plug into the block with a hammer and punch, then grasp them with large pliers and pull them back through the holes **(see illustration)**. **Caution:** *The core plugs (also known as freeze or soft plugs) may be difficult or impossible to retrieve if they're driven into the block coolant passages.*

2   Using a gasket scraper, remove all traces of gasket material from the engine block. Be very careful not to nick or gouge the gasket sealing surfaces.

3   Remove the main bearing caps and separate the bearing inserts from the caps and the engine block. Tag the bearings, indicating which cylinder they were removed from and whether they were in the cap or the block, then set them aside.

4   Remove all of the threaded oil gallery plugs from the block. The plugs are usually very tight - they may have to be drilled out and the holes retapped. Use new plugs when the engine is reassembled.

5   If the engine is extremely dirty it should be taken to an automotive machine shop to

be steam cleaned or hot tanked.

6   After the block is returned, clean all oil holes and oil galleries one more time. Brushes specifically designed for this purpose are available at most auto parts stores. Flush the passages with warm water until the water runs clear, dry the block thoroughly and wipe all machined surfaces with a light, rust preventive oil. If you have access to compressed air, use it to speed the drying process and to blow out all the oil holes and galleries. **Warning:** *Wear eye protection when using compressed air!*

7   If the block isn't extremely dirty or sludged up, you can do an adequate cleaning job with hot soapy water and a stiff brush. Take plenty of time and do a thorough job. Regardless of the cleaning method used, be sure to clean all oil holes and galleries very thoroughly, dry the block completely and coat all machined surfaces with light oil.

8   The threaded holes in the block must be clean to ensure accurate torque readings during reassembly. Run the proper size tap into each of the holes to remove rust, corrosion, thread sealant or sludge and restore damaged threads. If possible, use compressed air to clear the holes of debris produced by this operation. Now is a good time to clean the threads on the head bolts and the main bearing cap bolts as well.

9   Reinstall the main bearing caps and tighten the bolts finger tight.

10   After coating the sealing surfaces of the new core plugs with Permatex no. 2 sealant, install them in the engine block **(see illustration)**. Make sure they're driven in straight and seated properly or leakage could result. Special tools are available for this purpose, but a large socket, with an outside diameter that will just slip into the core plug, a 1/2-inch drive extension and a hammer will work just as well.

11   Apply non-hardening sealant (such as Permatex no. 2 or Teflon pipe sealant) to the new oil gallery plugs and thread them into the holes in the block. Make sure they're tightened securely.

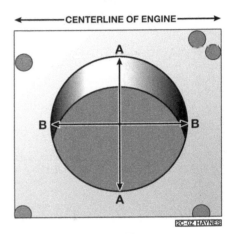

← CENTERLINE OF ENGINE →

16.4a Measure the diameter of each cylinder at a right angle to the engine centerline (A), and parallel to the engine centerline (B) - out-of-round is the difference between A and B - taper is the difference between A and B at the top of the cylinder and A and B at the bottom of the cylinder

12   If the engine isn't going to be reassembled right away, cover it with a large plastic trash bag to keep it clean.

16 Engine block - inspection

Refer to illustrations 16.4a, 16.4b and 16.4c

1   Before the block is inspected, it should be cleaned as described in Section 15.

2   Visually check the block for cracks, rust and corrosion. Look for stripped threads in the threaded holes. It's also a good idea to have the block checked for hidden cracks by an automotive machine shop that has the special equipment to do this type of work. If defects are found, have the block repaired, if possible, or replaced.

3   Check the cylinder bores for scuffing and scoring.

4   Check the cylinders for taper and out-

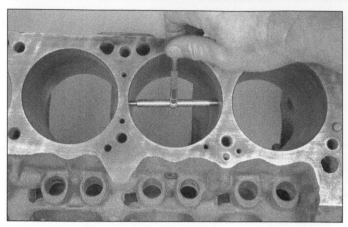

16.4b The ability to "feel" when the telescoping gauge is at the correct point will be developed over time, so work slowly and repeat the check until you are satisfied that the bore measurement is accurate

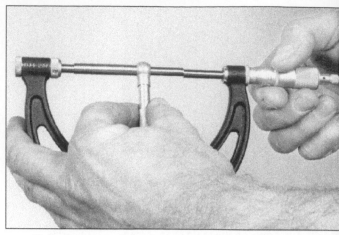

16.4c The gauge is then measured with a micrometer to determine the bore size

of-round conditions as follows **(see illustrations)**.

5   Measure the diameter of each cylinder at the top (just under the ridge area), center and bottom of the cylinder bore, parallel to the crankshaft axis.

6   Next, measure each cylinder's diameter at the same three locations across the crankshaft axis. Compare the results to the Specifications.

7   If the required precision measuring tools aren't available, the piston-to-cylinder clearances can be obtained, though not quite as accurately, using feeler gauge stock. Feeler gauge stock comes in 12-inch lengths and various thicknesses and is generally available at auto parts stores.

8   To check the clearance, select a feeler gauge and slip it into the cylinder along with the matching piston. The piston must be positioned exactly as it normally would be. The feeler gauge must be between the piston and cylinder on one of the thrust faces (90-degrees to the piston pin bore).

9   The piston should slip through the cylinder (with the feeler gauge in place) with moderate pressure.

10   If it falls through or slides through easily, the clearance is excessive and a new piston will be required. If the piston binds at the lower end of the cylinder and is loose toward the top, the cylinder is tapered. If tight spots are encountered as the piston/feeler gauge is rotated in the cylinder, the cylinder is out-of-round.

11   Repeat the procedure for the remaining pistons and cylinders.

12   If the cylinder walls are badly scuffed or scored, or if they're out-of-round or tapered beyond the limits given in the Specifications, have the engine block rebored and honed at an automotive machine shop. If a rebore is done, oversize pistons and rings will be required.

13   If the cylinders are in reasonably good condition and not worn to the outside of the limits, and if the piston-to-cylinder clearances can be maintained properly, then they don't

have to be rebored. Honing is all that's necessary (see Section 17).

17 Cylinder honing

Refer to illustrations 17.3a and 17.3b

1   Prior to engine reassembly, the cylinder bores must be honed so the new piston rings will seat correctly and provide the best possible combustion chamber seal. **Note:** *If you don't have the tools or don't want to tackle the honing operation, most automotive machine shops will do it for a reasonable fee.*

2   Before honing the cylinders, install the main bearing caps and tighten the bolts to the specified torque.

3   Two types of cylinder hones are commonly available - the flex hone, or "bottle brush," type and the more traditional surfacing hone with spring-loaded stones. Both will do the job, but for the less experienced mechanic the "bottle brush" hone will probably be easier to use. You'll also need some kerosene or honing oil, rags and an electric drill motor. Proceed as follows:

17.3a A "bottle brush" hone will produce good results even if you have never honed cylinders before

a)   *Mount the hone in the drill motor, compress the stones and slip it into the first cylinder* **(see illustration)**. *Be sure to wear safety goggles or a face shield!*

b)   *Lubricate the cylinder with plenty of honing oil, turn on the drill and move the hone up and down in the cylinder at a pace that will produce a fine crosshatch pattern on the cylinder walls. Ideally, the crosshatch lines should intersect at approximately a 60-degree angle* **(see illustration)**. *Be sure to use plenty of lubricant and don't take off any more material than is absolutely necessary to produce the desired finish.* **Note:** *Piston ring manufacturers may specify a smaller crosshatch angle than the traditional 60-degrees - read and follow any instructions included with the new rings.*

c)   *Don't withdraw the hone from the cylinder while it's running. Instead, shut off the drill and continue moving the hone up-and-down in the cylinder until it comes to a complete stop, then com-*

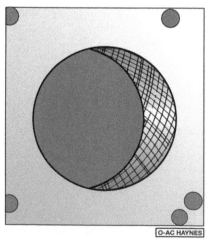

17.3b The cylinder hone should leave a smooth, crosshatch pattern with the lines intersecting at approximately a 45 to 60-degree angle

18.4a The piston ring grooves can be cleaned with a special tool, as shown here . . .

18.4b . . . or a section of a broken ring

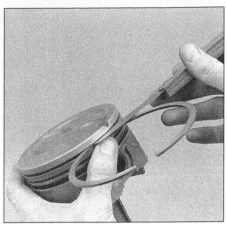

18.10 Check the ring side clearance with a feeler gauge at several points around the groove

press the stones and withdraw the hone. If you're using a "bottle brush" type hone, stop the drill motor, then turn the chuck in the normal direction of rotation while withdrawing the hone from the cylinder.

d) Wipe the oil out of the cylinder and repeat the procedure for the remaining cylinders.

4   After the honing job is complete, chamfer the top edges of the cylinder bores with a small file so the rings won't catch when the pistons are installed. Be very careful not to nick the cylinder walls with the end of the file.

5   The entire engine block must be washed again very thoroughly with warm, soapy water to remove all traces of the abrasive grit produced during the honing operation. **Note:** *The bores can be considered clean when a lint-free white cloth - dampened with clean engine oil - used to wipe them out doesn't pick up any more honing residue, which will show up as gray areas on the cloth. Be sure to run a brush through all oil holes and galleries and flush them with running water.*

6   After rinsing, dry the block and apply a coat of light rust preventive oil to all machined surfaces. Wrap the block in a plastic trash bag to keep it clean and set it aside until reassembly.

18 Piston/connecting rod assembly - inspection

Refer to illustrations 18.4a, 18.4b, 18.10 and 18.11

1   Before the inspection process can be carried out, the piston/connecting rod assemblies must be cleaned and the original piston rings removed from the pistons. **Note:** *Always use new piston rings when the engine is reassembled.*

2   Using a piston ring installation tool, carefully remove the rings from the pistons. Be careful not to nick or gouge the pistons in the process.

3   Scrape all traces of carbon from the top of the piston. A hand-held wire brush or a piece of fine emery cloth can be used once the majority of the deposits have been scraped away. Do not, under any circumstances, use a wire brush mounted in a drill motor to remove deposits from the pistons. The piston material is soft and may be eroded away by the wire brush.

4   Use a piston ring groove cleaning tool to remove carbon deposits from the ring grooves. If a tool isn't available, a piece broken off the old ring will do the job. Be very careful to remove only the carbon deposits - don't remove any metal and do not nick or scratch the sides of the ring grooves **(see illustrations)**.

5   Once the deposits have been removed, clean the piston/rod assemblies with solvent and dry them with compressed air (if available). Make sure the oil return holes in the back sides of the ring grooves are clear.

6   If the pistons and cylinder walls aren't damaged or worn excessively, and if the engine block is not rebored, new pistons won't be necessary. Normal piston wear appears as even vertical wear on the piston thrust surfaces and slight looseness of the top ring in its groove. New piston rings, however, should always be used when an engine is rebuilt.

7   Carefully inspect each piston for cracks around the skirt, at the pin bosses and at the ring lands.

8   Look for scoring and scuffing on the thrust faces of the skirt, holes in the piston crown and burned areas at the edge of the crown. If the skirt is scored or scuffed, the engine may have been suffering from overheating and/or abnormal combustion, which caused excessively high operating temperatures. The cooling and lubrication systems should be checked thoroughly. A hole in the piston crown is an indication that abnormal combustion (preignition) was occurring. Burned areas at the edge of the piston crown are usually evidence of spark knock (detonation). If any of the above problems exist, the causes must be corrected or the damage will occur again. The causes may include intake

air leaks, incorrect fuel/air mixture, incorrect ignition timing and EGR system malfunctions.

9   Corrosion of the piston, in the form of small pits, indicates that coolant is leaking into the combustion chamber and/or the crankcase. Again, the cause must be corrected or the problem may persist in the rebuilt engine.

10   Measure the piston ring side clearance by laying a new piston ring in each ring groove and slipping a feeler gauge in beside it **(see illustration)**. Check the clearance at three or four locations around each groove. Be sure to use the correct ring for each groove - they are different. If the side clearance is greater than specified, new pistons will have to be used.

11   Check the piston-to-bore clearance by measuring the bore (see Section 16) and the piston diameter. Make sure the pistons and bores are correctly matched. Measure the piston across the skirt, at a 90-degree angle to and in line with the piston pin **(see illustration)**. Subtract the piston diameter from the bore diameter to obtain the clearance. If it's greater than specified, the block will have to be rebored and new pistons and rings installed.

18.11 Measure the piston diameter at a 90-degree angle to the piston pin and in line with it

12   Check the piston-to-rod clearance by twisting the piston and rod in opposite directions. Any noticeable play indicates excessive wear, which must be corrected. The piston/connecting rod assemblies should be taken to an automotive machine shop to have the pistons and rods resized and new pins installed.

13   If the pistons must be removed from the connecting rods for any reason, they should be taken to an automotive machine shop. While they are there have the connecting rods checked for bend and twist, since automotive machine shops have special equipment for this purpose. **Note:** *Unless new pistons and/or connecting rods must be installed, do not disassemble the pistons and connecting rods.*

14   Check the connecting rods for cracks and other damage. Temporarily remove the rod caps, lift out the old bearing inserts, wipe the rod and cap bearing surfaces clean and inspect them for nicks, gouges and scratches. After checking the rods, replace the old bearings, slip the caps into place and tighten the nuts finger tight. **Note:** *If the engine is being rebuilt because of a connecting rod knock, be sure to install new or remanufactured rods.*

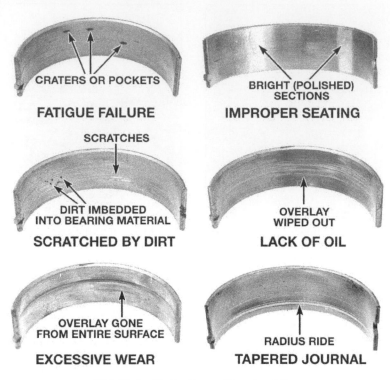

20.1 Typical indications of bearing failure

19   Crankshaft - inspection

1   Clean the crankshaft with solvent and dry it with compressed air (if available). Be sure to clean the oil holes with a stiff brush and flush them with solvent.

2   Check the main and connecting rod bearing journals for uneven wear, scoring, pits and cracks.

3   Rub a penny across each journal several times. If a journal picks up copper from the penny, it's too rough and must be reground.

4   Remove all burrs from the crankshaft oil holes with a stone, file or scraper.

5   Check the rest of the crankshaft for cracks and other damage. It should be magnafluxed to reveal hidden cracks - an automotive machine shop will handle the procedure.

6   Using a micrometer, measure the diameter of the main and connecting rod journals and compare the results to the Specifications. By measuring the diameter at a number of points around each journal's circumference, you'll be able to determine whether or not the journal is out-of-round. Take the measurement at each end of the journal, near the crank throws, to determine if the journal is tapered.

7   If the crankshaft journals are damaged, tapered, out-of-round or worn beyond the limits given in the Specifications, have the crankshaft reground by an automotive machine shop. Be sure to use the correct size bearing inserts if the crankshaft is reconditioned.

8   Check the oil seal journals at each end of the crankshaft for wear and damage. If the seal has worn a groove in the journal, or if it's

nicked or scratched, the new seal may leak when the engine is reassembled. In some cases, an automotive machine shop may be able to repair the journal by pressing on a thin sleeve. If repair isn't feasible, a new or different crankshaft should be installed.

9   Refer to Section 20 and examine the main and rod bearing inserts.

20   Main and connecting rod bearings - inspection

Refer to illustration 20.1

1   Even though the main and connecting rod bearings should be replaced with new ones during the engine overhaul, the old bearings should be retained for close examination, as they may reveal valuable information about the condition of the engine **(see illustration)**.

2   Bearing failure occurs because of lack of lubrication, the presence of dirt or other foreign particles, overloading the engine and corrosion. Regardless of the cause of bearing failure, it must be corrected before the engine is reassembled to prevent it from happening again.

3   When examining the bearings, remove them from the engine block, the main bearing caps, the connecting rods and the rod caps and lay them out on a clean surface in the same general position as their location in the engine. This will enable you to match any bearing problems with the corresponding crankshaft journal.

4   Dirt and other foreign particles get into the engine in a variety of ways. It may be left in the engine during assembly, or it may pass through filters or the PCV system. It may get into the oil, and from there into the bearings. Metal chips from machining operations and normal engine wear are often present. Abrasives are sometimes left in engine components after reconditioning, especially when parts are not thoroughly cleaned using the proper cleaning methods. Whatever the source, these foreign objects often end up embedded in the soft bearing material and are easily recognized. Large particles will not embed in the bearing and will score or gouge the bearing and journal. The best prevention for this cause of bearing failure is to clean all parts thoroughly and keep everything spotlessly clean during engine assembly. Frequent and regular engine oil and filter changes are also recommended.

5   Lack of lubrication (or lubrication breakdown) has a number of interrelated causes. Excessive heat (which thins the oil), overloading (which squeezes the oil from the bearing face) and oil leakage or throw off (from excessive bearing clearances, worn oil pump or high engine speeds) all contribute to lubrication breakdown. Blocked oil passages, which usually are the result of misaligned oil holes in a bearing shell, will also oil starve a bearing and destroy it. When lack of lubrication is the cause of bearing failure, the bearing material is wiped or extruded from the steel backing of the bearing. Temperatures may increase to the point where the steel backing turns blue from overheating.

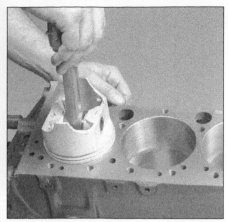

22.3 When checking piston ring end gap, the ring must be square in the cylinder bore (this is done by pushing the ring down with the top of a piston as shown)

22.4 With the ring square in the cylinder, measure the end gap with a feeler gauge

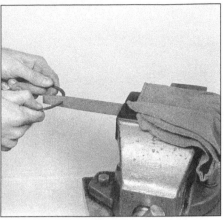

22.5 If the end gap is too small, clamp a file in a vise and file the ring ends (from the outside in only) to enlarge the gap slightly

6   Driving habits can have a definite effect on bearing life. Full throttle, low speed operation (lugging the engine) puts very high loads on bearings, which tends to squeeze out the oil film. These loads cause the bearings to flex, which produces fine cracks in the bearing face (fatigue failure). Eventually the bearing material will loosen in pieces and tear away from the steel backing. Short trip driving leads to corrosion of bearings because insufficient engine heat is produced to drive off the condensed water and corrosive gases. These products collect in the engine oil, forming acid and sludge. As the oil is carried to the engine bearings, the acid attacks and corrodes the bearing material.

7   Incorrect bearing installation during engine assembly will lead to bearing failure as well. Tight fitting bearings leave insufficient bearing oil clearance and will result in oil starvation. Dirt or foreign particles trapped behind a bearing insert result in high spots on the bearing which lead to failure.

21 Engine overhaul - reassembly sequence

1   Before beginning engine reassembly, make sure you have all the necessary new parts, gaskets and seals as well as the following items on hand:

Common hand tools
A 1/2-inch drive torque wrench
Piston ring installation tool
Piston ring compressor
Vibration damper installation tool
Short lengths of rubber or plastic hose to fit over connecting rod bolts
Plastigage
Feeler gauges
A fine-tooth file
New engine oil
Engine assembly lube or moly-base grease
Gasket sealant
Thread locking compound

2   In order to save time and avoid problems, engine reassembly must be done in the following general order:

New camshaft bearings (must be installed by an auto machine shop)
Piston rings
Crankshaft and main bearings
Piston/connecting rod assemblies
Oil pump - except external-mount type
Camshaft and lifters
Oil pan
Timing chain and sprockets
Cylinder heads, pushrods and rocker arms
Timing cover (and oil pump - external-mount type)
Intake and exhaust manifolds
Rocker arm covers
Engine rear plate (if equipped)
Flywheel/driveplate

22 Piston rings - installation

Refer to illustrations 22.3, 22.4, 22.5, 22.9a, 22.9b and 22.12

1   Before installing the new piston rings, the ring end gaps must be checked. It's assumed that the piston ring side clearance has been checked and verified correct (see Section 18).

2   Lay out the piston/connecting rod assemblies and the new ring sets so the ring sets will be matched with the same piston and cylinder during the end gap measurement and engine assembly.

3   Insert the top (number one) ring into the first cylinder and square it up with the cylinder walls by pushing it in with the top of the piston **(see illustration)**. The ring should be near the bottom of the cylinder, at the lower limit of ring travel.

4   To measure the end gap, slip feeler gauges between the ends of the ring until a gauge equal to the gap width is found **(see illustration)**. The feeler gauge should slide between the ring ends with a slight amount of drag. Compare the measurement to the Specifications. If the gap is larger or smaller

than specified, double-check to make sure you have the correct rings before proceeding.

5   If the gap is too small, it must be enlarged or the ring ends may come in contact with each other during engine operation, which can cause serious damage to the engine. The end gap can be increased by filing the ring ends very carefully with a fine file. Mount the file in a vise equipped with soft jaws, slip the ring over the file with the ends contacting the file face and slowly move the ring to remove material from the ends. When performing this operation, file only from the outside in **(see illustration)**.

6   Excess end gap isn't critical unless it's greater than 0.040-inch. Again, double-check to make sure you have the correct rings for your engine.

7   Repeat the procedure for each ring that will be installed in the first cylinder and for each ring in the remaining cylinders. Remember to keep rings, pistons and cylinders matched up.

8   Once the ring end gaps have been checked/corrected, the rings can be installed on the pistons.

9   The oil control ring (lowest one on the piston) is usually installed first. It's composed of three separate components. Slip the

22.9a Installing the spacer/expander in the oil control ring groove

22.9b DO NOT use a piston ring installation tool when installing the oil ring side rails

22.12 Installing the compression rings with a ring expander - the mark (arrow) must face up

23.11 Lay the Plastigage strips (arrow) on the main bearing journals, parallel to the crankshaft centerline

spacer/expander into the groove **(see illustration)**. If an anti-rotation tang is used, make sure it's inserted into the drilled hole in the ring groove. Next, install the lower side rail. Don't use a piston ring installation tool on the oil ring side rails, as they may be damaged. Instead, place one end of the side rail into the groove between the spacer/expander and the ring land, hold it firmly in place and slide a finger around the piston while pushing the rail into the groove **(see illustration)**. Next, install the upper side rail in the same manner.

10   After the three oil ring components have been installed, check to make sure that both the upper and lower side rails can be turned smoothly in the ring groove.

11   The number two (middle) ring is installed next. It's usually stamped with a mark which must face up, toward the top of the piston. **Note:** *Always follow the instructions printed on the ring package or box - different manufacturers may require different approaches. Do not mix up the top and middle rings, as they have different cross sections.*

12   Use a piston ring installation tool and make sure the identification mark is facing the top of the piston **(see illustration)**, then slip the ring into the middle groove on the piston. Don't expand the ring any more than necessary to slide it over the piston.

13   Install the number one (top) ring in the same manner. Make sure the mark is facing up. Be careful not to confuse the number one and number two rings.

14   Repeat the procedure for the remaining pistons and rings.

23   Crankshaft - installation and main bearing oil clearance check

Refer to illustrations 23.11 and 23.15

1   Crankshaft installation is the first step in engine reassembly. It's assumed at this point that the engine block and crankshaft have been cleaned, inspected and repaired or reconditioned.

2   Position the engine with the bottom facing up.

3   Remove the main bearing cap bolts and lift out the caps. Lay them out in the proper order to ensure correct installation.

4   If they're still in place, remove the original bearing inserts from the block and the main bearing caps. Wipe the bearing surfaces of the block and caps with a clean, lint-free cloth. They must be kept spotlessly clean.

Main bearing oil clearance check

5   Clean the back sides of the new main bearing inserts and lay one in each main bearing saddle in the block. If one of the bearing inserts from each set has a large groove in it, make sure the grooved insert is installed in the block. Lay the other bearing from each set in the corresponding main bearing cap. Make sure the tab on the bearing insert fits into the recess in the block or cap. **Caution:** *The oil holes in the block must line up with the oil holes in the bearing insert. Do not hammer the bearing into place and don't nick or gouge the bearing faces. No lubrication should be used at this time.*

6   The flanged thrust bearing must be installed in the proper cap and saddle.

7   Clean the faces of the bearings in the block and the crankshaft main bearing journals with a clean, lint-free cloth.

8   Check or clean the oil holes in the crankshaft, as any dirt here can go only one way - straight through the new bearings.

9   Once you're certain the crankshaft is clean, carefully lay it in position in the main bearings.

10   Before the crankshaft can be permanently installed, the main bearing oil clearance must be checked.

11   Cut several pieces of the appropriate size Plastigage (they must be slightly shorter than the width of the main bearings) and place one piece on each crankshaft main bearing journal, parallel with the journal axis **(see illustration)**.

12   Clean the faces of the bearings in the caps and install the caps in their respective

positions (don't mix them up) with the arrows pointing toward the front of the engine. Don't disturb the Plastigage.

13   Starting with the center main and working out toward the ends, tighten the main bearing cap bolts, in three steps, to the specified torque. Don't rotate the crankshaft at any time during this operation.

14   Remove the bolts and carefully lift off the main bearing caps. Keep them in order. Don't disturb the Plastigage or rotate the crankshaft. If any of the main bearing caps are difficult to remove, tap them gently from side to side with a soft-face hammer to loosen them.

15   Compare the width of the crushed Plastigage on each journal to the scale printed on the Plastigage envelope to obtain the main bearing oil clearance **(see illustration)**. Check the Specifications to make sure it's correct.

16   If the clearance is not as specified, the bearing inserts may be the wrong size (which means different ones will be required). Before

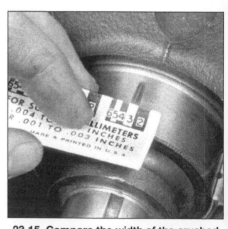

23.15 Compare the width of the crushed Plastigage to the scale on the container to determine the main bearing oil clearance (always take the measurement at the widest point of the Plastigage); be sure to use the correct scale - standard and metric scales are included

24.3 Use a wood hammer handle for installing the rear main bearing seal - fabric (rope) type seals

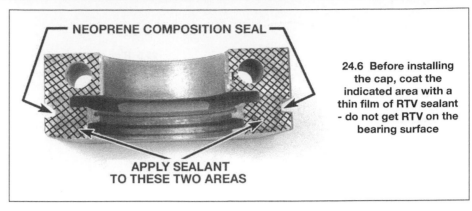

NEOPRENE COMPOSITION SEAL

24.6 Before installing the cap, coat the indicated area with a thin film of RTV sealant - do not get RTV on the bearing surface

APPLY SEALANT TO THESE TWO AREAS

deciding that different inserts are needed, make sure that no dirt or oil was between the bearing inserts and the caps or block when the clearance was measured. If the Plastigage was wider at one end than the other, the journal may be tapered (refer to Section 19).

17 Carefully scrape all traces of the Plastigage material off the main bearing journals and/or the bearing faces. Use your fingernail or the edge of a credit card - don't nick or scratch the bearing faces.

Final crankshaft installation

18 Carefully lift the crankshaft out of the engine. If the engine is equipped with a two piece rear main oil seal refer to Section 24 and install the seal halves into the cylinder block and the rear main bearing cap.

19 Clean the bearing faces in the block, then apply a thin, uniform layer of moly-base grease or engine assembly lube to each of the bearing surfaces. Be sure to coat the thrust faces as well as the journal face of the thrust bearing.

20 Make sure the crankshaft journals are clean, then lay the crankshaft back in place in the block.

21 Clean the faces of the bearings in the caps, then apply lubricant to them.

22 Install the caps in their respective positions with the arrows pointing toward the front of the engine.

23 Install the bolts.

24 Tighten all except the thrust bearing cap bolts to the specified torque (work from the center out and approach the final torque in three steps).

25 Tighten the thrust bearing cap bolts to 10-to-12 ft-lbs.

26 Tap the ends of the crankshaft forward and backward with a lead or brass hammer to line up the main bearing and crankshaft thrust surfaces.

27 Retighten all main bearing cap bolts to the specified torque, starting with the center main and working out toward the ends.

28 On manual transmission equipped models, install a new pilot bearing in the end of the crankshaft (see Chapter 8).

29 Rotate the crankshaft a number of times by hand to check for any obvious binding.

30 The final step is to check the crankshaft end play with a feeler gauge or a dial indicator as described in Section 14. The end play should be correct if the crankshaft thrust faces aren't worn or damaged and new bearings have been installed.

31 If you are working on an engine with a one-piece rear main oil seal, refer to Section 24 and install the new seal, then bolt the housing to the block.

24 Rear main oil seal installation

Two-piece fabric (rope) type seal

Refer to illustrations 24.3 and 24.6

1 Braided fabric seals pressed into grooves formed in the crankcase and rear bearing cap are used to seal against oil leakage around the crankshaft. The crankshaft must be removed for this operation.

2 With the bearing caps and the crankshaft removed place the new seals in the grooves of the cylinder block and the rear main bearing cap with both ends projecting above the cap parting surface.

3 Use a handle of a hammer or similar tool to force the seal into the groove by rubbing down until the seal projects above the groove not more then 1/16-inch **(see illustration)**. Cut the ends of the seal flush with the surface of the cap with a single-edged razor blade.

4 Soak the neoprene seals (if equipped), which go into the grooves in the sides of the bearing cap in kerosene for one to two minutes.

5 Install the neoprene seals (if equipped) in the groove between the bearing cap and the crankcase. The seals are slightly undersized and swell in the presence of heat and oil. They are slightly longer than the groove in the bearing cap and must not be cut to fit.

6 Apply a small amount of RTV sealer at the joint where the bearing cap meets the crankcase to help eliminate oil leakage **(see illustration)**. A very thin coat is all that is necessary.

7 Install the bearing cap in the crankcase. Force the seals up into the bearing cap with a blunt instrument to be sure of a good seal at

the upper parting line between the cap and case.

Two-piece neoprene type seal

Refer to illustration 24.9

8 Inspect the rear main bearing cap and engine block mating surfaces, as well as the seal grooves, for nicks, burrs and scratches. Remove any defects with a fine file or deburring tool.

9 Install one seal section in the block with the lip facing the front of the engine (if the seal has two lips, the one with the helix must face the front) **(see illustration)**. Leave one end protruding from the block approximately 1/4- to 3/8-inch and make sure it's completely seated.

10 Repeat the procedure to install the remaining seal half in the rear main bearing cap. In this case, leave the opposite end of the seal protruding from the cap the same distance the block seal is protruding from the block.

11 During final installation of the crankshaft (after the main bearing oil clearances have been checked with Plastigage) as described in Section 23, apply a thin, even coat of anaerobic-type gasket sealant to the area of the cap adjacent to the seal **(see illustration 24.6)**. Don't get any sealant on the bearing face, crankshaft journal, seal ends or seal lips. Also, lubricate the seal lips with moly-base grease or engine assembly lube.

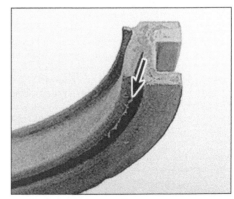

24.9 The rear main oil seal may have two lips - the oil seal (arrow) must point toward the front of the engine, which means that the dust seal will face out, toward the rear of the engine

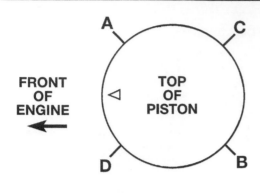

25.5 Ring end gap positions

A   Oil ring rail gap - lower
B   Oil ring rail gap - upper
C   Top compression ring gap
D   Second compression ring gap and oil ring spacer gap

O-BO HAYNES

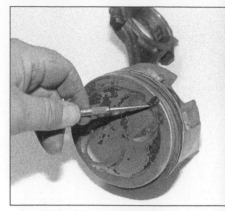

25.9 The notch in each piston must face the FRONT of the engine as the pistons are installed

One piece neoprene type seal

12   Some models are equipped with a one-piece seal that fits into a housing attached to the block. The crankshaft must be installed first and the main bearing caps bolted in place, then the new seal should be installed in the housing and the housing bolted to the block.

13   Before installing the crankshaft, check the seal contact surface very carefully for scratches and nicks that could damage the new seal lip and cause oil leaks. If the crankshaft is damaged, the only alternative is a new or different crankshaft.

14   The old seal can be removed from the housing with a screwdriver by prying it out from the front (see illustration 17.23 in Chapter 2A). Be sure to note how far it's recessed into the housing bore before removing it; the new seal will have to be recessed an equal amount. Be very careful not to scratch or otherwise damage the bore in the housing or oil leaks could develop.

15   Make sure the housing is clean, then apply a thin coat of engine oil to the outer edge of the new seal. The seal must be pressed squarely into the housing bore, so hammering it into place is not recommended. If you don't have access to a press, sandwich the housing and seal between two smooth pieces of wood and press the seal into place with the jaws of a large vise. The pieces of wood must be thick enough to distribute the force evenly around the entire circumference of the seal. Work slowly and make sure the seal enters the bore squarely.

16   The seal lips must be lubricated with moly-base grease or engine assembly lube before the seal/housing is slipped over the crankshaft and bolted to the block. Use a new gasket - no sealant is required - and make sure the dowel pins are in place before installing the housing.

17   Tighten the screws a little at a time until they're all at the specified torque.

25   Piston/connecting rod assembly - installation and rod bearing oil clearance check

Refer to illustrations 25.5, 25.9 and 25.11

1   Before installing the piston/connecting rod assemblies, the cylinder walls must be perfectly clean, the top edge of each cylinder must be chamfered, and the crankshaft must be in place.

2   Remove the cap from the end of the number one connecting rod (refer to the marks made during removal). Remove the original bearing inserts and wipe the bearing surfaces of the connecting rod and cap with a clean, lint-free cloth. They must be kept spotlessly clean.

Connecting rod bearing oil clearance check

3   Clean the back side of the new upper bearing insert, then lay it in place in the connecting rod. Make sure the tab on the bearing fits into the recess in the rod. Don't hammer the bearing insert into place and be very careful not to nick or gouge the bearing face. Don't lubricate the bearing at this time.

4   Clean the back side of the other bearing insert and install it in the rod cap. Again, make sure the tab on the bearing fits into the recess in the cap, and don't apply any lubricant. It's critically important that the mating surfaces of the bearing and connecting rod are perfectly clean and oil-free when they're assembled.

5   Position the piston ring gaps at 120-degree intervals around the piston (see illustration).

6   Slip a section of plastic or rubber hose over each connecting rod cap bolt.

7   Lubricate the piston and rings with clean engine oil and attach a piston ring compressor to the piston. Leave the skirt protruding about 1/4-inch to guide the piston into the cylinder. The rings must be compressed until they're flush with the piston.

8   Rotate the crankshaft until the number one connecting rod journal is at BDC (bottom dead center) and apply a coat of engine oil to the cylinder walls.

9   With the mark or notch on top of the piston (see illustration) facing the front of the engine, gently insert the piston/connecting rod assembly into the number one cylinder bore and rest the bottom edge of the ring compressor on the engine block.

10   Tap the top edge of the ring compressor to make sure it's contacting the block around its entire circumference.

11   Gently tap on the top of the piston with the end of a wooden hammer handle (see

illustration) while guiding the end of the connecting rod into place on the crankshaft journal. The piston rings may try to pop out of the ring compressor just before entering the cylinder bore, so keep some downward pressure on the ring compressor. Work slowly, and if any resistance is felt as the piston enters the cylinder, stop immediately. Find out what's hanging up and fix it before proceeding. Do not, for any reason, force the piston into the cylinder - you might break a ring and/or the piston.

12   Once the piston/connecting rod assembly is installed, the connecting rod bearing oil clearance must be checked before the rod cap is permanently bolted in place.

13   Cut a piece of the appropriate size Plastigage slightly shorter than the width of the connecting rod bearing and lay it in place on the number one connecting rod journal, parallel with the journal axis.

14   Clean the connecting rod cap bearing face, remove the protective hoses from the connecting rod bolts and install the rod cap. Make sure the mating mark on the cap is on the same side as the mark on the connecting rod.

15   Install the nuts and tighten them to the specified torque, working up to it in three

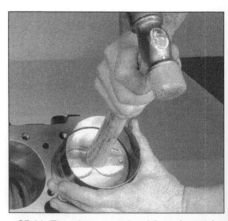

25.11 The piston can be driven (gently) into the cylinder bore with the end of a wooden hammer handle

steps. **Note:** *Use a thin-wall socket to avoid erroneous torque readings that can result if the socket is wedged between the rod cap and nut. If the socket tends to wedge itself between the nut and the cap, lift up on it slightly until it no longer contacts the cap. Do not rotate the crankshaft at any time during this operation.*

16   Remove the nuts and detach the rod cap, being very careful not to disturb the Plastigage.

17   Compare the width of the crushed Plastigage to the scale printed on the Plastigage envelope to obtain the oil clearance. Compare it to the Specifications to make sure the clearance is correct.

18   If the clearance is not as specified, the bearing inserts may be the wrong size (which means different ones will be required). Before deciding that different inserts are needed, make sure that no dirt or oil was between the bearing inserts and the connecting rod or cap when the clearance was measured. Also, recheck the journal diameter. If the Plastigage was wider at one end than the other, the journal may be tapered (refer to Section 19).

Final connecting rod installation

19   Carefully scrape all traces of the Plastigage material off the rod journal and/or bearing face. Be very careful not to scratch the bearing; use your fingernail or the edge of a credit card.

20   Make sure the bearing faces are perfectly clean, then apply a uniform layer of clean moly-base grease or engine assembly lube to both of them. You'll have to push the piston into the cylinder to expose the face of the bearing insert in the connecting rod - be sure to slip the protective hoses over the rod bolts first.

21   Slide the connecting rod back into place on the journal, remove the protective hoses from the rod cap bolts, install the rod cap and tighten the nuts to the specified torque. Again, work up to the torque in three steps.

22   Repeat the entire procedure for the remaining pistons/connecting rods.

23   The important points to remember are:

a)   *Keep the back sides of the bearing inserts and the insides of the connecting rods and caps perfectly clean when assembling them.*

b)   *Make sure you have the correct piston/rod assembly for each cylinder.*

c)   *The notch or mark on the piston must face the front of the engine.*

d)   *Lubricate the cylinder walls with clean oil.*

e)   *Lubricate the bearing faces when installing the rod caps after the oil clearance has been checked.*

24   After all the piston/connecting rod assemblies have been properly installed, rotate the crankshaft a number of times by hand to check for any obvious binding.

25   As a final step, the connecting rod end play must be checked. Refer to Section 13 for this procedure.

26   Compare the measured end play to the Specifications to make sure it's correct. If it was correct before disassembly and the original crankshaft and rods were reinstalled, it should still be right. If new rods or a new crankshaft were installed, the end play may be inadequate. If so, the rods will have to be removed and taken to an automotive machine shop for resizing.

26   Pre-oiling engine after overhaul

Refer to illustrations 26.3 and 26.5
Note: *This procedure does not apply to Buick V6 engines. For those engines, go on the Section 27.*

1   After an overhaul it is a good idea to pre-oil the engine before it is installed in the vehicle and started for the first time. Pre-oiling will reveal any problems with the lubrication system at a time when corrections can be made easily and will prevent major engine damage. It will also allow the internal engine parts to be lubricated thoroughly in the nor-

mal fashion without the heavy loads associated with combustion placed on them.

2   The engine should be completely assembled with the exception of the distributor and rocker arm covers. The oil filter and oil pressure sending unit must be in place and the specified amount of oil must be in the crankcase (see Chapter 1).

3   An old distributor will be needed for this procedure - a salvage yard should be able to supply one for a reasonable price. In order to function as a pre-oil tool, the distributor must have the gear on the lower end of the shaft ground off **(see illustration)** and, if equipped, the advance weights on the upper end of the shaft removed.

4   Install the pre-oil distributor in place of the original distributor and make sure the lower end of the shaft mates with the upper end of the oil pump driveshaft. Turn the distributor shaft until they are aligned and the distributor body seats on the block. Install distributor hold-down clamp and bolt.

5   Mount the upper end of the shaft in the chuck of an electric drill and use the drill to turn the pre-oil distributor shaft, which will drive the oil pump and circulate the oil throughout the engine **(see illustration)**. **Note:** *The drill must turn in a clockwise direction on Chevrolet built engines.*

6   It may take two or three minutes, but oil should soon start to flow out of all of the rocker arm holes, indicating that the oil pump is working properly. Let the oil circulate for several seconds, then shut off the drill motor.

7   Remove the pre-oil distributor, then install the rocker arm covers. The distributor should be installed after the engine is installed in the vehicle, so plug the hole with a clean cloth.

27   Initial start-up and break-in after overhaul

Warning: *Have a fire extinguisher handy when starting the engine for the first time.*

1   Once the engine has been installed in

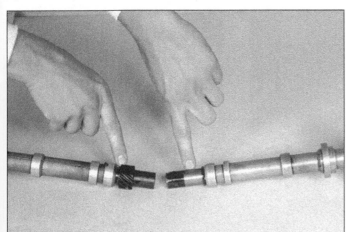

26.3 The pre-oil distributor (right) has the gear ground off and the advance weights (if equipped) removed

26.5 A drill motor connected to the modified distributor shaft drives the oil pump

the vehicle, double-check the engine oil and coolant levels.

2   With the spark plugs out of the engine and the ignition system disabled (see Section 3), crank the engine until oil pressure registers on the gauge or the indicator light goes out.

3   Install the spark plugs, hook up the plug wires and restore the ignition system functions (see Section 3).

4   Start the engine. It may take a few moments for the fuel system to build up pressure, but the engine should start without a great deal of effort. **Note:** *If backfiring occurs, recheck the valve timing and ignition timing.*

5   After the engine starts, it should be allowed to warm up to normal operating temperature. While the engine is warming up, make a thorough check for fuel, oil and coolant leaks.

6   Shut the engine off and recheck the engine oil and coolant levels.

7   Drive the vehicle to an area with minimum traffic, accelerate at full throttle from 30 to 50 mph, then allow the vehicle to slow to 30 mph with the throttle closed. Repeat the procedure 10 or 12 times. This will load the piston rings and cause them to seat properly against the cylinder walls. Check again for oil and coolant leaks.

8   Drive the vehicle gently for the first 500 miles (no sustained high speeds) and keep a constant check on the oil level. It is not unusual for an engine to use oil during the break-in period.

9   At approximately 500 to 600 miles, change the oil and filter.

10   For the next few hundred miles, drive the vehicle normally. Do not pamper it or abuse it.

11   After 2000 miles, change the oil and filter again and consider the engine broken in.

Chapter 3
Cooling, heating and air conditioning systems

Contents

Specifications

General
Radiator cap pressure rating .... 14 to 17 psi
Thermostat rating
   High altitude .... 180 to 185-degrees F
   Standard .... 192 to 198-degrees F

Torque specifications   **Ft-lbs** (unless otherwise indicated)
Water pump mounting bolts
  V6 engine
    200, 229 and 262 engines 30
    231 engine .... 84 in-lbs
  V8 engines 30
Thermostat housing cover
  V6 engine
  200, 229 and 262 engines 20
  231 engine .... 156 in-lbs
  V8 engines 30

1 General information

Engine cooling system

All vehicles covered by this manual employ a pressurized engine cooling system with thermostatically controlled coolant circulation. An impeller type water pump mounted on the front of the block pumps coolant through the engine. The coolant flows around each cylinder and toward the rear of the engine. Cast-in coolant passages direct coolant around the intake and exhaust ports, near the spark plug areas and in close proximity to the exhaust valve guides.

A wax pellet type thermostat is located in the thermostat housing near the front of the engine. During warm up, the closed thermostat prevents coolant from circulating through the radiator. When the engine reaches normal operating temperature, the thermostat opens and allows hot coolant to travel through the radiator, where it is cooled before returning to the engine.

The cooling system is sealed by a pressure type radiator cap. This raises the boiling point of the coolant, and the higher boiling point of the coolant increases the cooling efficiency of the radiator. If the system pressure exceeds the cap pressure relief value, the excess pressure in the system forces the spring-loaded valve inside the cap off its seat and allows the coolant to escape through the overflow tube into a coolant reservoir on 1972 and later models. When the system cools, the excess coolant is automatically drawn from the reservoir back into the radiator.

The coolant reservoir does double duty as both the point at which fresh coolant is added to the cooling system to maintain the proper fluid level and as a holding tank for overheated coolant.

This type of cooling system is known as a closed design because coolant that escapes past the pressure cap is saved and reused.

Heating system

The heating system consists of a blower fan and heater core located within the heater box, the inlet and outlet hoses connecting the heater core to the engine cooling system and the heater/air conditioning control head on the dashboard. Hot engine coolant is circulated through the heater core. When the heater mode is activated, a flap door opens

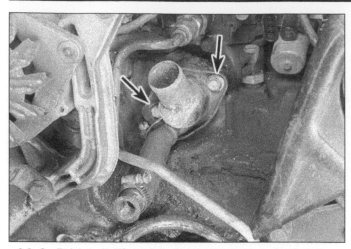

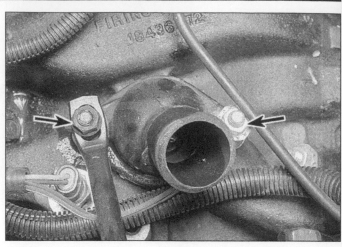

3.8 On Buick-built V6 models, the thermostat housing is below and behind the distributor (removed for clarity) - remove the bypass hose and the two housing bolts (arrows)

3.9 Remove the two thermostat housing mounting bolts (arrows) and the accessory support bracket if equipped

to expose the heater box to the passenger compartment. A fan switch on the control head activates the blower motor, which forces air through the core, heating the air.

Air conditioning system

The air conditioning system consists of a condenser mounted in front of the radiator, an evaporator mounted adjacent to the heater core, a compressor mounted on the engine, a filter-drier (accumulator) which contains a high pressure relief valve and the plumbing connecting all of the above.

A blower fan forces the warmer air of the passenger compartment through the evaporator core (sort of a radiator-in-reverse), transferring the heat from the air to the refrigerant. The liquid refrigerant boils off into low pressure vapor, taking the heat with it when it leaves the evaporator.

2   Antifreeze - general information

Warning: *Do not allow antifreeze to come in contact with your skin or painted surfaces of the vehicle. Rinse off spills immediately with plenty of water. Antifreeze is highly toxic if ingested. Never leave antifreeze lying around in an open container or in puddles on the floor; children and pets are attracted by it's sweet smell and may drink it. Check with local authorities about disposing of used antifreeze. Many communities have collection centers which will see that antifreeze is disposed of safely. Never dump used anti-freeze on the ground or into drains.*
Note: *Non-toxic coolant is available at most auto parts stores. Although the coolant is non-toxic, proper disposal is still required.*

The cooling system should be filled with a water/ethylene glycol based antifreeze solution, which will prevent freezing down to at least -20-degrees F, or lower if local climate requires it. It also provides protection against corrosion and increases the coolant

boiling point.

The cooling system should be drained, flushed and refilled at least every other year (see Chapter 1). The use of antifreeze solutions for periods of longer than two years is likely to cause damage and encourage the formation of rust and scale in the system. If your tap water is "hard", use distilled water with the antifreeze.

Before adding antifreeze to the system, check all hose connections, because antifreeze tends to leak through very minute openings. Engines do not normally consume coolant. Therefore, if the level goes down find the cause and correct it.

The exact mixture of antifreeze-to-water which you should use depends on the relative weather conditions. The mixture should contain at least 50 percent antifreeze, but should never contain more than 70 percent anti-freeze. Consult the mixture ratio chart on the antifreeze container before adding coolant. Hydrometers are available at most auto parts stores to test the ratio of antifreeze to water. Use antifreeze which meets the vehicle manufacturer's specifications.

3.10 Lift out the thermostat - note the location of the spring

3   Thermostat - check and replacement

Warning: *The engine must be completely cool when this procedure is performed.*

Check

1   Before assuming the thermostat is to blame for a cooling system problem, check coolant level (see Chapter 1), drivebelt tension (see Chapter 1) and temperature gauge (or light) operation.
2   If the engine takes a long time to warm up, the thermostat is probably stuck open. Replace the thermostat.
3   If the engine runs hot, use your hand to check the temperature of the upper radiator hose. If the hose is not hot, but the engine is, the thermostat is probably stuck in the closed position, preventing the coolant inside the engine from escaping to the radiator. Replace the thermostat.
4   If the upper radiator hose is hot, it means that the coolant is flowing and the thermostat is open. Consult the *Troubleshooting* section at the front of this manual for further diagnosis.

Replacement

Refer to illustrations 3.8, 3.9 and 3.10
5   Disconnect the negative battery cable.
6   Drain the cooling system (see Chapter 1) and remove the air cleaner housing.
7   Remove the upper radiator hose from the thermostat housing and disconnect any vacuum lines if the vehicle is equipped with a thermal vacuum switch.
8   On Buick V6 engines, remove the smaller coolant hose from the thermostat housing **(see illustrations)**.
9   Remove the bolts from the thermostat housing and detach the housing **(see illustration)**. Be prepared for some coolant to spill as the gasket seal is broken.
10   Remove the thermostat, noting the way it was installed **(see illustration)**.

4.4a On early models, remove the upper and lower screws securing the shroud to the radiator . . .

4.4b . . . then hang the shroud over the fan

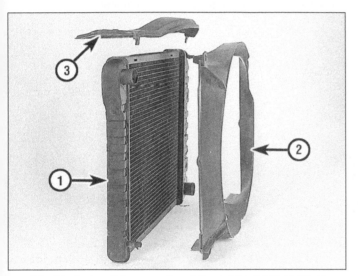

4.4c Typical radiator mount and shroud (early V8 engines)

1 Radiator
2 Radiator shroud
3 Mounting panel

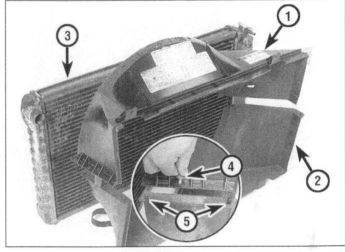

4.4d Typical radiator mount and shroud (later V8 engines)

1 Upper radiator shroud
2 Lower radiator shroud
3 Radiator
4 Bolt
5 Locating lugs

11   Remove all traces of gasket material from the sealing surfaces.
12   Apply gasket sealer to both sides of a new gasket and position it on the engine.
13   Install the thermostat, housing and bolts. Tighten the bolts to the specified torque.
14   Refill the cooling system (see Chapter 1).

4   Radiator - removal and installation

Warning: *The engine must be completely cool before beginning this procedure!*

Removal

Refer to illustrations 4.4a, 4.4b, 4.4c, 4.4d, 4.4e and 4.5
1   Disconnect the negative battery cable.
2   Drain the radiator, referring to Chapter 1.

3   Disconnect the radiator upper and lower hoses and the automatic transmission cooling lines if applicable.
4   On early models, remove the radiator shroud-to-radiator core support bolts and

move the shroud back over the fan **(see illustrations)**. On later models with a two-piece shroud, remove the screws attaching the upper and lower shroud sections along the sides **(see illustrations)**.

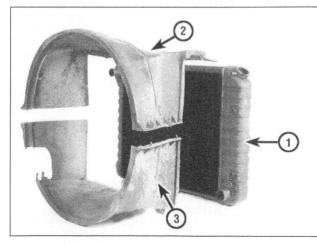

4.4e Typical radiator mount and shroud (later V6 engines)

1 Radiator
2 Upper radiator shroud
3 Lower radiator shroud

4.5 Remove the screws (arrows) retaining the upper shroud to the radiator support

5.10a On models with a clutch assembly, the fan is attached to the assembly or water pump hub with four nuts or bolts (arrow)

5   On early models, remove the upper metal panel at the top of the radiator. On later models remove the screws retaining the top of the shroud to the core support **(see illustration)**.

6   Lift the radiator straight up and out of the engine compartment. Be careful not to scratch the paint on the front panel. If coolant drips on any body paint, immediately wash it off with clear water as the antifreeze solution can damage the finish.

7   With the radiator removed, it can be inspected for leaks or damage. If in need of repairs, have a professional radiator shop or dealer perform the work, as special equipment and techniques are required.

8   Bugs and dirt can be cleaned from the radiator by using compressed air and a soft brush. Do not bend the cooling fins as this is done.

9   Inspect the rubber mounting pads which the radiator sits on and replace as necessary.

Installation

10   Lift the radiator into position making sure it is seated in the mounting pads.

11   Install the upper panel, shroud and hoses in the reverse order of removal.

12   Connect the negative battery cable and fill the radiator as described in Chapter 1.

13   Start the engine and check for leaks. Allow the engine to reach normal operating temperature (upper radiator hose hot) and add coolant until the level reaches the bottom of the filler neck.

14   Install cap with arrows aligned with the overflow tube.

5   Engine cooling fan and clutch - check and replacement

Check

1   Most vehicles covered by this manual are equipped with thermostatically controlled

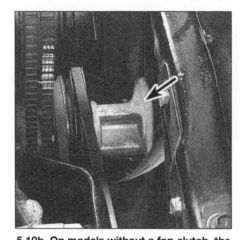

5.10b On models without a fan clutch, the fan and spacer are retained to the water pump hub by four long bolts (arrow)

fan clutches. Some non-air conditioned models are equipped with solid-hub type fans.

2   Begin the clutch check with a lukewarm engine (start it when cold and let it run for two minutes only).

3   Remove the key from the ignition switch for safety purposes.

4   Turn the fan blades and note the resistance. There should be moderate resistance, depending on temperature.

5   Drive the vehicle until the engine is warmed to operating temperature. Shut it off and remove the key.

6   Turn the fan blades and again note the resistance. There should be a noticeable increase in resistance.

7   If the fan clutch fails this check or is locked up, replacement is indicated. If excessive fluid is leaking from the hub or lateral play over 1/4-inch is noted, replace the fan clutch.

8   If any fan blades are bent, don't straighten them! The metal will be weakened and blades could fly off during engine operation. Replace fan with a new one.

5.12 Cooling fan clutch mounting details

Replacement

Refer to illustrations 5.10a, 5.10b and 5.12

Note: *A thermostatic fan clutch must remain in the "in car" position to prevent the fluid from leaking out.*

9   Remove the upper fan shroud.

10   Remove the fasteners holding the fan assembly to the water pump hub **(see illustrations)**.

11   Detach the fan and clutch assembly.

12   Unbolt the fan from the clutch **(see illustration)**.

13   Installation is the reverse of removal.

14   Tighten all fasteners securely.

6   Coolant temperature sending unit - check and replacement

Temperature warning light system check

Refer to illustrations 6.1a and 6.1b

1   If the light doesn't come on when the ignition switch is turned on, check the bulb. If the light stays on even with the engine cold,

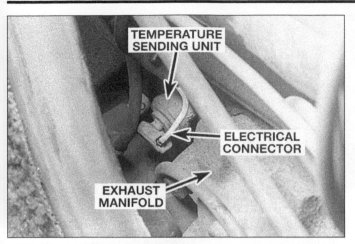

6.1a On most Chevrolet-built engines, the coolant temperature sending unit is located on the left side between the number one and number three spark plugs (arrow)

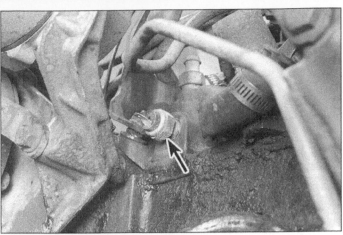

6.1b Coolant temperature sending unit location - Buick 231 V6 engine - at the front of the intake manifold, hidden by the distributor (distributor removed here for clarity)

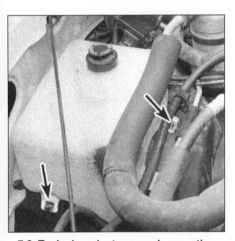

7.2 Typical coolant reservoir mounting details (late model shown)

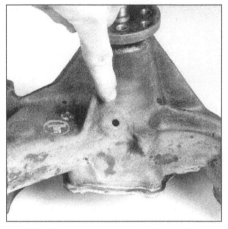

8.2 Water pump weep hole location

unplug the wire at the sending unit (see illustrations). If the light goes off, replace the sending unit. If the light stays on, the wire is grounded somewhere in the harness.

Temperature gauge system check

2   If the gauge is inoperative, check the fuse (see Chapter 12).
3   If the fuse is OK, unplug the wire connected to the sending unit and ground it with a jumper wire (see illustrations 6.1a and 6.1b). Turn on the ignition switch momentarily. The gauge should register at maximum. If it does, replace the sending unit. If it's still inoperative, the gauge or wiring may be faulty.

Sending unit replacement

4   Allow the engine to cool completely.
5   Unplug the wire connected to the sending unit.
6   Unscrew the sending unit and quickly install the new unit to prevent loss of coolant.
7   Connect the wire and check indicator operation.

7   Coolant reservoir - removal and installation

Refer to illustration 7.2
1   Disconnect the coolant overflow hose at the radiator neck.
2   Remove the screws attaching the reservoir to the inner fender (see illustration).
3   Lift the reservoir straight up, being careful not to spill any coolant on the paint.
4   Installation is the reverse of removal.

8   Water pump - check

Refer to illustration 8.2
1   Water pump failure can cause overheating of and serious damage to the engine. There are two ways to check the operation of the water pump while it's installed on the engine. If either of the following quick checks indicates water pump failure, it should be replaced immediately.
2   A seal protects the water pump impeller shaft bearing from contamination by engine coolant. If the seal fails, weep holes in the top

and bottom of the water pump snout (see illustration) will leak coolant under the vehicle. If the weep hole is leaking, shaft bearing failure will follow. Replace the water pump immediately.
3   Besides contamination by coolant after a seal failure, the water pump impeller shaft bearing can also be prematurely worn out by an improperly tensioned drivebelt. When the bearing wears out, it emits a high pitched squealing sound. If noise is coming from the water pump during engine operation, the shaft bearing has failed. Replace the water pump immediately.
4   To identify excessive bearing wear before the bearing actually fails, grasp the water pump pulley and try to force it up and down or from side to side. If the pulley can be moved either horizontally or vertically, the bearing is nearing the end of its service life. Replace the water pump.

9   Water pump - removal and installation

Warning: The engine must be completely cool before performing this procedure.

Removal

Refer to illustrations 9.6a, 9.6b, 9.7, 9.8a and 9.8b
1   Disconnect the negative battery cable.
2   Drain the radiator, referring to Chapter 1 if necessary.
3   Remove the radiator shroud for better access to the water pump.
4   Remove the engine cooling fan (see Section 5).
5   Remove the drivebelts from the water pump pulley. The number of belts will depend upon model, year, and equipment. Loosen the adjusting and pivot bolts of each component (air pump, alternator, air conditioning compressor, power steering pump) and push the component inward to loosen the belt

9.6a Note the location and position of all related mounting brackets for ease of reassembly

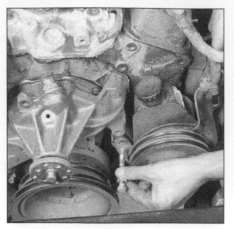

9.6b On some models, the long pivot bolt for the power steering pump may pass through the water pump

9.7 Remove the radiator and coolant hoses (arrows) attached to the water pump (Buick 231 V6 model shown)

enough to be removed from the water pump pulley (see Chapter 1).

6   Disconnect and remove all mounting brackets which are attached to the water pump. These may include the alternator, air conditioning compressor, and power steering brackets **(see illustrations)**.

7   Disconnect the lower radiator hose, heater hose, and bypass hose at the water pump **(see illustration)**.

8   Remove the remaining bolts which secure the water pump to the engine or front cover. Lightly tap on the water pump housing to break the gasket seal and remove it from the engine **(see illustrations)**.

Installation

9   If installing a new or rebuilt water pump, transfer all fittings and studs to the new water pump.

10   Clean the gasket surfaces of the front cover completely using a gasket scraper or putty knife.

11   Use a thin coat of gasket sealant on the new gaskets and install the new water pump. Place the pump into position on the front cover and secure loosely with the bolts. Do not tighten the bolts until all brackets have

been installed to their original position on the water pump.

12   Tighten the water pump bolts to the torque listed in this Chapter's Specifications.

13   Install the engine components in the reverse order of removal, tightening the appropriate fasteners securely.

14   Adjust all drivebelts to the proper tension (see Chapter 1).

15   Connect the negative battery cable and fill the radiator with the proper mixture of antifreeze and water. Start the engine and allow it to idle until the upper radiator hose gets hot. Check for leaks. With the engine hot, fill the coolant reservoir with coolant mixture to the correct level. Check coolant level periodically over the next few miles of driving.

10   Blower motor - removal and installation

1973 and earlier models

1   Disconnect the battery ground cable, then remove all wiring, hoses, etc., to the right-hand fender skirt.

2   Raise the front of the vehicle and sup-

port it securely on jackstands.

3   Remove all fender skirt attaching bolts except those fastened to the radiator support.

4   Pull outwards then down on the fender skirt and place a 2 x 4-inch wooden block between the skirt and fender.

5   Remove the blower motor to blower housing retaining screws and withdraw the blower motor through the fender skirt opening.

6   On some models it may be necessary to remove the blower fan retaining nut and separate the fan from the blower motor before the blower motor can be removed through the fender skirt opening.

7   Installation is the reverse of the removal procedure. Be sure the blower wheel is installed with the open end away from the motor and connect the cooling hose to the motor and housing.

1974 and later models

Refer to illustrations 10.10a, 10.10b and 10.10c

8   On 1974 and later models, the blower motor is accessible from under the hood.

9   Disconnect the electrical connector and

9.8a Remove the water pump bolts (arrows) noting the location of any special bolts or studs (Chevrolet built engine shown)

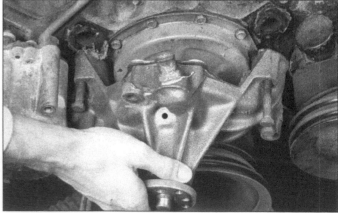

9.8b Tap on the water pump housing to break the gasket seal

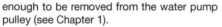

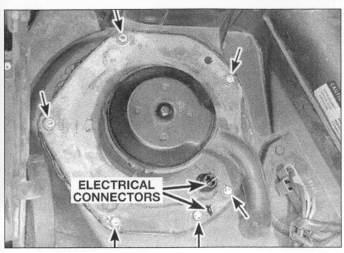

ELECTRICAL
CONNECTORS

10.10a Unplug the electrical connections and remove the screws (arrows) retaining the blower motor to the housing

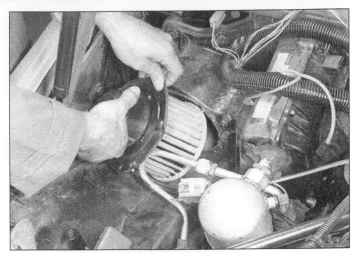

10.10b On early models, the blower motor is mounted horizontally

ground wire from the blower motor.
10   Remove the blower-to-case attaching screws and withdraw the blower motor **(see illustrations)**.
11   Installation is the reverse of removal.

11   Heater core - removal and installation

Warning: *The engine must be completely cool before performing this procedure.*

1977 and earlier models

1   Disconnect the negative battery cable.
2   Drain the cooling system (see Chapter 1).
3   Disconnect the heater hoses from the heater core tubes at the firewall.
4   Remove the heater case nuts from the engine side of the firewall.
5   Remove the glove box.
6   Detach the lower air duct and defroster duct from the heater unit.

7   Disconnect the electrical connectors, vacuum hoses and control cables from the heater unit. Mark the components so they can be installed in their original locations.
8   Remove the heater case-to-dash nuts/bolts and pull the heater case back until the studs clear the firewall. Remove the heater unit from the vehicle.
9   Separate the heater case, remove the heater core retaining screws and remove the core from the case.
10   Installation is the reverse of removal. Be sure to replace any damaged seals and seal the case with caulking.

1978 and later models
Refer to illustrations 11.13, 11.14, 11.17, 11.18, 11.19, and 11.20
11   Disconnect the negative battery cable.
12   Drain the cooling system (see Chapter 1).
13   Remove the five screws retaining the chrome trim strip at the base of the windshield, both wiper arms and the engine compartment seal **(see illustration)**.

10.10c Later model blower motors are mounted vertically near the right hood hinge

14   Remove the screws retaining the leaf screen to the cowl and the nuts retaining the trim brackets to the firewall **(see illustration)**.

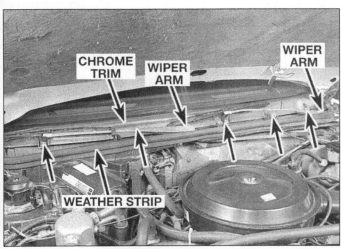

CHROME
TRIM

WIPER
ARM

WIPER
ARM

WEATHER STRIP

11.13 Remove the screws retaining the chrome trim at the bottom of the windshield, then remove the wiper arms and the hood seal

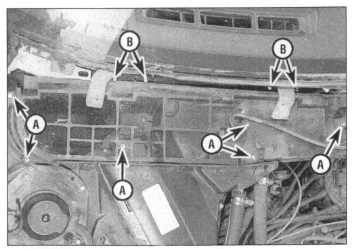

11.14 Remove the screws (A) retaining the leaf screen and the nuts (B) retaining the brackets

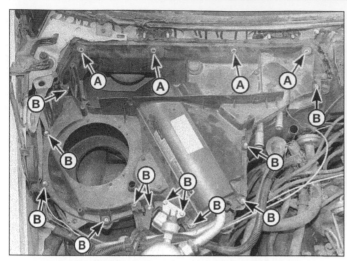

11.17 Remove the bolts (A), the screws (B) and separate the cover from the housing (non A/C equipped vehicles similar)

11.18 Pull the front of the cover up and guide it out toward the front of the engine

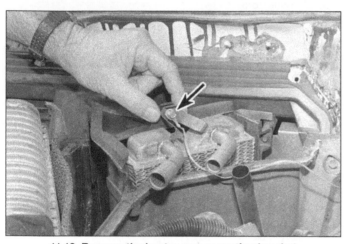

11.19 Remove the heater core mounting bracket and ground strap

11.20 Pull the heater core up and out of the case

15   Remove the blower motor (see Section 10).

16   Disconnect the heater hoses from the heater core tubes at the firewall.

17   Remove the mounting screws retaining the upper cover to the heater/air conditioning housing **(see illustration)**.

18   Separate the cover from the housing by pulling up on the front half of the cover and out toward the front of the car **(see illustration)**.

19   Disconnect the heater core retaining bolt and ground wire **(see illustration)**.

20   Remove the heater core by pulling it up and out of the housing **(see illustration)**.

21   Installation is the reverse of removal. Make sure to reseal the cover to the housing with body sealer. Refill the cooling system (see Chapter 1) and test for leaks.

---

12   Air conditioning and heating system - check and maintenance

---

Warning: *The air conditioning system is under high pressure. DO NOT loosen any fittings or remove any components until after the system has been discharged. Air conditioning refrigerant should be properly discharged into an EPA-approved recovery container at a dealer service department or an automotive air conditioning facility. Always wear eye protection when disconnecting air conditioning system fittings.*

Air conditioning system

Refer to illustrations 12.1a, 12.1b, 12.2, 12.8, 12.9 and 12.12

1   Two distinctly different types of air conditioning systems are used on the models covered by this manual. Very early models use the Pilot Operated Absolute (POA) valve system. The POA system consists of the compressor, condenser, receiver-drier, expansion valve, evaporator and POA valve **(see illustration)**. The POA valve controls the system pressure and evaporator temperature. On a POA system, the compressor is engaged anytime the system is operating. Later models use several variations of the Cycling Clutch Orifice Tube (CCOT) system. The

CCOT system consists of the compressor, condenser, accumulator, orifice tube, evaporator and pressure cycling switch. On a CCOT system, the compressor cycles on-and-off to control the system pressure and the orifice tube performs the duties of the expansion valve. The CCOT system can be identified by the presence of the accumulator near the evaporator case **(see illustration)**.

2   The following maintenance steps should be performed on a regular basis to ensure that the system continues to operate at peak efficiency:

   a) *Check the tension of the air conditioning compressor drivebelt and adjust it if necessary* (refer to Chapter 1).

   b) *Visually inspect the condition of the hoses, looking for any cracks, hardening and other deterioration.* **Note:** *Don't remove any hoses until the system has been discharged.*

   c) *Make sure the fins of the condenser aren't covered with foreign material, such as leaves or bugs. A soft brush and compressed air can be used to remove*

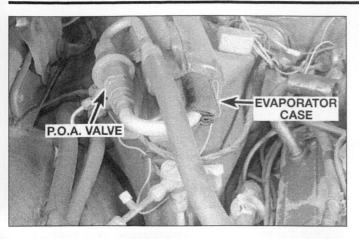

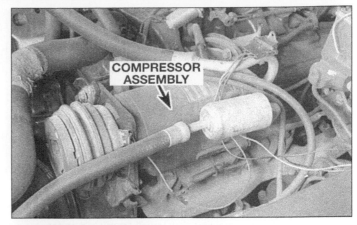

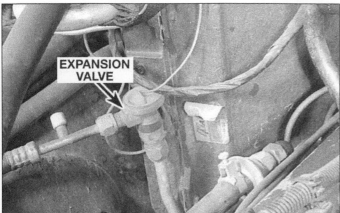

12.1a Components of a typical Pilot Operated Absolute (POA) valve system

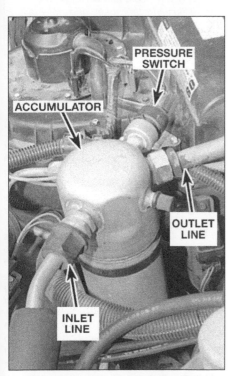

12.1b A Cycling Clutch Orifice Tube (CCOT) system can be identified by the presence of the accumulator near the evaporator housing

them. **Warning:** *Always wear eye protection when using compressed air!*

d) *Be sure the evaporator drain is open by slipping a wire into the evaporator housing drain occasionally* **(see illustration)**.

e) *Check the wiring for broken wires, damaged insulation, etc. Make sure the electrical connectors are clean and tight.*

f) *Maintain the correct refrigerant charge.*

g) *The air conditioning compressor should be run about 10 minutes at least once every month. This is especially important to remember during the winter months because long-term non-use can cause hardening of the seals.*

3   Leaks in the air conditioning system are best spotted when the system is brought up to operating temperature and pressure, by running the engine with the air conditioning ON for five minutes. Shut the engine off and inspect the air conditioning hoses and connections. Traces of oil usually indicate refrigerant leaks.

4   Because of the complexity of the air conditioning system and the special equipment required to effectively work on it, accurate troubleshooting of the system should be left to a professional technician.

5   The most common cause of poor cooling is simply a low system refrigerant charge. If a noticeable drop in cool air output occurs, the following quick check will

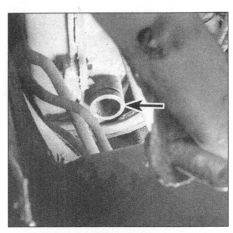

12.2 The evaporator drain (arrow) is located under the evaporator housing - occasionally check the drain for restrictions

help you determine if the refrigerant level is low. For more complete information on the air conditioning system, refer to the *Haynes Automotive Heating and Air Conditioning Manual.*

Checking the refrigerant charge

6   Warm the engine up to normal operating temperature.

7   Place the air conditioning temperature

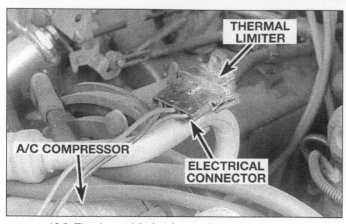

12.8 The thermal limiter is usually mounted on the compressor bracket

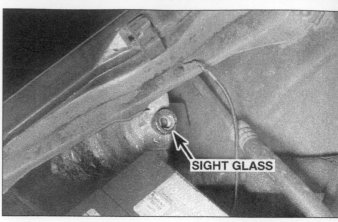

12.9 The sight glass is located on top of the receiver-drier

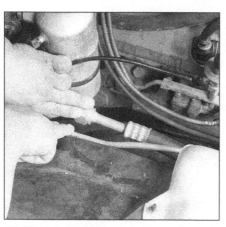

12.12 With the engine running and the air conditioning system On, feel the temperature of the evaporator inlet pipe and the evaporator outlet pipe/accumulator surface

selector at the coldest setting and put the blower at the highest setting. Open the doors (to make sure the air conditioning system doesn't cycle off as soon as it cools the passenger compartment).

Early models (POA system with receiver-drier and sight glass)

8   Most of the earlier POA systems use a thermal limiter circuit to protect the compressor **(see illustration)**. If the compressor clutch won't engage, check the thermal limiter fuse, if it's blown, replace it with a new one.
9   With the engine running and air conditioning ON, note the condition of the refrigerant in the sight glass **(see illustration)**. If the sight glass is clear and the system is cooling properly the system has the correct refrigerant charge. If the sight glass is clear and the system does not cool, the system is probably empty.
10   If occasional bubbles appear in the sight glass, this is an indication of either a low refrigerant charge or air in the refrigerant.
11   If a heavy stream of bubbles appear

in the sight glass, this indicates a large refrigerant leak. In such a case, do not operate the compressor until the fault has been corrected.

Later models (CCOT system with accumulator)

12   With the compressor engaged - the clutch will make an audible click and the center of the clutch will rotate. After the system reaches operating temperature, with your hands, feel the temperature of the evaporator inlet pipe and the evaporator outlet pipe/accumulator surface **(see illustration)**.
13   If the system is operating properly, the two surfaces will both feel about the same temperature and both should be colder than the outside air by the same amount.
14   If the evaporator inlet pipe feels cooler than the accumulator surface, the refrigerant charge is probably low.

All systems

15   Insert a thermometer in the center air distribution duct while operating the air conditioning system - the temperature of the output air should be 35 to 40-degrees F below the ambient air temperature (down to approximately 40-degrees F). If the air isn't as cold as it used to be, the system probably needs a charge. Further inspection or testing of the system is beyond the scope of the home mechanic and should be left to a professional.
16   Because of federal restrictions on the sale of R-12 refrigerant, it isn't practical anymore for refrigerant to be added by the home mechanic. When the system needs recharging, take the vehicle to a dealer service department or professional air conditioning shop for evacuation, leak-testing and recharging.

Heating system

17   If the carpet under the heater core is damp or smells of coolant, or if coolant vapors are coming through the vents, the heater core is leaking. Refer to Section 12 for the replacement procedure.
18   The air coming out of the heater vents isn't hot, the problem could stem from any of the following causes:

a) *The thermostat is stuck open, preventing the engine coolant from warming up enough to carry heat to the heater core. Replace the thermostat (see Section 3).*
b) *The heater water valve is blocked or malfunctioning. Replace the water valve.*
c) *A heater hose is blocked, preventing the flow of coolant through the heater core. Feel both hoses at the firewall. They should be hot, and when the heater is operating and the blower is on, the inlet hose should be hotter than the outlet hose. If one of the hoses is cool, there is an obstruction in either the hoses or the heater core. Detach the heater hoses and attach a garden hose to the inlet fitting of the core and backflush the core.*
d) *If flushing the hoses and core fails to remove the obstruction, replace the hoses and core.*

13 Air conditioning accumulator/receiver-drier - removal and installation

Refer to illustrations 13.3a and 13.3b
Warning: *The air conditioning system is under high pressure. DO NOT loosen any fittings or remove any components until after the system has been discharged. Air conditioning refrigerant should be properly discharged into an EPA-approved recovery container at a dealership service department or an automotive air conditioning repair facility. Always wear eye protection when disconnecting air conditioning system fittings.*
1   The accumulator/receiver-drier acts as a reservoir, dehumidifier and filter for the refrigerant. The receiver-drier (used on early POA systems) is the canister-shaped object mounted on the radiator support adjacent to the condenser. The accumulator (used on CCOT systems) is mounted near the evaporator case.
2   Have the air conditioning system discharged and the refrigerant recovered (see **Warning** above).
3   Disconnect the two refrigerant lines

13.3a The receiver-drier is mounted on the radiator support near the condenser - disconnect the refrigerant lines (arrows) from the unit

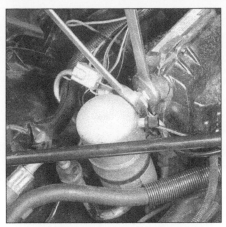

13.3b Use a back-up wrench when disconnecting the refrigerant lines from the accumulator to prevent damage to the lines or fittings

14.3 Disconnect the wiring connector at the front of the compressor

from the accumulator or receiver-drier **(see illustrations)**. Cap the open fittings immediately to prevent moisture from entering the system.

4   Remove the accumulator or receiver-drier from its bracket.

5   Installation procedures are the reverse of those for removal.

6   Have the system evacuated, charged and leak tested. If a new accumulator or receiver-drier was installed, add refrigerant oil according to the component manufacturer's instructions.

14 Air conditioning compressor - removal and installation

Refer to illustration 14.3

Warning: *The air conditioning system is under high pressure. DO NOT loosen any fittings or remove any components until after the system has been discharged. Air conditioning refrigerant should be properly discharged into an EPA-approved recovery container at a dealership service depart-*

ment or an automotive air conditioning repair facility. Always wear eye protection when disconnecting air conditioning system fittings.

1   Have the air conditioning system discharged and the refrigerant recovered (see **Warning** above).

2   Disconnect the negative battery cable.

3   Disconnect the electrical connector from the compressor clutch **(see illustration)**.

4   Remove the drivebelt (see Chapter 1).

5   Disconnect the refrigerant lines from the rear of the compressor. Plug the open fittings to prevent entry of dirt and moisture.

6   Unbolt the compressor from the mounting brackets and lift it out of the vehicle.

7   If a new compressor is being installed, follow the directions which come with the compressor regarding the draining of excess oil prior to installation.

8   Installation is the reverse of removal. Replace any O-rings with new ones specifically made for the purpose and lubricate them with refrigerant oil.

9   Have the system evacuated, recharged and leak tested by the shop that discharged it.

15 Air conditioning condenser - removal and installation

Refer to illustrations 15.5 and 15.6

Warning: *The air conditioning system is under high pressure. DO NOT loosen any fittings or remove any components until after the system has been discharged. Air conditioning refrigerant should be properly discharged into an EPA-approved recovery container at a dealership service department or an automotive air conditioning repair facility. Always wear eye protection when disconnecting air conditioning system fittings.*

1   Have the air conditioning system discharged and the refrigerant recovered (see **Warning** above).

2   Disconnect the negative battery cable.

3   Drain the cooling system (see Chapter 1).

4   Remove the radiator (see Section 4).

5   Disconnect the refrigerant lines from the condenser **(see illustration)**.

6   Remove the mounting bolts from the condenser brackets **(see illustration)**.

7   Lift the condenser out of the vehicle and plug the lines to prevent dirt and moisture

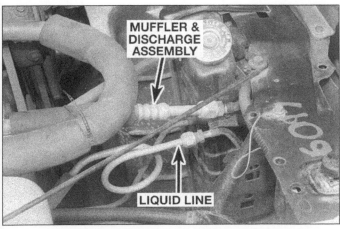

15.5 Typical air conditioning condenser refrigerant lines

15.6 Remove the condenser mounting bolts (arrows) and pull the condenser up and out - don't lose the rubber isolators at the bottom - some models may have bolts at the bottom as well

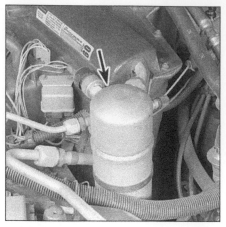

16.2 On a CCOT system, the expansion (orifice) tube is located in the evaporator inlet line (arrow) behind the accumulator

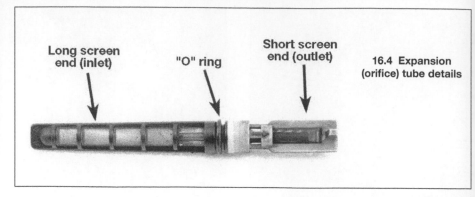

16.4 Expansion (orifice) tube details

from entering.

8   If the same condenser will be reinstalled, store it with the line fittings on top to prevent oil from draining.

9   If a new condenser is being installed, pour one ounce of refrigerant oil into the new condenser prior to installation.

10   Reinstall the components in the reverse order of removal. Be sure the rubber pads are in place under the condenser.

11   Have the system evacuated, recharged and leak tested by the shop that discharged it.

16   Air conditioning expansion (orifice) tube CCOT system - removal and installation

Refer to illustrations 16.2 and 16.4
Warning: *The air conditioning system is under high pressure. DO NOT loosen any fittings or remove any components until after the system has been discharged. Air conditioning refrigerant should be properly dis-* charged into an EPA-approved container at a dealership service department or an automotive air conditioning repair facility. Always wear eye protection when disconnecting air conditioning system fittings.

1   Have the air conditioning system discharged and the refrigerant recovered (see **Warning** above). Disconnect the negative battery cable.

2   The orifice tube is located in the evaporator inlet line at the fitting behind the accumulator **(see illustration)**.

3   Using a back-up wrench, loosen and separate the fittings.

4   The expansion (orifice) tube is a tube with a fixed-diameter orifice and a mesh filter at each end **(see illustration)**. When you separate the pipe at the fitting you will see one end of the orifice tube inside the pipe leading to the evaporator. Use needle-nose pliers to remove the orifice tube.

5   The orifice tube acts to meter the refrigerant, changing it from high-pressure liquid to low-pressure liquid. It is possible to reuse the orifice tube if:

a)   The screens aren't plugged with grit or foreign material

b)   Neither screen is torn

c)   The plastic housing over the screens is intact

d)   The brass orifice inside the plastic housing is unrestricted

6   Installation is the reverse of removal. Be sure to insert the expansion tube with the shorter end in first, toward the evaporator. **Caution:** *Always use a new O-ring when installing the orifice tube.*

7   Retighten the fitting and refrigerant line, then have the system evacuated, recharged and leak-tested by the shop that discharged it.

17   Air conditioner and heater control assembly -removal and installation

Refer to illustrations 17.3 and 17.5

1   Disconnect the negative battery cable.

2   Remove the trim bezel around the control panel (if equipped). On early models remove the ashtray.

3   Remove the screws and pull the control assembly out of the dashboard **(see illustration)**.

4   Label and then disconnect the control cables.

5   Disconnect the vacuum hose harness and the electrical connectors from the backside **(see illustration)**.

6   Remove the assembly from the dash.

7   Installation is the reverse of removal.

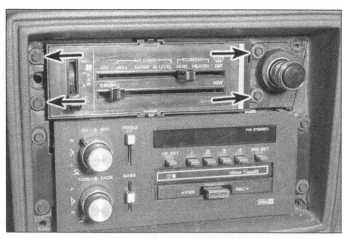

17.3 Remove the air conditioning and heater control assembly mounting screws (arrows)

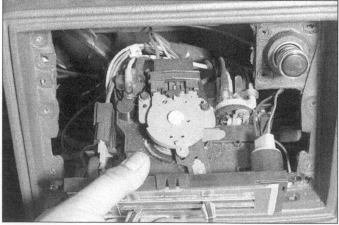

17.5 Pull the air conditioning and heater control assembly out from the dash and disconnect the vacuum harness, electrical connectors and control cables

Chapter 4 Part A
Fuel and exhaust systems

Contents

Specifications

Fuel pressure
Carbureted models
V6 and V8 engines ... 5 to 9 psi
Fuel injected (TBI) models ... 9 to 13 psi

Carburetor
Note 1: *The following specifications were compiled from the latest information available. If the specifications on the Vehicle Emissions Control Information label differ from those listed here, assume that the label is correct. Also, if the specifications in the rebuild kit for your particular carburetor differ from those listed here, assume that the specifications in the rebuild kit are correct.*
Note 2: *All dimensions listed are in inches.*

Applications
1970 through 1978	Rochester 2GV(2 barrel)
	Rochester 2GC and 2GE(2 barrel)
	Rochester 4MV(4 barrel)
	Rochester M4MC/CA(4 barrel)
1979 and later	Rochester M2MC, M2ME and E2ME (2 barrel)
	Rochester M4MC, M4ME and E4ME/E4MC (4 barrel)

Carburetor specifications - Rochester 2GV (1970)

Engine displacement (cu in)	350	400
Float level	23/32	23/32
Float drop	1-3/8	1-3/8
Accelerator pump stroke	1-17/32	1-17/32
Fast idle (rpm)	2200 to 2400 (one turn)	2200 to 2400 (one turn)
Choke rod	0.085	0.085
Choke vacuum break		
manual	M 0.215	0.215
automatic	A 0.200	
Choke unloader		
manual	M 0.275	0.325
automatic	A 0.325	
Main metering jet		
manual	0.059	0.069
auto	0.060	0.066
*Curb idle speed (rpm)		
manual	750	700
auto	600	600

M - Manual transmission
A - Automatic transmission

*Refer also to relevant text for intermediate speed levels to be set during adjustment procedure. Also see emissions label.

Note: *A twist drill of the required diameter can be used as a substitute for a gauge when making the carburetor adjustments.*

Carburetor specifications - Rochester 2GV (1971 through 1974)

		1971	1972	1973	1974
Engine displacement (cu in)		350	350	350, 400	350, 400
Float level		M 23/32 A 25/32	23/32	19/32	19/32
Float drop		1-3/8	1-9/32	1-9/32	1-9/32
Accelerator pump stroke		1-5/32	1-1/2	1-7/16	M 1-9/32 A 1-3/16
Fast idle (rpm)		2200 to 2400 (one turn)	M 2200 A 1850	1600	1600
Choke rod		0.100	0.100	M 0.200 A 0.245	M 0.200 A 0.245
Choke vacuum break		M 0.180 A 0.170	M 0.180 A 0.170	M 0.140 A 0.130	M 0.140 A 0.130
Choke unloader		0.325	0.325	M 0.250 A 0.325	M 0.250 A 0.325
Curb idle speed (rpm)	manual initial	700	1050	1000	1000
	final	600	900	900	900
	auto initial	580	650	630	650
	final	550	600	600	600
Maximum CO at idle		0.5%	0.5%	0.5%	0.5%

M - Manual transmission
A - Automatic transmission
*Refer also to relevant text for intermediate speed levels to be set during adjustment procedure. Also see emissions label.

Carburetor specifications - Rochester 2GC (1975 and 1976)

	1975	1976	
Engine displacement (cu in)	350	305	350
Float level	21/32	9/16	21/32
Float drop	31/32	1-9/32	31/32
Accelerator pump stroke	1-5/8	1-21/32	1-11/16
Fast idle (rpm)	1600	1600	1600
Choke rod	0.400	0.260	0.260
Choke vacuum break	0.130	0.140	0.130
Choke unloader	0.350	0.325	0.325

			1975	1976	
Engine displacement (cu in)			**350**	**305**	**350**
Curb idle (rpm)	manual	initial	1000	900	—
		final	900	800	—
	auto	initial	650	630	650
		final	600	600	600
Maximum CO at idle ....See emissions label				See emissions label	See emissions label

Carburetor specifications - Rochester 2GC (1977)

Float level			
Carburetor no.	17057108		19/32
	17057110		
	17057112		
	17057114		
All others			21/32
Float drop			1-9/32
Accelerator pump stroke			1-21/32
Choke rod			0.260
Choke unloader			0.325
Automatic choke coil			
Carburetor no.	17057108		Index
	17057110		
	17057112		
	17057114		
All others			1/2 notch lean
Choke vacuum break..			0.160

Carburetor specifications - Rochester 2GC/2GE (1978)

		231	305
Engine displacement (cu in)		**231**	**305**
Float level...			7/16
Carburetor no. 17058404		—	1/2
17058405		—	15/32
All others		—	
Float drop		1-5/32	1-9/32
Accelerator pump stroke		1-5/8	
Carburetor no. 17058104		—	1-21/32
17058105			
17058404			
17058405			
All others		—	1-17/32
Choke coil lever		0.120	0.120
Automatic choke coil		1 notch rich	
Carburetor no. 17058404		—	1/2 notch lean
17058405			
All others		—	Index
Choke rod		0.080	0.260
Choke vacuum break (choke side) (V8)			
Carburetor no. 17058404		—	0.140/0.160
17058405			
Carburetor no. 17058102		—	0.130/0.150
17058103			
All others		—	0.130/0.160
Choke vacuum break (choke side) (V6)			
Carburetor no. 17058447		0.110	—
Carburetor no. 17058143		0.040	—
Carburetor no. 17058147		0.100	—
All others		0.060	—
Choke vacuum break (throttle lever side)			
Carburetor no. 17058147		0.140	—
Carburetor no. 17058447		0.150	—
All others		0.110	—
Choke unloader		0.140	0.325
Idle speed		See emissions label	See emissions label

Carburetor specifications - Rochester M2MC/M2ME (1979)

Engine displacement (cu in)	200	267	231
Float level	9/32	9/32	11/32
Accelerator pump stroke	1/4 (inner)	1/4 (inner)	1/4 (inner)
Choke coil lever	0.120	0.120	0.120
Fast idle (bench setting)	2 turns	2 turns	2 turns
Fast idle cam	38°	38°	24.5°
Front vacuum break	27°	28°	—
Carburetor no. 17059196	—	—	23°
All others	—	—	19°
Rear vacuum break			
Carburetor no. 17059196	—	—	21°
All others	—	—	17°
Automatic choke coil	—	1 notch lean	—
Carburetor no. 17059130	Index	—	—
17059132			
All others	1 notch lean	—	—
Choke unloader	38°	38°	—
Carburetor no. 17059196	—	—	42°
All others	—	—	38°
Fast idle speed (rpm)**			
Manual	1300	1300	—
49-state auto	1600	1600	2200
Calif. auto	—	—	2200
Idle speed	See emissions label	See emissions label	See emissions label

***Note:** Recheck fast idle with carburetor in place on vehicle; readjust if necessary*
***If information on the Vehicle Emissions Control information label (in the engine compartment) differs from the information given here, assume that the label is correct*

Carburetor specifications - Rochester M2ME (1980)

Float level	
Carburetor no. 17080130	5/16
17080131	
17080132	
17080133	
All others	3/8
Accelerator pump stroke	5/16 (inner)
Choke coil lever	0.120
Fast idle (bench setting)*	2 turns
Fast idle cam	38°
Front vacuum break	25°
Choke unloader	38°
Idle speed	See emissions label
Fast idle speed	See emissions label

***Note:** Recheck fast idle with carburetor in place on vehicle; readjust if necessary*

Carburetor specifications - Rochester E2ME (1980)

Float level	5/16
Choke coil lever	0.120
Fast idle (bench setting)*	2 turns
Fast idle cam	24.5°
Front vacuum break	21°
Rear vacuum break	
Carburetor no. 17080403	30°
17080495	
All others	33°
Choke unloader	38°
Idle speed	See emissions label
Fast idle speed	See emissions label

***Note:** Recheck fast idle with carburetor in place on vehicle; readjust if necessary*

Carburetor specifications - Rochester E2ME (1981)

Float level		
Carburetor no. 17081194		5/16
All others		3/8
Choke coil lever		0.120
Fast idle (bench setting)*		2 turns
Fast idle cam		
Carburetor no. 17081194		24.5°
17081198		
All others		20°
Front vacuum break		
Carburetor no. 17081194		21°
17081198		28°
All others		25°
Rear vacuum break		
Carburetor no. 17081194		24°
17081198		
Choke unloader		
Carburetor no. 17081138		40°
17081140		
All others		38°
Idle speed		See emissions label
Fast idle speed		See emissions label

*__Note:__ *Recheck fast idle with carburetor in place on vehicle; readjust if necessary*

Carburetor specifications - Rochester 4MV (1970)

Engine displacement (cu in)	**350,400**	**454**
Float level	1/4	1/4
Accelerator pump stroke	5/16	5/16
Fast idle (rpm)	2400 (2 turns)	2400 (2 turns)
Choke rod	0.100	0.100
Choke vacuum break	M 0.275	M 0.275
	A 0.245	A 0.245
Choke unloader	0.450	0.450
Air valve spring	(350) 7/16	
	(400) 13/16	13/16
Air valve dashpot	0.020	0.020
*Curb idle manual	700	700
speed (rpm) auto	600	600

*__*Refer also to relevant text for intermediate speed levels to be set during adjustment procedure__*
All carburetors: Primary throttle bore 1-3/8   Secondary throttle bore 2-1/4   M - Manual transmission   A - Automatic transmission

Carburetor specifications - Rochester 4MV (1971 through 1974)

	1971	1972	1973		1974	
Engine displacement (cu in)	**350, 402, 454**	**350, 402, 454**	**350**	**454**	**350, 400**	**454**
Float level					(350) 1/4	3/8
	1/4	1/4	7/32	1/4	(400) 3/8	
Accelerator pump stroke	3/8	3/8	13/32	13/32	13/32	13/32
Fast idle (rpm)	M 1350	M 1350	M 1300	M 1300	M 1300	M 1300
	A 1500	A 1500	A 1600	A 1600	A 1600	A 1600
Choke rod	0.100	0.100	0.430	0.430	0.430	0.430
Choke vacuum break	M 0.275	(350) 0.215	0.250	0.250	0.230	0.250
	A 0.260	(others) 0.250				
Choke unloader	0.450	0.450	0.450	0.450	0.450	0.450
Air valve spring	13/16	13/16	—	—	—	—
Air wind up valve	—	—	1/2	11/16	7/8	7/16
Air valve dashpot	0.020	0.020	0.020	—	—	—
§ Curb Idle speed						
(rpm) manual initial	675	750*†	920*	925	950	850
final	600	750**	900‡	900‡	900‡	800‡
auto initial	630	(350) 630*	620*	625	650	630
		(400) 650*				
		(400) 650*				
		(454) 600*				
final	600	600**	600‡	600‡	600‡	600‡
Maximum CO at idle	1%	1%	1%	1%	0.5%	0.5%

__§ Refer to relevant text for intermediate speed levels to set during adjustment procedure__

Carburetor specifications - Rochester 4MV (1971 through 1974) (continued)

All carburetors: Primary throttle bore 1-3/8
                    Secondary throttle bore 2-1/4
M - Manual transmission
A - Automatic transmission
‡ With solenoid adjuster 500 rpm
† 350 cu in engine initial 1000 rpm, final 900 rpm
* On vehicles equipped with Air Injector Reactor system, turn mixture screw 1/4 turn from lean roll point for final carburetor setting
** Set low idle, using idle speed screw or solenoid Allen head screw (with solenoid de-energized), at 450 rpm

Carburetor specifications - Rochester M4MC/CA

		1975 M4MC/CA		1976 M4MC
Engine displacement (cu in)		**350 or 400**	**454**	**350 or 400**
Float level		15/32	17/32	13/32
Accelerator pump stroke		0.275	0.275	9/32
Fast idle (rpm)		1600	1000	1600
Choke rod		0.300 (350)	0.300	0.325
		0 325 (400)		
Choke vacuum break		Front 0.180	Front 0.200	0.185
		Rear 0.170	Rear 0.550	
Choke unloader		0.325	0.325	0.325
Spring wind up		7/8 (350)		7/8 (350)
		3/4 (400)	9/16	3/4 (400)
Air valve dashpot		0.015	0.015	0.015
Choke coil lever		0.120	0.120	0.120
Curb idle	manual initial	900	—	900
speed (rpm)*	final	800	—	800
	auto initial	650	650	650
	final	600	600	600

* If information on the Vehicle Emission Control Information label (in the engine compartment) differs from the information given here, assume that the label is correct

Carburetor specifications - Rochester M4MC (1977, 350 cu in)

Float level
    Carburetor no.   17057228 .... 13/32
    All others 15/32
Accelerator pump stroke
    Carburetor no.   17057582 .... 9/32 (outer)
                     17057584
    All others .... 9/32 (inner)
Choke coil lever 0.120
Fast idle (rpm) .... M 1300
                                                                        A 1600
Choke rod 0.325
Air valve dashpot 0.015
Choke vacuum break
    Carburetor no.   17057202 .... 0.160
                     17057203
                     17057204
    Carburetor no.   17057502 .... 0.165
                     17057504
    All others 0.180
Automatic choke coil
    Carburetor no.   17057203 .... 3 notches lean
                     17057211
    All others .... 2 notches lean
Choke unloader 0.280
Secondary lockout 0.015
Secondary closing 0.020
Spring wind up
    Carburetor no.   17057210 .... 1 turn
                     17057211
                     17057228
    All others .... 7/8 turn
Idle speed .... See emissions label

Carburetor specifications - Rochester M4MC (1979, 305 cu in)

Float level..	15/32
Accelerator pump stroke	
Carburetor no. 17059582	11/32
17059584	
All others	1/4
Choke coil lever	0.120
Fast idle (bench setting)*	2 turns
Fast idle cam	38°
Air valve dashpot	0.015
Front vacuum break	
Carburetor no. 17059502	28°
17059504	
17059222	
Carburetor no. 17059582	33°
17059584	
All others	27°
Choke unloader	
Carburetor no. 17059582	46°
17059584	
All others	38°

*Note: *recheck fast idle with carburetor in place on vehicle; readjust if necessary*

Carburetor specifications - Rochester M4ME (1980 and 1981)

Engine displacement (cu in)	231	305
Float level..	3/16	7/16
Accelerator pump stroke	9/32	1/4
Choke coil lever	0.120	0.120
Fast idle (bench setting)*	4 turns	4 turns
Fast idle cam	14.5°	20°
Air valve dashpot	0.025	0.025
Front vacuum break...	16°	27°
Rear vacuum break	16°	—
Choke unloader...	30°	38°
Secondary lockout..	0.015	0.015
Secondary closing	0.020	0.020
Spring wind up	9/16 turn	7/8 turn
Idle speed	See emissions label	See emissions label

*Note: *Recheck fast idle with carburetor in place on vehicle; readjust if necessary*

Carburetor specifications - Rochester E4ME (1980)

Engine displacement (cu in)	231	305
Float level..	3/8	1/2
Choke coil lever	0.120	0.120
Fast idle (bench setting)*	4 turns	4 turns
Fast idle cam	14.5°	20°
Air valve dashpot	0.015	0.015
Front vacuum break...	19°	24°
Rear vacuum break	23°	30°
Choke unloader...	38°	38°
Secondary lockout..	0.015	0.015
Secondary closing	0.020	0.020
Spring wind up	9/16 turn	7/8 turn
Idle speed	See emissions label	See emissions label

Note: *Recheck fast idle with carburetor in place on vehicle, readjust if necessary*

Carburetor specifications - Rochester E4ME (1981)

Engine displacement (cu in)	231	305
Float level..	1/4	11/32
Choke coil lever	0.120	0.120
Fast idle (bench setting)*	4-1/2 turns	4-1/2 turns
Fast idle cam	24.5°	20°
Air valve dashpot	0.025	0.025
Front vacuum break...	19°	26°
Rear vacuum break	17°	—
Choke unloader	38°	38°

Automatic choke coil

Carburetor no. 17059216	1 notch lean	
17059217		
17059582		
17059584		
All others	2 notches lean	
Secondary lockout..	0.015	0.015
Secondary closing	0.020	0.020
Spring wind up	9/16 turn	7/8 turn
Idle speed	See emissions label	See emissions label

Note: *Recheck fast idle with carburetor in place on vehicle readjust if necessary*

Carburetor specifications - Rochester E2ME, E2MC and E2SE (1982 and later)

Float level
   Carburetor number... 5/16 in
     17082497
     17083192
     17083193
     17083497
     17084191
     17084193
     17084194
     17084195
   Carburetor number... 3/8 in
     17082130
     17082132
     17082140
     17083130
     17083132
Choke rod
   Carburetor number 17082497... 24.5°
   Carburetor number... 20°
     17082131
     17082132
     17082138
     17082140
     17083130
     17083132
   Carburetor number... 18°
     17083190
     17083192
     17084191
   Carburetor number... 17°
     17083193
     17084193
     17084195
Vacuum break
   Carburetor number... 28° (front) 24° (rear)
     17082497
     17083190
     17083192
   Carburetor number... 27° (front only)
     17082130
     17082132
     17082138
     17084193
     17084194
     17084195
   Carburetor number 17083193... 23° (front) 28° (rear)
   Carburetor number 17082132... 27° (front) 24° (rear)
   Carburetor number 17084191... 28° (front) 25° (rear)
Choke unloader
   Carburetor number... 38°
     17082130
     17082132
     17082138
     17082140
     17083130

Carburetor number .... 32°
  17082497
  17083190
  17083192
Carburetor number 17083193 .... 27°
Carburetor number 17084191 .... 35°
Curb and fast idle speed .... See *Emissions Control Information* label

Carburetor specifications - Rochester M4ME, E4ME and E4MC (1982 and later)

Float level
  Carburetor number .... 11/32 in
    17082202
    17082204
    17083202
    17083204
    17083207
    17083210
    17083216
    17083218
    17083236
    17084201
    17084205
    17084208
    17084209
    17085202
    17085203
    17085204
    17085207
    17085218
    17086003
    17086004
    17086005
    17086006
    17086008
    17086040
    17086077
    17087129
    17087130
    17087132
    17087306
    17088115
  Carburetor number .... 7/16 in
    17083506
    17083508
    17083524
    17083526
    17084507
    17084509
    17084525
    17084527
    17085502
    17085503
    17086506
    17085508
    17085524
    17085526
Choke rod
  Carburetor number .... 20°
    17082202
    17082204
    17083202
    17083204
    17083216
    17083218
    17083506
    17083508
    17083524
    17083526
    17084201

Choke rod
    Carburetor number...   20°
        17084208
        17084210
        17084507
        17084509
        17084527
        17084529
        17085202
        17085203
        17085204
        17085218
        17085502
    Carburetor number...   20°
        17085503
        17085506
        17085508
        17085524
        17085526
        17086003
        17086004
        17086005
        17086006
        17086040
        17087129
        17087130
        17087132
        17087306
    Carburetor number...   38°
        17083203
        17083207
        17084205
        17084209
        17085207
    Carburetor number...   14°
        17086008
        17086077
        17088115
Vacuum break
    Carburetor number...   26° (front only)
        17082202
        17082204
    Carburetor number...   26° (front) 36° (rear)
        17085502
        17085503
    Carburetor number...   27° (front only)
        17085202
        17085203
        17085204
        17085207
        17085218
        17086003
        17086004
        17086005
        17086006
        17086040
        17087129
        17087130
        17087132
        17087306
    Carburetor number...   27° (rear only)
        17083202
        17083203
        17083207
        17083216
        17083218
        17083236

Carburetor number... 27° (front) 36° (rear)
    17083506
    17083508
    17085506
    17085508
Carburetor number... 25° (front) 36° (rear)
    17084525
    17084527
    17085524
    17085526
Carburetor number... 25° (front) 43° (rear)
    17086008
    17086077
    17088115
Choke unloader
Carburetor number... 39°
    17085502
    17085503
Carburetor number... 38°
    17082202
    17082204
    17083202
    17083204
    17083207
    17083216
    17083218
    17083230
    17084201
    17084205
    17084208
    17084209
    17084210
    17087129
    17087130
    17087132
Carburetor number... 36°
    17083506
    17083508
    17083524
    17083526
    17084507
    17084509
    17084525
    17084527
Carburetor number... 35°
    17086008
    17088115
Carburetor number... 32°
    17087129
    17087132
    17087306
Curb and fast idle speeds.. See *Emissions Control information label*

Torque specifications

In-lb (unless otherwise indicated)

Two barrel carburetor (Rochester)

Throttle body to bowl	72
Bowl cluster	46
Fast idle cam	58
Metering jet	40
Choke lever	14
Choke housing to throttle body	46
Choke housing cover	26
Air horn to bowl	46
Vacuum break unit	26
Choke shaft	14
Fuel inlet nut	33 ft-lbs
Fuel inlet needle seat	45

Four barrel carburetor (Rochester)

Throttle body to bowl	46
Choke lever	14
Choke housing	46
Choke housing cover	26
Air horn to bowl	
large	46
small	26
Air horn to throttle body	46
Choke lever	14
Vacuum break unit	26
Solenoid bracket	71
Fuel inlet nut	33 ft-lbs

TBI mounting bolts	132 in-lbs

Manifold bolts	**ft-lb**
Intake	
except 231 V6	30
231 V6	45
Exhaust	
except 350 engine inner bolts	20
350 engine inner bolts	30
Exhaust pipe to manifold bolts	15

Catalytic converter

Fill plug	50

1   General information

Fuel system

The fuel system consists of a rear-mounted gas tank, a fuel pump, a carburetor or fuel injection system and the fuel feed and return lines between the engine and the tank. On carburetor equipped models, the fuel pump is mounted on the engine; on fuel injected models it's mounted in the gas tank.

Several types of carburetors are used over the production life of these models. 1985 and later models equipped with a 4.3L V6 engine are equipped with Throttle Body Injection (TBI), type of fuel injection.

Exhaust system

The exhaust system consists of manifolds, pipes, mufflers and, on some models, resonators which direct the exhaust gases to the rear of the vehicle. Later models are equipped with a catalytic converter, which is part of the emissions system (see Chapter 6).

2   Fuel pump/fuel pressure - check

Warning: *Gasoline is extremely flammable, so take extra precautions when you work on any part of the fuel system. Don't smoke or allow open flames or bare light bulbs near the work area, and don't work in a garage where a natural gas-type appliance (such as a water heater or a clothes dryer) with a pilot light is present. Since gasoline is carcinogenic, wear latex gloves when there's a possibility of being exposed to fuel, and, if you spill any fuel on your skin, rinse it off immediately with soap and water. Mop up any spills immediately and do not store fuel-soaked rags where they could ignite. When you perform any kind of work on the fuel system, wear safety glasses and have a Class B type fire extinguisher on hand.*

Carbureted models

1   The fuel pump is mounted on the front of the engine. It's permanently sealed and isn't serviceable or rebuildable.
2   Before testing the fuel pump, check all fuel hoses and lines as well as the fuel filter (see Chapter 1).
3   Disconnect the primary (BAT) wire from the ignition coil or distributor (see Chapter 5). Isolate the wire aside so it does not contact any metal component (the wire is "hot" with the ignition key On).
4   Disconnect the fuel line at the carburetor.

Volume check

5   To check the fuel pump volume, connect a hose to the end of the fuel line and place the open end of the hose in a graduated plastic container. Volume marks should be clearly visible on the container.
6   Crank the engine and allow gasoline to flow into the container. At the end of 15 seconds, note the volume of fuel in the container. The fuel pump should output a minimum of 1/2-pint in 15 seconds.
7   If the volume is below the specified amount, attach an auxiliary fuel supply to the inlet side of the fuel pump. A small gas can with a hose forced tightly into the cap can be used as an auxiliary fuel supply. This will eliminate the possibility of a clogged tank and/or delivery line. Repeat the test and check the volume. If the volume has changed or is now normal, the fuel lines and/or tank are clogged. If the volume is still low, the fuel pump must be replaced with a new one.

Pressure check

8   Connect a fuel pressure gauge with a flexible hose to the end of the fuel line (the gauge must be held about 16-inches above the fuel pump, so make sure the hose is long enough). Make sure the inside diameter of the hose is no smaller than the diameter of the fuel line.
9   Crank the engine over several revolutions. Note the fuel pressure on the gauge. It should be within the Specifications listed at the front of this Chapter.

Fuel injected (TBI) models

10   TBI models are equipped with an electric fuel pump mounted inside the fuel tank. If the fuel system fails to deliver the proper amount of fuel, or any fuel at all, to the Throttle Body Injection system, inspect it as follows.
11   Always make certain there's fuel in the tank. Check the fuel filter and replace it if necessary.
12   Check for leaks at the threaded fittings at both ends of the fuel lines (see Chapter 1). Tighten any loose connections. Inspect all hoses for flat spots and kinks which would restrict the flow of fuel.

Fuel pump operational check

Refer to illustrations 2.15 and 2.18

13   If you suspect a problem with the fuel pump, verify that it actually runs. Have an assistant turn the ignition switch to On while you listen at the fuel tank - you should hear a whirring as the pump runs. The ECM will shut the pump off after 2 seconds, so listen carefully and cycle the key On and Off several times, if necessary.
14   If the pump doesn't run (makes no sound), continue with the Operational check. If the fuel pump runs, proceed to the Pressure check.

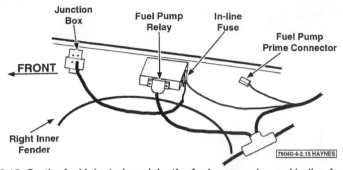

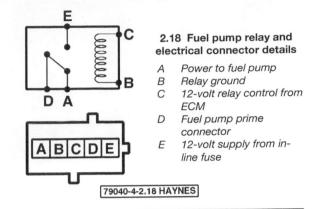

2.18 Fuel pump relay and electrical connector details

A   *Power to fuel pump*
B   *Relay ground*
C   *12-volt relay control from ECM*
D   *Fuel pump prime connector*
E   *12-volt supply from in-line fuse*

2.15 On the fuel injected models, the fuel pump relay and in-line fuse are mounted on a bracket on the right inner fender panel - the fuel pump prime connector is taped to the wiring harness

15   Check the fuel pump in-line fuse located in a fuse holder attached to a bracket on the right inner fender **(see illustration)**. If the fuse is blown, replace the fuse and see if the pump works.

16   Turn the ignition key off. Using a fused jumper wire, connect the positive battery terminal to the fuel pump test terminal **(see illustration 2.15)** and listen for the sound that indicates the fuel pump is running.

17   If the pump does not run, disconnect the fuel pump connector at the fuel tank and check for battery power at the tan wire terminal of the harness connector (with power applied to the test terminal). If no power is present at the fuel pump connector at the fuel tank, check for an open circuit between the relay and the fuel pump. Check for continuity to chassis ground at the fuel pump connector black wire terminal. If power is present, the ground circuit is good and the fuel pump does not run when connected, replace the fuel pump.

18   If the pump runs with power applied to the test terminal but not under normal operation, disconnect the fuel pump relay electrical connector and check for battery voltage on terminal E (hot at all times) and terminal C (for 2 seconds with the ignition key On) **(see illustration)**. Check for continuity to ground

on terminal B. If there is no voltage at terminal C for 2 seconds when the ignition key is turned On, the circuit from the relay to the ECM is open or the ECM is defective.

19   If the circuits are good, replace the fuel pump relay.

Pressure check

Note: *In order to perform the fuel pressure test, you will have to obtain a fuel pressure gauge and adapter set for the Throttle Body Injection system.*

20   Disconnect the fuel feed hose at the inlet fitting on the TBI unit and attach the fuel pressure gauge inline with a T-fitting.

21   With the ignition off, connect a fused jumper wire from the positive battery terminal to the fuel pump test terminal. The fuel pump test terminal is taped to the fuel pump relay wiring harness on the right inner fender **(see illustration 2.15)**.

22   Note the pressure reading. If the pressure is within specifications, no further testing is necessary.

23   If the pressure is higher than specified, Check for a restricted fuel return line. If the line is okay, replace the fuel pressure regulator (See Section 20).

24   If the pressure was less than specified, slowly pinch the return hose shut and note

the pressure. If the pressure rises above the specified pressure, replace the fuel pressure regulator (See Section 20).

25   If the pressure is still less than specified, check for a restricted fuel line, including the fuel filter and the pump inlet filter. If no restriction is found, replace the fuel pump.

26   After testing is complete, relieve the fuel system pressure and remove the fuel gauge.

3   Fuel pump - removal and installation

Warning: *Gasoline is extremely flammable, so extra precautions must be taken when working on any part of the fuel system. See the **Warning** in Section 2.*

Mechanical pump (carbureted models)

Refer to illustrations 3.1, 3.2a, 3.2b, 3.3 and 3.5

1   To remove the pump, remove the fuel inlet and outlet lines. Use two wrenches to prevent damage to the pump and connections **(see illustration)**.

2   Remove the fuel pump mounting bolts,

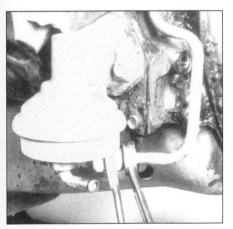

3.1 When detaching the fuel lines from a mechanical pump, use two wrenches so you don't bend or kink the metal fuel lines or the pipes protruding from the pump

3.2a To remove a mechanical pump from the engine, remove the two mounting bolts . . .

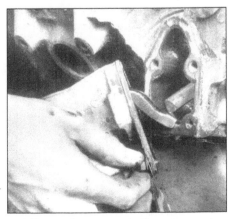

3.2b . . . then separate the pump and adapter plate from the block (be sure to remove all old gasket material from the pump and block mating surfaces) (Chevrolet V8 engine shown)

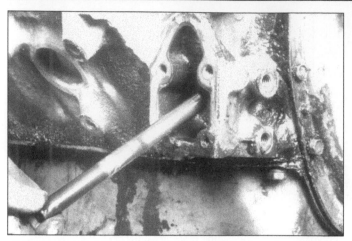

3.3 The pushrod (on smallblock Chevrolet-built engines) is easy to pull out once the pump and adapter plate are removed

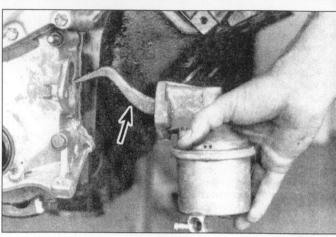

3.5 Make sure the rocker arm (arrow) contacts the drive eccentric properly (Buick 231 V6 engine)

the pump and the gasket **(see illustrations)**.
3   On Chevrolet built engines, remove the pipe plug or the pump adapter plate (if equipped) and remove the fuel pump pushrod **(see illustration)**.
4   Coat the pushrod with heavy grease to hold it in position and install it into the engine block. On Chevrolet built engines, install the the pipe plug or adapter plate with a new gasket.
5   Install the fuel pump using a new gasket. Make sure the rocker arm contacts the drive eccentric or pushrod properly **(see illustration)**. Use gasket sealant on the bolt threads.
6   Connect the fuel lines, start the engine and check for leaks.

Electric pump (fuel-injected models)

Refer to illustrations 3.10 and 3.12
7   Disconnect the cable from the negative battery terminal.
8   Remove the fuel tank (see Section 4).
9   Remove the fuel pump/sending unit assembly (see Section 6).

10   Remove and inspect the strainer on the lower end of the fuel pump **(see illustration)**. If it's dirty, replace it.
11   Disconnect the electrical connector from the fuel pump.
12   Remove the pump from the sending unit by pushing the fuel pump up into the rubber coupler and sliding the bottom of the pump away from the bottom support **(see illustration)**. After the pump is clear of the bottom support, pull it out of the rubber coupler. Care should be taken to prevent damage to the rubber insulator during removal.
13   Inspect the rubber coupler and sound insulator and replace them if necessary. Reassemble the fuel pump onto the sending unit assembly.
14   Insert the fuel pump/sending unit assembly into the fuel tank.
15   Turn the inner lock ring clockwise until the locking cams are fully engaged by the retaining tangs. **Note:** *If you have installed a new O-ring type rubber gasket, it may be necessary to push down on the inner lock ring until the locking cams slide under the retaining tangs.*
16   Install the fuel tank (see Section 5).

4   Fuel tank - removal and installation

Refer to illustrations 4.8a and 4.8b
Warning: *Gasoline is extremely flammable, so extra precautions must be taken when working on any part of the fuel system. See the* **Warning** *in Section 2.*
1   Remove the fuel filler neck cap to relieve fuel tank pressure.
2   Detach the cable from the negative terminal of the battery.
3   If the tank is full or nearly full, use a hand-operated pump to remove as much fuel through the filler tube as possible (if no such pump is available, you can siphon the tank at the fuel feed line after raising the vehicle - see Step 6).
4   Raise the vehicle and place it securely on jackstands.
5   Disconnect the fuel lines, the vapor return line and the fuel filler neck. **Note:** *The fuel feed and return lines and the vapor return line are three different diameters, so reattachment is simplified. If you have any doubts, however, clearly label the three lines and their*

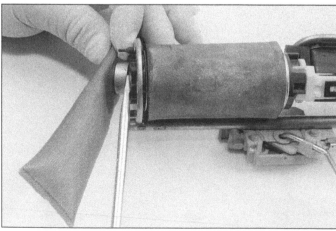

3.10 Use a small screwdriver to separate the strainer from the bottom of the fuel pump inlet

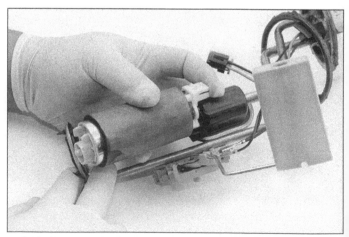

3.12 Separate the fuel pump from the sending unit assembly

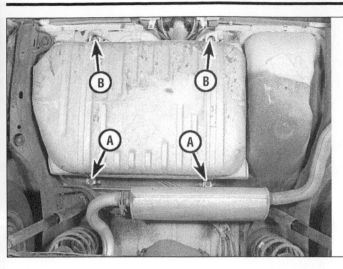

4.8a Typical fuel tank mounting

A   Mounting strap bolts
B   Mounting strap hinges

respective inlet or outlet pipes. Be sure to plug the hoses to prevent leakage and contamination of the fuel system.

6   If fuel remains in the tank, siphon the fuel from the tank at the fuel feed line (not the return line). **Warning:** *DO NOT start the siphoning action by mouth! Use a siphoning kit, available at most auto parts stores.*

7   Support the fuel tank with a floor jack or other suitable means. Place a sturdy plank between the jack head and the fuel tank to protect the tank.

8   Disconnect both fuel tank retaining straps and pivot them down until they are hanging out of the way **(see illustrations)**.

9   Lower the tank far enough to disconnect the electrical wires and ground strap from the fuel pump/fuel gauge sending unit. On early models, the wire should be disconnected from inside the trunk compartment and then fed through the trunk floorpan with the rubber grommet pushed out of place.

10   Remove the tank from the vehicle.

11   Installation is the reverse of removal.

---

5   Fuel tank cleaning and repair - general information

1   Any repairs to the fuel tank or filler neck should be carried out by a professional who has experience in this critical and potentially dangerous work. Even after cleaning and flushing of the fuel system, explosive fumes can remain and ignite during repair of the tank.

2   If the fuel tank is removed from the vehicle, it should not be placed in an area where sparks or open flames could ignite the fumes coming out of the tank. Be especially careful inside garages where a natural gas-type appliance is located, because the pilot light could cause an explosion.

---

6   Fuel level sending unit - check and replacement

Warning: *Gasoline is extremely flammable, so take extra precautions when you work on*

any part of the fuel system. See **Warning** in Section 2.

Check

Refer to illustration 6.2

1   Remove the fuel level sending unit (see below).

2   Using an ohmmeter, connect the positive probe onto the fuel level sending unit electrical connector terminal and the negative probe onto the sending unit body **(see illustration)**.

3   Check the resistance of the sending unit as you move the float from the empty position to the full position. The resistance of the sending unit should be approximately zero ohms empty and approximately 90 ohms full. The resistance should change smoothly with the movement of the float arm.

4   If the readings are incorrect or there is very little change in resistance as the float travels from full to empty, replace the fuel level sending unit assembly.

Replacement

Refer to illustrations 6.8 and 6.9

5   Disconnect the cable from the negative battery terminal.

6   Remove the fuel tank (see Section 4).

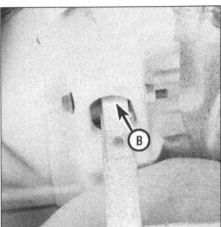

4.8b Close-up of mounting strap bolts (A) and hinges (B)

7   The fuel pump/sending unit assembly is located inside the fuel tank. It is held in place by a cam lock ring mechanism consisting of an inner ring with three locking cams and an outer ring with three tangs. The outer ring is welded to the tank and can't be turned.

8   To unlock the fuel pump/sending unit assembly, turn the inner ring counterclockwise until the locking cams are free of the

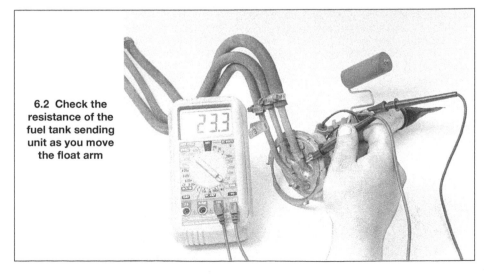

6.2 Check the resistance of the fuel tank sending unit as you move the float arm

6.8 Using a hammer and brass punch, tap the lock ring counterclockwise to loosen it

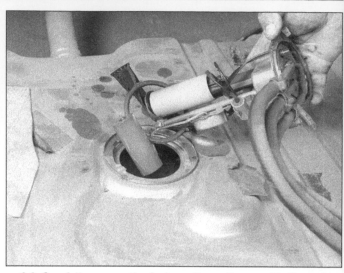

6.9 Carefully remove the fuel sending unit assembly through the opening

tangs **(see illustration)**. A special tool is available to loosen the lock ring, but if the tool is not available a hammer and brass punch can be used. **Warning:** *If using a punch to loosen the lock ring, use a brass punch. Do not use a steel punch - a spark could cause an explosion!*

9   Lift the fuel pump/sending unit assembly from the fuel tank **(see illustration)**. **Caution:** *The fuel level float and sending unit are delicate. Do not bump them against the tank during removal or the accuracy of the sending unit may be affected.*

10   Inspect the condition of the rubber gasket around the opening of the tank. If it is dried, cracked or deteriorated, replace it.

11   On TBI models, remove the fuel pump from the sending unit assembly (see Section 3).

12   Installation is the reverse of removal.

7   Fuel lines and fittings - general information

Warning: *Gasoline is extremely flammable, so extra precautions must be taken when working on any part of the fuel system. See the Warning in Section 2.*

1   The fuel lines on these vehicles are generally made of metal with short lengths of rubber hose connecting critical flex points such as the tank at the rear of the vehicle and fuel pump or TBI unit at the front of the vehicle.

2   The fuel lines extend from the fuel tank to the engine compartment. The lines are secured to the underbody with clip and screw assemblies. Both fuel feed lines must be occasionally inspected for leaks, kinks and dents.

3   If evidence of dirt is found in the system or fuel filter during disassembly, the line should be disconnected and blown out. Check the fuel strainer on the fuel gauge sending unit (see Section 6) for damage and deterioration.

Steel tubing

4   If replacement of a fuel line or emission line is called for, use welded steel tubing, available at most auto parts stores.

5   Don't use copper or aluminum tubing to replace steel tubing. These materials cannot withstand normal vehicle vibration.

6   Because fuel lines used on fuel injected vehicles are under high pressure, they require special consideration.

7   Most fuel lines have threaded fittings with O-rings. Any time the fittings are loosened to service or replace components:

a) *Use a back-up wrench while loosening and tightening the fittings.*

b) *Check all O-rings for cuts, cracks and deterioration. Replace any that appear worn or damaged.*

c) *If the lines are replaced, always use original equipment parts, or parts that meet the manufacturer's standards.*

Rubber hose

8   When rubber hose is replaced, use reinforced, fuel resistant hose with the word "Fluoroelastomer" imprinted on it. Hose(s) not clearly marked like this could fail prematurely and could fail to meet Federal emission standards. Hose inside diameter must match line outside diameter.

9   On fuel injected models, replace the rubber hose with the exact replacement part specified for the particular application, the metal fittings are permanently installed at each end.

10   Don't use rubber hose within four inches of any part of the exhaust system or within ten inches of the catalytic converter. Metal lines and rubber hoses must never be allowed to chafe against the frame. A minimum of 1/4-inch clearance must be maintained around a line or hose to prevent contact with the frame.

8   Carburetor - general information, diagnosis and overhaul

General information

Refer to illustrations 8.2, 8.4, 8.5, 8.6 and 8.9

1   Depending on model, year and engine size, carburetors of various types were used on these vehicles.

Rochester 2G/4G series

2   This Rochester carburetor is a 2-barrel (2GC) or 4-barrel (4GC), side bowl design **(see illustration)**. The units installed on manual and automatic transmission equipped vehicles are similar but vary in calibration. The main metering jets are fixed - calibration is handled by a system of air bleeds. A power enrichment valve is incorporated (power mixtures are controlled by air velocity past the boost venturi according to engine demand).

3   On later model vehicles, an electrically-operated throttle closing solenoid (controlled through the ignition switch) is used to ensure that the throttle valve closes completely after the ignition is turned off (to prevent dieseling). The choke is automatic and is operated by an exhaust manifold heated coil.

Rochester 4MC/4MV (Quadrajet) series

4   This is a 4-barrel two stage unit. The primary side uses a triple venturi system **(see illustration on page 4A-18)**. The secondary side has two large bores and one metering system which supplements the primary main metering system and receives fuel from a common float chamber.

Rochester M4M/E4M (Quadrajet) series

5   This is also a 4-barrel two stage unit and is very similar to the 4MV carburetor (see

8.2 Exploded view of a typical Rochester 2G series carburetor

1   Pump Rod Retainer
2   Pump Rod
3   Fast Idle Cam Screw
4   Fast idle Cam
5   Trip Lever Screw
6   Trip Lever
7   Choke Lever and Collar
8   Choke Rod
9   Idle Vent Valve Screw
10   Idle Vent Valve Shield
11   Idle Vent Valve
12   Fuel Inlet Fitting
13   Fuel Inlet Fitting Gasket
14   Fuel Filter Gasket
15   Fuel Inlet Filter
16   Fuel Filter Spring
17   Stat Cover Retainer Screw
18   Stat Cover Toothed Retainer
19   Stat Cover Plain Retainer
20   Stat Cover & Spring Assembly
21   Stat Cover Basket
22   Choke Baffle Plate
23   Choke Housing Basket (Not Shown)
24   Bowl Cover Screw & Lockwasher (Extra Long)
25   Bowl Cover Screw & Lockwasher
26   Bowl Cover Assembly
27   Float Hinge Pin
28   Float Assembly
29   Rotary Inlet Valve & Gasket
30   Bowl Cover Gasket
31   Pump Plunger Retainer
32   Pump Plunger Lock Ring
33   Pump Plunger Washer
34   Pump Plunger Spring
35   Pump Plunger Cup
36   Power Piston Assembly
37   Venturi Cluster Screw & Fiber Washer (Center)
38   Venturi Cluster Screw & Lockwasher
39   Venturi Cluster
40   Venturi Cluster Gasket
41   Main Well Inserts
42   Pump Return Spring
43   Pump Check Ball
44   Pump Discharge Ball Guide
45   Pump Discharge Ball Spring
46   Pump Discharge Ball
47   Power Valve Assembly
48   Power Valve Assembly Gasket
49   Main Metering Jets
50   Throttle Body To Bowl Screw & Lockwasher
51   Throttle Body Gasket
52   Throttle Body & Shaft Assembly
53   Idle Needles
54   Idle Needle Springs
55   Flange Gasket
56   Idle Adjust Screw
57   Idle Compensator Cover Screw
58   Idle Compensator Cover
59   Idle Compensator Valve Screw
60   Idle Compensator Valve
61   Idle Compensator Valve Gasket
62   Hot Water Choke Assembly
63   Hot Water Filter Retainer
64   Hot Water Filter
65   Vacuum Control Rod Retainer
66   Vacuum Control Rod

67   Stat Rod Lever Screw
68   Stat Rod Lever
69   Vacuum Control Attaching Screw
70   Vacuum Break Control Assembly

71   Vacuum Break Control Hose
72   Idle Air Adjust Needle
73   Idle Air Adjust Needle Spring
74   Bowl Assembly

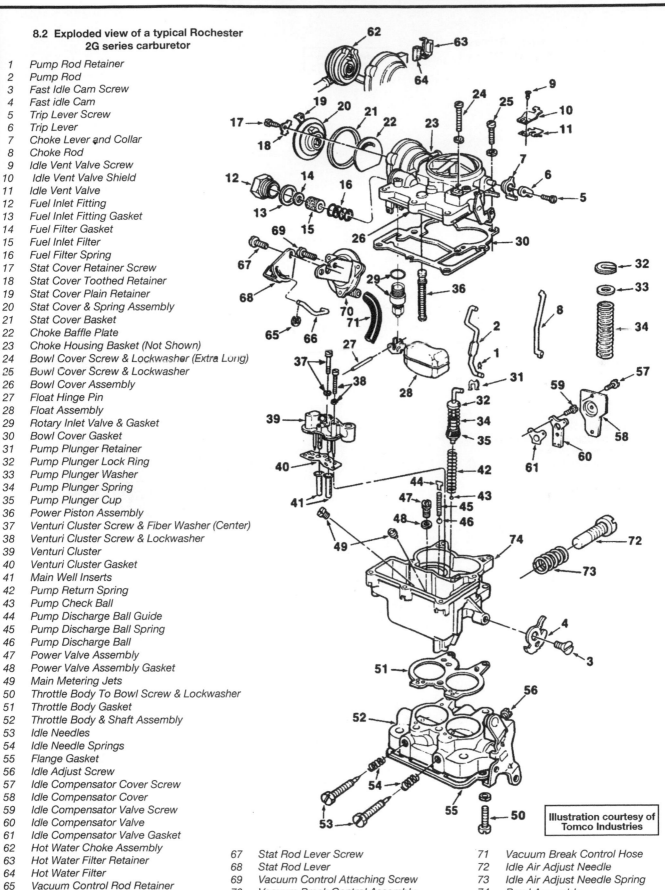

Illustration courtesy of Tomco Industries

8.4 Exploded view of a typical Rochester 4MC/4MV series carburetor

1   Fuel Filter Nut (Inlet)
2   Filter Nut Gasket
3   Fuel Filter Gasket
4   Fuel Filter
5   Fuel Filter Spring
6   Idle Vent Valve Screw
7   Idle Vent Valve
8   Pump Rod Retainer
9   Pump Rod
10 Idle Vent Valve Lever
11 Choke Rod Retainer
12 Choke Rod
13 Air Horn Screw (4)
14 Air Horn Screw (3)
15 Air Horn Screw (2)
16 Bowl Cover Assembly
17 Dashpot Piston And Rod Assembly
18 Secondary Metering Rod (2)
19 Pump Assembly
20 Pump Return Spring
21 Air Horn Gasket
22 Primary Power Piston Assembly
23 Primary Metering Rods (2)
24 Power Piston Spring
25 Float Bowl Insert
26 Float Hinge Pin
27 Float And Lever Assembly
28 Float Needle Full Clip
29 Needle Diaphragm Retainer Screw (2)
30 Needle Diaphragm Retainer
31 Needle Diaphragm Assembly
31A Needle Seat Gasket Assembly
32 Pump Discharge Ball Plug
33 Pump Discharge Ball
34 Primary Jets (2)
35 Idle Compensator Cover Screw (2)
36 Idle Compensator Cover
37 Idle Compensator Assembly
38 Idle Compensator Gasket
39 Throttle Body Screw (3)
40 Throttle Body Assembly
41 Idle Adjustment Needle (2)
42 Idle Adjustment Needle Spring (2)
43 Throttle Body Gasket
44 Vacuum Hose (4MV)
45 Vacuum Break Control Bracket
    Attaching Screw
46 Vacuum Break Control And Bracket
    Assembly (4MV)
47 Fast Idle Cam
48 Secondary Lockout Lever (4MV)
49 Intermediate Choke Lever
50 Stat Retainer Screw (3) (4MC)
51 Stat Cover Retainers (3) (4MC)
52 Stat Cover And Spring Assembly
    (4MC)
53 Stat Cover Gasket (4MC)
54 Choke Baffle Plate (4MC)
55 Stat Housing Attaching Screw
56 Choke Housing And Vacuum
    Break Assembly (4MC)
57 Choke Housing Gasket (4MC)
58 Float Bowl Assembly

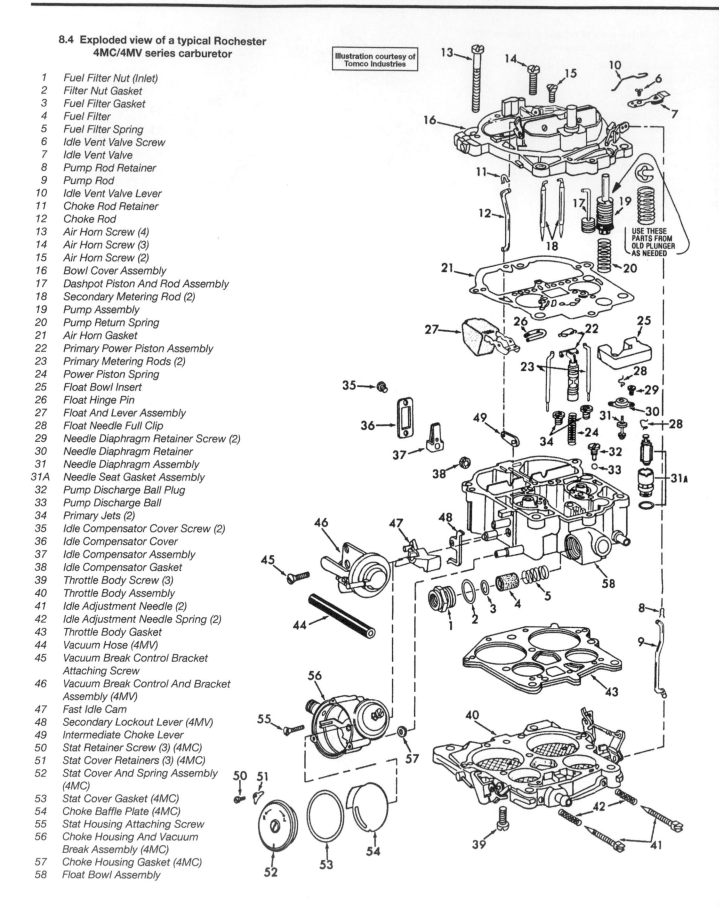

Illustration courtesy of Tomco Industries

USE THESE PARTS FROM OLD PLUNGER AS NEEDED

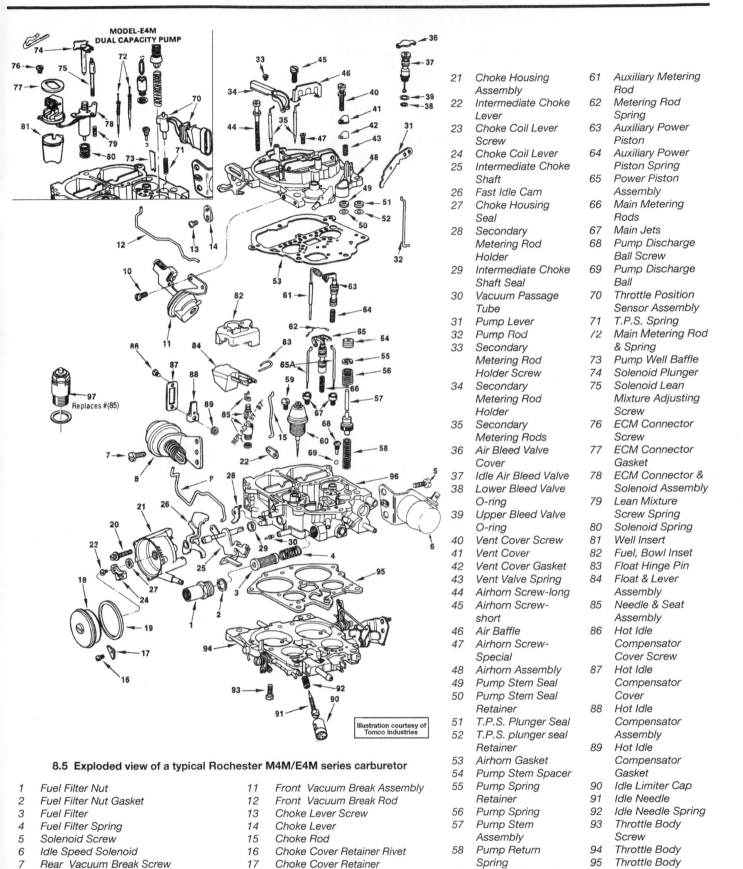

21	Choke Housing Assembly	61	Auxiliary Metering Rod
22	Intermediate Choke Lever	62	Metering Rod Spring
23	Choke Coil Lever Screw	63	Auxiliary Power Piston
24	Choke Coil Lever	64	Auxiliary Power Piston Spring
25	Intermediate Choke Shaft	65	Power Piston Assembly
26	Fast Idle Cam	66	Main Metering Rods
27	Choke Housing Seal	67	Main Jets
28	Secondary Metering Rod Holder	68	Pump Discharge Ball Screw
29	Intermediate Choke Shaft Seal	69	Pump Discharge Ball
30	Vacuum Passage Tube	70	Throttle Position Sensor Assembly
31	Pump Lever	71	T.P.S. Spring
32	Pump Rod	72	Main Metering Rod & Spring
33	Secondary Metering Rod Holder Screw	73	Pump Well Baffle
		74	Solenoid Plunger
34	Secondary Metering Rod Holder	75	Solenoid Lean Mixture Adjusting Screw
35	Secondary Metering Rods	76	ECM Connector Screw
36	Air Bleed Valve Cover	77	ECM Connector Gasket
37	Idle Air Bleed Valve	78	ECM Connector & Solenoid Assembly
38	Lower Bleed Valve O-ring	79	Lean Mixture Screw Spring
39	Upper Bleed Valve O-ring	80	Solenoid Spring
40	Vent Cover Screw	81	Well Insert
41	Vent Cover	82	Fuel, Bowl Inset
42	Vent Cover Gasket	83	Float Hinge Pin
43	Vent Valve Spring	84	Float & Lever Assembly
44	Airhorn Screw-long	85	Needle & Seat Assembly
45	Airhorn Screw-short	86	Hot Idle Compensator Cover Screw
46	Air Baffle	87	Hot Idle Compensator Cover
47	Airhorn Screw-Special	88	Hot Idle Compensator Assembly
48	Airhorn Assembly	89	Hot Idle Compensator Gasket
49	Pump Stem Seal	90	Idle Limiter Cap
50	Pump Stem Seal Retainer	91	Idle Needle
51	T.P.S. Plunger Seal	92	Idle Needle Spring
52	T.P.S. plunger seal Retainer	93	Throttle Body Screw
53	Airhorn Gasket	94	Throttle Body
54	Pump Stem Spacer	95	Throttle Body Gasket
55	Pump Spring Retainer	96	Main Body
56	Pump Spring	97	Rotary Inlet Valve
57	Pump Stem Assembly		
58	Pump Return Spring		
59	Aneroid Assembly Screw		
60	Aneroid Assembly		

8.5 Exploded view of a typical Rochester M4M/E4M series carburetor

1	Fuel Filter Nut	11	Front Vacuum Break Assembly
2	Fuel Filter Nut Gasket	12	Front Vacuum Break Rod
3	Fuel Filter	13	Choke Lever Screw
4	Fuel Filter Spring	14	Choke Lever
5	Solenoid Screw	15	Choke Rod
6	Idle Speed Solenoid	16	Choke Cover Retainer Rivet
7	Rear Vacuum Break Screw	17	Choke Cover Retainer
8	Rear Vacuum Break Assembly	18	Choke Cover Assembly
9	Rear Vacuum Break Rod	19	Choke Cover Gasket
10	Front Vacuum Break Screw	20	Choke Housing Screw

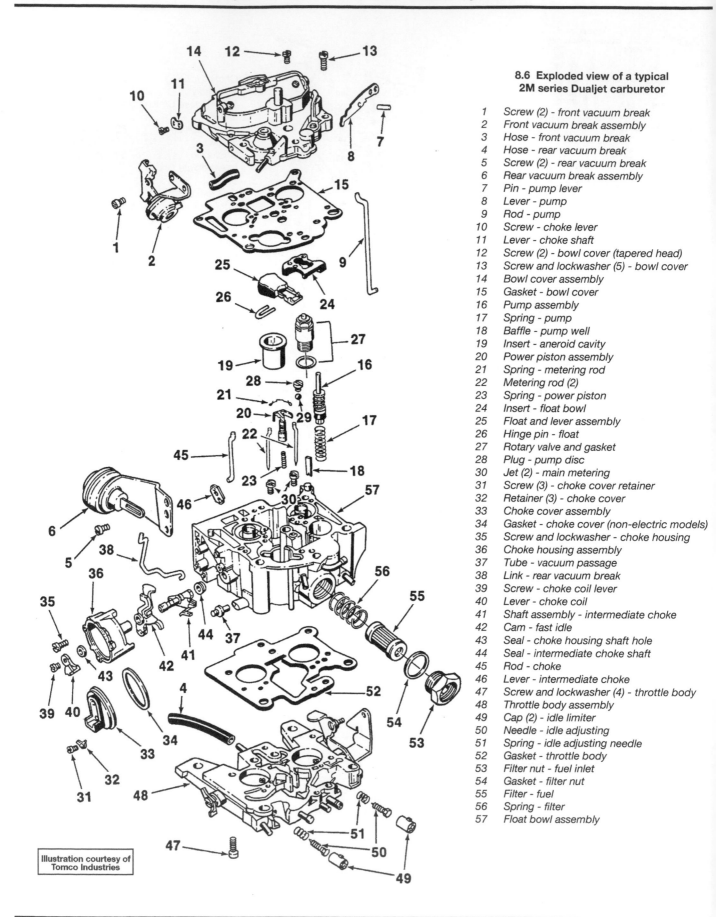

8.6 Exploded view of a typical 2M series Dualjet carburetor

1   Screw (2) - front vacuum break
2   Front vacuum break assembly
3   Hose - front vacuum break
4   Hose - rear vacuum break
5   Screw (2) - rear vacuum break
6   Rear vacuum break assembly
7   Pin - pump lever
8   Lever - pump
9   Rod - pump
10   Screw - choke lever
11   Lever - choke shaft
12   Screw (2) - bowl cover (tapered head)
13   Screw and lockwasher (5) - bowl cover
14   Bowl cover assembly
15   Gasket - bowl cover
16   Pump assembly
17   Spring - pump
18   Baffle - pump well
19   Insert - aneroid cavity
20   Power piston assembly
21   Spring - metering rod
22   Metering rod (2)
23   Spring - power piston
24   Insert - float bowl
25   Float and lever assembly
26   Hinge pin - float
27   Rotary valve and gasket
28   Plug - pump disc
29   Jet (2) - main metering
30   Jet (2) - main metering
31   Screw (3) - choke cover retainer
32   Retainer (3) - choke cover
33   Choke cover assembly
34   Gasket - choke cover (non-electric models)
35   Screw and lockwasher - choke housing
36   Choke housing assembly
37   Tube - vacuum passage
38   Link - rear vacuum break
39   Screw - choke coil lever
40   Lever - choke coil
41   Shaft assembly - intermediate choke
42   Cam - fast idle
43   Seal - choke housing shaft hole
44   Seal - intermediate choke shaft
45   Rod - choke
46   Lever - intermediate choke
47   Screw and lockwasher (4) - throttle body
48   Throttle body assembly
49   Cap (2) - idle limiter
50   Needle - idle adjusting
51   Spring - idle adjusting needle
52   Gasket - throttle body
53   Filter nut - fuel inlet
54   Gasket - filter nut
55   Filter - fuel
56   Spring - filter
57   Float bowl assembly

illustration). Later model vehicles use the Rochester E4ME carburetor, which is very similar to the earlier model except for the changes due to increased emission restrictions. These carburetors work together with the emission systems through the use of a mixture control solenoid mounted in the float bowl.

Rochester M2M/E2M (Dualjet) series

6   This is a 2-barrel unit with a triple venturi system **(see illustration)**. It is very similar to the primary side of the M4M carburetor ill. The M2ME carburetors used on air conditioned vehicles have an electrically-operated air conditioner idle speed solenoid which maintains proper idle speed when the air conditioner is being used. The model M2M carburetors have an identification number stamped vertically on the left rear corner of the float bowl.

7   Later model vehicles use the Rochester E2M Dualjet carburetor which is very similar to the earlier M2M model except for the changes due to increased emission restrictions. These carburetors work together with the emission systems through the use of a mixture control solenoid mounted in the float bowl.

Diagnosis

8   A thorough road test and check of the carburetor adjustments should be done before any carburetor service work is done. Specifications for some adjustments are listed on the Vehicle Emissions Control Information label found in the engine compartment.

9   Some performance complaints directed at the carburetor are actually a result of loose, out-of-adjustment or malfunctioning engine or electrical components. Others develop when vacuum hoses leak, are disconnected or are incorrectly routed. The proper approach to analyzing carburetor problems should include a routine check of the following:

a) *Inspect all vacuum hoses and actuators for leaks and correct installation.*
b) *Tighten the intake manifold and carburetor mounting nuts/bolts evenly and securely.*
c) *Perform a cylinder compression test (see Chapter 2).*
d) *Clean or replace the spark plugs as necessary (see Chapter 1).*
e) *Check the spark plug wires (see Chapter 1).*
f) *Inspect the ignition primary wires.*
g) *Check the ignition timing (follow the instructions printed on the Vehicle Emissions Control Information label).*
h) *Check the fuel pump pressure/volume (see Section 2).*
i) *Check the heat control valve in the air cleaner for proper operation (see Chapter 1).*
j) *Check/replace the air filter element (see Chapter 1).*

8.14 An identification number will be stamped on the side of the carburetor body - copy the numbers off the carburetor when purchasing a rebuild kit or a new or rebuilt carburetor

k) *Check the PCV system (see Chapters 1 and 6).*
l) *Check/replace the fuel filter (see Chapter 1). Also, the strainer in the tank could be restricted.*
m) *Check for a plugged exhaust system.*
n) *Check EGR valve operation (see Chapter 6).*
o) *Check the choke - it should be completely open at normal engine operating temperature (see Chapter 1).*
p) *Check for fuel leaks and kinked or dented fuel lines (see Chapters 1 and 4).*
q) *Check accelerator pump operation with the engine off (remove the air cleaner cover and operate the throttle as you look into the carburetor throat - you should see a stream of gasoline enter the carburetor).*
r) *Check for incorrect fuel or bad gasoline.*
s) *Check the camshaft lobe lift (see Chapter 2).*
t) *Have a dealer service department or repair shop check the electronic engine and carburetor controls (if equipped).*

10   Carburetor problems usually show up as flooding, hard starting, stalling, severe backfiring, poor acceleration and lack of response to idle mixture screw adjustments. A carburetor that is leaking fuel and/or covered with wet-looking deposits definitely needs attention.

11   Diagnosing carburetor problems may require that the engine be started and run with the air cleaner removed. While running the engine without the air cleaner it's possible that it could backfire. A backfiring situation is likely to occur if the carburetor is malfunctioning, but the removal of the air cleaner alone can lean the fuel/air mixture enough to produce an engine backfire. **Warning:** *DO NOT position your face directly over the carburetor throat during inspection and servicing procedures.*

Overhaul

Warning: *Gasoline is extremely flammable, so extra precautions must be taken when working on any part of the fuel system. See the* **Warning** *in Section 2.*

Refer to illustration 8.14

12   Once it's determined that the carburetor is in need of work or an overhaul, several alternatives should be considered. If you're going to attempt to overhaul the carburetor yourself, first obtain a good quality carburetor rebuild kit. You'll also need some special solvent and a means of blowing out the internal passages of the carburetor with air.

13   Due to the fact that several types of carburetors were used on the vehicles covered by this manual, it isn't possible to include a step-by-step overhaul of each one. You'll receive a detailed, well-illustrated set of instructions with any quality carburetor overhaul kit; they will apply in a more specific manner to the carburetor on your vehicle.

14   Another alternative, which is probably the best approach, is to obtain a new or rebuilt carburetor. They are readily available from dealers and auto parts stores. The most important fact to consider when purchasing one of these units is to make sure the exchange carburetor is identical to the original. In most cases an identification number that will help you determine which type you have will be stamped on the side of the carburetor **(see illustration)**. When obtaining a rebuilt carburetor or a rebuild kit, take time to ensure that the kit or carburetor matches your application exactly. Seemingly insignificant differences can make a considerable difference in the performance of your engine.

15   If you choose to overhaul your own carburetor, allow enough time to disassemble the carburetor carefully, soak the necessary parts in the cleaning solvent (usually for at least one half day or according to the instructions listed on the carburetor cleaner) and reassemble it, which will usually take much longer than disassembly. When you're disassembling the carburetor, match each part with the illustration in the carburetor kit and lay the parts out in order on a clean work surface. Overhauls by inexperienced mechanics can result in an engine which runs poorly, or not at all. To avoid this, use care and patience when disassembling the carburetor so you can reassemble it correctly.

16   When the overhaul is complete, adjustments which may be beyond the ability of the home mechanic may be required, especially on later models. If so, take the vehicle to a dealer service department or a reputable tune-up shop for final carburetor adjustments,

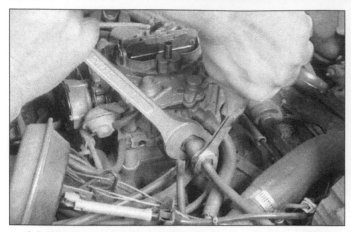

9.2 When removing the fuel line always use the correct size wrenches, preferably a "flare-nut" wrench to avoid stripping the tube nut, and a "backup" wrench to keep the carburetor inlet nut from turning

9.6 Remove the carburetor mounting bolts or nuts

which will ensure compliance with emissions regulations and acceptable performance.

9 Carburetor - removal and installation

Warning: *Gasoline is extremely flammable, so extra precautions must be taken when working on any part of the fuel system. See the **Warning** in Section 2.*

Removal

Refer to illustrations 9.2 and 9.6

1   Remove the air cleaner (see Chapter 1).
2   Disconnect the fuel and vacuum lines from the carburetor **(see illustration)**. Mark the hoses and fittings with pieces of numbered tape to ensure correct installation.
3   Mark and disconnect the wire harness connectors from the choke and carburetor.
4   Disconnect the throttle linkage and throttle return springs.
5   Disconnect the throttle valve or detent linkage (automatic transmission equipped models).
6   Remove the carburetor mounting nuts and bolts and detach the carburetor from the intake manifold **(see illustration)**. **Note:** *Quadrajet carburetors are attached to the manifold with two nuts and two long bolts. The bolts are located at the front and extend all the way through the air horn assembly, the main body and the throttle body of the carburetor.*
7   Remove the insulator and gaskets. On later models, remove the EFE heater.

Installation

8   Installation is the reverse of removal, but the following points should be noted:
   a) *By filling the float bowl with fuel, the initial start up will be easier and there will be less strain on the battery.*
   b) *New gaskets must be used.*
   c) *Idle speed and mixture should be checked and adjusted if necessary.*

10 Carburetor (Rochester 2G series) - adjustments

Refer to illustrations 10.6, 10.34, 10.39, 10.43, 10.47 and 10.49

Idle speed adjustment

1   Run the engine to normal operating temperature, make sure that the choke plate is completely open and have the air cleaner in position. If air conditioning is installed, switch it Off during adjustment.
2   Make sure the ignition timing is correct.
3   Disconnect the hose from the fuel tank connector on top of the charcoal canister.
4   Disconnect and plug the vacuum hose to the distributor.
5   Connect a tachometer to the engine according to the manufacturers instructions.
6   Set the curb idle speed. Place the idle stop screw on the low step of the fast idle cam and adjust the idle speed stop screw until the engine idles at the specified rpm **(see illustration)**. If the vehicle is equipped with an idle stop solenoid proceed to Step 7. If the vehicle is not equipped with an idle stop solenoid, turn off the engine and reconnect any vacuum hoses that were disconnected.

With idle stop solenoid

7   2G series carburetors are fitted with a idle stop solenoid when the vehicle is equipped with automatic transmission or air conditioning. This solenoid allows the carburetor to raise the idle speed during increased engine loads.
8   When an idle stop solenoid is equipped follow Steps 1 through 6 and set the curb idle speed first, then follow the remaining Steps and set the curb idle speed with the solenoid energized.
9   If the vehicle is equipped with an automatic transmission, set the parking brake and place the vehicle in drive. For safety reasons this will require the help of an assistant to apply the brake pedal during this procedure.

If the vehicle is equipped with a manual transmission, set the parking brake and place the vehicle in neutral.
10   Disconnect the lead from the air conditioner compressor and turn the air conditioner On.
11   Open the throttle to allow the solenoid plunger to extend fully.
12   Turn the solenoid plunger bolt until the specified idle speed is obtained with the solenoid energized. Reconnect the compressor lead on completion.
13   Turn off the engine and reconnect any vacuum hoses that were disconnected.

Idle mixture adjustment

14   The mixture screws on these carburetors are equipped with limiter caps. The caps must remain in place and the screws adjusted (lean drop method) only to the limits of the cap stops.
15   If the carburetor has been dismantled to

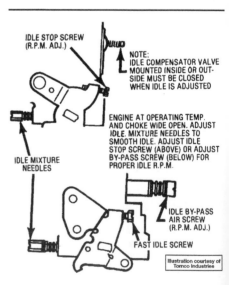

10.6 Rochester 2G carburetor idle speed and idle mixture adjustments

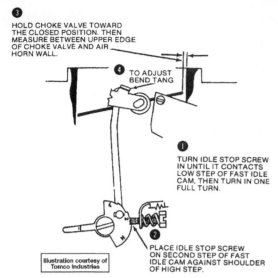

10.34 Rochester 2G carburetor fast idle (choke rod) adjustment details

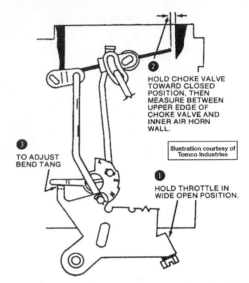

10.39 Rochester 2G carburetor choke unloader adjustment details

such an extent that the mixture screw limiter caps have been removed or, if after checks of all other engine tune-up specifications, the idle mixture seems to be off, adjust it as follows in order to maintain the exhaust emission levels set by the vehicle manufacturer.

Lean drop method

16 Disconnect the hose from the fuel tank connector on the charcoal canister.
17 Disconnect and plug the distributor vacuum hose.
18 With the parking brake applied, run the engine until normal operating temperature is reached. Make sure the air cleaner is installed and the air conditioner (if applicable) is Off.
19 On vehicles with an automatic transmission, place the selector lever in Drive.
20 On vehicles with a manual transmission, disconnect the wire harness from the carburetor idle stop solenoid.
21 If not already removed, break off the tab on the idle mixture screw limiter cap.
22 Adjust the curb idle speed as described in Steps 1 through 6. If an idle stop solenoid is installed, set the curb idle speed with the solenoid energized as described in Steps 7 through 12.
23 Unscrew each of the idle mixture screws an equal amount until the idle speed reaches its highest point and any further movement of the screws would cause it to decrease.
24 Readjust the idle stop screw if necessary to bring the engine speed back to that previously set.
25 Turn the mixture screws in until the final curb idle speed is set, based on the type of transmission, as indicated in the Specifications.
26 Reconnect all disconnected components and turn off the engine.

CO meter method

27 This method can be employed if a reliable CO meter can be obtained.

28 Carry out the procedure already described in Steps 16 through 21 of this Section.
29 Install the meter probe in the tailpipe.
30 Set the curb idle speed to the Specifications as described above.
31 If the engine idles smoothly and the CO level does not exceed 0.5, no further adjustment is necessary.
32 If the CO level is excessive, turn the idle mixture screws clockwise until the idle CO level is acceptable. Readjust the curb idle speed if necessary.

Choke rod adjustment

33 Turn the idle stop screw in until it just touches the bottom step of the fast idle cam, then screw it in exactly one full turn.
34 Position the fast idle screw on the second step of the fast idle cam against the shoulder of the high step **(see illustration)**.
35 Hold the choke plate in the closed position (using a rubber band to keep it in place) and check the gap or angle of the choke plate. **Note:** *The method for setting the adjustment depends on the specification given for the particular year and model. If the specification is listed in inches, use a drill bit between the upper edge of the choke plate and the air horn wall. If the specification is listed in degrees, use an angle gauge which measures the angle of the choke plate to the air horn wall.*
36 If the gap is not as specified, bend the tang to correct it.

Choke unloader adjustment

37 Hold the throttle plates all the way open.
38 Hold the choke plate in the closed position using a rubber band to keep it in place.
39 Check the gap between the upper edge of the choke plate and the inside wall of the air horn. The gap should be as shown in the Specifications **(see illustration)**. **Note:** *The*

method for setting the adjustment depends on the specification given for the particular year and model. If the specification is listed in inches, use a drill bit between the upper edge of the choke plate and the air horn wall. If the specification is listed in degrees, use an angle gauge which measures the angle of the choke plate to the air horn wall..
40 Bend the tang on the throttle lever to adjust the gap.

Vacuum break adjustment

41 Remove the air cleaner. On temperature controlled air cleaners, plug the sensor vacuum take-off port.
42 Apply vacuum to the vacuum break diaphragm until the plunger is fully seated.
43 Push the choke plate towards the closed position, then measure the gap between the upper edge of the choke plate and the air horn wall with a drill bit **(see illustration)**. **Note:** *The method for setting the*

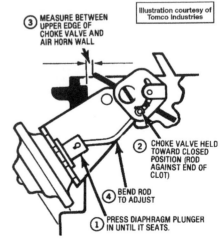

10.43 Rochester 2G carburetor vacuum break adjustment details

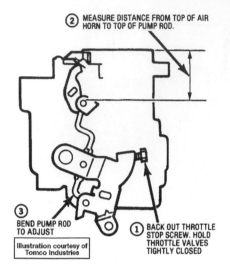

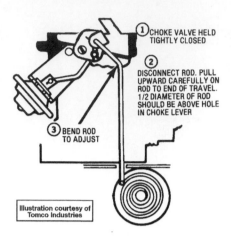

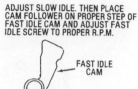

10.47 Rochester 2G carburetor accelerator pump rod adjustment details

10.49 Rochester 2G carburetor choke coil rod adjustment details

11.2 Rochester 4MV carburetor fast idle adjustment (typical)

adjustment depends on the specification given for the particular year and model. If the specification is listed in inches, use a drill bit between the upper edge of the choke plate and the air horn wall. If the specification is listed in degrees, use an angle gauge which measures the angle of the choke plate to the air horn wall.

44   The gap should be as specified. If it isn't, bend the vacuum break rod.

Accelerator pump rod adjustment

45   Back out the idle speed screw.
46   Close both throttle plates completely and measure from the top surface of the air horn ring to the tip of the pump rod.
47   This measurement should be as specified. Bend the rod as required **(see illustration)**.

Choke coil rod adjustment

48   Hold the choke plate open, then, with the thermostatic coil rod disconnected from the upper lever, push down on the rod to the end of its travel.

49   The bottom of the rod should now be level with the bottom of the elongated hole in the lever. If it isn't, bend the lever by inserting a screwdriver blade into the slot **(see illustration)**.

11   Carburetor (Rochester 4MV) - adjustments

Refer to illustrations 11.2, 11.7, 11.8, 11.13 and 11.15

Idle adjustment

1   The procedure is very similar to the one for the 2G series carburetor. Refer to Section 10, Steps 1 through 32.

Fast idle adjustment

2   Place the transmission in Neutral. Position the fast idle lever on the high step of the fast idle cam **(see illustration)**.
3   Make sure that the engine is at normal operating temperature with the choke fully open.

4   On vehicles equipped with a manual transmission, disconnect the vacuum advance hose.
5   Connect a tachometer to the engine in accordance with the manufacturer's instructions.
6   Turn the fast idle screw in or out as necessary to adjust the fast idle speed.

Choke rod (fast idle cam) adjustment

7   Set the cam follower on the second step of the fast idle cam and against the high step. Rotate the choke valve towards the closed position by turning the external choke lever counterclockwise. Use a drill bit as a gauge and measure the gap between the lower edge of the choke plate (lower end) and air horn wall. Bend the choke rod if necessary to achieve the specified clearance **(see illustration)**. **Note:** *The method for setting the adjustment depends on the specification given for the particular year and model. If the specification is listed in inches, use a drill bit between the upper edge of the choke plate and the air horn wall. If the specification is listed in degrees, use an angle gauge which measures the angle of the choke plate to the air horn wall.*

Choke vacuum break adjustment

8   Seat the choke vacuum break diaphragm with a vacuum pump. Open the

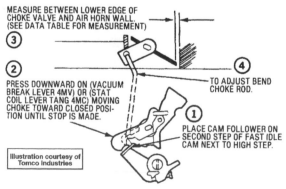

11.7 Rochester 4MV carburetor choke rod adjustment (typical)

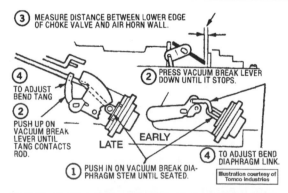

11.8 Rochester 4MV carburetor choke vacuum break adjustment (typical)

① ALL. HOLD CHOKE VALVE COMPLETELY CLOSED AND CHOKE ROD AT BOTTOM OF SLOT IN CHOKE LEVER.

② ALL. DISCONNECT CHOKE COIL ROD FROM LEVER.

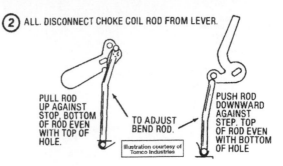

PULL ROD UP AGAINST STOP, BOTTOM OF ROD EVEN WITH TOP OF HOLE.

TO ADJUST BEND ROD.

PUSH ROD DOWNWARD AGAINST STEP. TOP OF ROD EVEN WITH BOTTOM OF HOLE

Illustration courtesy of Tomco Industries

11.13 Rochester 4MV carburetor choke coil rod adjustment (typical)

throttle plate slightly so the cam follower will clear the step of the fast idle cam. Rotate the vacuum break lever counterclockwise and use a rubber band to hold it in position **(see illustration)**.

9 Make sure that the end of the vacuum brake rod is in the outer slot of the diaphragm plunger.

10 Measure the gap between the lower edge of the choke plate and the inside of the air horn wall. Use a drill bit as a gauge. If the gap is not as specified, bend the link rod. **Note:** *The method for setting the adjustment depends on the specification given for the particular year and model. If the specification is listed in inches, use a drill bit between the upper edge of the choke plate and the air horn wall. If the specification is listed in degrees, use an angle gauge which measures the angle of the choke plate to the air horn wall.*

Choke coil rod adjustment

11 Hold the choke plate closed by turning the choke coil lever counterclockwise.

12 Disconnect the thermostatic coil rod and remove the coil cover, then push the coil rod down until the rod contacts the surface of the bracket.

13 Check that the coil rod fits in the notch in the choke lever. Bend the rod as necessary if it doesn't **(see illustration)**.

Air valve dashpot adjustment

14 Seat the choke vacuum break diaphragm, using an external vacuum source.

15 With the diaphragm seated and the air valve fully closed, measure the distance between the end of the slot in the vacuum break plunger lever and the air valve. If it's not as specified, bend the link rod as shown **(see illustration)**.

12 Carburetor (Rochester M4MC/M4MCA) - adjustments

Refer to illustrations 12.12, 12.18, 12.26, 12.30, 12.33, 12.38, 12.42, 12.48, 12.52, 12.57, 12.60 and 12.62

Mixture adjustment

1 Remove the air cleaner for access to the carburetor, keeping the vacuum hoses connected. Disconnect the remaining hoses from the air cleaner and plug them.

2 Prior to making the adjustment, the engine must be at normal operating temperature, the choke must be open and the air conditioning system (where applicable) off. The ignition timing must be correct. Make sure that the parking brake is applied.

3 Connect a tachometer to the engine in accordance with the manufacturer's instructions.

4 Remove the limiter caps from the idle mixture screws. Turn the screws in until they seat lightly, then unscrew them equally until the engine will just run.

5 Place the transmission in Neutral (manual transmission) or Drive (automatic transmission).

6 Back the mixture screws out equally 1/8-turn at a time until the maximum engine speed is obtained. Now turn the idle speed screw until the initial idle speed is as specified.

7 Repeat the adjustment to make sure that the maximum idle speed was obtained when unscrewing the mixture screws.

8 Now turn each of the idle mixture screws in 1/8-turn at a time until the idle speed is as shown for lean drop in the Specifications.

9 Check that the speed shown in the Specifications is the same as on the Vehicle Emission Control Information label. If necessary, readjust until the specified speed is achieved.

10 Reconnect the vacuum hoses, install the air cleaner and turn the engine off.

11 Install new limiter caps so future adjustment of one turn clockwise (leaner) can be done.

Pump rod adjustment

12 With the fast idle cam follower off the steps of the fast idle cam, back out the idle speed screw until the throttle valves are completely closed in the bore. Make sure that the secondary actuating rod is not restricting movement; bend the secondary closing tang if necessary, then readjust it after pump adjustment **(see illustration)**.

13 Place the pump rod in the specified hole in the lever. On all carburetors except numbers 17057262, 17057263 and 17057266, place the rod in the inner hole. On carburetors 17057262, 17057263 and 17057266, place the rod in the outer hole.

14 Measure from the top of the choke valve

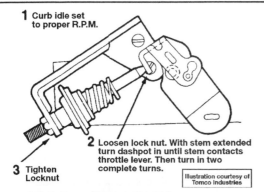

1 Curb idle set to proper R.P.M.

2 Loosen lock nut. With stem extended turn dashpot in until stem contacts throttle lever. Then turn in two complete turns.

3 Tighten Locknut

Illustration courtesy of Tomco Industries

11.15 Rochester 4MV carburetor air valve dashpot adjustment (typical)

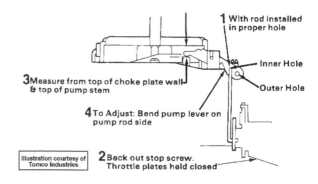

1 With rod installed in proper hole

Inner Hole

Outer Hole

3 Measure from top of choke plate wall & top of pump stem

4 To Adjust: Bend pump lever on pump rod side

Illustration courtesy of Tomco Industries

2 Back out stop screw. Throttle plates held closed

12.12 Rochester M4MC/M4MCA carburetor pump rod adjustment details

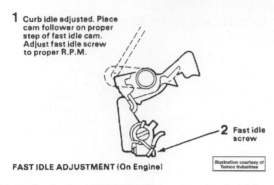

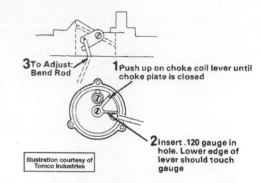

12.18 Rochester M4MC/M4MCA carburetor fast idle adjustment details

12.26 Rochester M4MC/M4MCA carburetor choke coil lever adjustment details

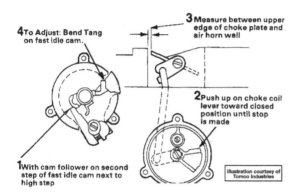

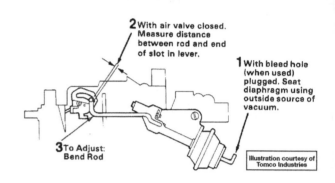

12.30 Rochester M4MC/M4MCA carburetor choke rod (fast idle cam) adjustment details

12.33 Rochester M4MC/M4MCA carburetor air valve rod adjustment details

wall (next to the vent stack) to the top of the pump stem.

15   If necessary, adjust to obtain the specified dimension by bending the lever while supporting it with a screwdriver.

16   Adjust the idle speed.

Fast idle adjustment

17   Place the transmission in Park (automatic) or Neutral (manual).

18   Hold the cam follower on the highest step of the fast idle cam **(see illustration)**.

19   Disconnect and plug the vacuum hose at the EGR valve (if applicable).

20   Turn the fast idle screw to obtain the correct specified idle speed.

Choke coil lever adjustment

21   Remove the screws and detach the cover and coil assembly from the choke housing.

22   Push up the thermostatic coil tang (counterclockwise) until the choke plate is closed.

23   Check that the choke rod is at the bottom of the slot in the choke lever.

24   Insert a plug gauge (or the equivalent size drill bit shank) of the specified size into the hole in the choke housing.

25   The lower edge of the choke coil lever should just contact the side of the plug gauge.

26   If necessary, bend the choke rod at the point shown to adjust **(see illustration)**.

Choke rod (fast idle cam) adjustment

27   Turn the fast idle cam screw in until it contacts the fast idle cam follower, then turn it in an additional three full turns. Remove the coil cover.

28   Place the lever on the second step of the fast idle cam, against the rise of the high step.

29   Push up on the choke coil lever inside the housing to close the choke plate.

30   Measure between the upper edge of the choke plate and the air horn wall with the shank of a drill bit of the specified size **(see illustration)**. **Note:** *The method for setting the adjustment depends on the specification given for the particular year and model. If the specification is listed in inches, use a drill bit between the upper edge of the choke plate and the air horn wall. If the specification is listed in degrees, use an angle gauge which measures the angle of the choke plate to the air horn wall.*

31   If necessary, bend the tang on the fast idle cam to adjust, but make sure that the tang lies against the cam after bending.

Air valve rod adjustment

32   Using an external source of vacuum,

seat the choke vacuum break diaphragm.

33   Make sure that the air valves are completely closed, then measure between the air valve dashpot and the end of the slot in the air valve lever **(see illustration)**.

34   Bend the air valve dashpot rod at the point shown, if adjustment is necessary.

Front vacuum break adjustment

35   Remove the thermostatic coil cover.

36   Place the cam follower on the highest step of the fast idle cam.

37   Use a vacuum pump to seat the vacuum diaphragm.

38   Push the inner choke coil lever counterclockwise until the tang on the vacuum break lever contacts the tang on the vacuum break plunger **(see illustration)**.

39   Check the gap between the upper edge of the choke plate and the inside wall of the air horn. Use a drill bit of the specified size to do this. **Note:** *The method for setting the adjustment depends on the specification given for the particular year and model. If the specification is listed in inches, use a drill bit between the upper edge of the choke plate and the air horn wall. If the specification is listed in degrees, use an angle gauge which measures the angle of the choke plate to the air horn wall.*

40   Turn the adjuster screw to achieve the correct gap.

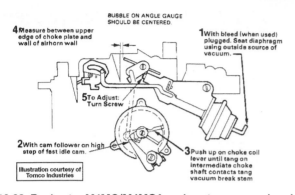

12.38 Rochester M4MC/M4MCA carburetor vacuum break adjustment details

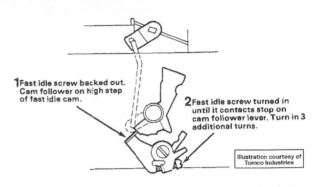

12.42 Rochester M4MC/M4MCA carburetor automatic choke coil adjustment details

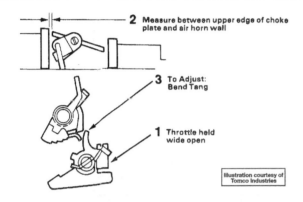

12.48 Rochester M4MC/M4MCA carburetor choke unloader adjustment details

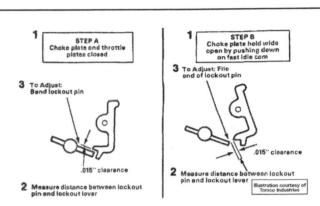

12.52 Rochester M4MC/M4MCA carburetor secondary throttle lockout adjustment details

41 Install the cover and check the automatic choke adjustment (Steps 42 through 45).

Automatic choke coil adjustment (hot air type)

42 Place the cam follower on the highest point of the cam **(see illustration)**.
43 Loosen the three retaining screws and rotate the coil cover until the choke plate just closes.
44 Align the mark on the cover with the specified point on the housing, depending on the carburetor type.
45 Tighten the cover screws.

Choke unloader adjustment

46 Make sure that the automatic choke housing cover is set to the specified position (Steps 42 through 45).
47 Hold the throttle plate wide open.
48 Close the choke plate by pushing up on the tang of the intermediate choke lever **(see illustration)**.
49 Check the gap between the upper edge of the choke plate and the air horn inner wall. Use a suitable size drill bit as a gauge. **Note:** *The method for setting the adjustment depends on the specification given for the particular year and model. If the specification is listed in inches, use a drill bit between the upper edge of the choke plate and the air*

horn wall. If the specification is listed in degrees, use an angle gauge which measures the angle of the choke plate to the air horn wall.
50 Bend the tang if necessary to adjust it.

Secondary throttle lockout adjustment

Lockout lever clearance

51 Hold the choke plates and secondary lockout plates closed, then measure the clearance between the lockout pin and lockout lever.
52 If adjustment is necessary, bend the lockout pin to obtain the specified clearance **(see illustration)**.

Opening clearance

53 Push down on the tail of the fast idle cam to hold the choke wide open.
54 Hold the secondary throttle plates partly open, then measure between the end of the lockout pin and the toe of the lockout lever.
55 If adjustment is necessary, file the end of the lockout pin, making sure that no burrs remain afterwards.

Secondary closing adjustment

56 Adjust the engine idle speed.
57 Hold the choke plate wide open, with the cam follower lever off the steps of the fast idle cam **(see illustration)**.

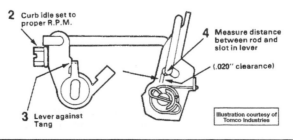

12.57 Rochester M4MC/M4MCA carburetor secondary closing adjustment details

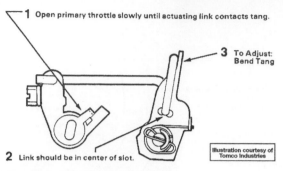

1 Open primary throttle slowly until actuating link contacts tang.

3 To Adjust: Bend Tang

2 Link should be in center of slot.

Illustration courtesy of Tomco Industries

12.60 Rochester M4MC/M4MCA carburetor secondary opening adjustment details

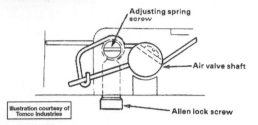

TO ADJUST LOOSEN ALLEN LOCK SCREW. AIR VALVES HELD CLOSED, TURN ADJUST-ING SCREW CLOCKWISE UNTIL TORSION SPRING CONTACTS PIN ON AIR VALVE SHAFT. THEN TURN ADDITIONAL SPECIFIED TURNS AND TIGHTEN ALLEN LOCK SCREW. (SEE DATA TABLE).

Adjusting spring screw

Air valve shaft

Allen lock screw

Illustration courtesy of Tomco Industries

12.62 Rochester M4MC/M4MCA carburetor air valve spring adjustment details

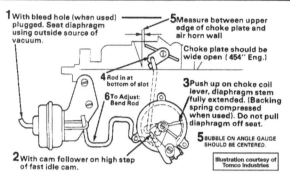

1 With bleed hole (when used) plugged. Seat diaphragm using outside source of vacuum.

5 Measure between upper edge of choke plate and air horn wall.

Choke plate should be wide open (454" Eng.)

4 Rod in at bottom of slot

6 To Adjust: Bend Rod

3 Push up on choke coil lever, diaphragm stem fully extended. (Backing spring compressed when used). Do not pull diaphragm off seat.

5 BUBBLE ON ANGLE GAUGE SHOULD BE CENTERED.

2 With cam follower on high step of fast idle cam.

Illustration courtesy of Tomco Industries

13.7 Rochester M4ME carburetor rear vacuum break adjustment details

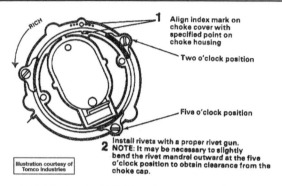

RICH

1 Align index mark on choke cover with specified point on choke housing

Two o'clock position

Five o'clock position

2 Install rivets with a proper rivet gun. NOTE: It may be necessary to slightly bend the rivet mandrel outward at the five o'clock position to obtain clearance from the choke cap.

Illustration courtesy of Tomco Industries

13.8 Rochester M4ME carburetor electric choke component layout

58   Measure the clearance between the slot in the secondary throttle plate pick-up lever and the secondary actuating rod.
59   If adjustment is necessary, bend the secondary closing tang on the primary throttle lever to obtain the specified clearance.

Secondary opening adjustment

60   Lightly open the primary throttle lever until the link just contacts the tang on the secondary lever **(see illustration)**.
61   Bend the tang on the secondary lever, if necessary, to position the link in the center of the secondary lever slot.

Air valve spring adjustment

62   Using an Allen wrench, loosen the lockscrew, then turn the tension adjusting screw counterclockwise until the air valve is partly open **(see illustration)**.
63   Hold the air valve closed, then turn the tension adjusting screw clockwise the speci-fied number of turns after the spring contacts the pin.
64   Tighten the lockscrew.

13 Carburetor (Rochester M4ME) - adjustments

Refer to illustrations 13.7, 13.8 and 13.9

1   The Rochester M4ME carburetor is very similar to M4MC and M4MCA carburetors, with the exception of an electrically-heated

(as opposed to hot air heated) choke and a rear vacuum break.
2   Except for the adjustments mentioned in the following Steps, all other on-vehicle adjustments are the same as the M4MC and M4MCA carburetors described in Section 12.

Rear vacuum break adjustment

3   Remove the thermostatic coil cover.
4   Place the cam follower on the highest step of the fast idle cam.
5   Use a vacuum pump to seat the vacuum diaphragm.
6   Push up on the choke coil lever, inside the choke housing, towards the closed choke position until the stem is pulled out and seated.
7   With the choke rod in the bottom of the

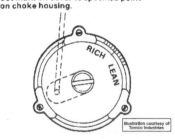

Rotate choke cover against spring tension. Set mark on cover to specified point on choke housing.

RICH LEAN

Illustration courtesy of Tomco Industries

13.9 Rochester M4ME carburetor automatic choke coil adjustment details

slot in the choke lever, place the specified size drill bit between the upper edge of the choke plate and the air horn wall. If the dimension is not as specified, bend the vac-uum break rod **(see illustration)**. **Note:** *The method for setting the adjustment depends on the specification given for the particular year and model. If the specification is listed in inches, use a drill bit between the upper edge of the choke plate and the air horn wall. If the specification is listed in degrees, use an angle gauge which measures the angle of the choke plate to the air horn wall.*

Automatic choke coil adjustment (electrically-heated type)

8   Place the cam follower on the highest step of the cam **(see illustration)**.
9   Loosen the retaining screws and rotate the coil cover until the choke plate just closes **(see illustration)**.
10   Align the index mark on the cover with the specified mark on the housing.
11   Tighten the cover screws.

14 Carburetor (Rochester M2M) - adjustments

Float adjustment

1   Hold the float retainer firmly in place and gently push down on the float against the needle.

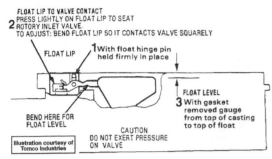

15.2 Rochester E4ME/E4MC carburetor float level adjustment details

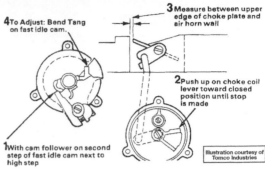

15.7 Rochester E4ME/E4MC carburetor choke rod adjustment details

2   Measure from the top of the casting to the top of the float, at a point 3/16-inch back from the end of the float, at the toe.
3   If adjustment is necessary, remove the float and bend the float arm up or down as necessary.
4   Reinstall the float, visually check the alignment and recheck the float setting.

Pump adjustment

5   With the throttle plates completely closed, make sure the fast idle cam follower lever is off the steps of the fast idle cam.
6   Insert the specified size drill bit into the specified hole in the pump lever.
7   Measure from the top of the choke plate wall, next to the vent stack, to the top of the pump stem as specified.
8   If adjustment is necessary, support the pump lever with a screwdriver while bending the lever.

Choke coil lever adjustment

9   The procedure is the same as the one for the M4MC carburetor (Section 12, Steps 21 through 26).

Fast idle adjustment - bench setting

10   Hold the cam follower on the highest step of the fast idle cam.
11   Turn the fast idle screw out until the primary throttle plates are closed. Turn the screw in to contact the lever, then turn it in an additional two complete turns.

Choke rod (fast idle cam) adjustment

12   This procedure is very similar to that used to adjust the M4MC carburetor (Section 12, Steps 27 through 31).
13   Close the choke by pushing up on the choke coil lever or vacuum break lever tang. Hold in that position with a rubber band.
14   To adjust, bend the tang on the fast idle cam until the specified clearance is obtained.
Note: *The method for setting the adjustment depends on the specification given for the particular year and model. If the specification is listed in inches, use a drill bit between the upper edge of the choke plate and the air horn wall. If the specification is listed in degrees,*

use an angle gauge which measures the angle of the choke plate to the air horn wall.
15   Remove the gauge.

Front vacuum break adjustment

16   This procedure can be carried out in the same manner as the front vacuum break adjustment described in Section 12, Steps 35 through 41 for the M4MC carburetor.

Automatic choke coil adjustment

17   Place the cam follower on the highest step of the cam.
18   Loosen the three retaining screws on the choke housing.
19   Rotate the cover and coil assembly counterclockwise until the choke plate just closes.
20   Align the mark on the cover with the point on the housing that is one notch to the lean side. **Note:** *Make sure the slot in the lever engages with the coil tang.*

Choke unloader adjustment

21   This procedure is very similar to the one for the M4MC carburetor (Section 12, Steps 46 through 50).
22   Hold the throttle plates wide open.
23   With the engine warm, close the choke plate by pushing up on the tang on the vacuum break lever. Hold in this position with a rubber band.
24   To adjust, bend the tang on the fast idle lever until the specified clearance is obtained.

Idle adjustment

25   The procedure is very similar to that described for the 2G carburetor. Refer to Section 10, Steps 1 through 32.

Fast idle adjustment - on vehicle

26   Place the transmission in Park or Neutral.
27   Place the cam follower on the specified step of the fast idle cam. Refer to the emission label.
28   Disconnect the vacuum hose from the EGR valve and plug the hose.
29   Start the engine and turn the fast idle

adjustment screw to obtain the rpm specified on the emission label.

15   Carburetor (Rochester E4ME/E4MC) - adjustments

Refer to illustrations 15.2, 15.7, 15.12, 15.14, 15.17, 15.23, 15.28, 15.29, 15.31 and 15.37
Note: *Refer to Section 16 for additional adjustments.*
1   Except where noted, adjustments are the same as for earlier M4MC models (see Section 12). Carburetors used with the Computer Command Control system do not require adjustment of the pump rod.

Float level adjustment - bench setting

2   Hold the float retainer firmly in place, push the float down until it lightly contacts the needle and measure the float level with the gauge. The gauging point is 3/16-inch back from the toe of the float **(see illustration)**.
3   On carburetors used with the Computer Command Control system, the float should be adjusted if the height varies from that shown in the Specifications Section of this Chapter by plus-or-minus 1/16-inch.
4   If the level is too high, hold the retainer in place and push down on the center of the float until the specified setting is obtained.
5   If the level is too low on non-solenoid carburetors, remove the power piston, metering rods, plastic filler block and the float. Bend the float arm up to adjust. Reinstall the parts and visually check the alignment of the float.
6   If the level is too low on solenoid equipped carburetors, remove the metering rods and the solenoid connector screw. Count and record for use at the time of reassembly, the number of turns necessary to lightly bottom the lean mixture screw. Back the screw out and remove it, the solenoid connector and float. Bend the float arm up to adjust. Install the parts and reset the mixture screw to the recorded number of turns.

Choke rod adjustment

7   Attach a rubber band to the green tang of the intermediate choke shaft **(see illustration)**.

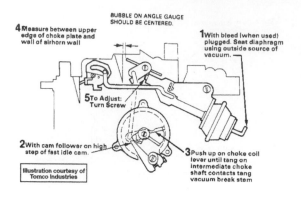

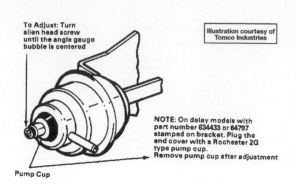

15.12 Rochester E4ME/E4MC carburetor front vacuum break adjustment details

15.14 Plug the vacuum delay air bleed

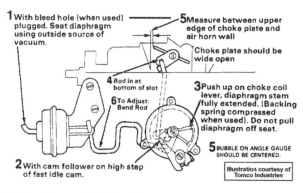

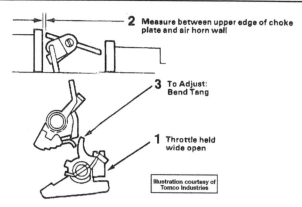

15.17 Rochester E4ME/E4MC carburetor rear vacuum break adjustment details

15.23 Rochester E4ME/E4MC carburetor choke unloader adjustment

8   Close the choke by opening the throttle.

9   Install a choke angle gauge and set the specified angle.

10   Place the cam follower on the second step of the cam, against the high step. If the follower does not contact the cam, turn the fast idle speed screw in until it does. The final fast idle adjustment must be made according to the information on the Emissions Control Information label under the hood.

11   Bend the fast idle cam tang until the bubble is centered.

Front vacuum break adjustment

12   Attach a rubber band to the green tang of the intermediate choke shaft and open the throttle to allow the choke to close (see illustration).

13   Install a gauge tool and set the vacuum break to the specified angle.

14   Apply at least 18-inches Hg of vacuum to retract the vacuum break plunger. Plug any air bleed holes (see illustration).

15   The air valve rod can sometimes restrict the plunger from retracting completely and it might be necessary to bend the rod slightly. The final rod clearance must be set after the vacuum break adjustment has been made.

16   With the vacuum applied, adjust the screw until the bubble is centered.

Rear vacuum break adjustment

17   Attach a rubber band to the green tang of the intermediate choke shaft (see illustration).

18   Open the throttle until the choke closes.

19   Set up an angle gauge tool on the carburetor and set the angle to specification. **Note:** *The method for setting the adjustment depends on the specification given for the particular year and model. If the specification is listed in inches, use a drill bit between the upper edge of the choke plate and the air horn wall. If the specification is listed in degrees, use an angle gauge which measures the angle of the choke plate to the air horn wall.*

20   Apply vacuum to retract the vacuum break plunger, making sure to plug any bleed holes.

21   The air valve rod can sometimes restrict the plunger from retracting completely. If necessary, bend the rod to allow full travel of the plunge.

22   To center the bubble, either of two methods can be used. With the vacuum applied, use a 1/8-inch Allen wrench to turn the adjustment screw. Alternatively, support the vacuum break rod and bend the rod with the vacuum applied.

Choke unloader adjustment

23   Attach a rubber band to the green tang of the intermediate choke shaft (see illustration) and open the throttle to allow the choke to close.

24   Hold the secondary lockout lever away from the pin.

25   Check the gap between the upper edge of the choke plate and the air horn wall using a suitable size drill bit as a gauge. **Note:** *The method for setting the adjustment depends on the specification given for the particular year and model. If the specification is listed in inches, use a drill bit between the upper edge of the choke plate and the air horn wall. If the specification is listed in degrees, use an angle gauge which measures the angle of the choke plate to the air horn wall.*

26   Hold the throttle lever in the wide open position and bend the fast idle lever tang to obtain the specified clearance.

Idle speed adjustment - preparation

27   Prior to idle speed adjustment, the engine must be at normal operating temperature and the ignition timing set to the specification on the Emissions Control Information label.

28   Some models equipped with the Com-

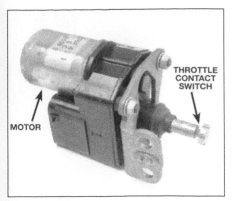

15.28 Rochester E4ME/E4MC idle speed control assembly used with the Computer Command Control system

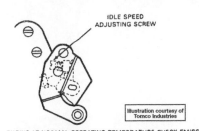

15.29 Rochester E4ME/E4MC non-solenoid equipped curb idle speed adjustment details

puter Command Control system use an idle speed control (ISC) assembly mounted on the carburetor, which is controlled by the ECM. Do not attempt to adjust the idle on the idle speed control **(see illustration)**. Adjustment of the idle speed control assembly should be left to a dealer service department or a repair shop.

Curb idle speed adjustment

Non-solenoid equipped
29   With the air conditioner off, adjust the curb idle screw to the specifications on the Emissions Control Information label **(see illustration)**.

Solenoid equipped
30   Adjust the curb idle as described in the previous Step.
31   With the air conditioner on, the compressor lead disconnected at the compressor, the solenoid energized and the transmission in Neutral (manual) or Drive (automatic), open the throttle slightly to completely extend the solenoid plunger **(see illustration)**.
32   Adjust the curb idle to the specified rpm by turning the solenoid screw.
33   Reconnect the air conditioner compressor wire after adjustment.

Fast idle speed adjustment

Non-solenoid equipped
34   With the transmission in Park (automatic) or Neutral (manual), hold the cam follower on the step specified on the Emissions Control Information label. Turn the screw to obtain the correct fast idle speed.

Solenoid equipped
35   With the transmission in Park (automatic) or Neutral (manual), hold the cam follower on the step of the fast idle cam specified on the Emissions Control Information label.
36   Disconnect the vacuum hose at the EGR valve and plug it.
37   To obtain the fast idle rpm specified on the label, turn the fast idle screw **(see illustration)**.

16   Electronic feedback carburetor system

General information
1   The electronic feedback carburetor system relies on an electronic signal, which is generated by an exhaust gas oxygen sensor, to control a variety of devices and keep emissions within limits. The system works in conjunction with a three-way catalyst to control the levels of carbon monoxide, hydrocarbons and oxides of nitrogen. The feedback carburetor system also works in conjunction with the computer. The two systems share certain sensors and output actuators; therefore, diagnosing this system will require a thor-

ough check of all the feedback carburetor components.
2   The system operates in two modes: open loop and closed loop. When the engine is cold, the air/fuel mixture is controlled by the computer in accordance with a program designed in at the time of production. The air/fuel mixture during this time will be richer to allow for proper engine warm-up. When the engine is at operating temperature, the system operates in closed loop and the air/fuel mixture is varied depending on the information supplied by the exhaust gas oxygen sensor.
3   Here is a list of the various sensors and output actuators involved with these feedback carburetor systems:

> *Coolant temperature sensor (ECT)*
> *Exhaust Gas Oxygen sensor (O2)*
> *High energy Electronic Ignition (HEI)*
> *Early Fuel Evaporation (EFE) system*
> *Mixture Control Solenoid*
> *Electronic Control Module (ECM)*
> *Throttle Position Sensor (TPS)*
> *Vehicle Speed Sensor (VSS)*
> *Barometric Pressure Sensor (BARO) or Manifold Absolute Pressure (MAP) sensor*
> *Idle Speed Control (ISC) switch*
> *Park/Neutral switch (see Chapter 7)*

Refer to Chapter 6 for the diagnostic checks for the feedback carburetor system sensors.

Mixture control (M/C) solenoid
Refer to illustrations 16.12, 16.16a and 16.16b
4   The mixture control (M/C) solenoid is a device that controls fuel flow from the bowl to the main well and at the same time controls the idle circuit air bleed.
5   The mixture control (M/C) solenoid is located in the float bowl where the power piston used to be. It is equipped with a spring loaded plunger that moves up and down like a power piston but more rapidly. Certain areas on the plunger head contact the metering rods and an idle air bleed valve. Plunger movement controls both the metering rods and the idle air bleed valve simultaneously.

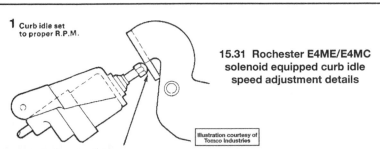

15.31 Rochester E4ME/E4MC solenoid equipped curb idle speed adjustment details

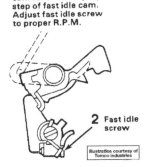

15.37 Rochester E4ME/E4MC fast idle speed adjustment details

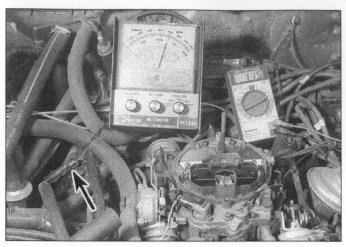

16.12 Connect the dwellmeter to the terminal in the green connector (arrow) near the carburetor

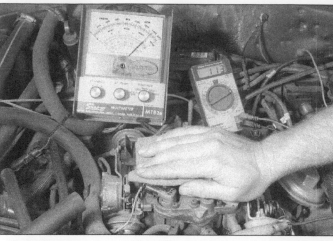

16.16a Place a rag over the carburetor air horn and observe the dwell indicate LEAN to compensate for a rich condition

6   When the mixture control solenoid is energized it moves down causing the metering rods to move into the jets and restrict the flow of fuel into the main well. The idle air bleed plunger opens the air bleed and allows air into the idle circuit. Both these movements reduce fuel flow and thereby LEANS out the system.

7   When the plunger is de-energized, the M/C solenoid moves up, causing the metering rods to move out of the jets and allow more fuel to the main well, less idle air and increased fuel flow. Here the solenoid is in the RICH position.

8   The mixture control solenoid varies the air/fuel ratio based on the electrical input from the ECM. When the solenoid is ON, the fuel is restricted and the air is admitted. This gives a lean air/fuel ratio (approximately 18:1). When the solenoid is OFF, fuel is admitted and the air/fuel ratio is approximately 13:1. During closed loop operation, the ECM controls the M/C solenoid to approximately 14.7:1 by controlling the ON and OFF time of the solenoid.

9   As the solenoid "on time" changes, the up time and down time of the metering rods also changes. When a lean mixture is desired, the M/C solenoid will restrict fuel flow through the metering jet 90-percent of the time, or, in other words, a lean mixture will be provided to the engine.

10   This lean command will read as 54-degrees on the dwell meter (54-degrees is 90-percent of 60-degrees), and means the M/C solenoid has restricted fuel flow 90-percent of the time. A rich mixture is provided when the M/C solenoid restricts it only 10-percent of the time and allows a rich mixture to flow to the engine. A rich command will have a dwellmeter reading of 6-degrees (10-percent of 60-degrees); the M/C solenoid has restricted fuel flow 10-percent of the time. On some engines dwellmeter readings can vary between 5-degrees and 55-degrees, rather than between 6-degrees and 54-degrees. The ideal mixture would be shown on the dwellmeter with the needle varying or swing-ing back and forth, anywhere between 10-degrees and 50-degrees. "Varying" means the needle continually moves up and down the scale. The amount it moves does not matter, only the fact that it does move. The dwell is being varied by the signal sent to the ECM by the oxygen sensor in the exhaust manifold.

11   The following checks assume the engine has been tuned and the ignition system is in order.

12   The dwellmeter is used to diagnose the M/C solenoid system. This is done by connecting a dwellmeter to the dwell test lead (a green single wire connector in the M/C solenoid wiring harness) **(see illustration)**.

13   In the older style contact-points ignition systems, the dwellmeter reads the period of time that the points were closed (or "dwelled" together). That period of time was when voltage flowed to the ignition coil. In the feedback carburetor system, the dwellmeter is used to read the time that the ECM closes the M/C solenoid circuit to ground, allowing voltage to operate the M/C solenoid. Dwell, as used in feedback system performance diagnosis, is the time that the M/C solenoid circuit is closed (energized). The dwellmeter will translate this time into degrees. The "6-cylinder" (0-degrees to 60-degrees) scale on the dwellmeter is used for this reading. The ability of the dwellmeter to make this kind of conversion makes it an ideal tool to check the amount of time the ECM internal switch is closed, thus energizing the M/C solenoid. The only difference is that the degree scale on the meter is more like percent of solenoid "on time" rather than "degrees of dwell".

14   First set the dwellmeter on the 6-cylinder position, then connect it to the M/C solenoid dwell test lead to measure the output of the ECM. Do not allow the terminal to touch ground, including hoses. You must use the 6-cylinder position when diagnosing all engines, whether the engine you're working on is a 4, 6, or 8 cylinder engine. **Note:** *Some older dwellmeters may not work properly on the feedback carburetor systems. Don't use any dwellmeter which causes a change in engine operation when it is connected to the solenoid lead.*

The 6-cylinder scale on the dwellmeter provides evenly divided points, for example:

   15-degrees = 1/4 scale
   30-degrees = midscale
   45-degrees = 3/4 scale

Connect the positive clip lead of the dwellmeter to the M/C solenoid dwell test connector. Attach the other dwellmeter clip lead to ground. Do not allow the clip leads to contact other conductive cables or hoses which could interfere with accurate readings.

15   After connecting the dwellmeter to a warm, operating engine the dwell at idle and part throttle will read between 5-degrees and 55-degrees and will be varying. That is, the needle will move continuously up and down the scale. What matters is that the needle does move, not how much it moves. Typical needle movement will occur between the 30 and 35-degree range in a normal operating feedback system. Needle movement indicates that the engine is in "closed-loop," and that the dwell is being varied by signals from the ECM. However, if the engine is cold, has just been restarted, or the throttle is wide open, the dwell will be fixed and the needle will be steady. Those are signs that the engine is in "open-loop."

16   Diagnostic checks to find a condition without a trouble code are usually made on a warm engine (in "closed-loop") easily checked by making sure the upper radiator hose is hot. There are three ways of distinguishing "open" from "closed-loop" operation.

   a) *A variation in dwell will occur only in "closed-loop."*
   b) *Test for closed loop operation by restricting airflow to the carburetor to choke the engine* **(see illustration)**. **Warning:** *Use a thick rag (not your bare hand) and keep your face away from the carburetor. If the dwellmeter moves up scale, that indicates "closed-loop."*

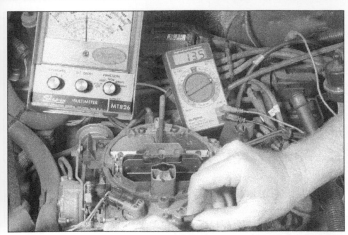

16.16b Disconnect a vacuum line and observe the dwell indicate RICH to compensate for a lean condition

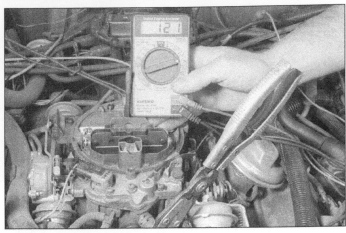

16.19a Clamp off the charcoal canister purge hose to prevent entrance of fuel vapors into the carburetor

c) *If you create a large vacuum leak and the dwell drops down, that also indicates "closed-loop"* **(see illustration)**.

17 Basically, "closed loop" indicates that the ECM is using information from the exhaust oxygen sensor to influence operation of the Mixture Control (M/C) solenoid. The ECM still considers other information such as engine temperature, RPM, barometric and manifold pressure and throttle position along with the exhaust oxygen sensor information.

18 During "open-loop" all information except the exhaust oxygen sensor input is considered by the ECM to control the M/C solenoid. Accurate readings from the oxygen sensor are not attained until the sensor is completely hot (600-degrees F).

Note: *The exhaust oxygen sensor may cool below its operational temperature during prolonged idling. This will cause an open loop condition, and make the diagnostic information not usable during diagnosis. Engine RPM must be increased to warm the exhaust oxygen sensor, and again re-establish "closed loop". Diagnosis should begin over at the first step after "closed-loop" is resumed.*

System performance test

Refer to illustration 16.19a and 16.19b

19 Start the engine and then ground the diagnostic terminal (see Chapter 6). This will allow the feedback system to run on the preset default values, thereby not allowing the computer to make any running adjustments. **Caution:** *Do not ground the diagnostic connector before starting the engine.* Disconnect the purge hose from the charcoal canister and plug it (or simply clamp it shut) **(see illustration)**. Disconnect the bowl vent hose at the carburetor and plug the hose on the canister side. These two steps will not allow any recirculated fuel or crankcase vapors to enter the carburetor during testing. Connect a tachometer according to the manufacturers' instructions. Disconnect the M/C solenoid and ground the M/C solenoid dwell test lead **(see illustration)**. Run the engine at 3,000 rpm and while holding the throttle steady,

reconnect the M/C solenoid and observe the rpm. Unground the dwell lead before returning the engine to idle.

20 The test results will be either:

a) *The engine speed will decrease at least 300 rpm*

b) *The engine speed does not decrease (at least 300 rpm) and may increase rpm*

21 If the engine speed does decrease (at least 300 rpm), check the wiring on the M/C solenoid for any damaged connectors and if they are OK, check the carburetor adjustments. Also check for fuel in the charcoal canister or crankcase or any other condition that would cause richness.

22 If the engine speed decrease's at least 300 rpm, connect a dwellmeter to the M/C solenoid dwell lead **(see illustration 16.12)**. Be sure to read the information on "closed loop" operation and dwellmeters. Set the carburetor on the high step of the fast idle cam and run the engine for one minute or until the dwell starts to vary (whichever happens first). Return the engine to idle and observe the dwell. In most cases the dwell should vary between 10 and 50-degrees but there are several types of problems that may occur. **Note:** *The following tests must be performed with the diagnostic connector still grounded unless otherwise indicated.*

Fixed dwell under 10-degrees

23 This condition indicates that the feedback system is responding RICH to offset a very lean condition in the engine. One way to separate the problem is to choke the carburetor with the engine at part throttle. This will either increase or decrease the dwell.

24 If the dwell increases, check for a vacuum leak in the hoses, gaskets, AIR system etc. Also, check for an exhaust leak near the oxygen sensor, any hoses that are misrouted and an EGR valve that is not operating or that is leaking.

25 If the dwell does not increase, check the oxygen sensor, the wiring harness from the ECM to the oxygen sensor, TPS, TPS voltage and/or the ECM. If necessary, have the sys-

16.19b With the dwell test lead grounded and a tachometer properly installed, disconnect the M/C solenoid connector and observe the engine speed

tem diagnosed by a dealer service department or other repair shop.

Fixed dwell between 10 and 50-degrees

26 This condition indicates that the feedback system is stuck in one mode (open loop) because of a faulty coolant temperature sensor, oxygen sensor or TPS. Start the diagnosis by running the engine at part throttle for one minute. Then with the engine idling, observe the dwell, remove the connector from the coolant sensor and jump the terminals on the connector. This will ground the signal and indicate to the ECM that the sensor is functioning. The dwell reading should not be fixed with the sensor grounded. Check each sensor for the correct resistance and voltage signal from its respective electrical connector.

Fixed dwell over 50-degrees

27 This condition indicates that the feedback system is responding LEAN to offset a very rich condition in the engine. Start the diagnosis by running the engine at fast idle

16.29 To measure plunger travel, first measure the plunger height with no pressure applied to the gauging tool, then . . .

16.32 . . . press lightly on the gauge until it bottoms the plunger and measure again. The difference between the two measurements is the travel

16.36a To adjust the lean-stop, place the gauging tool (being held at right) on the metering jet and under the solenoid plunger, as shown, then depress the plunger and turn the adjusting screw until the plunger bottoms against the lean stop and touches the top of the gauging tool simultaneously

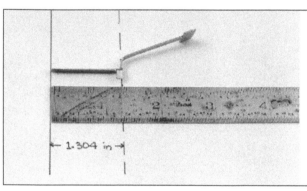

16.36b The gauging tool, shown here, allows you to precisely set the lean-stop at 1.304 inches

for about two minutes and then let it return to idle. This procedure makes sure that the feedback system is in closed-loop (warmed-up). Next, disconnect the large vacuum hose to the PCV valve and cause a major vacuum leak. Do not allow the engine to stall. The dwell should drop by approximately 20-degrees. If it does, check the carburetor adjustments. Also, check the evaporative canister for fuel overload or leaks in the purge control system that would cause an over-rich condition in the carburetor.

28   If the dwell does not drop, check the oxygen sensor, oxygen sensor signal voltage and their respective wiring circuits for any problems.

Feedback carburetor adjustments

Mixture control solenoid plunger adjustment

Refer to illustrations 16.29, 16.32, 16.36a, 16.36b, 16.41a, 16.41b and 16.43

Measuring plunger travel

Note: *If you're reassembling the carburetor after overhaul, just set the air horn in place when measuring the plunger travel (there's no need to install the gasket or screws, and you may need to remove the air horn again to adjust the travel).*

29   Insert a special carburetor float gauge tool (available from auto parts stores and tool

manufacturers) into the vertical "D"-shaped vent hole in the air horn casting next to the idle air bleed valve cover **(see illustration)**. It may be necessary to file material off the side of the gauge to make clearance for the tool to enter without binding.

30   First, press down on the gauge and make sure the gauge moves up and down freely without any binding.

31   With the gauge resting against the plunger (no pressure applied), sight across the air horn and record the mark on the gauge that lines up with the top of the air horn casting (upper edge).

32   Lightly press DOWN on gauge until the plunger bottoms. Record this measurement **(see illustration)**.

33   Now, subtract the first measurement from the second measurement and record the difference. This reading is the total plunger travel.

34   If the difference in travel is between 3/32 and 5/32-inch, no adjustment is necessary. If the difference is less than 3/32-inch (not enough travel) or greater than 5/32-inch (too much travel), adjust the plunger travel, as follows.

Adjusting plunger travel

Note 1: *This adjustment usually is not required after overhaul if you kept the adjustment at its original setting. It is only necessary if the travel is not correct or the duty cycle (percent "on time") is incorrect as determined*

by the System performance test.

Note 2: *The air horn must be removed from the carburetor during this procedure; however, the procedure can be performed with the carburetor on or off the vehicle.*

Note 3: *These carburetors are equipped with either a* **four-point** *mixture control solenoid adjustment system (earlier feedback systems) or a* **two-point** *mixture control solenoid adjustment system (mainly on 1985 and later models). The four-point system is equipped with separate lean-stop and rich-stop adjustments that correctly adjust the travel of the solenoid plunger from rich to lean. The* **two-point** *system uses an integral rich limit stop bracket that essentially limits the travel of the plunger going up (rich) as well as the plunger going down (lean). Therefore, only one adjustment is required.*

35   Remove the air horn from the carburetor. Also remove the solenoid plunger and the plastic float-bowl insert, then reinstall the plunger (removing the insert will allow access to the float-bowl area). **Note:** *On two-point systems, it will be necessary to remove the solenoid adjusting screw and rich-stop before the solenoid plunger and insert can be removed. Prior to removing the screw, turn it clockwise until the plunger bottoms, counting the number of turns. After the insert is removed, reassemble the plunger and stop, setting the adjusting screw the same number of turns from the lightly-bottomed position.*

36   Now install a special gauging tool (available from auto parts stores and automotive tool suppliers) onto the base of the metering jet and directly under the solenoid plunger. This tool is made specifically to measure the solenoid plunger travel **(see illustrations)**, which should be 1.304 inches (approximately 1-5/16 inches).

37   Apply very slight pressure to the solenoid plunger with the tip of your finger to bottom it. Slowly turn the lean-stop screw

16.41a On four-point systems, if the travel is not correct after setting the lean-stop, unscrew . . .

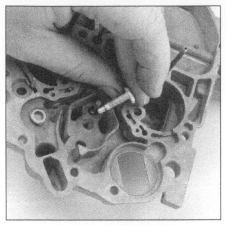

16.41b . . . and remove the rich-stop screw from the carburetor air horn

16.43 Use the special tool to adjust the rich-stop screw on a four-point system

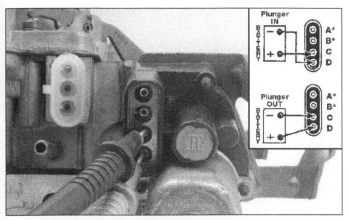

16.49 Using jumper wires, apply battery voltage to terminals C and D to extend or retract the ISC solenoid plunger

16.51 Location of the idle-stop screw (arrow) on feedback carburetors (on non-feedback carburetors, this is the curb-idle speed screw)

with the special adjustment tool until it (the solenoid plunger) just touches the gauging tool when it's bottomed against the lean stop. This is the factory-specified distance for the lean-stop adjustment.

38 If you have a four-point system, proceed to Step 39. If you have a two-point system, the adjustment is complete. Turn the adjusting screw clockwise to the lightly-bottomed position, counting the number of turns, then unscrew and remove the screw, rich stop and plunger. Install the insert and reassemble the parts, turning the adjusting screw in the same number of turns. Install the air horn and proceed to Step 44.

Models with four-point adjustment systems only

39 Reinstall the air horn, along with a new gasket, onto the carburetor main body, but attach it temporarily with two screws only.
40 Recheck the solenoid plunger travel as described earlier in this procedure. If the travel is correct, completely assemble the air horn and proceed to Step 44 to finish the job.
41 If the plunger travel is still not correct, remove the air horn, invert it and remove the rich-stop screw **(see illustrations)**.

42 With the rich-stop screw removed, drive the plug out of the air horn using a punch and a hammer. Reinstall the rich-stop screw.
43 Reinstall the carburetor air horn. Insert the special float gauge into the "D"-shaped vent hole (see *Measuring plunger travel* earlier in this procedure) and, using the special carburetor tool, turn the rich-stop screw **(see illustration)** until the total solenoid plunger travel is 1/8-inch.

All feedback models

44 With the plunger travel correctly set and the air horn reinstalled, install the replacement plugs into the carburetor body. Be sure to install the hollow end down.

Idle speed control (ISC) system

Refer to illustrations 16.49, 16.51, 16.53a and 16.53b

45 If the carburetor has just been overhauled, the ISC solenoid has just been replaced with a new one or the vehicle is having a fluctuating idle problem, it is necessary to check the ISC system adjustments.
46 First, place the transmission in Neutral, set the parking brake and block the drive wheels. Connect a tachometer according to the manufacturer's instructions.

47 With the A/C off (if equipped), start the engine and warm it up to normal operating temperature (closed loop).
48 Turn the ignition OFF and unplug the connector from the ISC motor.
49 Fully retract the ISC plunger by applying 12 volts to terminal C (run a fused jumper wire from the battery's positive terminal) of the ISC motor connector and grounding terminal D **(see illustration)**. **Note 1:** *Do not allow the battery voltage (12V) to contact terminal C any longer than necessary to retract the ISC plunger. Prolonged contact will cause damage to the motor. Also, it is very important to never connect a voltage source across terminals A and B, as it may damage the internal throttle contact switch.* **Note 2:** *If the ISC solenoid does not respond when battery voltage is applied to the C and D terminals, it is faulty and should be replaced.*
50 Start the engine and allow it to return to normal operation (closed loop). Have an assistant place his foot firmly on the brake and place the transmission in DRIVE.
51 With the plunger completely retracted (from Step 5), turn the idle stop screw **(see illustration)** to attain the minimum idle speed. The minimum base idle speed is usually about 450 rpm in DRIVE. Consult your

16.53a Adjust the ISC plunger by turning the plunger end with pliers (carburetor removed from engine for clarity), but . . .

16.53b . . . do not unscrew the plunger end more than 5/16-inch from its all-the-way-in setting

16.58 Be sure to protect the carburetor when drilling out the air-bleed rivets (otherwise, metal chips could fall into the carburetor!)

VECI label under the hood of the engine compartment to verify the idle specifications.

52   Now, with the transmission In DRIVE and an assistant holding his foot firmly on the brake pedal, fully extend the plunger by applying battery positive voltage (12V) to terminal D of the ISC connector and grounding terminal C **(see illustration 16.49)**. **Note:** *Do not allow the battery voltage (12V) to contact terminal D any longer than necessary to retract the ISC plunger. Prolonged contact will cause damage to the motor. Also, it is very important to never connect a voltage source across terminals A and B as it may damage the internal throttle contact switch.*

53   With the ISC plunger fully extended, check the idle speed, which should be approximately 900 rpm in DRIVE. If necessary, adjust the plunger length **(see illustrations)**. Consult your VECI label under the hood of the engine compartment to verify the idle specifications. **Note:** *When the ISC plunger is in the fully extended position, the plunger must also be in contact with the throttle lever (on carburetor) to prevent possible internal damage to the ISC solenoid.*

54   If the ISC motor was removed and the plunger length changed, double-check the maximum allowable speed (see Step 33). If necessary, readjust the plunger until the correct speed is obtained.

55   Reconnect the harness connector to the ISC motor.

Idle air bleed valve (E2ME and E4ME)

Refer to illustrations 16.58 and 16.61

56   To gain access to the idle air bleed valve, it is necessary to first remove the idle air bleed valve cover. Often, if the carburetor has been overhauled before, the cover has already been removed. If it has not been removed before, the air bleed is probably still at its factory setting and should not be tampered with unless it's known to be incorrect.

57   With the engine off, cover all the bowl vents, air inlets and air intakes with masking tape to prevent any metal chips from falling

into the carburetor.

58   Carefully align a Number 35 drill bit (0.110 inches) on one end of the steel rivet heads and drill only enough to remove the rivet head from the carburetor. Use a drift and hammer to remove the remaining rivet from the assembly **(see illustration)**.

59   Lift off the cover and remove any pieces of metal, rivets or debris from the carburetor body. Discard the idle air bleed valve cover.

60   Using compressed air (if available) blow out any remaining pieces of metal from the carburetor area. **Warning:** *Be sure to wear safety goggles whenever using compressed air.*

61   Install a special gauging tool (available from auto parts stores and automotive tool suppliers) into the "D"-shaped vent hole **(see illustration)**. The upper end of the tool should be positioned over the open cavity next to the idle air bleed valve.

62   While holding the gauging tool down lightly, engage the solenoid plunger against the solenoid stop. Adjust the idle air bleed valve so that the gauging tool will pivot over and just contact the top of the valve.

16.61 The special gauging tool shown is necessary to check the idle air bleed valve adjustment. Use a large screwdriver, as shown here, to adjust the valve

63   Remove the gauging tool. Next, it will be necessary to check the duty cycle (on-time) of the mixture control solenoid to verify that the air bleed is adjusted properly or if the idle mixture must be adjusted (see Section 12).

64   Disconnect the canister purge hose and plug the end (canister side). Start the engine and allow it to reach normal operating temperature (closed loop). Follow the procedure for checking the M/C solenoid duty cycle described earlier in this Section. If the dwell average is not within 25 to 35 degrees, then it will be necessary to adjust the idle mixture (refer to the procedure in this Chapter).

Idle load compensator (ILC) (some later engines)

Refer to illustrations 16.70 and 16.72

65   The idle load compensator (ILC) is used instead of an ISC solenoid on some later engines to control curb idle speed. The ILC uses manifold vacuum to sense changes in engine load and compensates by changing the idle speed. The idle load compensator is adjusted at the factory. It is not necessary to make any adjustments unless the curb idle speed is out of adjustment or the ILC solenoid was faulty and had to be replaced. Before adjusting the ILC, make sure the vacuum lines to the anti-diesel solenoid, vacuum regulator, ILC solenoid and all the components that require vacuum are in order and not leaking. Check the vacuum schematic on the VECI label under the hood for the correct vacuum routing.

66   Connect a tachometer according to the manufacturer's instructions. Remove the air cleaner and plug the hose to the thermal vacuum valve (TVV). Disconnect and plug the vacuum hose to the EGR valve. Disconnect and plug the vacuum hose to the canister purge port. Disconnect and plug the vacuum hose to the ILC.

67   Back out the idle-stop screw on the carburetor three turns **(see illustration 16.51)**.

68   Turn the A/C system OFF (if equipped). Place the transmission in PARK (automatic)

16.70 DO NOT adjust the ILC plunger to a point where the screw is extending more than 1-inch

16.72 Use a 3/32-inch Allen wrench to adjust the ILC solenoid (removed for clarity)

16.78 To adjust the TPS, drill out and remove the access plug (carburetor shown removed from vehicle for clarity) . . .

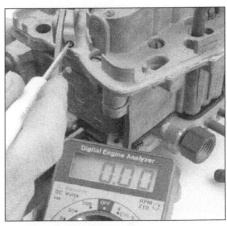

16.82 . . . then use a small screwdriver to turn the adjusting screw until the voltmeter readings are correct

or NEUTRAL (manual), apply the emergency brake and place a block under each of the drive wheels.

69   Start the engine and allow it to reach normal operating temperature. On automatic transmission models, have an assistant place his foot firmly on the brake pedal and place the transmission in DRIVE. The ILC plunger should be fully extended with no vacuum applied. Using back-up wrenches, adjust the ILC plunger to obtain 750 rpm, plus or minus 50 rpm.

70   Next, measure the distance from the jam nut to the tip of the plunger **(see illustration)**. It should not exceed 1-inch. If the plunger measures more than 1-inch, check for other carburetor problems.

71   Remove the plug from the vacuum hose and reconnect it to the ILC and observe the idle speed. Idle speed should be 450 rpm (vehicle in NEUTRAL (manual) or DRIVE with the brake applied). If the idle speed is correct, proceed to Step 74).

72   If the idle speed is not correct, it will be necessary to adjust the ILC diaphragm. Stop the engine and remove the ILC. Remove the center cap from the center outlet tube. Using a 3/32-inch Allen wrench, insert it through the open center tube to engage the idle speed adjusting screw inside **(see illustration)**. If the idle speed was low, turn the adjusting screw counterclockwise ONE turn to increase the idle speed approximately 75 rpm. Conversely, turn the screw clockwise ONE turn to lower the idle speed about 75 rpm. **Note:** *Turn the adjusting screw TWO full turns to raise or lower the idle speed approximately 150 rpm and consequently use the same ratio to calculate the necessary rpm change desired for the situation.*

73   Re-install the ILC onto the carburetor and attach the springs and other related parts. Recheck the idle speed. Make sure the engine is completely warmed up (closed loop). If it is not correct, repeat the ILC adjustment procedure.

74   The last adjustment must be performed on the engine after the TPS value has been reset by the ECM. This can be accomplished

by turning off the ignition for 10 seconds or more. Using a hand-held vacuum pump, apply vacuum to the ILC vacuum tube inlet to fully retract the plunger.

75   Adjust the idle stop screw on the carburetor to obtain 450 rpm with the vehicle in NEUTRAL (manual) or DRIVE with the brake applied.

76   Place the vehicle in PARK and stop the engine. Remove the plug from the ILC vacuum hose and install the hose onto the ILC. Reconnect all the vacuum hoses, install the air cleaner and gaskets.

Throttle Position Sensor (TPS)

Refer to illustrations 16.78 and 16.82

77   The Throttle Position Sensor is equipped with an adjustment screw that is covered over with a factory-installed plug. Do not remove the plug unless it has been determined with careful testing that the TPS is out of adjustment. Refer to Chapter 6 for complete diagnostic procedures for checking the TPS. This is a critical adjustment that must be performed correctly.

78   Using a 0.076-inch (5/64-inch) drill bit, carefully drill a hole in the aluminum plug covering the TPS adjustment screw **(see illustration)**. Drill only enough metal to start a self-tapping screw.

79   Start a number 8 X 1/2-inch long self-tapping screw into the drilled hole in the plug, turning the screw just enough to ensure good thread engagement in the hole.

80   Place a wide blade section of screwdriver between the screw head and the air horn casting. Carefully pry against the screw head to remove the plug. It is also possible to use a slide hammer.

81   Without disconnecting the TPS connector on the carburetor, backprobe the connector (using suitable probes) and connect the positive probe of the voltmeter to the dark blue wire and the negative probe on the black wire.

82   With the engine off and ignition on and the throttle closed, turn the screw until the voltage reading is approximately 0.48 volts **(see illustration)**.

83   Make sure the voltage is correct. Then

install a new plug into the air horn. Drive the plug into place until it is flush with the raised boss on the casting.

17   Throttle Body Injection (TBI) system - general information

Electronic fuel injection provides optimum fuel/air mixture ratios at all stages of combustion and offers better throttle response characteristics than carburetion. It also enables the engine to run at the leanest possible fuel/air mixture ratio, greatly reducing exhaust gas emissions.

All models employ a twin-injector Throttle Body Injection (TBI) unit, known as the Model 220.

The TBI system is controlled by an Electronic Control Module (ECM), which monitors engine performance and adjusts the fuel/air mixture accordingly during all engine operating conditions.

An electric fuel pump located in the fuel tank with the fuel gauge sending unit pumps fuel to the TBI unit through the fuel feed line

18.8 When disconnecting the fuel feed and return lines, be sure to use a back-up wrench to prevent damage to the lines

18.9 To detach the throttle body from the intake manifold, remove the bolts (arrows)

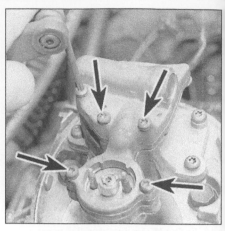

19.5 Unscrew the four cover retaining screws

19.6 Carefully peel away the old outlet passage and cover gaskets with a razor blade

19.7 DO NOT remove the four pressure regulator screws (arrows) from the fuel meter cover

and an in-line fuel filter. A pressure regulator in the TBI keeps fuel available to the injectors at a constant pressure between 9 and 13 psi. Fuel in excess of injector needs is returned to the fuel tank by a separate line.

18 Throttle Body Injection (TBI) unit - removal and installation

Warning: *Gasoline is extremely flammable, so extra precautions must be taken when working on any part of the fuel system. See the **Warning** in Section 2.*

Removal

Refer to illustrations 18.8 and 18.9
1   Disconnect the cable from the negative battery terminal.
2   Remove the air cleaner housing.
3   The fuel pressure regulator has an internal bleed valve to bleed-off fuel pressure, but a small amount of fuel will be released when the fuel lines are disconnected. Wrap a shop towel around the fuel inlet line fitting and slowly loosen the fitting. Dispose of the fuel soaked rag in an approved safety container.
4   Unplug the electrical connectors from the idle air control valve, throttle position sensor and fuel injectors.
5   Remove the wiring harness and insulating grommet from the throttle body.
6   Remove the throttle linkage and return spring, transmission control and cruise control cables (if applicable).
7   Using pieces of numbered tape, mark all of the vacuum hoses to the throttle body and disconnect them.
8   Disconnect the fuel inlet and return lines. Use a back-up wrench on the inlet and return fitting nuts to prevent damage to the throttle body and fuel lines **(see illustration)**. Remove the fuel line nut O-rings and discard them.
9   Remove the TBI assembly mounting bolts and lift the unit off the intake manifold **(see illustration)**. Stuff a rag into the intake manifold opening and carefully scrape all

traces of old gasket material off the intake manifold and throttle body mating surfaces. Scrape carefully so you don't damage the delicate aluminum surfaces.

Installation

10   Installation is the reverse of the removal procedure. Be sure to install a new throttle body-to-intake manifold gasket, new fuel line O-rings and tighten the mounting bolts to 132 in-lbs.
11   Turn the ignition switch on without starting the engine and check for fuel leaks.
12   Check to see if the accelerator pedal is free by depressing it to the floor and releasing it (with the ignition switch off).

19 Throttle Body Injection (TBI) unit - component replacement

Warning: *Gasoline is extremely flammable, so extra precautions must be taken when working on any part of the fuel system. See the **Warning** in Section 2.*
Note: *The fuel injectors, pressure regulator, throttle position sensor and idle air control valve can be replaced without removing the*

throttle body assembly.
1   Detach the cable from the negative terminal of the battery.
2   Remove the air cleaner housing assembly, adapter and gaskets.
3   The fuel pressure regulator has an internal bleed valve to bleed-off fuel pressure, but a small amount of fuel will be released when the fuel lines are disconnected. Wrap a shop towel around the fuel inlet line fitting and slowly loosen the fitting. Dispose of the fuel soaked rag in an approved safety container.

Fuel meter cover/fuel pressure regulator assembly

Refer to illustrations 19.5, 19.6 and 19.7
Note: *The fuel pressure regulator is housed in the fuel meter cover. Whether you are replacing the meter cover or the regulator itself, the entire assembly must be replaced. The regulator must not be removed from the cover.*
4   Unplug the electrical connectors to the fuel injectors.
5   Remove the cover screws **(see illustration)** and detach the fuel meter cover.
6   Remove the fuel meter outlet passage gasket, cover gasket and pressure regulator seal. Carefully remove old gasket material

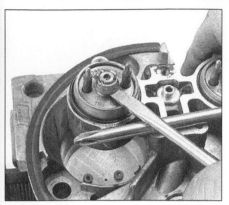

19.16 To remove an injector, slip the tip of a standard screwdriver under the lip of the lug on top of the injector and, using another screwdriver as a fulcrum, carefully pry the injector up and out

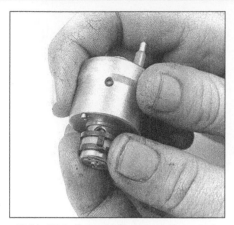

19.21 Slide the new filter onto the nozzle of the fuel injector

19.22 Lubricate the lower O-ring with transmission fluid, then place it on the shoulder in the bottom of the injector cavity

19.23 Place the steel back-up washer on the shoulder near the top of the injector cavity

19.24 Lubricate the upper O-ring with transmission fluid, then install it on top of the steel washer

19.25 Make sure the lug is aligned with the groove in the bottom of the fuel injector cavity

that's stuck with a razor blade **(see illustration)**. **Caution:** Do not attempt to reuse either of the gaskets.

7   Inspect the cover for dirt, foreign material and casting warpage. If it's dirty, clean it with a shop rag soaked in solvent. Do not immerse the fuel meter cover in solvent - it could damage the pressure regulator diaphragm and gasket. **Warning:** *Do not remove the four screws* **(see illustration)** *securing the pressure regulator to the fuel meter cover. The regulator contains a large spring under compression which, if accidentally released, could cause injury. Disassembly might also result in a fuel leak between the diaphragm and the regulator housing. The new fuel meter cover assembly will include a new pressure regulator.*

8   Install the new pressure regulator seal, fuel meter outlet passage gasket and cover gasket.

9   Install the fuel meter cover assembly using Loctite 262 or equivalent on the screws. **Note:** The short screws go next to the injectors.

10   Attach the electrical connectors to both injectors.

11   Attach the cable to the negative terminal of the battery.

12   With the engine off and the ignition on, check for leaks around the gasket and fuel line couplings.

13   Install the air cleaner, adapter and gaskets.

Fuel injector assembly

Refer to illustrations 19.16, 19.21, 19.22, 19.23, 19.24 and 19.25

14   To unplug the electrical connectors from the fuel injectors, squeeze the plastic tabs and pull straight up.

15   Remove the fuel meter cover/pressure regulator assembly. **Note:** *Do not remove the fuel meter cover assembly gasket - leave it in place to protect the casting from damage during injector removal.*

16   Use a screwdriver and fulcrum **(see illustration)** to pry out the injector.

17   Remove the upper (larger) and lower (smaller) O-rings and filter from the injector.

18   Remove the steel back-up washer from the top of the injector cavity.

19   Inspect the fuel injector filter for dirt and contamination. If present, check for the presence of dirt in the fuel lines and fuel tank.

20   Be sure to replace the fuel injector with an identical part. Injectors from other models can fit in the Model 220 TBI assembly but are calibrated for different flow rates.

21   Slide the new filter into place on the nozzle of the injector **(see illustration)**.

22   Lubricate the new lower (smaller) O-ring with automatic transmission fluid and place it on the small shoulder at the bottom of the fuel injector cavity in the fuel meter body **(see illustration)**.

23   Install the steel back-up washer in the injector cavity **(see illustration)**.

24   Lubricate the new upper (larger) O-ring with automatic transmission fluid and install it on top of the steel back-up washer **(see illustration)**. **Note:** *The back-up washer and large O-ring must be installed before the injector. If they aren't, improper seating of the large O-ring could cause fuel leakage.*

25   To install the injector, align the raised lug on the injector base with the groove in the fuel meter body cavity **(see illustration)**. Push down on the injector until it's fully seated in the fuel meter body. **Note:** *The electrical terminals should be parallel with the throttle shaft.*

26   Install the fuel meter cover assembly and gasket.
27   Attach the cable to the negative terminal of the battery.
28   With the engine off and the ignition on, check for fuel leaks.
29   Attach the electrical connectors to the fuel injectors.
30   Install the air cleaner housing assembly, adapter and gaskets.

Throttle Position Sensor (TPS)

31   For information on the TPS see Chapter 6.

Idle Air Control (IAC) valve

32   For information on the IAC valve see Chapter 6.

Fuel meter body assembly

Refer to illustrations 19.37 and 19.39
33   Unplug the electrical connectors from the fuel injectors.
34   Remove the fuel meter cover/pressure regulator assembly, fuel meter cover gasket, fuel meter outlet gasket and pressure regulator seal.
35   Remove the fuel injectors.
36   Unscrew the fuel inlet and return line threaded fittings, detach the lines and remove the O-rings.
37   Remove the fuel inlet and outlet nuts and gaskets from the fuel meter body assembly **(see illustration)**. Note the locations of the nuts to ensure proper reassembly. The inlet nut has a larger passage than the outlet nut.
38   Remove the gasket from the inner end of each fuel nut.
39   Remove the fuel meter body-to-throttle body screws and detach the fuel meter body from the throttle body **(see illustration)**.
40   Install the new throttle body-to-fuel meter body gasket. Match the cut-out portions in the gasket with the openings in the throttle body.
41   Install the fuel meter body on the throttle body. Coat the fuel meter body-to-throttle body screws with thread locking compound before installing them.
42   Install the fuel inlet and outlet nuts, with new gaskets, in the fuel meter body and tighten the nuts to 17 ft-lbs. Install the fuel inlet and return line threaded fittings with new O-rings. Use a back-up wrench to prevent the nuts from turning.
43   Install the fuel injectors.
44   Install the fuel meter cover/pressure regulator assembly.

19.37 Remove the fuel inlet and outlet nuts from the fuel meter body

45   Attach the cable to the negative terminal of the battery.
46   Attach the electrical connectors to the fuel injectors.
47   With the engine off and the ignition on, check for leaks around the fuel meter body, the gasket and around the fuel line nuts and threaded fittings.
48   Install the air cleaner housing assembly, adapters and gaskets.

---

20   Exhaust system servicing - general information

Warning: *Inspection and repair of exhaust system components should be done only after enough time has elapsed after driving the vehicle to allow the system components to cool completely. Also, when working under the vehicle, make sure it is securely supported on jackstands.*
Refer to illustration 20.1
1   The exhaust system consists of the exhaust manifold(s), the catalytic converter, the muffler, the tailpipe and all connecting pipes, brackets, hangers and clamps **(see illustration)**. The exhaust system is attached to the body with mounting brackets and rubber hangers. If any of these parts are improperly installed, excessive noise and vibration will be transmitted to the body.
2   Conduct regular inspections of the exhaust system to keep it safe and quiet. Look for any damaged or bent parts, open seams, holes, loose connections, excessive corrosion or other defects which could allow exhaust fumes to enter the vehicle. Deteriorated exhaust system components should

19.39 Once the fuel inlet and outlet nuts are off, pull the fuel meter body straight up to separate it from the throttle body

not be repaired; they should be replaced with new parts.
3   If the exhaust system components are extremely corroded or rusted together, they will probably have to be cut from the exhaust system. The convenient way to accomplish this is to have a muffler repair shop remove the corroded sections with a cutting torch. If, however, you want to save money by doing it yourself (and you don't have an oxy/acetylene welding outfit with a cutting torch), simply cut off the old components with a hacksaw. If you have compressed air, special pneumatic cutting chisels can also be used. If you do decide to tackle the job at home, be sure to wear OSHA-approved safety goggles to protect your eyes from metal chips and work gloves to protect your hands.
4   Here are some simple guidelines to apply when repairing the exhaust system:
   a) *Work from the back to the front when removing exhaust system components.*
   b) *Apply penetrating oil to the exhaust system component fasteners to make them easier to remove.*
   c) *Use new gaskets, hangers and clamps when installing exhaust system components.*
   d) *Apply anti-seize compound to the threads of all exhaust system fasteners during reassembly.*
   e) *Be sure to allow sufficient clearance between newly installed parts and all points on the underbody to avoid overheating the floor pan and possibly damaging the interior carpet and insulation. Pay particularly close attention to the catalytic converter and its heat shield.*

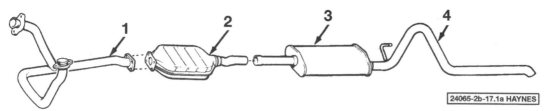

20.1 Typical exhaust system components

1   *Exhaust crossover pipe*
2   *Catalytic converter*
3   *Muffler*
4   *Exhaust pipe*

24065-2b-17.1a HAYNES

Chapter 4 Part B
Turbocharger (231 V6)

Contents

Specifications

Torque specifications

Ft-lbs (unless other wise indicated)

Exhaust outlet pipe-to-elbow	14
Elbow-to compressor housing	15
Exhaust inlet pipe-to-turbine housing	14
Exhaust inlet pipe-to-right manifold	14
Oil feed pipe fitting (both ends)	13
Oil feed pipe fitting-to-CHRA	7
CHRA-to-turbine housing	15
CHRA backplate-to-compressor housing	13
Compressor housing-to-plenum	20
Compressor housing-to-intake manifold	35
Oil drain-to-CHRA	15
EGR valve-to-manifold	15
EGR valve manifold-to-intake manifold	15
EGR valve manifold-to-plenum	15
Carburetor-to-plenum	21
Plenum front bracket-to-intake manifold	20
Plenum front bracket-to-plenum	21
PEVR-to-intake manifold	25
Turbine housing bracket-to-intake manifold	20
Turbine housing bracket-to-housing	18
Power brake vacuum line-to-plenum	10
Plenum side support bracket bolts	21
Linkage bracket-to-plenum	20

1 Turbocharger - general information

Turbocharging offers a way of raising the horsepower and torque output of an engine without the power-robbing belts, drives, and gears of a supercharger. The turbo-charger, a snail-shaped device that mounts on the 231 V6 engine between the carburetor and the exhaust manifold, is capable of increasing horsepower by approximately 35% and torque by about 25%. This increase in power is not constant, however. The unit works on an as needed basis.

An Electronic Spark Control (ESC) computer controls exactly when the turbocharger works and when it doesn't. This unit senses engine rpms (through a pickup in the HEI distributor) and engine detonation (via a detonation sensor on the engine block). When the unit determines that additional power is needed, it sends a signal to the turbo and additional boost is gained. When boost is not required, the control center makes sure that the engine remains normally aspirated.

The turbocharger is made up of two turbine wheels mounted to a common shaft. Each wheel is enclosed by a shroud which directs air flow. One shroud connects to the exhaust manifold. This is the turbine unit. The other shroud is linked to the carburetor and the intake manifold. This unit is called a compressor. The unit works as follows:

Compression of fuel and air in the combustion chamber of an engine is not just a mechanical action of the piston; when fuel/air mixtures are compressed, they gain heat and pressure due to the heat gain. When this pressurized gas is routed through a small exhaust port and down a narrow exhaust pipe, it gains velocity. Directing this pressurized exhaust into the turbine unit provides power for the turbine. The snail-shaped shroud (housing) keeps the gases tightly compressed but allows them to expand and cool as they leave the chamber. The change in heat and velocity is what powers the turbine wheel.

The compressor wheel is driven by the turbine. The fuel/air mixture is compressed as it enters the unit and a charge of compressed gases exits the compressor. From the compressor, the charge enters each cylinder and is burned as in a normally aspirated engine.

Turbine speeds can reach 50 000 rpm (or higher) during normal operation of the engine. Since engine speed determines both the speed of the turbine and the size of the charge, it is necessary that the system be protected from an overspeed condition, which could damage both the engine and the turbo. A wastegate, which controls speed and pressure in the turbine, is installed in the system as a safety device to prevent such overspeed conditions from occurring.

The wastegate, which opens and allows exhaust gas to bypass the turbine, slows the compressor and turbine wheels.

2 Turbocharger - cautions

Note: *Because of the critical nature of the working conditions required for repairs to be effectively performed on the turbocharger, it is recommended that such repairs be left to a Dealer service department or other qualified repair shop.*

1   Driven by superheated exhaust gases and routinely operating at extremely high temperatures, the turbocharger castings retain engine heat for a very long time. **Caution:** *It is very important to let the engine cool for a period of at least three hours after it has been run. Even then, it is prudent to wear heavy gloves when handling the turbo unit to prevent serious burns.*

2   The high-speed operation of the turbo dictates that bearing life is dependent upon a constant flow of clean engine oil. Careful attention should always be paid to the condition of the oil lines and to the tightness of their fittings. Overtightening will cause deformation and will lead to leaks.

3   Always change the engine oil and filter after any time that the turbo unit is removed.

4   When dismantling the turbo, take care that no bearings, washers, nuts or screws fall into the turbine. In a unit that operates at speeds up to 50 000 rpms, even a small amount of grit entering the turbo can cause severe damage to both the turbo and the engine. If an object is suspected of falling into the turbine or one of the passageways, flush the object out. After all servicing is done on the unit, flush all passageways out with clean oil. The best defense against dirt is to work on the turbo only after the engine has been steam cleaned.

5   Cover all inlets and pipes while the unit is dismantled. Account for every nut, screw, bolt and washer before and after the unit is reassembled. Don't try to rush your repair work. This will only lead to mistakes that could be very costly.

6   Be exceptionally careful when dismantling the turbocharger. Take special care not to bend, scratch or nick the compressor or turbine wheels. Even minor scratches on the blades can result in an imbalance that will result in failure of the unit.

7   Before disassembly of the turbocharger, scribe the center housing rotating assembly where it bolts up to the compressor and the turbine. The unit should be reassembled in the same basic position that it was in before disassembly.

8   If any sealer is found at any point in turbo disassembly (between the center housing rotating assembly backplate and the compressor housing, for example), it should be replaced when the unit is put back together.

3 Wastegate actuator boost pressure - testing

Note: *Read Section 2 before proceeding.*
1   Inspect all connections for proper hookup and check hoses for leaks and cracks. Inspect the mechanical linkage for damage.

2   Disconnect the hose that runs from the actuator to the compress housing at the actuator. Attach a hand-operated vacuum/pressure pump with gauge and apply 9 psi to the actuator. Somewhere between 8.5 and 9.5 psi, the actuator rod should move 0.015 inch and actuate the wastegate linkage.

3   If it does not work as outlined, replace the actuator and calibrate the linkage for 9 psi by crimping the threads on the rod once it has been turned for the correct setting. Reconnect the compressor housing-to-actuator hose.

4 Wastegate actuator assembly - removal and installation

Note: *Read Section 2 before proceeding.*
1   Disconnect the two hoses from the unit and remove the retaining clip from the actuator assembly rod.
2   Remove the two bolts that hold the actuator to the compressor.
3   Installation is the reverse of removal.

5 Power enrichment vacuum regulator (PEVR) - testing

Note: *Read Section 2 before proceeding.*
1   The power enrichment vacuum regulator is located in front of and between the turbocharger and the carburetor and threads into the intake manifold.
2   Check the PEVR and hoses for proper installation, cracks and other damage.
3   Connect 1 hose from a manometer (available at specialty tool manufacturers) between the yellow-striped input hose and the input port (use a 'T' fitting). Connect the remaining manometer hose directly to the PEVR output port.
4   Start the engine and allow, it to idle while observing the manometer. There should be no more than a 14 inch H20 difference. If there is, replace the PEVR with a new one.
5   If the preceding test proves inconclusive, remove the PEVR from the intake manifold and plug the manifold port on the valve, then reconnect the hoses to the PEVR.
6   Connect a pressure/vacuum gauge to the PEVR output hose (use a 'T' fitting), then start the engine and allow it to idle. The gauge reading should be 7 to 9 in Hg (vacuum).
7   Using a hand operated vacuum/pressure pump, apply 3 psi to the manifold signal port of the PEVR. The gauge reading at the output hose should now be 1.4 to 2.6 in Hg. If it is difficult to measure such a low level of vacuum with the gauge in use, apply at least 5 psi to the manifold signal port and check for a gauge reading of zero at the output hose.

8   If the PEVR does not check out as indicated, replace it with a new one.

6   Turbocharger - removal and installation

Note: *Read Section 2 before proceeding.*
1   Disconnect the exhaust outlet and inlet pipes at the turbo.
2   Unhook the oil pipe from the CHRA (Center Housing Rotating Assembly). Wipe up any spilled oil with a rag.
3   Undo the nut that attaches the air intake elbow to the carburetor and remove the elbow from the carburetor. Leave it attached to the flex tube.
4   Unhook the throttle, detent and cruise linkages from the carburetor. Disconnect the linkage bracket at the plenum. The plenum is the mixing box located underneath the carburetor. Take care not to lose any clips, screws, or nuts.
5   Remove the two bolts that attach the plenum to the side bracket.
6   Unhook the fuel line from the carburetor and plug the end. Take care to disconnect any necessary vacuum hoses and mop up any spilled fuel.
7   Drain the cooling system and disconnect the coolant hoses at the front and rear of the plenum. Unhook the power brake vacuum line at the plenum.
8   Disconnect the front plenum bracket by removing the bolt that attaches the bracket to the intake manifold. Leave the plenum attached to the bracket.
9   Remove the two bolts that attach the turbine housing to the bracket on the intake manifold and the two bolts that hold the EGR valve manifold to the plenum. Loosen the bolts that attach the EGR valve manifold to the intake manifold.
10   Next, loosen the clamp that attaches the hose from the .41R by-pass pipe to the check valve and remove the hose from the pipe.
11   Remove the three bolts that hold the compressor housing to the manifold and remove the turbocharger assembly. The actuator will be attached to the turbo. The turbo will still be attached to the carburetor and the plenum. Unhook any hoses still connected to the turbo assembly.
12   Now, unbolt the turbo/actuator assembly from the carburetor and plenum.
13   Take out the oil drain from the center

housing rotating assembly and let any oil in the unit drain out.
14   Assembly and installation of the unit is the exact reverse of the removal procedure. However, before the turbocharger is made totally operational, it is a good idea to change the engine oil and filter. This will insure a constant supply of clean oil to the unit.

7   Plenum - replacement

1   Steps 1 through 12 in the previous section cover removal of the plenum. If, however, you intend to replace the unit, it will be necessary to transfer all fittings and hoses to the new unit. Check back to Step 3 for a list of the parts that belong on the plenum. Installation of the plenum is covered in the previous section. Reverse the turbo removal procedure.

8   EGR valve manifold - removal and installation

Note: *Read Section 2 before proceeding.*
1   There are six bolts to unfasten in order to remove the EGR valve manifold; two attach the EGR valve to the valve manifold, two attach the valve manifold to the plenum, and two attach the valve manifold to the intake manifold. More information on the EGR is in Chapter 6.
2   Remove these bolts and unhook the vacuum line that runs to the EGR valve.
3   When installing, first loosely install the bolts that attach the valve manifold to the intake manifold. Next, loosely install the bolts that attach the valve manifold to the plenum.
4   Tighten the valve manifold-to-intake manifold bolts, then tighten the valve manifold-to-plenum bolts.
5   Bolt up the EGR valve to the valve manifold and install the vacuum line.

9   Turbocharger elbow assembly - removal and installation

1   Jack up the car and block the rear wheels. Set the parking brake.
2   Unbolt the turbo exhaust outlet from the catalytic converter.

3   Lower the car.
4   Remove the clip that attaches the wastegate linkage to the actuator rod and disconnect the turbocharger exhaust outlet from the elbow assembly.
5   Remove the bolts that attach the elbow assembly to the turbine housing.
6   Installation is the reverse of removal.

10   Turbocharger - disassembly and inspection

Note: *Read Section 2 before proceeding.*
1   Removal of the turbocharger is covered in Section 6. If the actuator is still connected to the turbo (you might have taken it off in Section 4), it will be necessary to remove the two bolts that hold the actuator to the turbo and disconnect the hose that runs from the compressor housing to the actuator at the housing.
2   Spin the compressor wheel in the Center Housing Rotating Assembly (CHRA) gently. If there is any binding, replace the CHRA.
3   Unbolt the drain from the CHRA and inspect for carbon build-up and oil sludging. If debris and dirt are minor, clean the area with a commercial solvent cleaner. Do NOT use gasoline or other fuels. If the unit is severely plugged, replace it.
4   Inspect the CHRA compressor wheel for signs of oil leakage. If there is leakage, replace the CHRA.
5   With the turbo on a bench, remove the six bolts and three clamps that hold the compressor housing to the turbine housing.' Be careful not to bang the CHRA when taking the compressor off the turbine.
6   Look at the CHRA wheels carefully. If there are any broken blades, scratches or other damage, replace the CHRA.
7   If the CHRA is being replaced, lubricate all fitting surfaces and the center shaft with clean engine oil. Carefully install the unit into the turbine housing and using all the bolts and clamps, bolt the turbocharger back together. Tighten the bolts in three steps, using a criss-cross pattern, to the specified torque.
8   If the CHRA unit seems OK and you wish to install it back into the turbocharger assembly, take the turbo to your Buick dealer or a competent machine shop and have the journal bearings inspected for proper clearance.

Notes

Chapter 5
Engine electrical systems

Contents

Specifications

Distributor type
1974 and earlier .... Mechanical contact breaker point
1975 and later .... Breakerless. Designated High Energy Ignition (HEI)

Distributor direction of rotation .... Clockwise

Point gap .... 0.019 in

Firing order
V6 1-6-5-4-3-2
V8 1-8-4-3-6-5-7-2

Condenser capacity (1974 and earlier) .... 0.18 to 0.23 mfd

Coil (with breaker type distributor)
Primary resistance .... 1.77 to 2.05 Ohms
Secondary resistance .... 3000 to 20 000 Ohms
Resistor .... 1.35 Ohms

Coil in cap (HEI distributor)
Primary resistance .... 0.41 to 0.51 Ohms
Secondary resistance .... 3000 to 20 000 Ohms
Resistor .... 0.43 to 0.68 Ohms

Alternator
Type (according to year of manufacture)... Delcotron Series 1 D, 10SI or 10DN Series 100B
Field current.. 2.2/2.6 amps at 80°F (pre-1973)
    4.0/4.5 amps at 80°F (1973 onward)
Output current .... Varies according to vehicle specification and alternator type

Voltage regulator (external type)
Field relay
    Air gap .... 0.015 in
    Point opening .... 0.030 in
    Closing voltage.. 1.5/3.2 volts
Regulator
    Air gap .... 0.067 in
    Point opening .... 0.014 in
    Voltage setting .... 13.8/14.8 V at 85° F

Spark plugs type and gap
Spark plugs type and gap .... see *Specifications* Section in Chapter 1 or vehicle emissions label

Ignition timing
Ignition timing .... see *Specifications* Section in Chapter 1 or vehicle emissions label

Distributor application chart (1981 and earlier)

Year and engine	Distributor No.	Centrifugal advance crank degrees @ engine rpm	Vacuum advance crank degrees @ in hg	Dwell angle degrees	Ignition timing at engine idle degrees BTDC	Transmission type
1970						
350 cu in 250 HP	1112001	0 @ 800 3 @ 1000 15 @ 1800 36 @ 4100	0 @ 7 24 @ 17	29 to 31	TDC	Manual
350 cu in 250 HP	1112002	0 @ 900 2 @ 1100 8 @ 1400 32 @ 4400	0 @ 7 24 @ 17	29 to 31	4	Automatic
350 cu in 300 HP	1111996	0 @ 950 14 @ 1400 20 @ 1800 30 @ 4700	0 @ 10 15 @ 17	29 to 31	TDC	Manual
350 cu in 300 HP	1111997	0 @ 900 9 @ 1300 15 @ 1700 26 @ 4700	0 @ 10 20 @ 17	29 to 31	4	Automatic
400 cu in 265 HP	1111492	0 @ 800 2 @ 1000 20 @ 2450 32 @ 4400	0 @ 6 15 @ 12	29 to 31	4	Manual
400 cu in 265 HP	1111494	0 @ 700 2 @ 1100 12 @ 2000 28 @ 4400!	0 @ 6 15 @ 12	29 to 31	8	Automatic
400 cu in 330 HP	1111998	0 @ 900 9 @ 1250 17 @ 2D00 32 @ 5000	0 @ 8 15 @ 15.5	29 to 31	4	Automatic and Manual
454 cu in 360 HP	1111436	0 @ 815 2 @ 1085 17 @ 2100 26 @ 4000	0 @ 8 15 @ 15.5	29 to 31	6	Automatic and Manual

Year and engine	Distributor No.	Centrifugal advance crank degrees @ engine rpm	Vacuum advance crank degrees @ in hg	Dwell angle degrees	Ignition timing at engine idle degrees BTDC	Transmission type
1971						
350 cu in 245 HP	1112042	0 @ 880 2 @ 1120 10 @ 1600 15 @ 2200 28 @ 4300	0 @ 8 20 @ 17	29 to 31	2	Manual
350 cu in 245 HP	1112005	0 @ 800 2 @ 1200 12 @ 2200 24 @ 4300	0 @ 8 20 @ 17	29 to 31	6	Automatic
350 cu in 270 HP	1112044	0 @ 840 2 @ 1160 10 @ 1800 15 @ 2400	0 @ 8 15 @ 15.5	29 to 31	4	Manual
350 cu in 270 HP	1112045	0 @ 865 2 @ 1335 11 @ 2400 18 @ 4200	0 @ 8 15 @ 15.5	29 to 31	8	Automatic
402 cu in 300 HP	1112057	0 @ 930 2 @ 1260 16 @ 2400 30 @ 4400	0 @ 8 20 @ 17	29 to 31	8	Automatic or Manual
454 cu in 365 HP	1112052	0 @ 857 2 @ 1143 14 @ 2000 22 @ 3900	0 @ 8 20 @ 17	29 to 31 Manual	8	Automatic or
1972						
350 cu in 165 HP	1112005	0 @ 800 2 @ 1200 12 @ 2200 24 @ 4300	0 @ 8 20 @ 17	29 to 31	6	Automatic or Manual
350 cu in 175 HP	1112044	0 @ 840 2 @ 1160 10 @ 1800 15 @ 2400	0 @ 8 15 @ 15.5	29 to 31	4	Manual
350 cu in 175 HP	1112045	0 @ 865 2 @ 1335 11 @ 2400 18 @ 4200	0 @ 8 15 @ 15.5	29 to 31	8	Automatic
402 cu in 240 HP	1112057	0 @ 930 2 @ 1260 16 @ 2400 30 @ 4400	0 @ 8 20 @ 17	29 to 31	8	Automatic or Manual
454 cu in 270 HP	1112051	0 @ 857 2 @ 1143 14 @ 2000 22 @ 3900	0 @ 8 20 @ 17	29 to 31	8	Automatic or Manual
1973						
350 cu in 145 HP	1112168	0 @ 1000 2 @ 1300 10 @ 2600 20 @ 4200	0 @ 4 14 @ 7	29 to 31	8	Automatic or Manual
350 cu in 175 HP	1112094	0 @ 1100 2 @ 1550 6 @ 2410 12 @ 3300 14 @ 4200	0 @ 6 15 @ 14	29 to 31	12	Automatic

Year and engine	Distributor No.	Centrifugal advance crank degrees @ engine rpm	Vacuum advance crank degrees @ in hg	Dwell angle degrees	Ignition timing at engine idle degrees BTDC	Transmission type
1973 (continued)						
350 cu in 175 HP	1112093	0 @ 1100 2 @ 1320 6 @ 1800 11 @ 2400 18 @ 4200	0 @ 6 15 @ 14	29 to 31	8	Manual
454 cu in 245 HP	1112113	0 @ 1100 2 @ 1320 11 @ 2400 18 @ 4200	0 @ 6 20 @ 15	29 to 31	10	Automatic or Manual
1974						
350 cu in 145 HP	1112168	0 @ 1000 20 @ 2400	0 @ 2 to 4 14 @ 7.5 to 8.5	29 to 31	TDC (Manual) 8 (Automatic)	Automatic or Manual
350 cu in 160 HP	1112847	0 @ 1100 11 @ 2400 18 @ 4200	0 @ 2 to 4 14 @ 7.5 to 8.5	29 to 31	4 (Manual) 8 (Automatic)	Automatic
400 cu in 150 HP	1112846	0 @ 1000 20 @ 4200	0.3 to 5 15 @ 9.5 to 10.5	29 to 31	8	Automatic
400 cu in 180 HP	1112854	0 @ 1000 20 @ 4200	0 @ 3 to 5 15 @ 9.5 to 10.5	29 to 31	8	Automatic
400 cu in 180 HP	1112854	0 @ 1000 20 @ 4200	0 @ 3 to 5 15 @ 9.5 to 10.5	29 to 31	8	Automatic (California)
454 cu in 235 HP	1112113	0 @ 1100 11 @ 2400 18 @ 4200	0 @ 5 to 7 20 @ 14.2 to 15.7	29 to 31	10	Automatic or Manual (California)
454 cu in 235 HP	1112504	0 @ 1100 11 @ 2400 18 @ 4200	0 @ 7 to 9 16 @ 15 to 16	29 to 31	10	Automatic (Federal)
1975						
350 cu in 145 HP	1112880	0 @ 1200 12 @ 2000 22 @ 4200	0 @ 4 18 @ 12	HEI	6	Automatic or Manual
350 cu in 155 HP	1112880	0 @ 1200 12 @ 2000 22 @ 4200	0 @ 4 18 @ 12	HEI	6	Automatic or Manual
400 cu in 175 HP	1112882	0 @ 1000 8 @ 1600 15 @ 2800	0 @ 8 15 @ 15.5	HEI	8	Automatic or Manual
454 cu in 215 HP	1112886	0 @ 1800 12 @ 4200	0 @ 4 18 @ 7	HEI	16	Automatic or Manual
1976						
305 cu in 140 HP	1112977	0 @ 1000 10 @ 1700 20 @ 3800	0 @ 4 18 @ 12	HEI	6 (Manual) 8 (Automatic)	Automatic or Manual
305 cu in 140 HP	1112999	0 @ l000 l0 @ 1700 20 @ 3800	0 @ 4 10 @ 8	HEI	0	Automatic (California)
350 cu in 145 HP	1112880	0 @ 1200 12 @ 2000 22 @ 4200	0 @ 4 18 @ 12	HEI	6	Automatic
350 cu in 165 HP	1112888	0 @ 1100 12 @ 1600 16 @ 2400 22 @ 4600	0 @ 4 18 @ 12	HEI	8	Automatic or Manual

Year and engine	Distributor No.	Centrifugal advance crank degrees @ engine rpm	Vacuum advance crank degrees @ in hg	Dwell angle degrees	Ignition timing at engine idle degrees BTDC	Transmission type
350 cu in 165 HP	1112905	0 @ 1200 12 @ 2000 22 @ 4200	0 @ 6 154 @ 12	HEI	6	Automatic or Manual (California)
400 cu in 175 HP	1103203	0 @ 1000 8 @ 1600 15 @ 2800	0 @ 6 15 @ 12	HEI	8	Automatic
400 cu in 175 HP	1112882	0 @ 1000 8 @ 1600 15 @ 2800	0 @ 8 15 @ 15.5	HEI	8	Automatic (California)

1977

Year and engine	Distributor No.	Centrifugal advance crank degrees @ engine rpm	Vacuum advance crank degrees @ in hg	Dwell angle degrees	Ignition timing at engine idle degrees BTDC	Transmission type
305 cu in 145 HP	1103239	0 @ 1200 15 @ 2700 20 @ 4200	0 @ 4 15 @ 10	HEI	8	Automatic
305 cu in 145 HP	1103244	0 @ 1000 10 @ 1700 20 @ 3800	0 @ 4 20 @ 10	HEI	6	Automatic (California)
350 cu in 170 HP	1103246	0 @ 1200 12 @ 2000 22 @ 4200	0 @ 4 18 @ 12	HEI	8	Automatic
350 cu in 170 HP	1103248	0 @ 1200 12 @ 2000 22 @ 4200	0 @ 4 10 @ 8	HEI	8	Automatic (California)

1978

Year and engine	Distributor No.	Centrifugal advance crank degrees @ engine rpm	Vacuum advance crank degrees @ in hg	Dwell angle degrees	Ignition timing at engine idle degrees BTDC	Transmission type
231 cu in	1110731	0-4 @ 2000 12-18 @ 3600	0 @ 4-6 16.@7-9	HEI	15	Automatic (California)
231 cu in	1110695	0-6 @ 2000 12-18 @ 3600	0 @ 7-9 24 @ 10-13	HEI	15	Automatic
305 cu in	1103281	0 @ 1000 10 @ 1700 20 @ 3800	0 @ 4 18 @ 12	HEI	4	Manual
305 cu in	1103282	0 @ 1000 10 @ 1700 20 @ 3800	0 @ 4 20 @ 10	HEI	4	Automatic

1979

Year and engine	Distributor No.	Centrifugal advance crank degrees @ engine rpm	Vacuum advance crank degrees @ in hg	Dwell angle degrees	Ignition timing at engine idle degrees BTDC	Transmission type
200 cu in	1110756	0 @ 1400 4 @ 1700 14 @ 3700	0 @ 2 24 @ 10	HEI	12	Automatic
200 cu in	1110696	0 @ 1400 4 @ 1700 14 @ 3800	0 @ 2 16 @ 7.5	HEI	8	Manual
231 cu in	111766	0-4 @ 2000 13-17 @ 3600	0-2 @ 4 24 @ 12	HEI	15	Automatic
231 cu in	1110767	0-4 @ 2000 13-17 @ 3600	0-2 @ 3 20 @ 12	HEI	15	Automatic (California)
267 cu in	1103371	0 @ 1000 8 @ 1700 22 @ 4400	0 @ 3 24 @ 10	HEI	8	Manual
267 cu in	1103370	0 @ 1300 16 @ 4200	0 @ 3 24 @ 10	HEI	8	Automatic
305 cu in	1103282	0 @ 1000 10 @ 1700 20 @ 3800	0 @ 4 20 @ 11	HEI	4	Manual
305 cu in	1103379	0 @ 1000 10 @ 1700 20 @ 3800	0 @ 3 20 @ 8.5	HEI	4	Automatic

Year and engine	Distributor No.	Centrifugal advance crank degrees	Vacuum advance crank degrees	Dwell angle degrees	Ignition timing at engine idle	Transmission type

1979 (continued)

Year and engine	Distributor No.	@ engine rpm	@ in hg		degrees BTDC	
305 cu in	1103282	0 @ 1000 10 @ 1700 20 @ 3800	0 @ 4 20 @ 11	HEI	4	Automatic
305 cu in	1103368	0 @ 1000 10 @ 1700 20 @ 3800	0 @ 4 10 @ 8	HEI	4	Automatic (California)

1980

Year and engine	Distributor No.	Centrifugal advance	Vacuum advance	Dwell	Ignition timing	Transmission
229 cu in	1110696	0 @ 1000 10 @ 1700 10 @ 3800	0 @ 2 16 @ 7.5	HEI	8	Manual
229 cu in	1110756	0 @ 1400 4 @ 1700 14 @ 3800	0 @ 2 24 @ 10	HEI	12	Automatic
231 cu in	1110766	0-4 @ 2000 13-17 @ 3600	0-2 @ 4 24 @ 12	HEI	15	Automatic
231 cu in	1110767	0-4 @ 2000 13-17 @ 3600	0-2 @ 3 20 @ 12	HEI	15	Automatic (California)
231 cu in	1110766	0-4 @ 2000 13-17 @ 3600	0-2 @ 4 24 @ 12	HEI	15	Automatic (California)
267 cu in	1103371	0 @ 1000 8 @ 1700 22 @ 4400	0 @ 3 24 @ 10	HEI	4	Automatic
305 cu in	1103282	0 @ 1000 10 @ 1700 20 @ 3800	0 @ 4 2 0 @ 11	HEI	4	Manual
305 cu in	1103379	0 @ 1000 10 @ 1700 20 @ 3800	0 @ 3 20 @ 8.5	HEI	4	Automatic
305 cu in	1103368	0 @ 1000 10 @ 1700 20 @ 3800	0 @ 4 10 @ 8	HEI	4	Automatic (California)

1981

Year and engine	Distributor No.	Centrifugal advance	Vacuum advance	Dwell	Ignition timing	Transmission
229 cu in	1110754	EST	EST	HEI	6	Automatic
229 cu in	1103434	EST	EST	HEI	6	Manual
231 cu in	1111386	EST	EST	HEI	15	Automatic
267 and 305 cu in	1103443	EST	EST	HEI	6	Automatic

* Electronic Spark Timing

1 General information and precautions

The engine electrical systems include all ignition, charging and starting components. Because of their engine-related functions, these components are discussed separately from chassis electrical devices such as the lights, the instruments, etc. (which are included in Chapter 12).

Always observe the following precautions when working on the electrical systems:

a) Be extremely careful when servicing engine electrical components. They are easily damaged if checked, connected or handled improperly.

b) The alternator is driven by an engine drivebelt which could cause serious injury if your hands, hair or clothes become entangled in it with the engine running.

c) Both the alternator and the starter are connected directly to the battery and could arc or even cause a fire if mishandled, overloaded or shorted out.

d) Never leave the ignition switch on for long periods of time with the engine off.

e) Don't disconnect the battery cables while the engine is running.

f) Maintain correct polarity when connecting a battery cable from another source, such as a vehicle, during jump starting.

g) Always disconnect the negative cable first and hook it up last or the battery may be shorted by the tool being used to loosen the cable clamps.

It's also a good idea to review the safety-related information regarding the

2.2 Typical battery hold-down clamp

engine electrical systems located in the *Safety First* section near the front of this manual before beginning any operation included in this Chapter.

2 Battery - removal and installation

Refer to illustration 2.2

1   **Caution:** *Always disconnect the negative cable first and hook it up last or the battery may be shorted by the tool being used to loosen the cable clamps. Disconnect both cables from the battery terminals.*
2   Remove the battery hold-down clamp **(see illustration)**.
3   Lift out the battery. Use the proper lifting technique - the battery is heavy.
4   While the battery is out, inspect the battery carrier (tray) for corrosion (see Chapter 1).
5   If you are replacing the battery, make sure that you get an identical battery, with the same dimensions, amperage rating, "cold cranking" rating, etc.
6   Installation is the reverse of removal.

3 Battery - emergency jump starting

Refer to the *Booster battery (jump) starting* procedure at the front of this manual.

4 Battery cables - check and replacement

1   Periodically inspect the entire length of each battery cable for damage, cracked or burned insulation and corrosion. Poor battery cable connections can cause starting problems and decreased engine performance.
2   Check the cable-to-terminal connections at the ends of the cables for cracks, loose wire strands and corrosion. The presence of white, fluffy deposits under the insulation at the cable terminal connection is a

sign that the cable is corroded and should be replaced. Check the terminals for distortion, missing mounting bolts and corrosion.
3   When removing the cables, always disconnect the negative cable first and hook it up last or the battery may be shorted by the tool used to loosen the cable clamps. Even if only the positive cable is being replaced, be sure to disconnect the negative cable from the battery first.
4   Disconnect the old cables from the battery, then trace each of them to their opposite ends and detach them from the starter solenoid and ground. Note the routing of each cable to insure correct installation.
5   If you are replacing either or both of the old cables, take them with you when buying new cables. It is vitally important that you replace the cables with identical parts. Cables have characteristics that make them easy to identify: positive cables are usually red, larger in cross-section and have a larger diameter battery post clamp; ground cables are usually black, smaller in cross-section and have a slightly smaller diameter clamp for the negative post.
6   Clean the threads of the solenoid or ground connection with a wire brush to remove rust and corrosion. Apply a light coat of battery terminal corrosion inhibitor, or petroleum jelly, to the threads to prevent future corrosion.
7   Attach the cable to the solenoid or ground connection and tighten the mounting nut/bolt securely.
8   Before connecting a new cable to the battery, make sure that it reaches the battery post without having to be stretched.
9   Connect the positive cable first, followed by the negative cable.

5 Ignition system - general information

1   In order for the engine to run correctly, it is necessary for an electrical spark to ignite the fuel/air mixture in the combustion chamber at exactly the right moment in relation to engine speed and load. The ignition coil converts low tension (LT) voltage from the battery into high tension (HT) voltage, powerful enough to jump the spark plug gap in the cylinder, providing that the system is in good condition and that all adjustments are correct.
2   The ignition system fitted to all pre-1975 cars as standard equipment is a conventional distributor with mechanical contact breaker points. On 1975 and later models, a breakerless high energy ignition (HEI) system is used.

Pre-1975 ignition systems

3   The ignition system is divided into the primary (low tension) circuit and the secondary (high tension) circuit.
4   The primary circuit consists of the battery lead to the starter motor, the lead to the ignition switch, the calibrated resistance wire

from the ignition switch to the primary coil winding and the lead from the low tension coil windings to the contact breaker points and condenser in the distributor.
5   The secondary circuit consists of the secondary coil winding, the high tension lead from the coil to the distributor cap, the rotor and the spark plug leads and spark plugs.
6   The system functions in the following manner. Low tension voltage in the coil is converted into high tension voltage by the opening and closing of the contact breaker points in the distributor. This high tension voltage is carried via the brush in the center of the distributor cap contacts to the rotor arm of the distributor cap. Every time the rotor contacts one of the spark plug terminals on the cap, it jumps the gap from the rotor arm to the terminal and is carried by the spark plug lead to the spark plug, where it jumps the spark plug gap to ground.
7   Ignition advance is controlled by both mechanical and vacuum operated systems. The mechanical governor mechanism consists of two weights, which due to centrifugal force move out from the distributor shaft as the engine speed rises. As they move outwards they rotate the cam relative to the distributor shaft, advancing spark timing. The weights are held in position by two light springs. It is the tension of these springs which determines correct spark advancement.
8   The vacuum control system consists of a diaphragm, one side of which is connected via a vacuum line to the carburetor, the other side to the contact breaker plate. Vacuum in the intake manifold and carburetor varies with engine speed and throttle opening. As the vacuum changes, it moves the diaphragm, which rotates the contact breaker plate slightly in relation to the rotor, thus advancing or retarding the spark. Control is fine-tuned by a spring in the vacuum assembly.
9   On some models, a Transmission Controlled Spark (TCS) system eliminates vacuum advance (see Chapter 6).

HEI ignition systems (1975 through 1980)

10   The high energy ignition (HEI) system is a pulse triggered, transistor controlled, inductive discharge system.
11   A magnetic pick-up inside the distributor contains a permanent magnet, pole-piece and pick-up coil. A time core, rotating inside the pole piece, induces a voltage in the pick-up coil. When the teeth on the timer and pole piece line up, a signal passes to the electronic module to open the coil primary circuit. The primary circuit current collapses and a high voltage is induced in the coil secondary winding. This high voltage is directed to the spark plugs by the distributor rotor in a manner similar to a conventional system. A capacitor suppresses radio interference.
12   The HEI system features a longer spark duration than a conventional breaker point

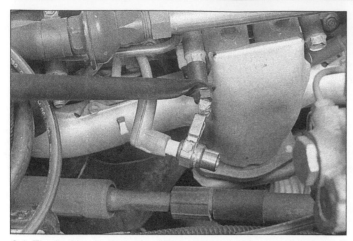

6.1 The ignition system should be checked with a spark tester - if the ignition system produces a spark that will jump the tester gap, it's functioning properly

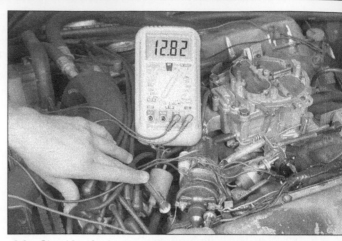

6.6a Checking for battery voltage on the positive terminal of the ignition coil (points-type ignition)

ignition system, and the dwell period increases automatically with engine speed. These characteristics are desirable for lean firing and EGR-diluted mixtures (see Chapter 6).

13   The ignition coil and the electronic module are both housed in the distributor cap on the HEI system. The distributor does not require routine servicing.

14   Spark timing is advanced by mechanical and vacuum devices similar to those used on conventional breaker point distributors (described above). The TCS system is eliminated.

HEI systems (1981 and later)

15   Since 1981, HEI systems have been equipped with electronic spark timing (EST). All spark timing changes are carried out by the Electronic Control Module (ECM), which monitors data from various engine sensors, computes the desired spark timing and signals the distributor to alter spark timing accordingly. Vacuum and mechanical advance is eliminated.

16   An Electronic Spark Control (ESC) system utilizes a knock sensor, and the ECM, to allow maximum spark advance without spark knock. The ESC system improves driveability and fuel economy.

6   Ignition system - check

Refer to illustrations 6.1, 6.6a and 6.6b
Warning: *Because of the very high secondary voltage generated by the ignition system - particularly the High Energy Ignition (HEI) system - extreme care should be taken whenever this check is performed.*

1   If the engine will not start even though it turns over, check for spark at the spark plug by installing a calibrated ignition system tester to one of the spark plug wires **(see illustration)**. **Note:** *The tool is available at most auto parts stores. Be sure to get the correct tool for your particular ignition system*

(breaker points or HEI).
2   Connect the clip on the tester to a ground such as a metal bracket or valve cover bolt, crank the engine and watch the end of the tester for bright blue, well defined sparks.
3   If sparks occur, sufficient voltage is reaching the plugs to fire the engine. However the plugs themselves may be fouled, so remove and check them as described in Chapter 1 or replace them with new ones.
4   If no sparks occur, detach the coil secondary wire from the distributor cap, connect the tester to the coil wire and repeat the test (breaker point systems only).
5   If sparks now occur, the distributor cap, rotor or spark plug wires may be defective. Remove the distributor cap and check the cap, rotor and spark plug wires as described in Chapter 1. Replace defective parts as necessary. If moisture is present in the distributor cap, use WD-40 or something similar to dry out the cap and rotor, then reinstall the cap and repeat the spark test at the spark plug wire.
6   If no sparks occur at the coil wire, check the primary wire connections at the coil to make sure they are clean and tight. Check for voltage to the ignition coil **(see illustrations)**. Battery voltage should be available to the ignition coil with the ignition switch ON. **Note:** *If the reading is 7 volts or less, repair the primary circuit from the ignition switch to the ignition coil.*
7   The coil-to-cap wire may be bad (breaker point systems only); check the resistance with an ohmmeter and compare it to the Specifications. Make any necessary repairs and repeat the test.
8   If there's still no spark, check the ignition module (HEI system) (see Section 11), the ignition coil (see Section 8) or other internal components that may be defective.
9   If there is still no spark (see Step 1), check the ignition points (refer to Chapter 1). If the points appear to be in good condition (no pits or burned spots on the point surface) and the primary wires are hooked up cor-

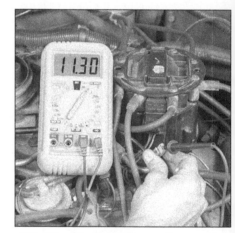

6.6b Checking for battery voltage on the BAT terminal of the ignition coil (HEI ignition)

rectly, adjust the points as described in Chapter 1.
10   If there is still no spark at the plugs, check for a ground or open circuit in the distributor/points circuit. There may be a damaged points terminal causing the ignition voltage to become shorted or diminished.

7   Distributor - removal and installation

Refer to illustrations 7.5, 7.6a and 7.6b

Removal

1   After disconnecting the negative battery terminal cable, unplug the primary lead from the coil.
2   Unplug or detach all electrical leads from the distributor. To find the connectors, trace the wires from the distributor.
3   Look for a raised "1" on the distributor cap This marks the location for the number one cylinder spark plug wire terminal. If the cap does not have a mark for the number one

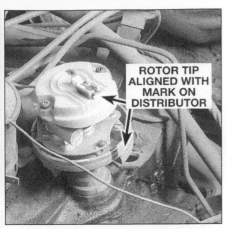

7.5 After turning the rotor until it is pointing at the terminal for the number 1 spark plug, paint or scribe a mark on the edge of the distributor base directly beneath it

7.6a Some distributor hold-down bolts can be removed with an open-end wrench . . .

7.6b . . . others may require a special distributor wrench

spark plug, locate the number one spark plug and trace the wire back to its corresponding terminal on the cap.

4   Remove the distributor cap (see Chapter 1) and turn the engine over until the rotor is pointing toward the number one spark plug terminal (see locating TDC procedure in Chapter 2).

5   Make a mark on the edge of the distributor base directly below the rotor tip and in line with it **(see illustration)**. Also, mark the distributor base and the engine block to ensure that the distributor is installed correctly.

6   Remove the distributor hold-down bolt and clamp **(see illustrations)**, then pull the distributor straight up to remove it. Be careful not to disturb the intermediate driveshaft. **Caution:** *DO NOT turn the engine while the distributor is removed, or the alignment marks will be useless.*

Installation

Note: *If the crankshaft has been moved while the distributor is out, locate Top Dead Center (TDC) for the number one piston (see Chapter 2) and position the distributor and rotor accordingly.*

7   Insert the distributor into the engine in exactly the same relationship to the block that it was in when removed.

8   To mesh the helical gears on the camshaft and the distributor, it may be necessary to turn the rotor slightly. If the distributor doesn't seat completely, the hex shaped recess in the lower end of the distributor shaft is not mating properly with the oil pump shaft. Recheck the alignment marks between the distributor base and the block to verify that the distributor is in the same position it was in before removal. Also check the rotor to see if it's aligned with the mark you made on the edge of the distributor base.

9   Place the hold-down clamp in position and loosely install the bolt.

10   Install the distributor cap and tighten the

cap screws securely.

11   Plug in the module electrical connector.

12   Reattach the spark plug wires to the plugs (if removed)

13   Connect the cable to the negative terminal of the battery.

14   Check the ignition timing (see Chapter 1) and tighten the distributor hold-down bolt securely.

8   Distributor (1970 through 1974) - overhaul

Refer to illustration 8.2

1   Remove the distributor (See Section 7).

2   Remove the rotor (two screws), the advance weight springs and the weights **(see illustration)**. Where applicable, also remove the radio frequency interference (RFI) shield.

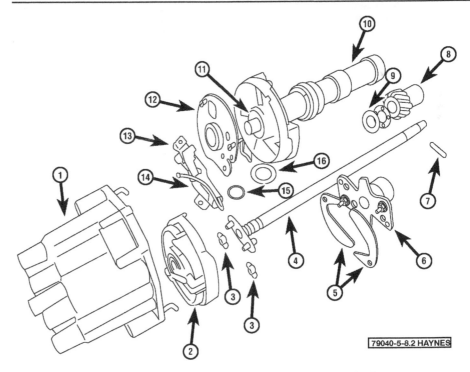

8.2 Exploded view of a typical contact breaker type distributor

1	Distributor cap	6	Cam and advance weight base	12	Breaker plate
2	Rotor			13	Contact point assembly
3	Advance weight springs	7	Drive gear roll pin	14	Condenser
4	Mainshaft	8	Distributor drive gear	15	Retaining ring
5	Advance weights	9	Washer and shim	16	Felt washer
		10	Distributor housing		
		11	Plastic washer		

79040-5-8.2 HAYNES

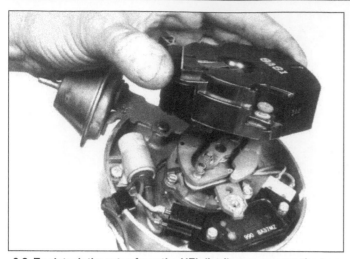

9.2 To detach the rotor from the HEI distributor, remove the two screws on top that attach it to the centrifugal advance mechanism

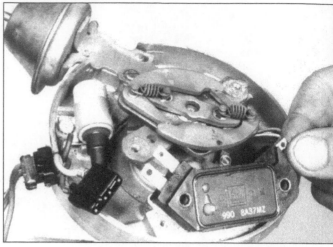

9.3 To detach the ignition module from the HEI distributor, remove the two mounting screws and unplug the connector from the B and C terminals

3   Drive out the roll pin retaining the gear to the shaft then pull off the gear and spacers.
4   Ensure that the shaft is not burred, then slide it from the housing.
5   Remove the cam weight base assembly.
6   Remove the screws retaining the vacuum unit and lift off the unit itself.
7   Remove the spring retainer (snap-ring) then remove the breaker plate assembly.
8   Remove the contact points and condenser, followed by the felt washer and plastic seal located beneath the breaker plate.
9   Wipe all components clean with a solvent-moistened cloth and examine them for wear, distortion and scoring. Replace parts as necessary. Pay particular attention to the rotor and distributor cap to ensure that they are not cracked.
10   Fill the lubricating cavity in the housing with general purpose grease, then fit a new plastic seal and felt washer.
11   Install the vacuum unit and the breaker plate in the housing, and the spring retainer on the upper bushing.
12   Lubricate the cam weight base and slide it on the mainshaft; install the weights and springs.
13   Insert the mainshaft in the housing then fit the shims and drivegear. Install a new roll pin.
14   Install the contact point set (see Chapter 1).
15   Install the rotor, aligning the round and square pilot holes.
16   Install the distributor (see Section 7).

9   HEI distributor (1975 through 1981) - overhaul

Refer to illustrations 9.2, 9.3, 9.5 and 9.9
Note: *To check the pick-up coil, see Steps 4 through 8 in Section 10.*
1   Remove the distributor (see Section 7).
2   Remove the rotor (two screws) **(see illustration)**.

9.5 To remove the distributor shaft, drive the roll pin out of the shaft with a hammer and punch, remove the gear, shim and tanged washer from the shaft, inspect the shaft for any burrs that might prevent its removal, then pull it out (be careful not to lose the washer at the upper end of the shaft)

3   Remove the two screws retaining the module. Move the module aside and remove the connector from the 'B' and 'C' terminals **(see illustration)**.
4   Remove the connections from the 'W' and 'G' terminals.
5   Carefully drive out the roll pin from the drive gear **(see illustration)**.
6   Remove the gear, shim and tanged washer from the distributor shaft.
7   Ensure that the shaft is not burred, then remove it from the housing.
8   Remove the washer from the upper end of the distributor housing.
9   Pry off the retaining ring and remove pick-up coil and pole piece assembly **(see illustration)**.

9.9 The pick-up coil and pole piece assembly can be removed after prying out the retaining ring

10   Remove the lock ring, then take out the pick-up coil retainer, shim and felt washer.
11   Remove the vacuum unit (two screws).
12   Disconnect the capacitor lead and remove the capacitor (one screw).
13   Disconnect the wiring harness from the distributor housing.
14   Wipe all components clean with a solvent-moistened cloth and examine them for wear, distortion and other damage. Replace parts as necessary.
15   To assemble, position the vacuum unit to the housing and secure with the two screws.
16   Position the felt washer over the lubricant reservoir at the top of the housing, then position the shim on top of the felt washer.
17   Install the pick-up coil and pole-piece assembly. Make sure it engages the vacuum advance arm properly. Install the retaining ring.
18   Install the washer to the top of the hous-

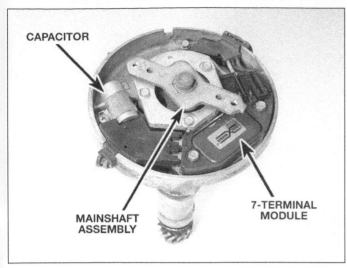

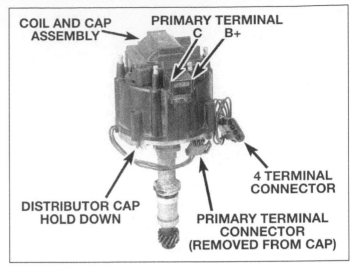

10.1a Typical late model HEI distributor details

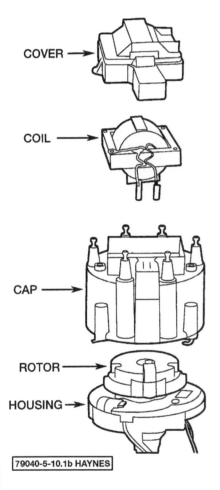

79040-5-10.1b HAYNES

10.1b Typical late-model HEI coil installation details

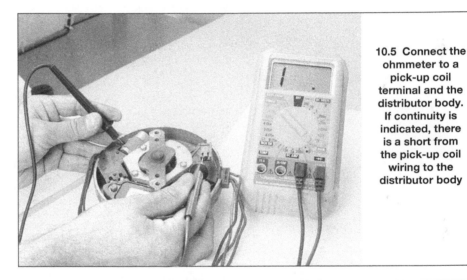

10.5 Connect the ohmmeter to a pick-up coil terminal and the distributor body. If continuity is indicated, there is a short from the pick-up coil wiring to the distributor body

ing. Install the distributor shaft, then rotate it and check for equal clearance all round between the shaft projections and pole-piece.

19   Install the tanged washer, shim and drivegear. Align the gear and install a new roll pin.

20   Loosely install the capacitor with one screw.

21   Install the connector to the 'B' and 'C' terminals on the module, with the tab at the top.

22   Apply silicone grease to the base of the module and secure it with two screws. The grease is essential to ensure good heat conduction.

23   Position the wiring harness, with the grommet in the housing notch, then connect the pink wire to the capacitor stud and the black wire to the capacitor mounting screw. Tighten the screw.

24   Connect the white wire from the pick-up coil to the module "W" terminal and the green to the 'G' terminal.

25   Install the advance weights, weight retainer (dimple downwards), and springs.

26   Install the rotor and secure with the two screws. Ensure that the notch on the side of the rotor engages with the tab on the cam weight base.

27   Install the distributor (see Section 7).

10   HEI distributor (1981 and later) - overhaul

Refer to illustrations 10.1a, 10.1b, 10.5, 10.6, 10.10 and 10.11

1   Later model HEI distributors **(see illustrations)** vary somewhat from those described in Section 9. Most later distributors are not equipped with vacuum advance units, as advance is controlled by the Electronic Control Module (ECM).

2   Unplug the electrical connector(s), disengage the latches and remove the distributor cap.

3   Remove the rotor.

4   If the distributor is equipped with a vacuum advance unit, connect a vacuum source or pump to it.

5   To test the pick-up coil, remove the pick-up coil leads from the module. Connect the negative ohmmeter probe to the distributor body and the positive probe to the pick-up coil electrical connector **(see illustration)**. Resistance should be infinite. If continuity is indicated, there is a short to ground in the pick-up coil. Replace the pick-up coil.

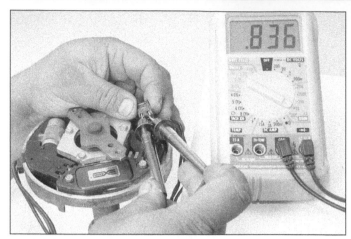

10.6 Measure the resistance of the pick-up coil. It should be between 500 and 1,500 ohms

10.10 To remove the distributor shaft, mark the relative positions of the gear and shaft, drive out the pin retaining the distributor gear to the shaft, remove the gear and pull out the shaft

6   Next, measure the pick-up coil resistance across the two pick-up coil terminals. It should be 500 to 1,500 ohms **(see illustration)**. If the resistance is incorrect, replace the pick-up coil.

10.11 The pick-up coil and pole piece assembly can be removed after carefully prying out the C-clip

7   Apply vacuum to the advance unit and verify that the indicated resistance remains steady as vacuum is applied. Replace the advance unit if it is inoperative or the ohmmeter reading changes. Make sure the application of vacuum does not cause the teeth to align (indicated by a jump in the ohmmeter reading).
8   Flex the pick-up coil leads and pull on the connectors (with the ohmmeter connected to the pick-up coil leads). Any change in resistance would indicate a broken wire in the pick-up coil, replace the pick-up coil.
9   Place the distributor in a vise, using blocks of wood to protect it.
10   Mark the relative positions of the gear and shaft. Drive the roll pin out **(see illustration)**. Remove the gear and pull the shaft from the distributor housing.
11   Remove the aluminum shield for access to the pick-up coil and module. The pick-up coil can be lifted out after removal of the C-clip **(see illustration)**. Remove the two screws and lift the module, capacitor and harness assembly from the distributor base.
12   Wipe the distributor base and module with a clean cloth and inspect it for cracks and damage.

13   Reassembly is the reverse of the disassembly procedure. Be sure to apply a coat of silicone grease to the distributor base under the module.

---

11   Ignition coil - check and replacement

Check

1   If the engine is hard to start (particularly when it's already hot), misses at high speed or cuts out during acceleration, the coil may be faulty. First, make sure that the battery and the distributor are in good condition, the points (pre-1975 vehicles) are properly adjusted and the plugs and plug wires are in good shape. If the problem persists, perform the following test:

Breaker points ignition system
Refer to illustrations 11.3a and 11.3b
2   Before performing any of the following coil electrical checks, make sure that the coil is clean, free of any carbon tracks and that all connections are tight and free of corrosion. Also make sure that both battery terminals

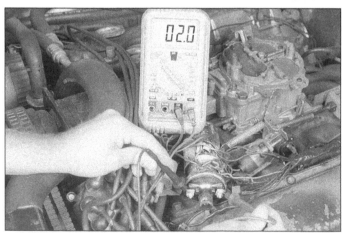

11.3a Measure the primary resistance between the ignition coil positive (+) and negative (-) terminals (points-type ignition)

11.3b Measure the secondary resistance between the positive (+) terminal and the ignition coil tower (points-type ignition)

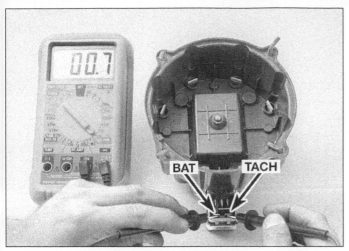

11.6 To test the HEI type coil-in-cap, attach the leads of an ohmmeter to the primary terminals and verify that the indicated resistance is zero or very near zero . . .

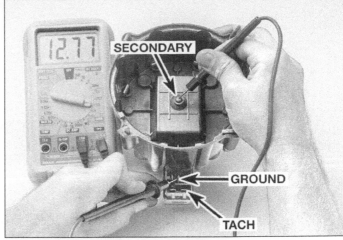

11.7 . . . then, using the high scale, attach one lead to the high tension terminal and the other to the GROUND terminal and then the TACH terminal. Verify that both of the readings are not infinite - if the indicated resistance is not as specified, replace the coil

are clean and that the cables are securely attached (especially the ground strap at the negative terminal).

3　Detach the coil high tension cable from the distributor cap. Using an insulated tool, hold the end of the cable about 3/16-inch away from some grounded part of the engine, turn the ignition on and operate the starter. A bright blue spark should jump the gap.

a)　*If the spark is weak, yellowish or red, spark voltage is insufficient. If the points, condenser and battery are in good condition, the coil is probably weak. Check the coil primary and secondary resistance* **(see illustrations)**. *Refer to the Specifications listed in the beginning of the Chapter. If the coil windings are damaged the resistance will be excessively high. Replace the coil.*

b)　*If there is no spark at all, try to locate the trouble before replacing the coil. Remove the distributor cap. Turn the engine until the points are open, or separate the points with a small piece of cardboard. Turn on the ignition switch. Using a 12-volt bulb with two test leads, attach one lead to ground somewhere on the engine and the other lead to first one of the coil's primary terminals and then the other:*

1)　*If the bulb lights when touched to the primary terminal that leads to the distributor, the coil is getting current and the primary windings are okay.*

2)　*If the bulb lights when touched to the other primary terminal but not when attached to the one leading to the distributor, the primary windings are faulty and the coil is no good.*

3)　*If the light does not go on when connected to either primary connection, the coil is not the problem. Check the ignition switch and starter solenoid.*

4)　*If the bulb lights when touched to both primary terminals (the coil is*

receiving current at both primary terminals), remove the high-tension cable from the center distributor cap tower and try shorting across the open distributor points with the tip of a clean (no oil) screwdriver.

(a)　*If a spark jumps from the coil's high tension secondary wire to a grounded point on the engine as the screwdriver is removed, the points are either contaminated by oil, dirt or water, or they're burned.*

(b)　*If the screwdriver fails to produce a spark at the high tension wire, disconnect the primary wire that passes between the coil and the distributor and attach a test wire to the coil in its place. Ground the other end of the wire against the engine block, then pull it away (the test wire is simulating the points: grounding the wire is just like closing the points; pulling it away creates the same effect - producing a spark from the coil's high tension wire - as opening the points).*

(1)　*If a spark jumps from the high-tension cable when the test wire is removed from the ground, the coil is okay. Either the points are grounded or the condenser is shorted.*

(2)　*If a spark does not jump during this test, the secondary windings of the coil are faulty. Replace the coil.*

4　Sometimes, a coil checks out perfectly but the engine is still hard to start and misses at higher speeds. The problem may be inadequate spark voltage caused by reversed coil polarity. If you have recently tuned up the engine or performed any service work involving the coil, it's possible that the primary

leads to the coil were accidentally reversed. To check for reversed polarity, remove one of the spark plug leads and, using an insulated tool, hold it about 1/4-inch from the spark plug terminal or any ground point. Then insert the point of a pencil between the ignition lead and the plug while the engine is running (if the plug connector terminals are deeply recessed in a boot or insulating shield, straighten all but one bend in a paper clip and insert the looped end into the plug connector).

a)　*If the spark flares on the ground or spark plug side of the pencil, the polarity is correct.*

b)　*If the spark flares between the ignition lead and the pencil, however, the polarity is wrong and the primary wires should be switched at the coil.*

HEI ignition

HEI-type coil-in-cap

Refer to illustrations 11.6 and 11.7

5　Remove the distributor cap (see Chapter 1).

6　Attach the leads of an ohmmeter across the two primary coil terminals as shown **(see illustration)**. The indicated resistance should be less than one ohm.

7　Using the high scale, attach one lead of the ohmmeter to the SECONDARY terminal in the middle of the distributor and the other lead to the GROUND terminal **(see illustration)**.

8　Using the high scale, attach one lead of the ohmmeter to the SECONDARY terminal in the middle of the distributor and the other lead to the TACH terminal.

9　If both of the above readings indicate infinite resistance, replace the coil.

Separately mounted coil

Refer to illustration 11.11

10　Disconnect the electrical connectors from the coil.

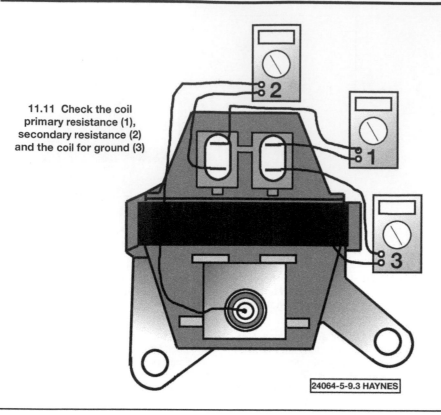

11.11 Check the coil primary resistance (1), secondary resistance (2) and the coil for ground (3)

24064-5-9.3 HAYNES

11.15 To remove a conventional coil, simply detach the high tension cable and the two primary leads and remove the mounting bracket screws

11   Connect an ohmmeter to the coil terminals as shown and perform the three tests **(see illustration)**. The test results should be as follows:

a) **Test 1** - Zero to one ohm.
b) **Test 2** - 6,000 to 30,000 ohms.
c) **Test 3** - Infinity.

12   If the coil fails any of the tests, replace the coil.

Replacement
Separately mounted coil
Refer to Illustration 11.15

13   Detach the cable from the negative terminal of the battery.

14   Disconnect the high tension cable from the coil.

15   Detach the electrical connections from the coil primary and secondary terminals. Be sure to mark the connections before removal to ensure that they are re-installed correctly. Remove the coil **(see illustration)**.

16   Installation is the reverse of removal

HEI-type coil-in-cap
Refer to Illustrations 11.19, 11.21 and 11.22

17   Detach the cable from the negative terminal of the battery.

18   Disconnect the battery wire and harness connector from the distributor cap.

19   Remove the coil cover screws and the cover **(see illustration)**.

20   Remove the coil assembly screws.

21   Note the position of each wire, marking them if necessary. Remove the coil ground wire, then push the leads from the underside of the connectors. Remove the coil from the distributor cap **(see illustration)**.

22   Installation is the reverse of the removal procedure. Be sure that the center electrode is in good shape **(see illustration)** and that the leads are connected to their original positions.

12   Charging system - general information and precautions

Refer to illustrations 12.2a and 12.2b

The charging system includes the alternator, either an internal or an external voltage regulator, a charge indicator, the battery, a

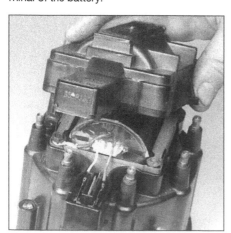

11.19 To get at the coil, remove the coil cover screws and the cover

11.21 To separate the coil from the distributor cap, clearly mark the wires, detach the coil ground wire and push the leads from the underside of the connectors

11.22 Before installing a new coil, make sure that the center electrode is in good condition

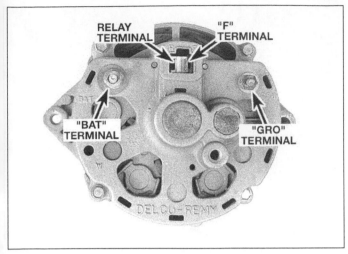

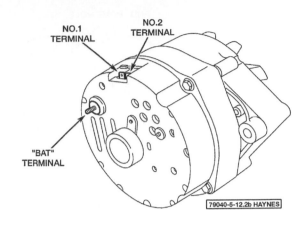

12.2a A Series 1D Delcotron alternator, which uses an external voltage regulator (Series 10DN and 100B types are similar)

12.2b A Series 10SI Delcotron alternator, which uses an internal voltage regulator (Series CS-121, 130 and 144 types are similar, but are not rebuildable)

fusible link and the wiring between all the components. The charging system supplies electrical power for the ignition system, the lights, the radio, etc. The alternator is driven by a drivebelt at the front (right end) of the engine.

The alternator fitted to all models is a Delco-Remy Delcotron. Several units have been used. Early units **(see illustration)**, such as the Series 1D, 10DN and 100B, are fitted with an external voltage regulator. Later units **(see illustration)**, such as the Series 10SI and the CS-121, CS-130 and CS-144, are fitted with an internal voltage regulator. Finally, it should be noted that the CS series alternators cannot be rebuilt. If one becomes inoperative, it must be replaced.

The purpose of the voltage regulator is to limit the alternator's voltage to a preset value. This prevents power surges, circuit overloads, etc., during peak voltage output.

The fusible link is a short length of insulated wire integral with the engine compartment wiring harness. The link is four wire gauges smaller in diameter than the circuit it protects. Production fusible links and their identification flags are identified by the flag color. See Chapter 12 for additional information regarding fusible links.

The charging system doesn't ordinarily require periodic maintenance. However, the drivebelt, battery and wires and connections should be inspected at the intervals outlined in Chapter 1.

The dashboard warning light should come on when the ignition key is turned to Start, then should go off immediately. If it remains on, there is a malfunction in the charging system (see Section 13). Some vehicles are also equipped with a voltage gauge. If the voltage gauge indicates abnormally high or low voltage, check the charging system (see Section 13).

Be very careful when making electrical circuit connections to a vehicle equipped with an alternator and note the following:

a) When reconnecting wires to the alternator from the battery, be sure to note the polarity.
b) Before using arc welding equipment to repair any part of the vehicle, disconnect the wires from the alternator and the battery terminals.
c) Never start the engine with a battery charger connected.
d) Always disconnect both battery leads before using a battery charger.
e) The alternator is driven by an engine drivebelt which could cause serious injury if your hands, hair or clothes become entangled in it with the engine running.
f) Because the alternator is connected directly to the battery, it could arc or cause a fire if overloaded or shorted out.
g) Wrap a plastic bag over the alternator and secure it with rubber bands before steam cleaning the engine.

13 Charging system - check

Refer to illustration 13.4
1   If a malfunction occurs in the charging circuit, don't automatically assume that the alternator is causing the problem. First check the following items:

a) Check the drivebelt tension and its condition. Replace it if worn or deteriorated.
b) Make sure that the alternator mounting and adjustment bolts are tight.
c) Inspect the alternator wiring harness and the connectors at the alternator and voltage regulator. They must be in good condition and tight.
d) Check the fusible link (if equipped) located between the starter solenoid and the alternator. If it's burned, determine the cause, repair the circuit and replace the link (the vehicle won't start and/or the accessories won't work if the fusible link blows).

e) Start the engine and check the alternator for abnormal noises (a shrieking or squealing sound indicates a bad bushing).
f) Check the specific gravity of the battery electrolyte. If it's low, charge the battery (doesn't apply to maintenance free batteries).
g) Make sure that the battery is fully charged (one bad cell in a battery can cause overcharging by the alternator).
h) Disconnect the battery cables (negative first, then positive). Inspect the battery posts and the cable clamps for corrosion. Clean them thoroughly if necessary (see Chapter 1). Reconnect the cable to the positive terminal.
i) With the key off, insert a test light between the negative battery post and the disconnected negative cable clamp.
   1) If the test light does not come on, reattach the clamp and proceed to the next Step.
   2) If the test light comes on, there is a short in the electrical system of the vehicle. The short must be repaired before the charging system can be checked.
   3) Disconnect the alternator wiring harness.
      a) If the light goes out, the alternator is bad.
      b) If the light stays on, pull each fuse until the light goes out (this will tell you which component is shorted).

2   Using a voltmeter, check the battery voltage with the engine off. It should be approximately 12 volts.
3   Start the engine and check the battery voltage again. It should now be approximately 13.5 to 14.5 volts.
4   Locate the D-shaped test hole in the back of the alternator. Ground the tab inside the hole to the alternator body by inserting a screwdriver blade through the hole and touching the tab and the body at the same

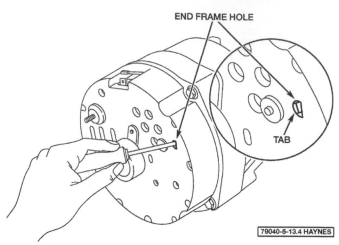

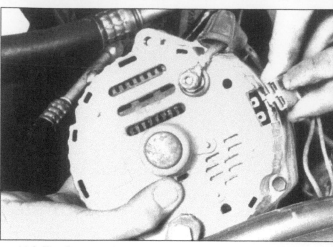

13.4 To "full-field" the alternator for testing, insert a small screwdriver through the D-shaped hole and ground the tab to the alternator body

14.2 The first step in alternator removal is disconnecting all electrical leads (which will vary somewhat with alternator type and year of manufacture)

14.3a To remove the drivebelt, loosen the alternator mounting bolt . . .

14.3b . . . then loosen the adjustment bolt and slip off the belt - finally, remove the mounting and adjusting bolts and remove the alternator

time **(see illustration)**. **Caution:** *Don't run the engine with the tab grounded any longer than necessary to obtain a voltmeter reading. If the alternator is charging, it's running unregulated during this test, which may overload the electrical system and cause damage to the components.*

5   The reading on the voltmeter should be 15-volts or higher with the tab grounded.

6   If the voltmeter indicates low voltage, the alternator is faulty and should be rebuilt or replaced with a new one.

7   If the voltage reading is 15-volts or higher and a "no charge" condition exists, the regulator or field circuit is the problem. Remove the alternator and have it load tested and repaired or replaced.

14   Alternator - removal and installation

Refer to illustrations 14.2, 14.3a, 14.3b

1   Detach the cable from the negative terminal of the battery.

2   Detach the electrical connectors from

the alternator and the voltage regulator **(see illustration)**.

3   Loosen the alternator adjustment and pivot bolts and detach the drivebelt **(see illustrations)**.

4   Remove the adjustment and pivot bolts and separate the alternator from the engine.

5   If you are replacing the alternator, take the old alternator with you when purchasing a replacement unit. Make sure that the new/rebuilt unit is identical to the old alternator. Look at the terminals - they should be the same in number, size and location as the terminals on the old alternator. Finally, look at the identification markings - they will be stamped in the housing or printed on a tag or plaque affixed to the housing. Make sure that these numbers are the same on both alternators.

6   Many new/rebuilt alternators do not have a pulley installed, so you may have to switch the pulley from the old unit to the new/rebuilt one. When buying an alternator, find out the shop's policy regarding pulleys - some shops will perform this service free of charge.

7   Installation is the reverse of removal.

8   After the alternator is installed, adjust

the drivebelt tension (see Chapter 1).

9   Check the charging voltage to verify proper operation of the alternator (see Section 13).

15   Alternator components - check and replacement

Refer to illustrations 15.1, 15.2, 15.3, 15.4, 15.5, 15.6, 15.7, 15.8, 15.9, 15.10, 15.13, 15.14a, 15.14b, 15.14c, 15.14d, 15.14e, 15.15a, 15.15b, 15.16a, 15.16b, 15.17a, 15.17b and 15.17c

Note: *This procedure does not apply to the CS-series alternator. If the CS-series alternator fails it must be replaced as an assembly; it cannot be rebuilt.*

1   Remove the alternator (see Section 14) and place it on a clean workbench **(see illustration)**.

2   Paint an alignment mark on each side of the end frame to insure correct reassembly **(see illustration)**.

3   Remove the four bolts that secure the two end frames together **(see illustration)**.

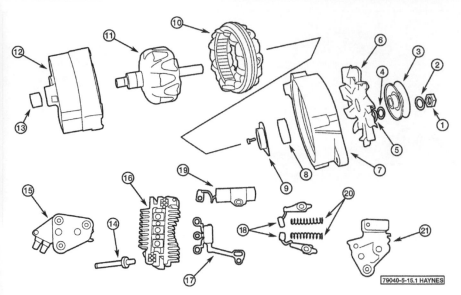

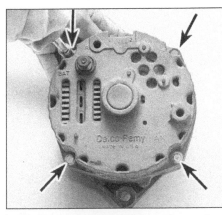

15.1 An exploded view of the 10SI alternator

1	Nut	9	Plate	15	Voltage regulator
2	Washer	10	Stator	16	Rectifier bridge
3	Pulley	11	Rotor	17	Diode trio
4	Washer	12	Slip-ring end frame	18	Brushes
5	Collar	13	Bearing	19	Capacitor
6	Fan	14	Terminal component	20	Brush springs
7	Drive end frame		stud	21	Brush holder
8	Bearing				

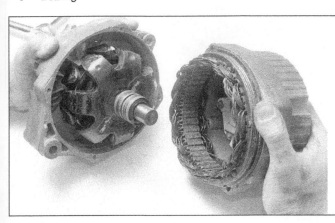

15.4 Separate the two halves

15.2 Paint an alignment mark on the alternator body between each end frame

15.3 Remove the bolts (arrows) securing the rear end frame to the drive end frame

4 Separate the two halves. The rotor will remain in one half while the stator will remain in the other half **(see illustration)**.
5 Remove the three nuts retaining the stator windings to the rectifier bridge **(see illustration)**.
6 Remove the stator from the end frame **(see illustration)**.
7 Remove the screw retaining the diode trio to the brush assembly **(see illustration)**.

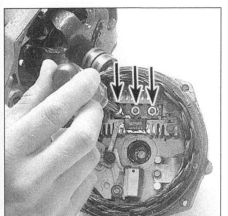

15.5 Remove the three nuts (arrows) retaining the stator windings to the rectifier bridge

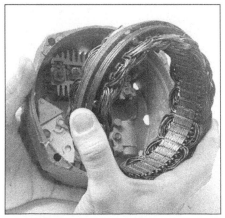

15.6 Separate the stator from the end frame

15.7 Remove the screw (arrow) retaining the diode trio to the regulator/brush assembly

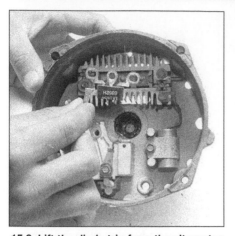

15.8 Lift the diode trio from the alternator

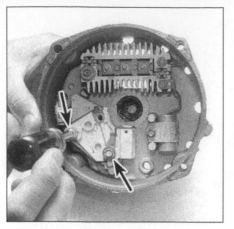

15.9 Remove the two remaining screws (arrows) and remove the brush assembly and voltage regulator

15.10 Remove the rectifier assembly from the end frame

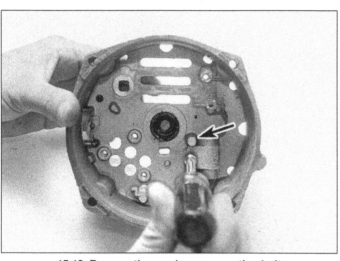

15.13 Remove the condenser mounting bolt

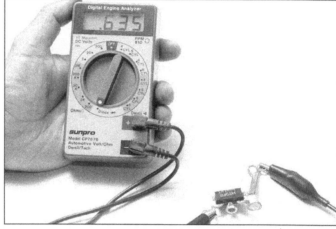

15.14a To test the diode trio, install the negative probe on the extension leg and the positive probe onto each diode (in turn). Continuity should exist through each diode

8   Lift the diode trio from the alternator assembly **(see illustration)**.
9   Remove the brush assembly and voltage regulator **(see illustration)**.
10   Remove the nut and washer from the output stud. Remove the screws retaining the rectifier bridge to the rear end frame and lift the assembly out **(see illustration)**.
11   Remove the nut and washer and separate the front pulley and fan from the drive end frame.
12   Remove the rotor from the drive end frame.
13   Remove the condenser mounting bolt and condenser **(see illustration)**.
14   Test the diodes as follows:

a)   *Test the diode trio by installing the negative probe of the multimeter onto the leg and the positive probe onto each terminal of the diode trio. Using the diode testing function on the multimeter, continuity should exist in one position only **(see illustrations)**. Reverse the polarity of the test by changing the position of the probes. Continuity should NOT exist. This checks the unidirectional capability*

of the operating diode. Next, repeat this test for each of the other two diodes (total six checks). Each diode should have the exact same characteristics. It will be necessary to use a multimeter that includes a diode check function. This function allows slight current application to the diode to assist in opening the voltage gate.

b)   *To test the rectifier assembly, follow the exact same procedure as detailed above. Be sure to use a multimeter with a diode check function. Use a screwdriver to lift the tabs off the contact surface to separate the negative and positive diode rectifiers. Make sure the tabs do not touch the posts, the heat sinks or each other. Install the multimeter probes on the tabs and the cooling fins on one side of the rectifier, then the other, to obtain results **(see illustrations)**. If the test results are incorrect for any one of the diodes, replace the complete unit.*

15   Test the rotor as follows:

a)   *Check for an open between the two slip rings **(see illustration)**. There should be*

1 to 5 ohms resistance between the slip rings.

b)   *Check for grounds between each slip ring and the rotor **(see illustration)**. There should be no continuity (infinite*

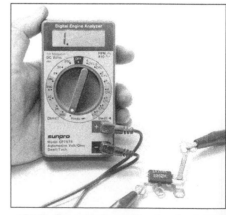

15.14b Reverse the probes to check the continuity in the other direction. The diode should allow continuity in one direction only

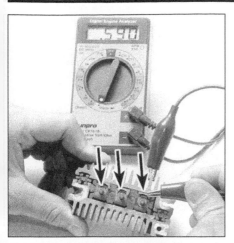

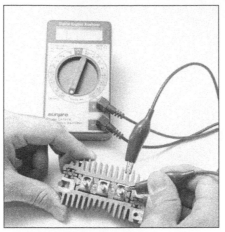

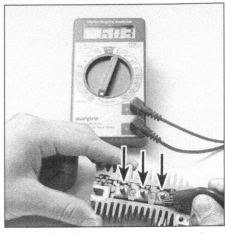

15.14c To test the three diodes (arrows) on the top side of the rectifier bridge, install the positive probe onto the cooling fin and touch the negative probe to each metal tab. Continuity should exist through each diode

15.14d Reverse the probes to check the continuity in the other direction. Continuity should NOT exist. The rectifier bridge diodes should allow continuity in one direction only

15.14e Test the three diodes (arrows) on the bottom row in the same manner

resistance) between the rotor and either slip ring. If the rotor fails either test, or if the slip rings are excessively worn, the rotor is defective.

16 Test the stator as follows:

a) *Check for opens between each end terminal of the stator windings* **(see illustration)**. *If either reading is high (infinite resistance), the stator is defective.*

b) *Check for a grounded stator winding between each stator terminal and the frame* **(see illustration)**. *If there's continuity between any stator winding and the frame, the stator is defective.*

17 Reassembly is a reversal of disassembly. Observe the following points:

a) *Take great care to position the insulating washers and sleeves correctly on the rectifier bridge and brush assembly screws.*

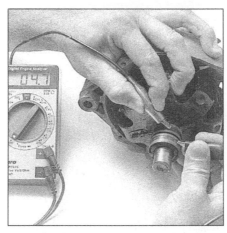

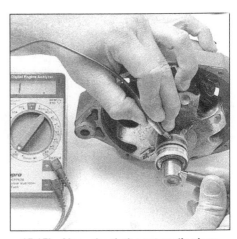

15.15a To test the rotor, check for an open between the two slip rings. There should be 1 to 5 ohms resistance

15.15b Also, check the rotor slip rings and the rotor for possible grounds. Continuity should NOT exist between the slip ring and the shaft or frame

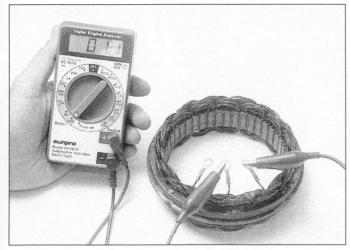

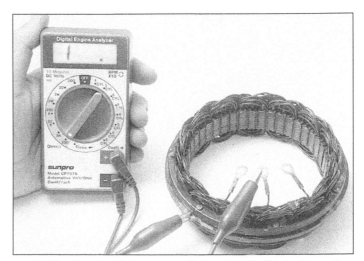

15.16a To test the stator, check for opens between each end terminal of the stator windings. Continuity should exist between all the terminals

15.16b Also, check for a grounded stator winding to the end frame. Continuity should NOT exist between any terminal and the frame

15.17a   Apply a small amount of grease to the bearing surface

15.17b   Press the two brushes into the holder and slide a paper clip through the eyelet to keep the brushes in the holder assembly

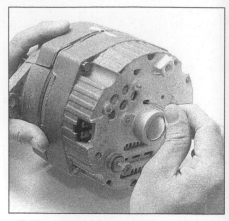

15.17c   After the alternator is completely assembled remove the wire or paper clip to release the brushes

b)   *Place a small amount of grease onto the bearing surface before installing the alternator halves together* **(see illustration)**.

c)   *Press the brushes into the holder and insert a stiff wire (such as a straightened-out paper clip) through the small hole in the back of the alternator to hold the brushes in a retracted position* **(see illustration)**. *This will prevent them from catching on the slip rings as the alternator halves are assembled.*

d)   *Clean the brush contact surfaces on the slip rings before installing the end-frame.*

e)   *Make sure that the marks on the rear end frame and drive end frame (which were made before dismantling) are in alignment.*

f)   *Tighten the pulley nut securely.*

g)   *Remove the wire or paper clip from the end-frame to permit the brushes to release onto the slip rings* **(see illustration)**.

---

16   External voltage regulator - check and replacement

Refer to illustration 16.4

1   A discharged battery will normally be due to a fault in the voltage regulator, but before testing the unit, do the following:

a)   *Check the drivebelt tension.*

b)   *Test the condition of the battery.*

c)   *Check the charging circuit for loose connections and broken wires.*

d)   *Made sure that lights or other electrical accessories have not been switched on inadvertently.*

e)   *Check the generator indicator lamp for normal illumination with the ignition switched on and off, and with the engine idling and stationary.*

2   Disconnect the battery ground cable. Disconnect the harness connector from the regulator.

3   Under no circumstances should the voltage regulator or field relay contacts be cleaned since any abrasive materials will destroy the contact material.

4   Voltage regulator point gap and air gap adjustments can be checked with a feeler gauge of the specified thickness (point gap should be 0.014-inch and the air gap should be 0.067-inch). Check the voltage regulator point opening of the upper contacts with the lower contacts just touching. Adjustments are made by carefully bending the upper contact arm. Check the voltage regulator air gap with the lower contacts touching and adjust it, if necessary **(see illustration)**.

5   The field relay point opening may be adjusted by bending the armature stop. The air gap is checked with the points just touch-

ing and is adjusted by bending the flat contact support spring. **Note:** *The field relay will normally operate satisfactorily even if the air gap is outside the specified limits, and should not be adjusted when the system is functioning satisfactorily.*

6   If the regulator must be replaced, simply remove the mounting screws.

7   Installation is the reverse of the removal procedure. Ensure that the rubber gasket is in place on the regulator base.

---

17   Starting system - general information and precautions

The sole function of the starting system is to turn over the engine quickly enough to allow it to start.

The starting system consists of the battery, the starter motor, the starter solenoid and the wires connecting them. The solenoid is mounted directly on the starter motor or is a separate component located in the engine compartment.

The solenoid/starter motor assembly is installed on the lower part of the engine, next to the transmission bellhousing.

When the ignition key is turned to the Start position, the starter solenoid is actuated through the starter control circuit. The starter solenoid then connects the battery to the starter. The battery supplies the electrical energy to the starter motor, which does the actual work of cranking the engine.

The starter motor on a vehicle equipped with a manual transmission can only be operated when the clutch pedal is depressed; the starter on a vehicle equipped with an automatic transmission can only be operated when the transmission selector lever is in Park or Neutral.

Always observe the following precautions when working on the starting system:

a)   *Excessive cranking of the starter motor can overheat it and cause serious damage. Never operate the starter motor for more than 15 seconds at a time without*

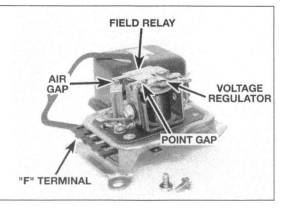

16.4   The voltage regulator point and air gap adjustments can be checked with a feeler gauge of the specified thickness (point gap = 0.014-inch; air gap = 0.067-inch)

FIELD RELAY

AIR GAP

VOLTAGE REGULATOR

POINT GAP

"F" TERMINAL

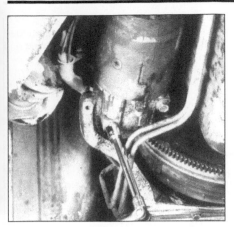

19.4a Working from underneath the vehicle (supported securely on jackstands), remove the starter motor mounting bolts . . .

19.4b . . . and remove the starter and solenoid (arrow) as an assembly

11 Crank the engine and take the voltmeter readings as soon as a steady figure is indicated. Do not allow the starter motor to turn for more than 15 seconds at a time. A reading of nine volts or more, with the starter motor turning at normal cranking speed, is normal. If the reading is nine volts or more but the cranking speed is slow, the motor is faulty. If the reading is less than nine volts and the cranking speed is slow, the solenoid contacts are probably burned, the starter motor is bad, the battery is discharged or there is a bad connection.

19 Starter motor - removal and installation

Refer to illustrations 19.4a and 19.4b
1   Detach the cable from the negative terminal of the battery.
2   Raise the vehicle and support it securely on jackstands.
3   Clearly label, then disconnect the wires from the terminals on the starter motor and solenoid.
4   Remove the starter motor mounting bolts **(see illustrations)**. Remove the starter.
5   Installation is the reverse of removal.

20 Starter motor solenoid - removal, repair and installation

Refer to Illustrations 20.3 and 20.4
1   After removing the starter/solenoid unit (see Section 19), disconnect the connector strap from the solenoid MOTOR terminal.
2   Remove the two screws which secure the solenoid housing to the end-frame assembly.
3   Twist the solenoid in a clockwise direction to disengage the flange key and then withdraw the solenoid **(see illustration)**.
4   Remove the nuts and washers from the solenoid terminals and then unscrew the two solenoid end-cover retaining screws and washers and pull off the end-cover **(see illustration)**.
5   Unscrew the nut washer from the bat-

---

pausing to allow it to cool for at least two minutes.
b) *The starter is connected directly to the battery and could arc or cause a fire if mishandled, overloaded or shorted out.*
c) *Always detach the cable from the negative terminal of the battery before working on the starting system.*

18 Starter motor - testing in vehicle

Note: *Before diagnosing starter problems, make sure that the battery is fully charged.*
1   If the starter motor does not turn at all when the ignition switch is operated, make sure that the shift lever is in Neutral or Park (automatic transaxle) or that the clutch pedal is depressed (manual transaxle).
2   Make sure that the battery is charged and that all cables, both at the battery and starter solenoid terminals, are clean and secure.
3   If the starter motor spins but the engine is not cranking, the overrunning clutch in the starter motor is slipping and the starter motor must be replaced.
4   If, when the switch is actuated, the starter motor does not operate at all but the

solenoid clicks, then the problem lies with either the battery, the main solenoid contacts or the starter motor itself, or the engine is seized.
5   If the solenoid plunger cannot be heard when the switch is actuated, the battery is bad, the fusible link is burned (the circuit is open) or the solenoid itself is defective.
6   To check the solenoid, connect a jumper lead between the battery (positive terminal) and the ignition switch terminal (the small terminal) on the solenoid. If the starter motor now operates, the solenoid is OK and the problem is in the ignition switch, neutral start switch or in the wiring.
7   If the starter motor still does not operate, remove the starter/solenoid assembly for disassembly, testing and repair.
8   If the starter motor cranks the engine at an abnormally slow speed, first make sure that the battery is charged and that all terminal connections are tight. If the engine is partially seized, or has the wrong viscosity oil in it, it will crank slowly.
9   Run the engine until normal operating temperature is reached, then disconnect the coil wire from the distributor cap and ground it on the engine.
10   Connect a voltmeter positive lead to the battery positive post and then connect the negative lead to the negative post.

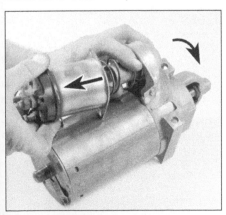

20.3 To disengage the solenoid from the starter, turn it in a clockwise direction

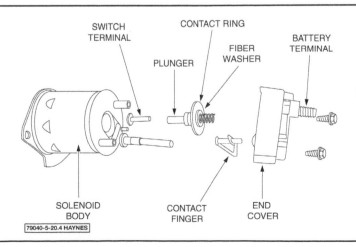

20.4 An exploded view of a typical starter solenoid assembly

SWITCH TERMINAL

CONTACT RING

FIBER WASHER

BATTERY TERMINAL

PLUNGER

SOLENOID BODY

CONTACT FINGER

END COVER

79040-5-20.4 HAYNES

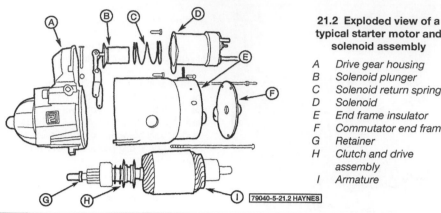

21.2 Exploded view of a typical starter motor and solenoid assembly

A Drive gear housing
B Solenoid plunger
C Solenoid return spring
D Solenoid
E End frame insulator
F Commutator end frame
G Retainer
H Clutch and drive assembly
I Armature

79040-5-21.2 HAYNES

tery terminal on the end-cover and remove the terminal.

6   Remove the resistor bypass terminal and contactor.

7   Remove the motor connector strap terminal and solder a new terminal in position.

8   Use a new battery terminal and install it to the end-cover. Install the bypass terminal and contactor.

9   Install the end-cover and the remaining terminal nuts.

10   Install the solenoid to the starter motor by first checking that the return spring is in position on the plunger and then insert the solenoid body into the drive housing and turn the body counter clockwise to engage the flange key.

11   Install the two solenoid securing screws and connect the MOTOR connector strap.

21 Starter motor - overhaul

Refer to Illustration 21.2
Note: *Due to the critical nature of the disassembly and testing of the starter motor it may be advisable for the home mechanic to simply purchase a new or factory-rebuilt unit. If it is decided to overhaul the starter, check on the availability of singular replacement components before proceeding.*

1   Disconnect the starter motor field coil connectors from the solenoid terminals.

2   Unscrew and remove the through-bolts **(see illustration)**.

3   Remove the commutator end-frame, field frame assembly and the armature from the drive housing.

4   Slide the two-section thrust collar off the end of the armature shaft and then, using a piece of suitable tube, drive the stop/retainer up the armature shaft to expose the snap-ring.

5   Extract the snap-ring from its shaft groove and then slide the stop/retainer and overrunning clutch assembly from the armature shaft.

6   Dismantle the brush components from the field frame.

7   Release the V-shaped springs from the brush holder supports.

8   Remove the brush holder support pin

and then lift the complete brush assembly upwards.

9   Disconnect the leads from the brushes if they are worn down to half their original length and they are to be replaced.

10   The starter motor is now completely dismantled except for the field coils. If these are found to be defective during the tests described later in this Section, removal of the pole shoe screws is best left to a service facility which has the necessary pressure driver.

11   Clean all components and replace any obviously worn components.

12   Never attempt to undercut the insulation between the commutator segments on starter motors having the molded type commutators. On commutators of conventional type, the insulation should be undercut (below the level of the segments) by 1/32-inch. Use an old hacksaw blade to do this, and make sure that the undercut is the full width of the insulation and the groove is quite square at the bottom. When the undercutting is completed, brush away all dirt and dust.

13   Clean the commutator by spinning it while a piece of '00' sandpaper is wrapped around it. Never use any other type of abrasive material for this work.

14   If necessary, because the commutator is in such bad shape, it may be turned down in a lathe to provide a new surface. Make sure to undercut the insulation when the turning is completed.

15   To test the armature for ground: use a lamp-type circuit tester. Place one lead on the armature core or shaft and the other on a segment of the commutator. If the lamp lights, then the armature is grounded and must be replaced.

16   To test the field coils for open circuit: place one test probe on the insulated brush and the other on the field connector bar. If the lamp does not light, the coils are open and must be replaced.

17   To test the field coils for ground: place one test probe on the connector bar and the other on the grounded brush. If the lamp lights, then the field coils are grounded.

18   The overrunning clutch cannot be repaired, and if faulty, it must be replaced as a complete assembly.

19   Install the brush assembly to the field frame as follows:

20   Install the brushes to their holders.

21   Assemble the insulated and grounded brush holders together with the V-spring and then locate the unit on its support pin.

22   Push the holders and spring to the bottom of the support and then rotate the spring to engage the V in the support slot.

23   Connect the ground wire to the grounded brush and the field lead wire to the insulated brush.

24   Repeat the operations for the second set of brushes.

25   Smear silicone oil onto the drive end of the armature shaft and then slide the clutch assembly (pinion to the front) onto the shaft.

26   Slide the pinion stop/retainer onto the shaft so that its open end is facing away from the pinion.

27   Stand the armature vertically on a piece of wood and then position the snap-ring on the end of the shaft. Using a hammer and a piece of hardwood, drive the snap-ring onto the shaft.

28   Slide the snap-ring down the shaft until it drops into its groove.

29   Install the thrust collar on the shaft so that the shoulder is next to the snap-ring. Using two pairs of pliers, squeeze the thrust collar and stop/retainer together until the snap-ring fully enters the retainer.

30   Lubricate the drive housing bushing with silicone oil and after ensuring that the thrust collar is in position against the snap-ring, slide the armature and clutch assembly into the drive housing so that, at the same time, the shift lever engages with the clutch.

31   Position the field frame over the armature and apply sealing compound between the frame and the solenoid case.

32   Position the field frame against the drive housing, taking care not to damage the brushes.

33   Lubricate the bushing in the commutator end-frame using silicone oil; place the leather brake washer on the armature shaft and then slide the commutator end-frame onto the shaft.

34   Reconnect the field coil connectors to the MOTOR terminal of the solenoid.

35   Now check the pinion clearance. To do this, connect a 12-volt battery between the solenoid S terminal and ground and at the same time fix a heavy connecting cable between the MOTOR terminal and ground (to prevent any possibility of the starter motor rotating). As the solenoid is energized it will push the pinion forward into its normal cranking position and retain it there. With the fingers, push the pinion away from the stop/retainer in order to eliminate any slack, then check the clearance between the face of the pinion and the face of stop/retainer using a feeler gauge. The clearance should be between 0.010 and 0.140-inch to ensure correct engagement of the pinion with the flywheel (or driveplate - automatic transmission) ring-gear. If the clearance is incorrect, the starter will have to be dismantled again and any worn or distorted components replaced, no adjustment being provided for.

Chapter 6
Emissions control systems

Contents

Specifications

Torque specifications

Ft-lbs (unless otherwise indicated)

Oxygen sensor	30
Coolant temperature switch (EFE system)	96 in-lbs
Air pump pulley bolts	24
Air pump mounting bolts	20 to 35
Catalytic converter fill plug	28
Diverter valve-to-air pump	90 in-lbs
Actuator mounting bolts	25 in-lbs
Thermal vacuum switches	15

1 General information

Refer to illustration 1.8

To prevent pollution of the atmosphere from incompletely burned and evaporating gases and to maintain good driveability and fuel economy, a number of emission control devices are incorporated on later models. They include the:

Fuel control system
Electronic Spark Timing (EST)
Electronic Spark Control (ESC) system
Exhaust Gas Recirculation (EGR) system
Evaporative Emission Control System (EECS)
Positive Crankcase Ventilation (PCV) system
Thermostatic air cleaner (THERMAC)
Air Injection Reaction (AIR) system
Early Fuel Evaporation (EFE) system
Transmission Controlled Spark (TCS) system
Catalytic converter

These systems were installed during different years for different models and engine options. Which systems were installed on which engines was also dependent upon the area where the vehicle was first sold, with 49-state, California and Canadian models having different combinations of emission control devices.

The fuel system installed (two-barrel carburetor, four-barrel carburetor, throttle body fuel injection) also caused variances in the systems used. Refer to the systems described and illustrated in this Chapter and compare them to the equipment on your engine to determine just what systems have been installed on your vehicle.

On later models some of these systems are linked, directly or indirectly, to the Computer Command Control (CCC) system.

The Sections in this Chapter include general descriptions, checking procedures within the scope of the home mechanic and component replacement procedures (when possible) for each of the systems listed above.

Before assuming that an emissions control system is malfunctioning, check the fuel and ignition systems carefully. The diagnosis of some emission control devices requires specialized tools, equipment and training. If checking and servicing become too difficult or if a procedure is beyond the scope of your skills, consult your dealer service department.

This doesn't mean, however, that emission control systems are particularly difficult to maintain and repair. You can quickly and easily perform many checks and do most (if not all) of the regular maintenance at home with common tune-up and hand tools. **Note:** *The most frequent cause of emissions problems is simply a loose or broken vacuum hose or wiring connection, so always check the hose and wiring connections first.*

Pay close attention to any special precautions outlined in this Chapter. It should be noted that the illustrations of the various systems may not exactly match the system installed on your vehicle because of changes made by the manufacturer during production or from year-to-year.

On most later model vehicles a Vehicle Emissions Control Information label is

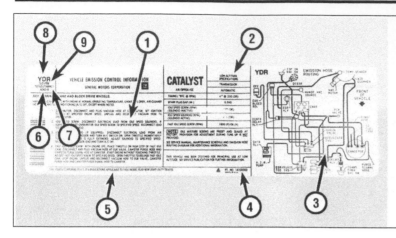

1.8 The vehicle Emission Control Information (VECI) label provides information regarding engine size, exhaust emission system used, engine adjustment procedures and specifications and an emission component and vacuum hose schematic diagram

1 *Adjustment procedures*
2 *Engine adjustment specifications*
3 *Emission component and vacuum hose routing*
4 *Label part number*
5 *Area of certification (California, Federal or Canada)*
6 *Evaporative emission system*
7 *Exhaust emission system*
8 *Label code*
9 *Engine size*

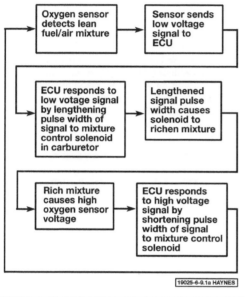

2.1a Typical Computer Command Control (CCC) system operation cycle (carbureted model)

located in the engine compartment **(see illustration)**. This label contains important emissions specifications and setting procedures, as well as a vacuum hose schematic with emissions components identified. When servicing the engine or emissions systems, the VECI label in your particular vehicle should always be checked for up-to-date information.

2   Computer Command Control (CCC) system and trouble codes

General description

Refer to illustrations 2.1a, 2.1b, 2.2a, 2.2b and 2.2c

The Computer Command Control (CCC) system first became available on 1980 models. In 1982, the system was upgraded. Both systems control exhaust emissions while retaining driveability by maintaining a continuous interaction between all of the emissions systems on your vehicle **(see illustrations)**.

The Computer Command Control (CCC) system is used on later models and consists of an Electronic Control Module (ECM) and information sensors which monitor various functions of the engine and send data back to the ECM **(see illustrations)**.

The CCC system is analogous to the central nervous system in the human body: The sensors (nerve endings) constantly relay information to the ECM (brain), which processes the data and, if necessary, sends out a command to change the operating parameters of the engine (body).

Here's a specific example of how one portion of this system operates: An oxygen sensor, located in the exhaust manifold, constantly monitors the oxygen content of the exhaust gas. If the percentage of oxygen in the exhaust gas is incorrect, indicating a too rich or too lean condition, an electrical signal is sent to the ECM. The ECM takes this information, processes it and then sends a command to the fuel injection system or carburetor mixture control solenoid, telling it to change the air/fuel mixture. This happens in a fraction of a second and it goes on continuously when the engine is running. The end

2.1b Typical Computer Command Control (CCC) system component guide (carbureted model)

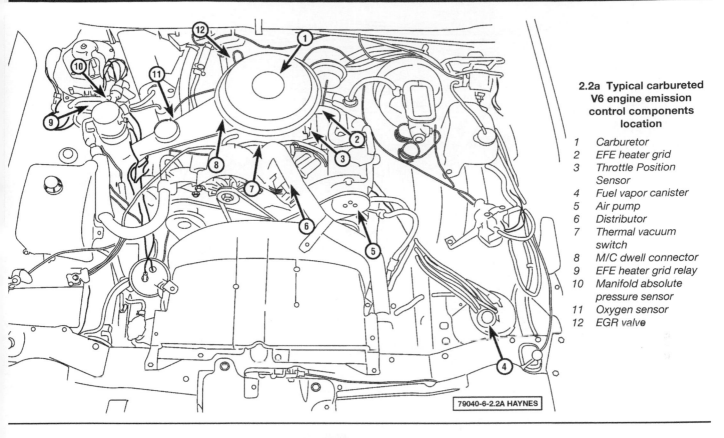

2.2a Typical carbureted V6 engine emission control components location

1. Carburetor
2. EFE heater grid
3. Throttle Position Sensor
4. Fuel vapor canister
5. Air pump
6. Distributor
7. Thermal vacuum switch
8. M/C dwell connector
9. EFE heater grid relay
10. Manifold absolute pressure sensor
11. Oxygen sensor
12. EGR valve

79040-6-2.2A HAYNES

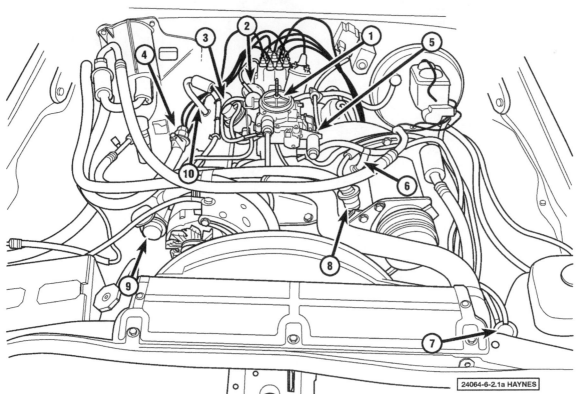

24064-6-2.1a HAYNES

2.2b Component locations on a typical early 2bbl carburetor equipped V8 engine with California emissions package

1	Carburetor	4	AIR check valve	7	EVAP canister
2	Vacuum advance unit	5	Idle stop solenoid	8	AIR check valve
3	EGR valve	6	PCV valve	9	Diverter valve

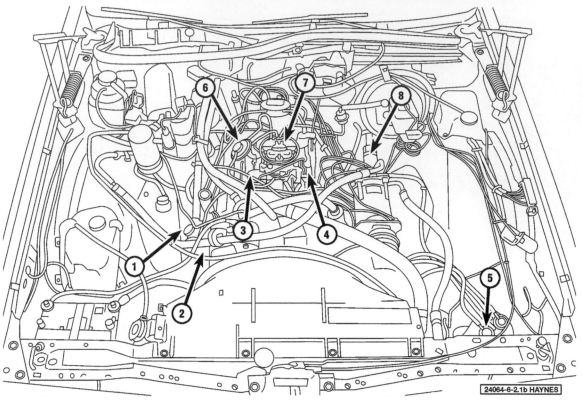

2.2c Component locations on a typical 4 bbl carburetor equipped V8 engine with Federal (49-state) emissions package

1   Diverter valve
2   AIR pump
3   Decel valve
4   Stepped speed control actuator
5   EVAP canister
6   EGR valve
7   Carburetor
8   EFE valve

24064-6-2.1b HAYNES

result is an air/fuel mixture ratio which is constantly maintained at a predetermined ratio, regardless of driving conditions.

One might think that a system which uses an on-board computer and electrical sensors would be difficult to diagnose. This is not necessarily the case. The system has a built-in diagnostic feature which indicates a problem by flashing a CHECK ENGINE light on the instrument panel. On the early system, the CHECK ENGINE light will remain on as long as the engine is running. With the ECM diagnostic system activated, this same lamp will flash the trouble code related to the cause of the malfunction. On the later system, the CHECK ENGINE light will remain on until the problem is identified and repaired and the code is erased from memory. In other words, the later system stores trouble codes in its memory, but the early system doesn't.

Diagnostic tool information

Refer to illustrations 2.6, 2.7 and 2.8

A digital multimeter is a necessary tool for checking fuel and emission related components **(see illustration)**. A digital volt-ohmmeter is preferred over the older style analog multimeter for several reasons. The analog multimeter cannot display the volts-ohms or amps measurement in hundredths and thousandths increments. When working with electronic circuits which are often very low voltage, this accurate reading is most important. Another good reason for the digital multimeter is the high impedance circuit. The digital multimeter is equipped with a high resistance

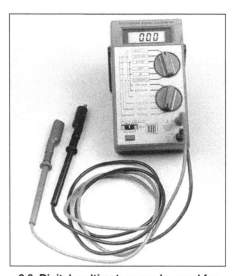

2.6 Digital multimeters can be used for testing all types of circuits; because of their high impedance, they are much more accurate than analog meters for measuring millivolts in low-voltage computer circuits

internal circuitry (10 million ohms). Because a voltmeter is hooked up in parallel with the circuit when testing, it is vital that none of the voltage being measured should be allowed to travel the parallel path set up by the meter itself. This dilemma does not show itself when measuring larger amounts of voltage (9 to 12 volt circuits) but if you are measuring a low voltage circuit such as the oxygen signal voltage, a fraction of a volt may be a sig-

2.7 Scanners like the Actron Scantool and the AutoXray XP240 are powerful diagnostic aids - programmed with comprehensive diagnostic information, they can tell you just about anything you want to know about your engine management system

nificant amount when diagnosing a problem.

Hand-held scanners are the most powerful and versatile tools for analyzing engine management systems used on later model vehicles **(see illustration)**. Each brand scan tool must be examined carefully to match the year, make and model of the vehicle you are working on. Often interchangeable cartridges are available to access the particular manufacturer; Ford, GM, Chrysler, etc.). Some manufacturers will specify by continent; Asia, Europe, USA, etc.

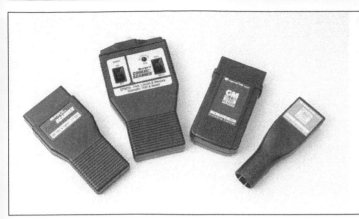

2.8 Trouble code tools simplify the task of extracting the trouble codes

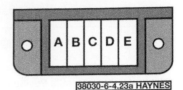

2.9a On 1981 models, connect a jumper wire between terminals D and E of the Assembly Line Data Link (ALDL) connector to enter the diagnostic mode (with the engine running)

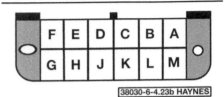

2.9b 1982 and later models, connect a jumper wire between terminals A and B of the Assembly Line Data Link (ALDL) connector to enter the diagnostic mode (with the key On, engine Off)

Another type of code reader is available at parts stores **(see illustration)**. These tools simplify the procedure for extracting codes from the engine management computer by simply "plugging in" to the diagnostic connector on the vehicle wiring harness.

Obtaining trouble codes

Refer to illustrations 2.9a and 2.9b

To retrieve trouble code information from the ECM memory, you must use a short jumper wire to ground a diagnostic terminal. **Note:** *The computer on 1980 and 1981 models does not possess long term memory. Once the CHECK ENGINE light appears on the instrument panel, it will be necessary to access the trouble code without turning the engine off or the trouble code will be lost.*

a) On **1980 models**, *locate the single wire green connector taped to the wiring harness near the ECM (the ECM is located under the dash, on the passenger side, behind the glove box). With the engine running, connect the terminal in this connector to a good chassis ground.*

b) On **1981 models**, *locate the 5-pin Assembly Line Diagnostic Link under the dash, just below the instrument panel* **(see illustration)**. *Attach one end of a short jumper wire to terminal D (diagnostic terminal) and the other end to terminal E (ground) while the engine is running.*

c) On **1982 and later models**, *locate the 12-pin Assembly Line Diagnostic Link under the dash, just below the instru-*

ment panel **(see illustration)**. *Turn the ignition Key On (engine not running) and verify the CHECK ENGINE light is on steady. Attach one end of a short jumper wire to terminal B (diagnostic terminal) and the other end to terminal A (ground).*

When the diagnostic terminal is grounded with the key On, engine Off, the ECM will display a "Code 12" by flashing the CHECK ENGINE light. A Code 12 is simply one flash, followed by a brief pause, then two flashes in quick succession. This indicates the system is operating properly and capable of trouble code output.

On 1982 and later models, this code will be flashed three times. If no other codes are stored, Code 12 will continue to flash until the diagnostic terminal ground is removed. After flashing Code 12 three times, the ECM will display any stored trouble codes. Each code will be flashed three times, then Code 12 will be flashed again, indicating that the display of any stored trouble codes has been completed.

When the ECM sets a trouble code, the CHECK ENGINE light will come on and a trouble code will be stored in memory. If the problem is intermittent, the light will go out after 10 seconds or when the fault goes away. However, the trouble code will stay in the ECM memory until the battery voltage to the ECM is interrupted. Removing battery voltage for 10 seconds will clear all stored trouble codes. Trouble codes should always be cleared after repairs have been completed. **Caution:** *To prevent damage to the*

ECM, the ignition switch must be off when disconnecting power to the ECM.

Clearing trouble codes

To clear the diagnostic trouble codes from ECM memory, disconnect the negative terminal from the battery for at least 10 seconds. Note any radio presets before disconnecting the battery so they may be reprogrammed into the radio memory when battery power is restored.

Trouble codes

Following is a list of the typical trouble codes which may be encountered while diagnosing the Computer Command Control System.

Also included are the probable causes for each code. Refer to the appropriate Section in this Chapter for sensor checks. If the problem persists after these checks have been made, more detailed service procedures will have to be done by a dealer service department.

Trouble code	Circuit or system	Probable cause *
Code 12	No distributor reference pulse	This code will flash whenever the diagnostic terminal is grounded with the ignition On and the engine not running. If this code appears while the engine is running, no reference pulses from the distributor are reaching the ECM.
Code 13	Oxygen sensor circuit	Sticking Throttle Position Sensor; poor electrical connection, open or short in circuit; defective oxygen sensor; defective ECM
Code 14	Coolant temperature sensor circuit	Engine overheating; poor electrical connection, open or short in circuit; defective coolant temperature sensor; defective ECM

Trouble code	Circuit or system	Probable cause *
Code 15	Coolant temperature sensor circuit	Poor electrical connection, open or short in circuit; defective coolant temperature sensor; defective ECM
Code 21	Throttle Position Sensor	Sticking or misadjusted TPS; poor electrical connection, open or short in circuit; defective TPS; defective ECM
Code 22	Throttle Position Sensor	Sticking or misadjusted TPS; poor electrical connection, open or short in circuit; defective TPS; defective ECM
Code 23	Mixture control solenoid	Open circuit or short to ground in the mixture solenoid circuit; defective mixture control solenoid; defective ECM - If the mixture control solenoid is shorted to ground, the ECM may have been damaged.
Code 24	Vehicle Speed Sensor	Poor electrical connection, open or short in circuit; defective VSS; defective ECM - A fault in this circuit should be indicated only when the vehicle is in motion. Disregard Code 24 if it is set when the drive wheels are not turning.
Code 32 (carbureted models)	BARO sensor	Defective BARO sensor; poor electrical connection, open or short in circuit; defective ECM
Code 32 (TBI models)	EGR system	Restricted vacuum hose to EGR solenoid or valve; poor electrical connection, open or short in circuit; defective EGR solenoid; defective EGR valve; defective ECM
Code 33 (TBI models)	MAP sensor	Leaking or restricted vacuum hoses; poor electrical connection, open or short in circuit; defective Map sensor; defective ECM
Code 34 (carbureted models)	Differential pressure sensor	Leaking or restricted vacuum hoses; poor electrical connection, open or short in circuit; defective differential pressure sensor; defective ECM
Code 34 (TBI models)	MAP sensor	Leaking or restricted vacuum hoses; poor electrical connection, open or short in circuit; defective Map sensor; defective ECM
Code 35	Idle speed control	TPS sticking or misadjusted; poor electrical connection, open or short in circuit; defective ISC motor; defective ECM
Code 41	No distributor reference pulses to ECM with engine running	Poor electrical connection, open or short in circuit; defective distributor pick-up coil; fault in the MAP or VAC sensor circuit
Code 42	Electronic Spark Timing	Poor electrical connection, open or short in circuit; defective ignition module; defective ECM
Code 43	Electronic Spark Control	Poor electrical connection, open or short in circuit; defective ESC module; defective knock sensor; defective ECM
Code 44	Lean exhaust	Lean condition caused by malfunctioning carburetor/fuel injector, vacuum leak, low fuel pressure, etc.; poor electrical connection, open or short in circuit; defective MAP sensor; defective oxygen sensor; defective ECM
Code 45	Rich exhaust	Rich condition caused by restricted air filter, fuel in evaporative charcoal canister or crankcase, malfunctioning carburetor/fuel injector, high fuel pressure, etc.; defective TPS; defective MAP; defective oxygen sensor; defective ECM
Code 51	PROM	Defective PROM; PROM not properly installed; incorrect PROM installed; defective ECM
Code 52 (TBI models)	Fuel CALPAK	Defective CALPAK; CALPAK not properly installed; incorrect CALPAK installed
Code 53 (carbureted models)	EGR control	Leaking, restricted or improperly routed vacuum line; EGR valve stuck or leaking; poor electrical connection, open or short in circuit; defective EGR solenoid; defective ECM

Trouble code	Circuit or system	Probable cause *
Code 54 (carbureted models)	Mixture control solenoid	Open circuit or short to ground in the mixture solenoid circuit; defective mixture control solenoid; defective ECM - If the mixture control solenoid is shorted to ground, the ECM may have been damaged.
Code 54 (TBI models)	Fuel pump circuit	Defective fuel pump; defective fuel pump relay; defective oil pressure switch; poor electrical connection, open or short in circuit; defective ECM
Code 55 (TBI models)	ECM	Poor electrical connection, open or short in circuit; defective ECM

Component replacement may not cure the problem in all cases. For this reason, you may want to seek professional advice before purchasing replacement parts.

3   Electronic Control Module/PROM - removal and installation

Caution: *To avoid electrostatic discharge damage to the ECM, handle the ECM only by its case. Do not touch the electrical terminals during removal and installation. If available, ground yourself to the vehicle with an anti-static ground strap, available at computer supply stores.*

Removal

Refer to illustrations 3.4, 3.6a, 3.6b, 3.8 and 3.9

1   The Electronic Control Module (ECM) is located under the instrument panel.
2   Disconnect the negative cable at the battery. Place the cable out of the way so it cannot accidentally come in contact with the negative terminal of the battery, as this would once again allow power into the electrical system of the vehicle.
3   Disconnect the electrical connectors from the ECM.
4   Remove the bolts and retainer and carefully detach the ECM **(see illustration)**.
5   Turn the ECM so that the bottom cover is facing up and carefully place it on a clean

3.4 Typical ECM installation

work surface.
6   Remove the screw(s) and lift off the PROM access cover **(see illustrations)**.
7   If you are replacing the ECM itself, the new ECM will not contain a PROM. It will be necessary to remove the old PROM from the old ECM and install it in the new one.
8   Using a PROM removal tool (available with a new PROM), grasp the PROM carrier

3.6a On early models the PROM cover is retained by one screw . . .

at the narrow ends **(see illustration)**. Gently rock the carrier from end to end while applying firm upward force. The PROM carrier and PROM should lift off the PROM socket easily. **Caution:** *The PROM carrier should only be removed with the special rocker-type PROM removal tool. Removal without this tool or with any other type of tool may damage the*

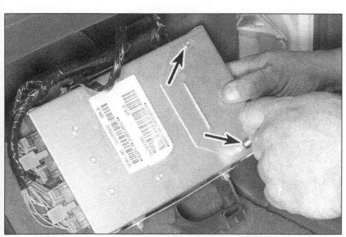

3.6b . . . while later models have two screws

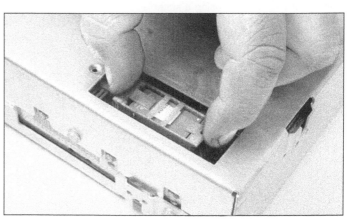

3.8 Grasp the PROM carrier at its narrow ends with the PROM removal tool (available with a new PROM) and gently rock the removal tool until the PROM is disconnected

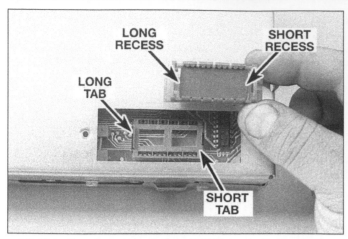

3.9 Note the reference end of the PROM carrier before setting it aside

3.10 Make sure the service numbers on the ECM and the PROM are the same - otherwise, depending on what you are replacing, you either have the wrong ECM or the wrong PROM

PROM or the PROM socket.

9   Note the reference end of the PROM carrier **(see illustration)** before setting it aside.

Installation

Refer to illustration 3.10

10   If you are replacing the ECM, remove the new ECM from its container and check the service number to make sure that it is the same as the number on the old ECM **(see illustration)**.

11   If you are replacing the PROM, remove the new PROM from its container and check the service number to make sure that it is the same as the number of the old PROM.

12   Position the PROM/PROM carrier assembly squarely over the PROM socket with the small notched end of the carrier aligned with the small notch in the socket at the pin 1 end. **Caution:** *If the PROM is installed backwards and the ignition switch is turned on, the PROM will be destroyed.*

13   Using the tool, press on the PROM carrier until it is firmly seated in the socket. **Caution:** *Do not press on the PROM - press only on the carrier.* On early models, further seat the PROM by pressing down on the carrier with your fingers while using a suitable narrow blunt tool to press down on the body of the PROM. Press alternately at either end of the PROM to fully seat it.

14   Attach the access cover to the ECM and tighten the two screws.

15   Install the ECM in the support bracket, plug in the electrical connectors to the ECM and install the hush panel.

16   Start the engine.

17   Enter the diagnostic mode by grounding the diagnostic lead of the ALDL (see Section 2). If the CHECK ENGINE light flashes Code 12 and no trouble codes appear, the PROM is correctly installed.

18   If Trouble Code 51 appears, or if the CHECK ENGINE light comes on and remains constantly lit, the PROM is not fully seated, is installed backwards, has bent pins or is defective.

19   If the PROM is not fully seated, pressing firmly on both ends of the carrier should correct the problem.

20   It is possible to install the PROM backwards. If this occurs, and the ignition key is turned to ON, the PROM circuitry will be destroyed and the PROM will have to be replaced.

21   If the pins have been bent, remove the PROM in accordance with the above procedure, straighten the pins and reinstall the PROM. If the bent pins break or crack when you attempt to straighten them, discard the PROM and replace it with a new one.

22   If careful inspection indicates that the PROM is fully seated, has not been installed backwards and has no bent pins, but the CHECK ENGINE light remains lit, the PROM is probably faulty and must be replaced.

---

4   Barometric pressure (BARO) sensor - check and replacement

1   The BARO sensor detects ambient pressure changes that occur as the result of changes in the weather and/or the altitude of the vehicle. It then sends an electronic signal to the ECM that is used to adjust the air fuel ratio and spark timing.

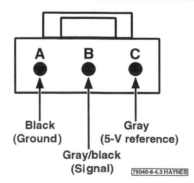

Black (Ground)

Gray/black (Signal)

Gray (5-V reference)

`79040-6-4.3 HAYNES`

4.3 Barometric pressure (BARO) sensor terminal guide

Check

Refer to illustration 4.3

2   A problem with the BARO sensor will usually set a Code 32. To check the sensor, begin by checking the reference voltage from the ECM. Disconnect the electrical connector and connect a voltmeter to the gray (+) and black (-) wire terminals of the harness connector. Turn the ignition key On, approximately 5.0 volts should be present. If reference voltage is not present, check the circuits from the sensor to the ECM. If the circuits are good the ECM is defective.

3   If reference voltage is available, turn the ignition key Off and reconnect the connector to the sensor. Using suitable probes, backprobe the connector and connect the voltmeter to terminals A and B at the sensor **(see illustration)**. Turn the ignition Key ON (engine not running) and compare your voltage reading with the following chart:

Altitude (in feet)	Voltage range
Below 1000	3.8 to 5.5
1000 to 2000	3.6 to 5.3
2000 to 3000	3.5 to 5.1
3000 to 4000	3.3 to 5.0
4000 to 5000	3.2 to 4.8
5000 to 6000	3.0 to 4.6
6000 to 7000	2.9 to 4.5
7000 to 8000	2.8 to 4.3
8000 to 9000	2.6 to 4.2
9000 to 10000	2.5 to 4.0

3   If the voltage reading is not within the correct range, replace the sensor.

Replacement

Refer to illustration 4.4

4   To replace the sensor, detach the vacuum hose, disconnect the electrical connector and remove the mounting screws **(see illustration)**.

5   Installation is reverse of removal.

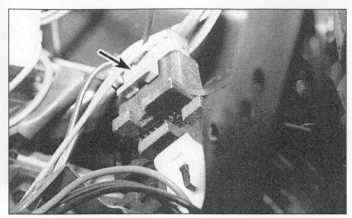

4.4 The BARO sensor is located under the passengers side of the dash or in the engine compartment on the passenger side fenderwell, to remove the sensor disconnect the electrical connector (arrow) and remove the mounting screw

5.2 The coolant temperature sensor (arrow) is normally located near the thermostat housing

5   Coolant temperature sensor - check and replacement

1   The coolant temperature sensor is a thermistor (a resistor which varies the value of its resistance in accordance with temperature changes). The coolant sensor is mounted in a coolant passage, usually in the front of the intake manifold. As the sensor temperature INCREASES, the resistance will DECREASE. The change in the resistance value will directly affect the voltage signal from the coolant sensor to the ECM. A failure in the coolant sensor circuit should set either a Code 14 or a Code 15. These codes indicate a failure in the coolant temperature sensor or circuit, so the appropriate solution to the problem will be either repair of a wire or replacement of the sensor. The sensor can be checked with an ohmmeter, by measuring its resistance when cold, then warming up the engine and taking another measurement.

Check

Refer to illustration 5.2
2   To check the sensor resistance, discon-

5.5 Wrap the threads of the coolant temperature sensor with Teflon tape to prevent leaks and thread corrosion

nect the harness connector from the coolant temperature sensor and measure the resistance across the two terminals on the sensor **(see illustration)**. Check its resistance with the engine completely cold, then start the engine and monitor the resistance change as the engine warms-up. Compare your readings with the following chart: **Note:** *In the event the coolant temperature sensor is difficult to position the ohmmeter probes, it will be necessary to remove the sensor and perform the tests in a pan of heated water to simulate the conditions.*

Temperature	Resistance
210-degrees F	185 ohms
160-degrees F	450 ohms
100-degrees F	1,800 ohms
70-degrees F	3,400 ohms
40-degrees F	7,500 ohms
20-degrees F	13,500 ohms

3   If the resistance reading is not within the correct range, replace the sensor.
4   Check the reference voltage from the ECM (computer) to the ECT with the ignition key ON (engine not running). Connect the positive probe of the ohmmeter onto the yellow (+) wire and the negative probe onto the black (-) wire. The reading should be approximately 5.0 volts.

Replacement

Refer to illustration 5.5
Warning: *Wait until the engine is completely cool before beginning this procedure.*
5   Drain the cooling system to a level just below the sensor. Disconnect the electrical connector and remove the sensor. **Caution:** *Handle the coolant sensor with care. Damage to this sensor will affect the operation of the entire fuel system.* Before installing the new sensor, wrap the threads with Teflon sealing tape to prevent leakage and thread corrosion **(see illustration).**
6   Installation is the reverse of removal.

6   Differential pressure sensor - check and replacement

Refer to illustration 6.1
1   The differential pressure sensor is used on some carbureted models. The differential pressure sensor measures the difference between atmospheric pressure and manifold vacuum. This information is used by the ECM to determine engine load. The sensor is located in the engine compartment on the firewall adjacent to the brake master cylinder or right fender near the coolant recovery bottle **(see illustration)**.
2   A failure in the differential pressure sensor should set a Code 34.

Check

Refer to illustration 6.5
3   Start the engine and allow it to reach normal operating temperature. Disconnect the vacuum hose at the sensor and measure the vacuum at the hose, it should be a minimum of 10 in-Hg at idle. If it isn't repair the

6.1 On this model, the differential pressure sensor (arrow) is located on a passenger side inner fenderwell of the engine compartment

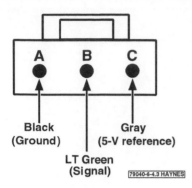

Black
(Ground)

LT Green
(Signal)

Gray
(5-V reference)

79040-6-4.3 HAYNES

**6.5 Differential pressure sensor
terminal guide**

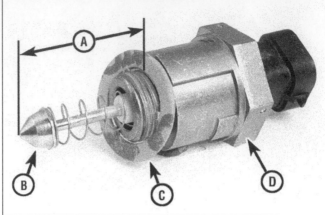

**7.10 Typical
IAC valve**

A   *Pintle extension
distance*
B   *Pintle*
C   *Gasket*
D   *IAC valve body*

problem before proceeding.

4   To check the sensor, begin by checking the reference voltage from the ECM. Disconnect the electrical connector and connect a voltmeter to the gray (+) and black (-) wire terminals of the harness connector. Turn the ignition key On, approximately 5.0 volts should be present. If reference voltage is not present, check the circuits from the sensor to the ECM. If the circuits are good the ECM is defective.

5   If reference voltage is available, turn the ignition key Off and reconnect the connector to the sensor. Using suitable probes, back-probe the connector and connect the voltmeter to terminals A and B at the sensor **(see illustration)**. Turn the ignition Key ON (engine not running) and note the voltage reading. It should be approximately 0.5 volt.

6   Start the engine and allow it to idle. The voltage should increase to over 2.0 volts. If it doesn't, check for damage to the wiring from the ECM or faulty connectors. If the circuits and connections are good, replace the sensor.

Replacement

7   Disconnect the Differential Pressure Sensor electrical connector and vacuum hose.

8   Remove the sensor retaining screws and remove the sensor.

9   Installation is reverse of removal.

---

7   Idle Air Control (IAC) valve - check and replacement

1   On TBI models, the engine idle speed is controlled by the ECM through the Idle Air Control (IAC) valve. The ECM sends voltage pulses to the IAC motor windings causing the IAC motor shaft and pintle to move in-and-out a given distance for each pulse. This movement controls airflow around the throttle plate, which in turn, controls engine idle speed, either hot or cold. The IAC valve is located on the throttle body.

Check

2   Start engine and allow the engine to

reach normal operating temperature, turn the ignition off and connect a tachometer.

3   Disconnect the electrical connector from the IAC valve and start the engine. If the idle speed increases, turn the ignition off and reconnect the IAC valve electrical connector. Restart the engine and check the idle, if the idle speed returns to normal, the IAC valve is functioning properly.

4   If the idle speed didn't respond as described, turn the engine Off and ground the diagnostic test terminal at the ALDL **(see illustration 2.9b)**. Disconnect the IAC valve electrical connection and turn the ignition switch On (engine not running).

5   Using a test light connected to ground, probe each of the four terminals in the harness connector. The test light should light at each IAC valve terminal indicating the ECM is attempting to control the IAC valve. If there is no control signal at the connector, check for an open wire, short to ground, faulty connection at the ECM or possible faulty ECM.

6   Finally, measure the resistance across each pair of terminals in the IAC valve. There should be a minimum of 20 ohms between the IAC valve terminals opposite the connector terminals A to B and C to D (the terminals are marked on the harness connector). If the resistance is not as specified, replace the IAC valve.

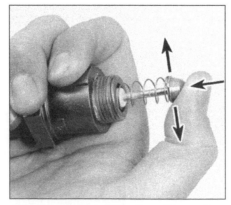

7.11 To adjust an IAC valve with a collar, retract the pintle by exerting firm pressure while moving it from side-to-side very slightly

Replacement

Refer to illustrations 7.10 and 7.11

7   Remove the air cleaner housing assembly and disconnect the IAC electrical connector.

8   Unscrew the IAC valve and withdraw the valve from the throttle body.

9   Clean the mounting surface and discard the old gasket.

10   Before installing the valve, measure the distance between the tip of the pintle and the housing mounting surface **(see illustration)**. If the distance is greater than 1-1/8 inch, it must be reduced to prevent damage to the valve.

11   The valve is adjusted by one of two methods, depending on the type of valve. If the valve has a collar at the end where the electrical connector attaches, retract the pintle by exerting firm pressure while moving it from side-to-side slightly **(see illustration)**. If the valve has no collar, pull the spring back and turn the pintle clockwise. After adjustment, make sure the straight section on the spring is engaged in the flat on the underside of the pintle.

12   Installation is the reverse of removal.

13   Start the engine and allow the engine to reach operating temperature, then turn it off. The IAC valve is reset by the ECM when the engine is turned off.

---

8   Manifold Absolute Pressure (MAP) sensor - check and replacement

Refer to illustration 8.1

1   The Manifold Absolute Pressure (MAP) sensor **(see illustration)** is mounted to the passenger side inner fenderwell on carbureted models or mounted on a bracket attached the air cleaner assembly on TBI models. The MAP sensor monitors the intake manifold pressure changes resulting from changes in engine load and speed and converts the information into a voltage output. The ECM uses the MAP sensor to control fuel delivery and ignition timing.

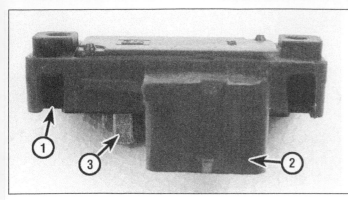

8.1 Manifold Absolute Pressure (MAP) sensor

1   MAP sensor
2   ECM harness connection
3   Manifold vacuum connection

Check

2   A failure in the MAP sensor circuit should set a Code 33 or Code 34.

3   To check the sensor, begin by checking the reference voltage from the ECM. Disconnect the electrical connector and connect a voltmeter to the gray (+) and black (-) wire terminals of the harness connector. Turn the ignition key On, approximately 5.0 volts should be present. If reference voltage is not present, check the circuits from the sensor to the ECM. If the circuits are good the ECM is defective.

4   If reference voltage is available, turn the ignition key Off and reconnect the connector to the sensor. Using suitable probes, backprobe the connector and connect the voltmeter to the light green (+) and black (-) wire terminals at the sensor. Turn the ignition Key ON (engine not running) and compare your voltage reading with the following chart:

Altitude (in feet)	Voltage range
Below 1000	3.8 to 5.5
1000 to 2000	3.6 to 5.3
2000 to 3000	3.5 to 5.1
3000 to 4000	3.3 to 5.0
4000 to 5000	3.2 to 4.8
5000 to 6000	3.0 to 4.6
6000 to 7000	2.9 to 4.5
7000 to 8000	2.8 to 4.3
8000 to 9000	2.6 to 4.2
9000 to 10000	2.5 to 4.0

5   If the voltage reading is within the correct range, disconnect the vacuum hose at the sensor, connect a hand held vacuum pump to the sensor fitting and apply 10 to 15 in-Hg of vacuum to the sensor (or start the engine). With vacuum applied to the sensor, the voltage should decrease sharply to 1.2 to 2.3 volts less than the voltage recorded in Step 4. If the voltage readings are not correct, replace the sensor.

Replacement

6   To replace the sensor, detach the vacuum hose, disconnect the electrical connector and remove the mounting screws.

7   Installation is reverse of removal.

9   Oxygen sensor

Refer to illustration 9.1

1   The oxygen sensor, which is located in the exhaust pipe **(see illustration)**, monitors the oxygen content of the exhaust gas stream. The oxygen content in the exhaust reacts with the oxygen sensor to produce a voltage output from 100 millivolts (high oxygen, lean mixture) to 900 millivolts (low oxygen, rich mixture). The ECM monitors this voltage output to determine optimum the air/fuel mixture. The ECM alters the air/fuel mixture ratio by controlling the pulse width (open time) of the mixture control solenoid (carbureted models) or fuel injectors (TBI models). A mixture of 14.7 parts air to 1 part fuel is the ideal ratio for minimizing exhaust emissions, thus allowing the catalytic converter to operate at maximum efficiency. It is this ratio of 14.7 to 1 which the ECM and the oxygen sensor attempt to maintain at all times.

2   The oxygen sensor must be hot to operate properly (approximately 600-degrees F). During the initial warm-up period, the ECM operates in open loop mode - that is, it controls fuel delivery in accordance with a programmed default value instead of feedback information from the oxygen sensor.

3   The proper operation of the oxygen sensor depends on four conditions:

9.1 Typical oxygen sensor installation

a)   **Electrical** - The low voltages and low currents generated by the sensor depend upon good, clean connections which should be checked whenever a malfunction of the sensor is suspected or indicated.

b)   **Outside air supply** - The sensor is designed to allow air circulation to the internal portion of the sensor. Whenever the sensor is removed and installed or replaced, make sure the air passages are not restricted.

c)   **Proper operating temperature** - The ECM will not react to the sensor signal until the sensor reaches approximately 600-degrees F. This factor must be taken into consideration when evaluating the performance of the sensor.

d)   **Unleaded fuel** - The use of unleaded fuel is essential for proper operation of the sensor. Make sure the fuel you are using is of this type.

4   In addition to observing the above conditions, special care must be taken whenever the sensor is serviced.

a)   The oxygen sensor has a permanently attached pigtail and connector which should not be removed from the sensor. Damage or removal of the pigtail or connector can adversely affect operation of the sensor.

b)   Grease, dirt and other contaminants should be kept away from the electrical connector and the louvered end of the sensor.

c)   Do not use cleaning solvents of any kind on the oxygen sensor.

d)   Do not drop or roughly handle the sensor.

e)   The silicone boot must be installed in the correct position to prevent the boot from being melted and to allow the sensor to operate properly.

Check

4   An open in the oxygen sensor circuit should set a Code 13. A low voltage in the circuit should set a Code 44. A high voltage in the circuit should set a Code 45. Codes 44 and 45 may also be set as a result of fuel system problems.

5   The sensor can also be checked with a high-impedance digital voltmeter. Warm up the engine to normal operating temperature, then turn the engine off.

6   Without disconnecting the oxygen sensor electrical connector, backprobe the sensor wire with a suitable probe and connect the positive lead of the voltmeter to the probe. **Caution:** *Don't let the sensor wire or the voltmeter lead touch the exhaust pipe or manifold.* Connect the negative lead of the meter to a good chassis ground. Place the meter to the millivolt setting.

7   Start the engine and monitor the voltage output of the sensor. When the engine is cold the sensor should produce a steady voltage of approximately 100 to 200 millivolts. After a period of approximately two minutes the

11.2 Disconnect the TPS electrical connector and check for a 4.5 to 5.0 volt reference voltage on the gray (+) and black (-) wire terminals of the harness connector

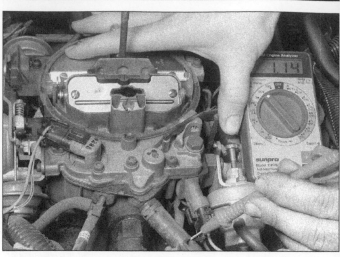

11.3a Backprobe the TPS connector with a voltmeter and check the signal voltage with the throttle closed . . .

engine should reach operating temperature and the voltage should begin to fluctuate between 100 and 900 millivolts. This indicates the system has reached closed loop and the computer is controlling fuel delivery. If the system fails to reach closed loop mode in a reasonable amount of time, the sensor may be defective or a fuel system problem could be the cause. If the voltage is fixed (or fluctuating very slowly), create a rich condition by opening the accelerator sharply and snapping it shut and then a lean condition by disconnecting a vacuum hose, if the sensor voltage does not respond quickly with the proper voltage fluctuations, the sensor is probably defective.

Replacement

Note: *Because the oxygen sensor is located in the exhaust pipe, it may be too tight to remove when the engine is cold. If you find it difficult to loosen, start and run the engine for a minute or two, then shut it off. Be careful not to burn yourself during the following procedure.*

8   Disconnect the cable from the negative terminal of the battery.
9   If necessary for clearance, raise the vehicle and place it securely on jackstands.
10   Disconnect the electrical connector from the sensor.
11   Note the position of the silicone boot, if equipped, and carefully unscrew the sensor from the exhaust manifold. **Caution:** *Excessive force may damage the threads.*
12   Anti-seize compound must be used on the threads of the sensor to facilitate future removal. The threads of a new sensor will already be coated with this compound, but if an old sensor is removed and reinstalled, recoat the threads.
13   Install the sensor and tighten it to the specified torque.
14   Reconnect the electrical connector of the pigtail lead to the main engine wiring harness.

15   Lower the vehicle and reconnect the cable to the negative terminal of the battery.

10   Park/Neutral start switch

1   The Park/Neutral start switch, located on the automatic transmission on early models and on the steering column on later models, indicates to the ECM when the transmission is in Park or Neutral. This information is used for Transmission Converter Clutch (TCC), Exhaust Gas Recirculation (EGR) and Idle Speed Control (ISC) motor operation. **Caution:** *The vehicle should not be driven with the Park/Neutral start switch disconnected because idle quality will be adversely affected.*
2   For more information regarding the Park/Neutral start switch, which is part of the Neutral start and back-up light switch assembly, see Chapter 7B.

11   Throttle Position Sensor - check and replacement

1   The Throttle Position Sensor (TPS) is mounted in the carburetor body (carbureted models) or on the side of the throttle body (TBI models). The TPS output voltage varies according to the angle of the throttle. Its job is to inform the computer about the rate of throttle opening and relative throttle position (on some early models, a Wide Open Throttle (WOT) switch may be used to signal the computer when the throttle is wide open). The TPS consists of a variable resistor that changes resistance as the throttle changes its opening. By signaling the computer when the throttle opens, the computer can richen the fuel mixture to retain the proper air/fuel ratio. The initial setting of the TPS is very important because the voltage signal the computer receives tells the computer the

exact position of the throttle at idle. A problem in any of the TPS circuits will set a Code 21 or 22.

Check

Refer to illustrations 11.2, 11.3a and 11.3b
2   Disconnect the Throttle Position Sensor (TPS) electrical connector. Using a voltmeter, check the reference voltage from the ECM. Connect the positive probe of the meter onto the gray wire terminal of the connector and negative probe onto the black wire terminal. Turn the ignition ON (engine not running). The voltage should read approximately 5.0 volts **(see illustration)**. If reference voltage is not present, check the circuits from the sensor to the ECM. If the circuits are good the ECM is defective.
3   Next, check the TPS signal voltage. Turn the ignition key Off and reconnect the connector to the TPS. Using suitable probes, backprobe the TPS connector and connect the positive probe of the voltmeter onto the dark blue wire and the negative probe onto the black wire. Turn the ignition key ON

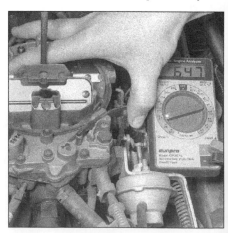

11.3b . . . gradually open the throttle and confirm that the voltage increases smoothly

11.6 Remove the Throttle Position Sensor mounting screws (arrows) - TBI model

(engine not running). The TPS voltage at closed throttle should be 0.5 to 1.5 volts. Gradually open the throttle valve and observe the TPS sensor voltage. The voltage should increase smoothly to approximately 4.5 volts at wide-open throttle **(see illustrations)**. If the readings are incorrect, replace the TPS.

Replacement

Refer to illustration 11.6

Carbureted models

4   TPS replacement on carbureted models requires partial carburetor disassembly (see Chapter 4). Remove the carburetor and have the TPS replaced by a competent carburetor technician.

TBI models

5   Disconnect the electrical connector from the TPS.
6   Remove the mounting screws and remove the TPS from the throttle body **(see illustration)**.
7   Install the TPS onto the throttle body with the throttle in the closed position. Make sure the arm on the TPS is above the actuating lever on the throttle body. Install the screws and tighten them securely.

12   Vehicle Speed Sensor (VSS) - check and replacement

1   The Vehicle Speed Sensor (VSS) sends a pulsing voltage signal to the ECM. The ECM uses this data for various functions, including the operation of the Transmission Converter Clutch. The VSS is located in the instrument cluster and is part of the speedometer. On some models a VSS buffer is taped to the wiring harness behind the instrument cluster.
2   Raise the rear of the vehicle and support it securely on jackstands. Place the transmission in the neutral position.
3   Backprobe the VSS brown wire with a voltmeter positive lead at the rear of the

speedometer, at the buffer or at the ECM connector. Connect the negative lead to a good chassis ground. **Caution:** *If you backprobe the ECM connector make sure you are connected to the correct brown wire, there may be more than one. Make sure the meter is set on the volts scale - DO NOT measure resistance at the ECM connector or damage to the ECM may result.*
4   Turn the ignition switch On and rotate a rear wheel. Voltage should vary form 1 to 6 volts as the wheel is turned. If the voltage doesn't vary check the circuits from the speed sensor to the buffer and ECM. If the circuits are good, remove the instrument cluster and have the VSS and buffer checked by a certified speedometer repair shop.

13   Distributor reference signal

The distributor sends a signal to the ECM to tell it both engine rpm and crankshaft position. See Electronic Spark Timing (EST), Section 14, for further information.

14   Electronic Spark Timing (EST) system

General description

Refer to illustration 14.4

1   To provide improved engine performance, fuel economy and control of exhaust emissions, the Electronic Control Module (ECM) controls distributor spark advance (ignition timing) with the Electronic Spark Timing (EST) system.
2   The ECM receives a reference pulse from the distributor, which indicates both engine rpm and crankshaft position. The ECM then determines the proper spark advance for the engine operating conditions and sends an EST pulse to the distributor.

Check

3   The ECM will set EST at a specified

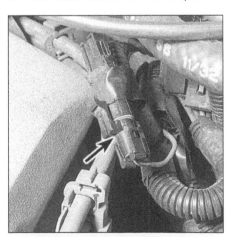

14.4 The EST connector (arrow) is located in the wiring harness near the rear of the engine

value when the diagnostic test terminals on the ALDL connector are grounded **(see illustrations 2.9a and 2.9b)**. To check for EST operation, the timing should be checked at 2000 rpm with the terminal ungrounded. Then ground the test terminals. If the timing changes at 2000 rpm, the EST is operating. A fault in the EST system will usually set Trouble Code 42.
4   Check the distributor 4-wire connector **(see illustration)**. If the connector is open, it will set a Code 42.
5   For further information regarding the testing and component replacement procedures for the HEI/EST distributor, refer to Chapter 5.

15   Electronic Spark Control (ESC) system

General description

1   Irregular octane levels in modern gasoline can cause detonation in an engine. Detonation is sometimes referred to as "spark knock."
2   The Electronic Spark Control (ESC) system is designed to retard spark timing up to 20-degrees to reduce spark knock in the engine. This allows the engine to use maximum spark advance to improve driveability and fuel economy.
3   The ESC module sends a voltage signal of 8 to 10-volts to the ECM when no spark knock is occurring and the ECM provides normal advance. When the knock sensor detects abnormal vibration (spark knock), the ESC module turns off the circuit to the ECM. The ECM then retards the EST distributor until spark knock is eliminated.
4   Failure of the ESC knock sensor signal or loss of ground at the ESC module will cause the signal to the ECM to remain high. This condition will result in the ECM controlling the EST as if no spark knock is occurring. Therefore, no retard will occur and spark knock may become severe under heavy engine load conditions. At this point, the ECM will set a Code 43. Loss of the ESC signal to the ECM will cause the ECM to constantly retard EST. This will result in sluggish performance and cause the ECM to set a Code 43.

Check

5   Connect a timing light to the engine according to the manufacturers instructions. With the engine running at approximately 1500 rpm carefully tap on the engine block with a hammer near the knock sensor while monitoring the ignition timing with the light. The timing should retard with each tap of the hammer. If it doesn't, disconnect the electrical connector to the knock sensor. Connect the lead of a 12-volt test light to the positive terminal of the battery and touch the terminal of the knock sensor harness connector with the tip of the test light while monitoring the

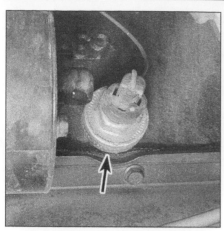

15.6 On TBI models, the knock sensor (arrow) is located on the side of the engine block just above the oil pan

15.12 To detach the ESC module, disconnect the electrical connector (B) and remove the mounting screws (A)

16.8 Push up on the diaphragm (arrow) when checking the EGR valve - if the diaphragm is stuck or binds in the up position, replace the valve with a new one

ignition timing with the timing light. The timing should retard each time the terminal is touched with the test light. If it does, the knock sensor is defective. If timing still did not retard, the ESC module may be defective. Before replacing the ESC module, check the following wiring and connections at the ESC harness connector and repair as necessary:

a) *Check for battery voltage with the key On at the pink/black wire terminal.*
b) *Check for continuity to ground at the brown wire terminal.*
c) *Check for continuity to the knock sensor at the dark blue or white wire terminal.*
d) *Check for continuity to the ECM at the black or yellow wire terminal.*

Component replacement

Knock sensor

Refer to illustration 15.6
6   On carbureted models, the knock sensor is located at the rear of the engine block, behind the intake manifold. On TBI models, it is located at the lower left side of the engine block, just above the oil pan **(see illustration)**. On TBI models, the threaded boss is open to the coolant passage.
7   Detach the cable from the negative terminal of the battery. On TBI models, drain the cooling system (see Chapter 1).
8   Disconnect the electrical connector and remove the knock sensor from the block.
10   Installation is the reverse of the removal procedure.

ESC module

Refer to illustration 15.12
11   Detach the cable from the negative terminal of the battery.
12   Locate the module on the right front inner fenderwell of the engine compartment **(see illustration)**.
13   Disconnect the electrical connector, remove the module mounting screws and remove the module.
14   Installation is the reverse of removal.

16   Exhaust Gas Recirculation (EGR) system

General description

1   The EGR system meters exhaust gases into the intake manifold. From there the exhaust gases pass into the combustion chambers for the purpose of lowering combustion temperature, thereby reducing the amount of oxides of nitrogen (NOx) formed.
2   The main element of the system is the EGR valve, mounted on the intake manifold, which feeds small amounts of exhaust gas back into the combustion chamber.
3   The amount of exhaust gas admitted is regulated by a vacuum or backpressure controlled EGR valve in response to engine operating conditions.
4   On earlier models, a Thermal Vacuum Switch (TVS) prevents the EGR valve from operating when the engine is cold.
5   On later models the EGR system is managed by the ECM. An ECM controlled solenoid is used in the vacuum line in order to maintain finer control of EGR flow. The ECM uses information from the coolant temperature, throttle position and manifold pressure sensors to regulate the vacuum solenoid.
6   Common engine problems associated with the EGR system are rough idle, stalling at idle after deceleration, surging at cruising speeds and stalling after cold starts. Engine overheating and detonation can also be caused by EGR system problems.

Check

EGR valve

Refer to illustration 16.8
7   Make a physical inspection of the hoses and electrical connections to ensure that nothing is loose.
8   With the engine off, place your finger under the EGR valve and push up on the underside of the diaphragm **(see illustration)** - if the engine is hot, wear a glove to prevent

burns! The diaphragm should move freely and not stick or bind. If it doesn't move or if it sticks, replace the valve with a new one.
9   With the engine running at normal operating temperature, push up on the diaphragm. The engine speed should drop and/or the engine should stall.
10   If it doesn't, clean the EGR valve and passages or replace the EGR valve if necessary.
11   If the engine speed drops, increase the engine speed from idle to about 2000 rpm. The EGR valve diaphragm should move as this is done.
12   If it moves the EGR system is operating correctly.
13   If it doesn't move, attach a vacuum gauge to the vacuum hose connected to the EGR valve (use a T-fitting to connect the gauge). Start the engine and increase the speed from idle to 2000 rpm, then let it return to idle. At idle, approximately 6-inches or more of vacuum should be indicated.
14   If vacuum is indicated but the valve does not move when connected to vacuum, the EGR valve is bad.
15   If no vacuum is indicated, check the vacuum hose to make sure it isn't leaking, restricted or disconnected.
16   Make the necessary repairs to the hose. If the hose is not a problem, check the thermal vacuum switch (non-ECM models) or EGR control solenoid (ECM models) as follows.

Thermal vacuum switch

Refer to illustration 16.17
17   Remove the carburetor-to-switch hose from the switch and check for vacuum at about 2000 rpm **(see illustration)**. The gauge should indicate approximately 10-inches of vacuum. If it doesn't, check for a plugged or cracked hose and plugged carburetor passage. If vacuum is indicated, reconnect the hose to the switch.
18   With the engine at operating temperature, check for vacuum at the EGR port of the

16.17 The thermal vacuum switch is threaded into the thermostat housing cover or intake manifold

1 EGR
2 Vacuum supply

switch. If no vacuum is indicated, replace the switch with a new one.

EGR control solenoid

Note: *A Code 53 (carbureted models) or a code 32 (TBI models) indicates a problem in the EGR system.*

19 Disconnect the electrical connector from the solenoid. Turn the ignition key On and check for battery voltage at the pink/black wire terminal of the harness connector. If battery voltage is not present, check the circuit from the EGR solenoid to the ignition switch (check the fuse first).

20 Turn the ignition key Off. Using jumper wires, connect positive battery power to terminal A (pink/black wire terminal) and momentarily ground terminal B (gray wire terminal). The solenoid should operate each time the ground wire is touched to terminal B. If the solenoid does not operate replace it.

21 Disconnect the vacuum harness from the EGR solenoid and connect a hand-held vacuum to the manifold vacuum port on the solenoid. Apply vacuum to the solenoid valve as you energize and de-energize the solenoid.

22 On carbureted models, the valve is normally open. It should hold vacuum when energized and allow air to pass when de-energized.

23 On TBI models, the valve is normally closed, it should hold vacuum when de-energized and allow air to pass when energized.

24 If the solenoid fails the above tests, replace it.

Component replacement

EGR valve

Refer to illustrations 16.26 and 16.29

25 Remove the air cleaner (see Chapter 1).
26 Disconnect the EGR valve vacuum line from the valve and remove the EGR valve mounting bolts **(see illustration).**

16.26 The EGR valve is mounted on the intake manifold by two nuts or bolts (arrows)

27 Remove the EGR valve from the manifold.
28 With the EGR valve removed, inspect the passages for excessive deposits. Scrape them clean and use a vacuum cleaner to remove the debris. **Caution:** *Do not wash the EGR valve in solvents or degreaser- permanent damage to the valve diaphragm may result. Sand blasting of the valve is also not recommended since it can affect the operation of the valve.*

29 The EGR valve can be cleaned by tapping the end of the pintle with a soft-face hammer **(see illustration).**
30 Look for exhaust deposits in the valve outlet. Remove deposit buildup with a screwdriver.
31 Clean the mounting surfaces of the intake manifold and the valve assembly.
32 Clean the mounting surface of the valve with a wire wheel or wire brush and the pintle with a wire brush.
33 Depress the valve diaphragm and check the seat area for cleanliness and wear by looking through the valve outlet. If the pintle or the seat are not completely clean, repeat the procedure in Step 32.
34 Hold the bottom of the valve securely and try to rotate the top of the valve back-and-forth. Replace the valve if any looseness is felt.
35 Inspect the valve outlet for deposits. Remove any deposit buildup with a screwdriver or other suitable sharp tool.
36 Using a new gasket, install the old (cleaned) or new EGR valve on the intake manifold.
37 Install the EGR valve mounting bolts and tighten them securely.
38 Attach the vacuum hose to the valve.
39 Install the air cleaner (see Chapter 4).

Thermal vacuum switch

40 Refer to Chapter 1 and drain about 1 gallon of coolant from the system.
41 Remove the air cleaner and detach the hoses from the switch (since the switch may be used to control vacuum to other compo-

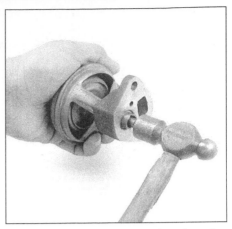

16.29 A hammer can be used to clean the EGR valve pintle

nents as well, more than two hoses may be attached - label the hoses and ports to ensure correct reinstallation).
42 Unscrew the switch and remove it from the engine.
43 Apply a non-hardening sealant to the switch threads (don't get it on the end of the switch) and install the switch. Tighten it securely, then reattach the hoses and add coolant as necessary.

EGR control solenoid

44 Disconnect the cable from the negative terminal of the battery.
45 Remove the air cleaner (see Chapter 4).
46 Disconnect the electrical connector at the solenoid.
47 Disconnect the vacuum hoses from the solenoid.
48 Remove the mounting nut and detach the solenoid.
49 Install the new solenoid and tighten the nut securely. The remainder of the installation procedure is the reverse of removal.

17 Air Injection Reaction (AIR) system

General description

1 The AIR system helps reduce hydrocarbons and carbon monoxide levels in the exhaust by injecting air into the exhaust ports of each cylinder during cold engine operation, or directly into the catalytic converter during normal operation. It also helps the catalytic converter reach proper operating temperature quickly during warm-up. On early models, the air is injected only into the exhaust ports.
2 The AIR system uses an air pump to force the air into the exhaust stream. An air management valve, controlled by the vehicle's Electronic Control Module (ECM) on later models, directs the air to the correct location, depending on engine temperature and driving conditions. During certain situations, such as deceleration, the air is diverted

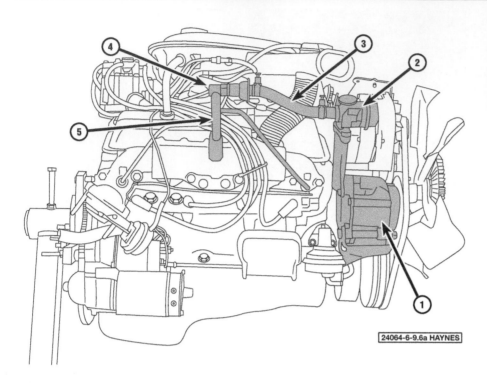

24064-6-9.6a HAYNES

17.6a Typical Air Injection Reaction (AIR) system component layout - (Right side)

1	Air pump	3	High temperature	4	One way check valve
2	Diverter valve		hose	5	Air injection pipe

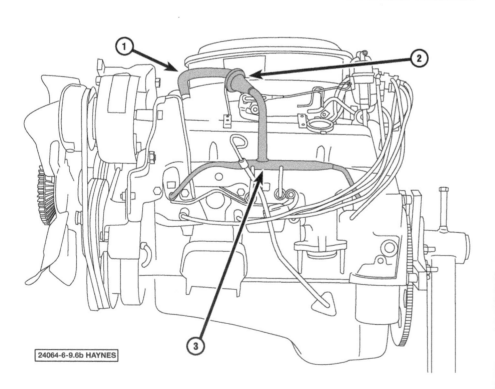

24064-6-9.6b HAYNES

17.6b Typical Air Injection Reaction (AIR) system component layout - (Left side)

1	High temperature hose	2	One way check valve	3	Air injection pipe

to the air cleaner to prevent backfiring from too much oxygen in the exhaust stream. One-way check valves are also used in the AIR system tubing to prevent exhaust gases from being forced back through the system.

3   The AIR system components include an engine driven air pump, air control, air switching and divert management valves, air flow and control hoses, check valves and a catalytic converter.

Check

Refer to illustrations 17.6a, 17.6b and 17.7

4   Because of the complexity of this system, it is difficult for the home mechanic to make an accurate diagnosis. If the system is malfunctioning, individual components can be checked.

5   Begin any inspection by carefully checking all hoses, vacuum lines and wires. Be sure they are in good condition and that all connections are tight and clean. Also make sure that the pump drivebelt is in good condition and properly adjusted (see Chapter 1).

6   To check the pump, allow the engine to reach normal operating temperature and run it at about 1500 rpm. Locate the hose running from the air pump **(see illustrations)** and squeeze to check for pulsations. Have an assistant increase the engine speed and check for a parallel increase in air flow. If an increase is noted, the pump is functioning properly. If not, there is a fault in the pump.

7   Each check valve can be inspected after removing it from the hose **(see illustration)**. Make sure the engine is completely cool before attempting this. Try to blow through the check valve from both ends. Air should pass through it only in the direction of normal air flow. If it is stuck either open or closed, the valve should be replaced.

8   To check the air control valve, disconnect the vacuum signal hose at the valve. With the engine running, check for vacuum at the hose. If none is felt, the line is clogged.

9   To check the deceleration valve, plug the air cleaner vacuum source, then reconnect the signal hose and listen for air flow through the ventilation pipe and into the

17.7 Typical one way check valve (arrow)

deceleration valve. There should also be a noticeable engine speed drop when the signal hose is reconnected. If the air flow does not continue for at least one second, or the engine speed does not drop noticeably, check the deceleration valve hoses for restrictions and leaks. If no restrictions or leaks are found, replace the deceleration valve.

Component replacement

Drivebelt
10   Loosen the pump mounting bolt and the pump adjustment bracket bolt.
11   Move the pump in until the belt can be removed.
12   Install the new belt and adjust it (refer to Chapter 1).

AIR pump pulley and filter
Refer to illustration 17.16
13   Compress the drivebelt to keep the pulley from turning and loosen the pulley bolts.
14   Remove the drivebelt as described above.
15   Remove the mounting bolts and lift off the pulley.
16   If the fan-like filter must be removed, grasp it firmly with a pair of needle-nose pliers **(see illustration)** and pull it from the pump. **Note:** *Do not insert a screwdriver between the filter and pump housing as the edge of the housing could be damaged. The filter will usually be distorted when pulled off. Be sure no fragments fall into the air intake hose.*
17   The new filter is installed by placing it in position on the pump, placing the pulley over it and tightening the pulley bolts evenly to draw the filter into the pump. Do not attempt to install a filter by pressing or hammering it into place. **Note:** *It is normal for the new filter to have an interference fit with the pump housing and, upon initial operation, it may squeal until worn in.*
18   Install the drivebelt and, while compressing the belt, tighten the pulley bolts securely.
19   Adjust the drivebelt tension.

17.16 The filter can be pulled out of the air pump with a pair of needle-nose pliers

Hoses and tubes
20   To replace a tube or hose, always note how it is routed first, either with a sketch or with numbered pieces of tape.
21   Remove the defective hose or tube and replace it with a new one of the same material and size and tighten all connections.

Check valve
22   Disconnect the pump outlet hose at the check valve.
23   Unscrew the check valve from the pipe assembly. Be careful not to bend or twist the assembly.
24   Install a new valve after making sure that it is a duplicate of the part removed, then tighten all connections.

Air control valve
25   Disconnect the negative battery cable at the battery. Remove the air cleaner.
26   Disconnect the vacuum signal line from the valve. Disconnect the air hoses and wiring connectors.
27   If the mounting bolts are retained by tabbed lockwashers, bend the tabs back, then remove the mounting bolts and lift the valve off the adapter or bracket.
28   Installation is the reverse of the removal procedure. Be sure to use a new gasket when installing the valve.

Air pump
29   Remove the air management valve and adapter, if so equipped.
30   If the pulley must be removed from the pump it should be done prior to removing the drivebelt.
31   If the pulley is not being removed, remove the drivebelt.
32   Remove the pump mounting bolts and separate the pump from the engine.
33   Installation is the reverse of removal. **Note:** *Do not tighten the pump mounting bolts until all components are installed.*
34   Following installation, adjust the drivebelt tension (see Chapter 1).

Deceleration valve
35   Disconnect the vacuum hoses from the valve.
36   Remove the screws retaining the valve to the engine bracket (if equipped) and detach the valve.
37   Install a new valve and reconnect all hoses.

18   Evaporative Emission Control System (EECS)

General description
Refer to illustration 18.2
1   This system is designed to trap and store fuel vapors that evaporate from the fuel tank, carburetor or throttle body and intake manifold.
2   The Evaporative Emission Control System (EECS) consists of a charcoal-filled can-

18.2 Typical late model charcoal - filled canister

ister and the lines connecting the canister to the fuel tank, ported vacuum and intake manifold vacuum **(see illustration)**.
3   Fuel vapors are transferred from the fuel tank, carburetor or throttle body and intake manifold to a canister where they are stored when the engine is not operating. When the engine is running, the fuel vapors are purged from the canister by intake air flow and consumed in the normal combustion process.
4   On later models there is a purge valve in the canister and, on some models, a tank pressure control valve. On some models the purge valve is operated by the ECM.
5   An indication that the system is not operating properly is a strong fuel odor. Poor idle, stalling and poor driveability can be caused by an inoperative purge valve, a damaged canister, split or cracked hoses or hoses connected to the wrong tubes.

Check
6   Checking and maintenance procedures for the EECS system canister and hoses are included in Chapter 1.
7   To check the purge valve, apply a short length of hose to the lower tube of the purge valve assembly and attempt to blow through it. Little or no air should pass into the canister (a small amount of air will pass because the canister has a constant purge hole).
8   With a hand vacuum pump, apply vacuum through the control vacuum signal tube to the purge valve diaphragm. If the diaphragm does not hold vacuum for at least 20 seconds, the diaphragm is leaking and the canister must be replaced.
9   If the diaphragm holds vacuum, again try to blow through the hose while vacuum is still being applied. An increased flow of air should be noted. If it isn't, replace the canister.

Canister replacement
10   Disconnect the hoses from the canister.
11   Remove the pinch bolt and loosen the clamp.
12   Remove the canister.
13   Installation is the reverse of removal.

19.1 Typical late model PCV valve (arrow)

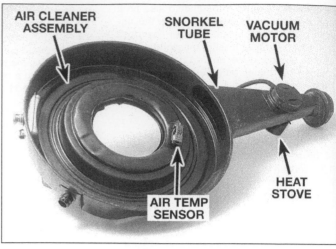

20.2 Typical THERMAC system components

19 Positive Crankcase Ventilation (PCV) system

General description

Refer to illustration 19.1

1   The Positive Crankcase Ventilation (PCV) system reduces hydrocarbon emissions by circulating fresh air through the crankcase to pick up blow-by gases, which are then rerouted through a PCV valve to the carburetor or intake manifold **(see illustration)** to be burned in the engine.
2   The main components of the PCV system are vacuum hoses and the PCV valve, which regulates the flow of gases according to engine speed and manifold vacuum.

Check

3   The PCV system can be checked quickly and easily for proper operation. It should be checked regularly because carbon and gunk deposited by blow-by gases will eventually clog the PCV valve and system hoses. The common symptoms of a plugged or clogged PCV valve include a rough idle, stalling or a slow idle speed, oil leaks, oil in the air cleaner or sludge in the engine.
4   Refer to Chapter 1 for PCV valve checking and replacement procedures.

20 Thermostatic Air Cleaner (THERMAC) system

General description

Refer to illustration 20.2

1   The thermostatic air cleaner (THERMAC) system improves engine efficiency and driveability under varying climatic conditions by controlling the temperature of the air coming into the air cleaner. A uniform incoming air temperature allows leaner air/fuel ratios during warm-up, which reduces hydrocarbon emissions.

2   The system uses a damper assembly, located in the snorkel of the air cleaner housing, to control the ratio of cold and warm air directed into the carburetor or throttle body. This damper is controlled by a vacuum motor which is, in turn, modulated by a temperature sensor in the air cleaner **(see illustration)**. On some engines a check valve is used in the sensor, which delays the opening of the damper when the engine is cold and the vacuum signal is low.
3   It is during the first few miles of driving (depending on outside temperature) that this system has its greatest effect on engine performance and emissions output. When the engine is cold, the damper blocks off the air cleaner inlet snorkel, allowing only warm air from around the exhaust manifold to enter the engine. As the engine warms up, the damper gradually opens the snorkel passage, increasing the amount of cold air allowed into the air cleaner. By the time the engine has reached its normal operating temperature, the damper opens completely, allowing only cold, fresh air to enter.
4   Because of this "cold engine only" function, it is important to periodically check this system to prevent poor engine performance when cold, or overheating of the fuel mixture once the engine has reached operating temperatures. If the air cleaner damper sticks in the "no heat" position the engine will run poorly, stall and waste gas until it has warmed up on its own. A valve sticking in the "heat" position causes the engine to run as if it is out of tune due to the constant flow of hot air to the carburetor or throttle body.

Check

5   Refer to Chapter 1 for maintenance and checking procedures for the THERMAC system. If any of the problems mentioned above are discovered while performing routine maintenance checks, refer to the following procedures.
6   If the damper does not close off snorkel air when the cold engine is started, disconnect the vacuum hose at the snorkel vacuum

motor and place your thumb over the hose end, checking for vacuum. Replace the vacuum motor if the hose routing is correct and the damper moves freely.
7   If there was no vacuum going to the motor in the above test, check the hoses for cracks, crimped areas and proper connection. If the hoses are clear and in good condition, replace the temperature sensor inside the air cleaner housing.

Component replacement

Note: *Check on parts availability before attempting component replacement. Some components may no longer be available for all models, requiring replacement of the complete air cleaner assembly.*

Air cleaner vacuum motor

8   Remove the air cleaner assembly from the engine and disconnect the vacuum hose from the motor.

Carbureted models

Refer to illustration 20.9

9   Drill out the two spot welds which secure the vacuum motor retaining strap to the snorkel tube **(see illustration)**.
10 Remove the motor retaining strap.
11 Lift up the motor, cocking it to one side to unhook the motor linkage at the damper assembly.
12 To install, drill a 7/64-inch hole in the snorkel tube at the center of the vacuum motor retaining strap.
13 Insert the vacuum motor linkage into the damper assembly.
14 Using the sheet metal screw supplied with the motor service kit, attach the motor and retaining strap to the snorkel. Make sure the sheet metal screw does not interfere with the operation of the damper door. Shorten the screw if necessary.

TBI models

Refer to illustration 20.15

15 Drill out the rivets retaining the wax pellet actuator and detach the actuator from the

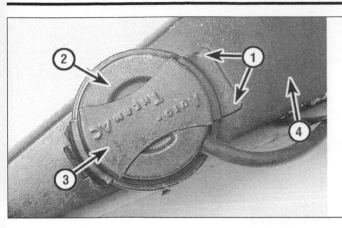

20.9 To remove the vacuum motor, drill out the spot welds and detach the retaining strap

1   Spot welds
2   Vacuum diaphragm motor
3   Retaining strap
4   Snorkel

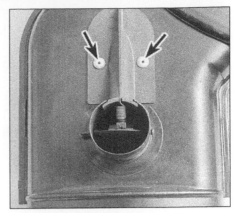

20.15 Fuel injection THERMAC actuator mounting rivets (arrows)

20.20 The sensor retainer must be pried off with a screwdriver

air cleaner housing **(see illustration)**.

16   To install, place the actuator in position in the housing, engage the spring and install new rivets.

17   Connect the vacuum hose to the motor and install the air cleaner assembly.

Air cleaner temperature sensor

Refer to illustration 20.20

18   Remove the air cleaner from the engine and disconnect the vacuum hoses at the sensor.

19   Carefully note the position of the sensor. The new sensor must be installed in exactly the same position.

20   Pry up the tabs on the sensor retaining clip and remove the sensor and clip from the air cleaner **(see illustration)**.

21   Install the new sensor with a new gasket in the same position as the old one.

22   Press the retaining clip onto the sensor. Do not damage the control mechanism in the center of the sensor.

23   Connect the vacuum hoses and attach the air cleaner to the engine.

---

21   Early Fuel Evaporation (EFE) system

Refer to illustration 21.1

General description

1   This system rapidly heats the intake air supply to promote evaporation of the fuel when the engine is started cold. On early models a valve in the exhaust pipe is opened and closed by a spring which is actuated by exhaust heat. Refer to Chapter 1 for maintenance operations for this type of valve. On later models a vacuum operated heat valve located in the exhaust pipe directs exhaust gas to the intake manifold to heat it when the engine is cold **(see illustration)**.

2   The components involved in the EFE's operation include the exhaust heat valve, the vacuum power actuator and rod or linkage and a coolant temperature switch and vacuum hoses.

3   The coolant temperature sensor is closed when the engine is cold, allowing vacuum to close the servo and direct exhaust gases to the manifold. As the engine warms up, the vacuum switch closes, cutting off vacuum to the servo so that it opens and exhaust gas is no longer directed to the manifold.

4   If the EFE system is not functioning, poor cold engine performance will result. If the system is not shutting off when the engine is warmed up, the engine will run as if it is out of tune and also may overheat.

Check

Refer to illustration 21.6

5   With the engine cold, the transmission in Neutral (manual) or Park (automatic), start the engine. Observe the operation of the actuator rod which leads to the heat valve inside the exhaust pipe. It should immediately move the valve to the closed position. If this is the case, the system is operating correctly.

6   If the actuator rod does not move, disconnect the vacuum hose at the actuator and place your thumb over the open end **(see illustration)**. With the engine cold and at idle, you should feel a suction indicating proper vacuum. If there is vacuum at this point, replace the actuator with a new one.

7   If there is no vacuum in the line, this is an indication that either the hose is crimped

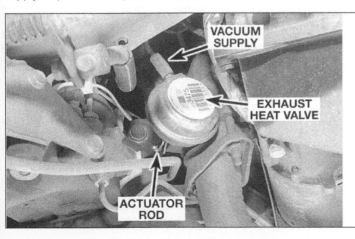

21.1 Typical EFE heat valve and actuator

VACUUM SUPPLY

EXHAUST HEAT VALVE

ACTUATOR ROD

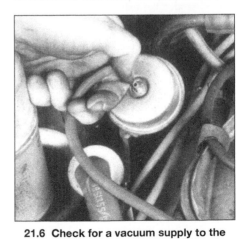

21.6 Check for a vacuum supply to the actuator with the engine idling

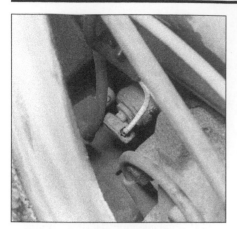

22.3 Typical TCS coolant temperature switch location in the cylinder head

22.4 The idle stop solenoid is mounted on the carburetor

23.5 Remove the catalytic converter U-bolts and flange bolts (arrows)

or plugged or the thermal vacuum switch threaded into the water passage is not functioning properly. Replace the hose or switch as necessary.

8   To make sure the system is disengaging once the engine has warmed up, make sure that the rod moves the exhaust heat valve to the open position as the engine reaches normal operating temperature (approximately 180 degrees).

9   If after the engine has warmed, the valve does not open, pull the vacuum hose at the actuator and check for vacuum with your thumb. If there is no vacuum, replace the actuator, If there is vacuum, replace the water temperature switch.

Component replacement

Actuator and rod assembly

10   Disconnect the vacuum hose from the actuator.

11   Remove the nuts which attach the actuator to the bracket.

12   Disconnect the rod from the heat valve and remove the actuator and rod from the engine.

13   Installation is the reverse of removal.

Exhaust heat valve

14   Remove the exhaust crossover pipe, if required for clearance.

15   If the actuator is part of the heat valve, disconnect the vacuum hose. If the actuator is remotely mounted, disconnect the rod from the heat valve.

16   Remove the exhaust pipe-to-manifold nuts and tension springs.

17   Remove the heat valve from inside the exhaust pipe.

18   Installation is the reverse of removal.

Coolant temperature thermal vacuum switch

19   Refer to Chapter 1 and drain about 1 gallon of coolant from the system.

20   Detach the vacuum hose from the temperature switch (located in the thermostat housing cover).

21   Unscrew the switch.

22   Apply non-hardening sealant to the threads of the new switch (don't get sealant on the end of the switch) and install it into the thermostat housing cover. Tighten it securely.

23   Reattach the vacuum hose and add coolant as required (see Chapter 1).

---

22 Transmission Controlled Spark (TCS) system

General description

Refer to illustrations 22.3 and 22.4

1   The purpose of this system is to reduce the emission of certain exhaust gases by eliminating ignition vacuum advance when the vehicle is operating in low forward gears on some models.

2   The vacuum advance is controlled by a solenoid-operated switch which is actuated by a transmission switch. When the switch is in operation, the vacuum normally applied to the distributor is vented to atmosphere and ignition advance is controlled by the advance mechanism in the distributor.

3   A coolant temperature switch is wired into the solenoid circuit to prevent vacuum cut-off to the distributor at engine temperatures below 93-degrees F **(see illustration)**.

4   In order to compensate for the retarded spark and to prevent possible engine run-on (dieseling) when the engine is turned off, an idle stop solenoid is provided on some carburetors to permit the throttle valve to close further than the normal idling position of the throttle valve plate **(see illustration)**.

5   Symptoms of faults in the TCS system are:

   a) *Slow idling or dieseling when the engine is shut off (faulty idle stop solenoid).*

   b) *Poor high gear performance (blown fuse).*

   c) *Excessive fuel consumption (faulty temperature switch).*

   d) *Backfiring during deceleration (transmission switch fault).*

   e) *Hard starting when cold (vacuum advance fault).*

Check

6   Checking of this system is limited to inspecting the security of the electrical and vacuum connections.

---

23 Catalytic converter - check and replacement

Refer to illustration 23.5

1   The catalytic converter is an emission control device added to the exhaust system to reduce pollutants in the exhaust gas. The converter contains platinum and rhodium, which lowers the levels of oxides of nitrogen (NOx) as well as hydrocarbons (HC) and carbon monoxide (CO).

Check

2   The test equipment for a catalytic converter is expensive and highly sophisticated. If you suspect that the converter on your vehicle is malfunctioning, take it to a dealer or authorized emissions inspection facility for diagnosis and repair.

3   The converter is located underneath the passenger compartment. Therefore, whenever the vehicle is raised for servicing of underbody components, check the converter for leaks, corrosion and other damage. If damage is discovered, the converter can simply be unbolted from the exhaust system and replaced.

Replacement

4   Raise the vehicle and place it securely on jackstands.

5   Remove the U-bolt nuts and flange bolts from both ends of the converter **(see illustration)**.

6   Detach the converter from the exhaust pipes and lower it from the vehicle.

7   Installation is the reverse of removal.

Chapter 7 Part A
Manual transmission

Contents

Specifications

Transmission oil capacity

3 speed units	2.0 US pts
4 speed units	3.0 US pts

Torque specifications — ft-lb

Three-speed Saginaw

Side cover to case bolts	15
Extension to case bolts	45
Shift lever to shifter shaft bolts	25
Lubrication filler plug	18
Transmission case to clutch housing bolts	75
Crossmember to frame nuts	25
Crossmember to mount bolts	40
Mount to transmission bolts	32

Torque specifications   ft-lb

Four-speed Saginaw

Side cover to case bolts	15
Extension to case bolts	45
Shift lever to shifter shaft bolts	25
Lubrication filler plug	18
Transmission case to clutch housing bolts	75
Crossmember to frame nuts	25
Crossmember to mount and mount to extension bolts	40
Mount to transmission bolts	32

Four-speed Muncie

Cover to case bolts	20
Extension and retainer to case bolts	
Upper	20
Lower	30
Lubrication filler plug	30
Shift lever to shifter shaft nut	20
Mount to transmission bolts	32

1   General information

All vehicles covered in this manual come equipped with a 3-speed or 4-speed manual transmission or an automatic transmission. All information on the manual transmission is included in this Part of Chapter 7. Information on the automatic transmission can be found in Part B of this Chapter.

The manual transmissions used in these models are synchromesh units.

Due to the complexity, unavailability of replacement parts and the special tools necessary, internal repair by the home mechanic is not recommended. The information in this Chapter is limited to general information and removal and installation of the transmission.

Depending on the expense involved in having a faulty transmission overhauled, it may be a good idea to replace the unit with either a new or rebuilt one. Your local dealer or transmission shop should be able to supply you with information concerning cost, availability and exchange policy. Regardless of how you decide to remedy a transmission problem, you can still save a lot of money by removing and installing the unit yourself.

2   Shift linkage - adjustment

Column shift

1   Place the shift lever in 'Reverse' and the ignition switch in the 'Lock' position.
2   Raise the vehicle for access beneath. then loosen the shift control swivel locknuts. Pull down slightly on the 1st/Reverse control rod attached to the column lever to remove any slackness. then tighten the locknut at the transmission lever.
3   Unlock the ignition switch and shift the column lever to Neutral. Position the column lower levers in the 'Neutral' position, align the

gauge holes in the levers and insert a 3/16 inch diameter gauge pin.
4   Support the rod and swivel to prevent movement, then tighten the 2nd/3rd shift control rod locknut.
5   Remove the alignment tool from the column lower levers and check the operation. Place the column shift lever in 'Reverse' and check the interlock control. It must not be possible to obtain 'Lock' except in 'Reverse'.
6   Lower the vehicle to the ground.

Floor shift

7   Switch the ignition to 'Off' then raise the vehicle for access beneath.
8   Loosen the swivel locknuts on the shift rods. Check that the rods pass freely through the swivels.
9   Set the shift levers to 'Neutral' at the transmission.
10   Move the shift control lever into the 'Neutral' detent position, align the control assembly levers, and insert the locating gauge into the lever alignment slot.
11   Tighten the shift rod swivel locknuts then remove the gauge.
12   Shift the transmission control lever into 'Reverse' and place the ignition switch in the 'Lock' position. Loosen the locknut at the back drive control rod swivel. then pull the rod down slightly to remove any slack in the column mechanism. Tighten the clevis jam nut.
13   Check the interlock control; the key should move freely to, and from, the 'Lock' position when the adjustment is correct.
14   Check the transmission shift control and readjust if necessary.
15   Lower the vehicle to the ground.

3   Backdrive linkage - adjustment

1   Shift the transmission into 'Reverse' and turn the ignition switch to the 'Lock' position.

2   Raise the vehicle for access beneath.
3   Loosen the backdrive control rod swivel locknut, pull down on the column linkage to remove any slackness, then tighten the clevis jam nut.
4   Check that the ignition key moves freely through the 'Lock' position; readjust if necessary at the bellcrank.
5   Lower the vehicle to the ground.

4   Transmission mount - check and replacement

1   Insert a large screwdriver or pry bar into the space between the transmission extension housing and the crossmember and pry up.
2   The transmission should not spread excessively away from the insulator.
3   To replace, remove the nuts attaching the insulator to the crossmember and the bolts attaching the insulator to the transmission.
4   Raise the transmission slightly with a jack and remove the insulator, noting which holes are used in the crossmember for proper alignment during installation.
5   Installation is the reverse of the removal procedure.

5   Transmission oil seal - replacement

Refer to illustrations 5.5 and 5.7
1   Oil leaks frequently occur due to wear of the extension housing oil seal and bushing (if equipped), and/or the speedometer drive gear oil seal and O-ring. Replacement of these seals is relatively easy, since the repairs can usually be performed without removing the transmission from the vehicle.
2   The extension housing oil seal is located

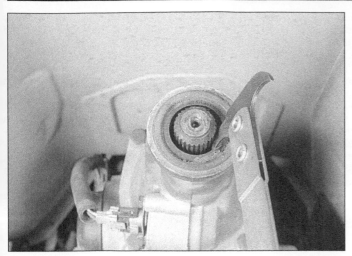

5.5 Use a seal removal tool or a long screwdriver to carefully pry the seal out of the end of the transmission

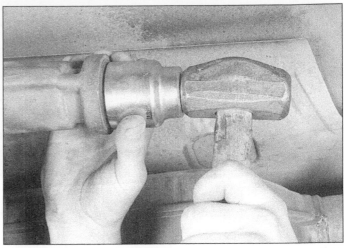

5.7 A large socket works well for installing the seal - the socket should come in contact with the outer edge of the seal

at the extreme rear of the transmission, where the driveshaft is attached. If leakage at the seal is suspected, raise the vehicle and support it securely on jackstands. If the seal is leaking, transmission lubricant will be built up on the front of the driveshaft and may be dripping from the rear of the transmission.

3   Refer to Chapter 8 and remove the driveshaft.

4   Using a soft-faced hammer, carefully tap the dust shield (if equipped) to the rear and remove it from the transmission. Be careful not to distort it.

5   Using a screwdriver or pry bar, carefully pry the oil seal and bushing (if equipped) out of the rear of the transmission **(see illustration)**. Take care not to damage the splines on the transmission output shaft.

6   If the oil seal and bushing cannot be removed with a screwdriver or pry bar, a special oil seal removal tool (available at auto parts stores) will be required.

7   Using a large section of pipe or a very large deep socket as a drift, install the new oil seal **(see illustration)**. Drive it into the bore squarely and make sure it's completely seated. Install a new bushing using the same method.

8   Reinstall the dust shield by carefully tapping it into place. Lubricate the splines of the transmission output shaft and the outside of the driveshaft sleeve yoke with lightweight grease, then install the driveshaft. Be careful not to damage the lip of the new seal.

9   The speedometer cable and driven gear housing is located on the side of the extension housing. Look for transmission oil around the cable housing to determine if the seal and O-ring are leaking.

10   Disconnect the speedometer cable.

11   Using a hook, remove the seal.

12   Using a small socket as a drift, install the new seal.

13   Install a new O-ring in the driven gear housing and reinstall the driven gear housing and cable assembly on the extension housing.

6   Manual transmission - removal and installation

Removal

1   Disconnect the negative cable at the battery. Place the cable out of the way so it cannot accidentally come in contact with the negative terminal of the battery, as this would once again allow power into the electrical system of the vehicle.

2   Disconnect the shift linkage.

3   Raise the vehicle and support it securely on jackstands.

4   Disconnect the speedometer cable and wire harness connectors from the transmission.

5   Remove the driveshaft (Chapter 8). Use a plastic bag to cover the end of the transmission to prevent fluid loss and contamination.

6   Remove the exhaust system components as necessary for clearance (Chapter 4).

7   Support the engine. This can be done from above with an engine hoist, or by placing a jack (with a block of wood as an insulator) under the engine oil pan. The engine should remain supported at all times while the transmission is out of the vehicle.

8   Support the transmission with a jack - preferably a special jack made for this purpose. Safety chains will help steady the transmission on the jack.

9   Remove the rear transmission support-to-crossmember nuts/bolts.

10   Remove the nuts from the crossmember bolts. Raise the transmission slightly and remove the crossmember.

11   Remove the bolts securing the transmission to the clutch housing.

12   Make a final check that all wires and hoses have been disconnected from the transmission and then move the transmission and jack toward the rear of the vehicle until the transmission input shaft is clear of the clutch housing. Keep the transmission level

as this is done.

13   Once the input shaft is clear, lower the transmission and remove it from under the vehicle. **Caution:** *Do not depress the clutch pedal while the transmission is out of the vehicle.*

14   The clutch components can be inspected by removing the clutch housing from the engine (Chapter 8). In most cases, new clutch components should be routinely installed if the transmission is removed.

Installation

15   If removed, install the clutch components (Chapter 8).

16   If removed, attach the clutch housing to the engine and tighten the bolts securely (Chapter 8).

17   With the transmission secured to the jack, as on removal, raise the transmission into position behind the clutch housing and then carefully slide it forward, engaging the input shaft with the clutch plate hub. Do not use excessive force to install the transmission - if the input shaft does not slide into place, readjust the angle of the transmission so it is level and/or turn the input shaft so the splines engage properly with the clutch.

18   Install the transmission-to-clutch housing bolts. Tighten the bolts to the specified torque.

19   Install the crossmember and transmission support. Tighten all nuts and bolts securely.

20   Remove the jacks supporting the transmission and the engine.

21   Install the various items removed previously, referring to Chapter 8 for the installation of the driveshaft and Chapter 4 for information regarding the exhaust system components.

22   Make a final check that all wires, hoses and the speedometer cable have been connected and that the transmission has been filled with lubricant to the proper level (Chapter 1). Lower the vehicle.

23   Connect and adjust the shift linkage.
24   Connect the negative battery cable. Road test the vehicle for proper operation and check for leakage.

7   Manual transmission overhaul - general information

Overhauling a manual transmission is a difficult job for the do-it-yourselfer. It involves the disassembly and reassembly of many small parts. Numerous clearances must be precisely measured and, if necessary, changed with select fit spacers and snap-rings. As a result, if transmission problems arise, it can be removed and installed by a competent do-it-yourselfer, but overhaul should be left to a transmission repair shop. Rebuilt transmissions may be available - check with your dealer parts department and auto parts stores. At any rate, the time and money involved in an overhaul is almost sure to exceed the cost of a rebuilt unit.

Nevertheless, it's not impossible for an inexperienced mechanic to rebuild a transmission if the special tools are available and the job is done in a deliberate step-by-step manner so nothing is overlooked.

The tools necessary for an overhaul include internal and external snap-ring pliers, a bearing puller, a slide hammer, a set of pin punches, a dial indicator and possibly a hydraulic press. In addition, a large, sturdy workbench and a vise or transmission stand will be required.

During disassembly of the transmission, make careful notes of how each piece comes off, where it fits in relation to other pieces and what holds it in place.

Before taking the transmission apart for repair, it will help if you have some idea what area of the transmission is malfunctioning. Certain problems can be closely tied to specific areas in the transmission, which can make component examination and replacement easier. Refer to the *Troubleshooting* section at the front of this manual for information regarding possible sources of trouble.

Chapter 7 Part B
Automatic transmission

Contents

Specifications

Fluid capacities

Powerglide	
*Routine fluid change	2 US qts
Filling from dry (new or overhauled unit)	9.0 US qts
Turbo Hydra-Matic 250 and 350	
*Routine fluid change	2.5 US qts
Filling from dry (new or overhauled unit)	10.0 US qts
Turbo Hydra-Matic 375/400 and 400	
*Routine fluid change	3.75 US qts
Filling from dry (new or overhauled unit)	8.5 US qts
THM 200	
*Routine fluid change	1.5 US qts
Filling from dry (new or overhauled unit)	6 US qts
CBC 350	
*Routine fluid change	2.5 US qts
Filling from dry (new or overhauled unit)	10 US qts
THM 400	
*Routine fluid change	3.75 US qts
Filling from dry (new or overhauled unit)	11 US qts

*The small quantity required at routine fluid changing is due to the fact that the fluid in the torque converter cannot be drained unless the converter is dismantled

Torque specifications

Ft-lbs (unless otherwise indicated)

Powerglide

Transmission case to engine	35
Transmission oil pan to case	8
Transmission extension to case	25
Speedometer driven gear fitting retainer	4
Filter screen attaching screws	2.5
Upper valve body plate bolts	5
Lower to upper body valve attaching bolts	15
Inner control lever Allen head screw	2.5
Parking lock pawl reaction bracket attaching bolts	10
Oil cooler plugs at transmission case	5
Pressure test point plugs	5
Low band adjustment locknut	15
Converter to flexplate bolts	35
Under pan to transmission case	7.5
Oil cooler pipe connectors to transmission case or radiator	10
Oil cooler pipe to connectors	10
Vacuum modulator to transmission case	15
Oil pan drain plug	20
Parking brake lock and range selector inner lever Allen head screw	2.5

Turbo Hydra-Matic 250 and 350 and CBC 350

Valve body and support plate	130 in-lbs
Parking lock bracket	29
Oil suction screen	40 in-lbs
Oil pan to case	130 in-lbs
Extension to case	25
Modulator retainer to case	130 in-lbs
Inner selector lever to shaft	25
Detent valve actuating bracket	52 in-lbs
Converter to flexplate bolts	35
Fluid pan to transmission case	110 in-lbs
Transmission case to engine	35
Oil cooler pipe connectors to transmission case or radiator	120 in-lbs
Oil cooler pipe to connectors	10
Gearshift bracket to frame	15
Gearshift shaft to swivel	20
Manual shaft to bracket	20
Detent cable to transmission	75 in-lbs
Intermediate band adjuster nut	15

Turbo Hydra-Matic 375/400 and THM 400

Center support bolts	23
Pump to case attaching bolts	18
Extension housing to case attaching bolts	23
Rear servo cover bolts	18
Detent solenoid bolts	7
Control valve body bolts	8
Fluid pan attaching screws	12
Modulator retainer bolt	18
Governor cover bolts	18
Manual lever to manual shaft nut	8
Manual shaft to inside detent lever	18
Linkage swivel clamp nut	43
Converter dust shield screws	93 in-lbs
Transmission to engine mounting bolts	35
Converter to flexplate bolts	32
Rear mount to transmission bolts	40
Rear mount to cross-member bolt	40
Cross-member mounting bolts	25
Line pressure take-off plug	13
Strainer retainer bolt	10
Oil cooler pipe connectors to transmission case or radiator	14
Oil cooler pipe to connector	10
Gearshift bracket to frame	15
Gearshift shaft to swivel	20
Manual shaft to bracket	20
Downshift switch to bracket	30

Turbo Hydra-Matic 200

Control valve body bolts.. 11
Oil screen retaining bolts... 11
Oil pan bolts 12
Torque converter to flexplate bolts.. 35
Transmission to engine bolts... 25
Converter dust shield screws 8
Speedometer driven gear bolts... 23
Fluid cooler line to transmission 25
Fluid cooler line to radiator 20
Linkage swivel clamp nut 30
Converter bracket to adapter 13
Transmission rear support bolts.. 40
Rear mounting to support nuts... 21
Support center nut... 33
Adapter to transmission bolts.. 33

1 General information

All vehicles covered in this manual come equipped with either a 3-speed or 4-speed manual transmission or an automatic transmission. All information on the automatic transmission is included in this Part of Chapter 7. Information on the manual transmission can be found in Part A of this Chapter.

Due to the complexity of the automatic transmissions covered in this manual and the need for specialized equipment to perform most service operations, this Chapter contains only general diagnosis, routine maintenance, adjustment and removal and installation procedures.

If the transmission requires major repair work, it should be left to a dealer service department or an automotive or transmission repair shop. You can, however, remove and install the transmission yourself and save the expense, even if the repair work is done by a transmission shop.

2 Diagnosis - general

Note: *Automatic transmission malfunctions may be caused by five general conditions: poor engine performance, improper adjustments, hydraulic malfunctions, mechanical malfunctions or malfunctions in the computer or its signal network. Diagnosis of these problems should always begin with a check of the easily repaired items: fluid level and condition (see Chapter 1), shift linkage adjustment and throttle linkage adjustment. Next, perform a road test to determine if the problem has been corrected or if more diagnosis is necessary. If the problem persists after the preliminary tests and corrections are completed, additional diagnosis should be done by a dealer service department or transmission repair shop. Refer to the* Troubleshooting *section at the front of this manual for information on symptoms of transmission problems.*

Preliminary checks

1 Drive the vehicle to warm the transmission to normal operating temperature.
2 Check the fluid level as described in Chapter 1:

a) *If the fluid level is unusually low, add enough fluid to bring the level within the designated area of the dipstick, then check for external leaks (see below).*

b) *If the fluid level is abnormally high, drain off the excess, then check the drained fluid for contamination by coolant. The presence of engine coolant in the automatic transmission fluid indicates that a failure has occurred in the internal radiator walls that separate the coolant from the transmission fluid (see Chapter 3).*

c) *If the fluid is foaming, drain it and refill the transmission, then check for coolant in the fluid or a high fluid level.*

3 Check the engine idle speed. **Note:** *If the engine is malfunctioning, do not proceed with the preliminary checks until it has been repaired and runs normally.*
4 Check the throttle valve cable for freedom of movement. Adjust it if necessary (see Section 9). **Note:** *The throttle cable may function properly when the engine is shut off and cold, but it may malfunction once the engine is hot. Check it cold and at normal engine operating temperature.*
5 Inspect the shift control linkage (see Sections 4 and 5). Make sure that it's properly adjusted and that the linkage operates smoothly.

Fluid leak diagnosis

6 Most fluid leaks are easy to locate visually. Repair usually consists of replacing a seal or gasket. If a leak is difficult to find, the following procedure may help.
7 Identify the fluid. Make sure it's transmission fluid and not engine oil or brake fluid (automatic transmission fluid is a deep red color).
8 Try to pinpoint the source of the leak. Drive the vehicle several miles, then park it over a large sheet of cardboard. After a minute or two, you should be able to locate the leak by determining the source of the fluid dripping onto the cardboard.
9 Make a careful visual inspection of the suspected component and the area immediately around it. Pay particular attention to gasket mating surfaces. A mirror is often helpful for finding leaks in areas that are hard to see.
10 If the leak still cannot be found, clean the suspected area thoroughly with a degreaser or solvent, then dry it.
11 Drive the vehicle for several miles at normal operating temperature and varying speeds. After driving the vehicle, visually inspect the suspected component again.
12 Once the leak has been located, the cause must be determined before it can be properly repaired. If a gasket is replaced but the sealing flange is bent, the new gasket will not stop the leak. The bent flange must be straightened.
13 Before attempting to repair a leak, check to make sure that the following conditions are corrected or they may cause another leak. **Note:** *Some of the following conditions cannot be fixed without highly specialized tools and expertise. Such problems must be referred to a transmission repair shop or a dealer service department.*

Gasket leaks

14 Check the pan periodically. Make sure the bolts are tight, no bolts are missing, the gasket is in good condition and the pan is flat (dents in the pan may indicate damage to the valve body inside).
15 If the pan gasket is leaking, the fluid level or the fluid pressure may be too high, the vent may be plugged, the pan bolts may be too tight, the pan sealing flange may be warped, the sealing surface of the transmission housing may be damaged, the gasket may be damaged or the transmission casting may be cracked or porous. If sealant instead of gasket material has been used to form a seal between the pan and the transmission housing, it may be the wrong sealant.

A   **B**   **C**   **D**   **E**

3.1 The oil pan gasket shape can help you determine which transmission your vehicle is equipped with

| A | THM700-R4 | B | THM400 | C | THM250/350 * | D | THM200-4R | E | Powerglide |

*THM200 and 200C look the same as the THM250/350 but with fewer bolts front and rear.

Seal leaks

16   If a transmission seal is leaking, the fluid level or pressure may be too high, the vent may be plugged, the seal bore may be damaged, the seal itself may be damaged or improperly installed, the surface of the shaft protruding through the seal may be damaged or a loose bearing may be causing excessive shaft movement.

17   Make sure the dipstick tube seal is in good condition and the tube is properly seated. Periodically check the area around the speedometer gear or sensor for leakage. If transmission fluid is evident, check the O-ring for damage.

Case leaks

18   If the case itself appears to be leaking, the casting is porous and will have to be repaired or replaced.

19   Make sure the oil cooler hose fittings are tight and in good condition.

Fluid comes out vent pipe or fill tube

20   If this condition occurs, the transmission is overfilled, there is coolant in the fluid, the case is porous, the dipstick is incorrect, the vent is plugged or the drain back holes are plugged.

3   Transmission identification

Refer to illustration 3.1

1   Besides checking the transmission serial number, there is a quick way to determine which of the several automatic transmissions a particular vehicle is equipped with. Read the following transmission oil pan descriptions and **refer to accompanying illustration 3.1** to identify the various models.

Powerglide

2   This two-speed transmission case is made of either cast iron or aluminum. The word 'powerglide' is plainly stamped on the case. The shift quadrant indicator is set up in one of two ways: P-N-D-L-R or P-R-N-D-L.

Turbo Hydra-Matic 200/200C

3   This three-speed transmission is a two-piece design with a downshift cable running from the accelerator to the right side of the transmission case. The oil pan has 11 bolts and is square-shaped with one corner angled.

Turbo Hydra-Matic 200-4R

4   This four-speed transmission has a pan with a unique shape, secured by 16 bolts.

Turbo Hydra-Matic 250/350

5   This closely resembles the THM 200 but the oil pan has 13 bolts.

Turbo Hydra-Matic 400

6   The case of the three-speed THM400 is also two-piece, but the downshifting is electrically controlled from a switch at the carburetor to the left side of the transmission. The oil pan also has 13 bolts. The shape of the pan is elongated and irregular.

4   Column shift linkage - checking and adjustment

1   The selector linkage will be in need of adjustment if at any time "Low" or "Reverse" can be obtained without first having to lift the shift control lever to enable it to pass over the mechanical stop.

1970 through 1973 models

2   If adjustment is required, release the control rod swivel or clamp and set the lever on the side of the transmission in the "Drive" or L2 detent. This can be clearly defined by placing the lever in L or L1 and moving the lever back one detent (click).

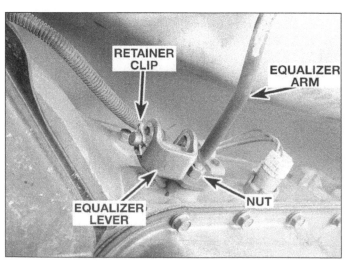

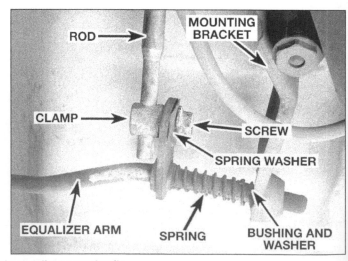

4.7 1974 and later shift linkage adjustment details

3   Position the shift control lever up against the "Drive" stop and then tighten the swivel or clamp on the control rod.
4   Check all selector positions, especially "Park." In some cases, especially with worn linkage, it may be necessary to readjust slightly in order to ensure that the "Park"detent is fully engaged.

1974 and later models

Refer to illustration 4.7
5   Place the shift lever in the Neutral position of the shift indicator.
6   Position the transmission shift lever in the Neutral detent.
7   Install the clamp spring and screw assembly on the equalizer lever control rod **(see illustration)**.
8   Hold the clamp flush against the equalizer lever and tighten the clamp screw finger tight. Make sure that no force is exerted in either direction on the rod or equalizer lever while the screw is tightened.
9   Tighten the screw securely.
10   Check that the ignition key can be moved freely to the "Lock" position when the shift lever is in "Park" and not in any other position.

5   Floor shift linkage - check and adjustment

1   If the engine can be started in any of the Drive positions and the Neutral start switch is properly adjusted (Section 11), the shift linkage must be adjusted.

Powerglide

2   Set the shift lever in "Drive".
3   Working under the vehicle, disconnect the selector cable from the lever on the side of the transmission.
4   Move the lever on the side of the transmission to the "Drive" detent.
5   Measure the distance from the rear face of the cable mounting bracket to the center of the cable pivot stud. This should be 5-1/2 inches. Adjust the position of the stud if necessary to achieve this measurement.
6   Adjust the cable in its mounting bracket so that the cable end fits freely onto the pivot stud. Install the cable retaining clip.
7   Working inside the vehicle, remove the shift quadrant cover, plate and illumination bulbs.
8   Remove the selector cable clip and disconnect the cable from the shift lever.
9   Insert a gauge (0.07-inch thick) between the pawl and the detent plate, then measure the distance between the front face of the shifter assembly bracket and the center of the cable pivot pin. This should be 6-1/4 inches. If it is not, loosen bolt A and move the lever as necessary.
10   Adjust the cable mounting to the shifter bracket until the cable eye freely enters over the pivot pin. If at any time depressing the handle button does not clear the cut-outs in

the detent plate, or conversely if the handle can be moved to P and R positions without depressing the button, raise or lower the detent plate after loosening the retaining bolt.

Turbo Hydra-Matic

Note: *Apply the parking brake and block the wheels to prevent the vehicle from rolling.*
11   Working under the vehicle, loosen the nut attaching the shift lever to the pin on the shift cable assembly.
12   Place the console shifter (inside the vehicle) in Park.
13   Place the manual shifting shaft on the transmission in Park.
14   Move the pin to give a 'free pin' fit, making sure that the console shifter is still in the Neutral position, and tighten the pin-to-lever retaining nut to the specified torque.
15   Make sure the engine will start in the Park and Neutral positions only.
16   If the engine can be started in any of the drive positions (as indicated by the shifter inside the vehicle), repeat the steps above or have the vehicle examined by a dealer, because improper linkage adjustment can lead to band or clutch failure and possible personal injury.

6   Powerglide - on-vehicle adjustments

Low band adjustment

1   This adjustment should normally be carried out at the time of the first oil change and thereafter only when unsatisfactory performance indicates it to be necessary (see *Troubleshooting*).
2   Raise the vehicle to provide access to the transmission, making sure to secure the vehicle on jackstands.
3   Place the selector lever in the Neutral position.
4   Remove the protective cap from the transmission adjusting screw.
5   Release and unscrew the adjusting screw locknut one-quarter turn and hold it in this position with a wrench.
6   Using a torque wrench and adapter, tighten the adjusting screw to 70 in-lbs., then back off the screw the exact number of turns as follows:

a) *For a band which has been in operation for less than 6000 miles- three complete turns.*
b) *For a band which has been in operation for more than 6000 miles- four complete turns.*

7   Tighten the adjusting screw locknut.

Throttle valve/linkage adjustment

8   Remove the air cleaner. Disconnect the accelerator linkage at the carburetor, and the accelerator and throttle valve return springs.
9   Pull the throttle valve upper rod forward

with the right hand then open the carburetor throttle side with the left hand.
10   Adjust the swivel on the end of the upper throttle valve rod so that the ball stud contacts the end of the slot in the upper throttle valve rod as the carburetor reaches the wide open throttle (WOT) position.
11   Connect and adjust the accelerator linkage.

Neutral start switch alignment

12   This operation will only be required if a new switch is being installed.
13   Set the shift lever in Neutral and locate the lever tang against the transmission selector plate.
14   Align the slot in the contact support with the hole in the switch by inserting a 3/32-inch diameter pin.
15   Place the contact support drive slot over the shifter tube drive tang and tighten the screws. Withdraw the pin.
16   Connect the switch wires and check that the operation of the switch is correct when the ignition is switched on.

7   Turbo Hydra-Matic 250 - on-vehicle adjustments

Intermediate band adjustment

1   This adjustment should be carried out every 24,000 miles or if the performance of the transmission indicates the need for it.
2   Raise the vehicle to gain access to the transmission, making sure to secure the vehicle on jackstands.
3   Place the speed selector lever in Neutral.
4   The adjusting screw and locknut for the intermediate band is located on the right-hand side of the transmission case.
5   Loosen the locknut 1/4-turn using a wrench or special tool, available at most auto parts stores. Hold the locknut in this position and tighten the adjusting screw to a torque of 30 in-lbs. Now back off the screw three complete turns exactly.
6   Without moving the adjusting screw, tighten the locknut to 15 ft-lbs.

Downshift (detent) cable adjustment

7   The cable will normally only require adjustment if a new one has been installed.
8   Depress the accelerator pedal fully. The ball will slide into the cable sleeve and automatically pre-set the cable tension.

8   Turbo Hydra-Matic 350 downshift (detent) cable - adjustment

Refer to illustration 8.1

1970 through 1980 models

1   Insert a screwdriver on each side of the

snap-lock and pry out to release **(see illustration)**.

2   Compress the locking tabs and disconnect the snap-lock assembly from its bracket.

3   Manually set the carburetor in the fully open position with the throttle lever fully against its stop.

4   With the carburetor in the fully open position, push the snap-lock on the detent cable into the locked position and release the throttle lever.

1981 and later models

5   This will normally be required only after installation of a new cable.

6   Depress the accelerator pedal to the fully open position. The cable ball will slide into the sleeve of the cable and automatically adjust the setting of the detent cable.

---

9   Throttle valve (TV) cable - description, inspection and adjustment

Description

1   The throttle valve cable used on these transmissions should not be thought of as merely a "downshift" cable, as in earlier transmissions. The TV cable controls line pressure, shift points, shift feel, part throttle downshifts and detent downshifts.

2   If the TV cable is broken, sticky, misadjusted or is the incorrect part, the vehicle will experience a number of problems.

Inspection

Refer to illustration 9.4

3   Inspection should be made with the engine running at idle speed with the selector lever in Neutral. Set the parking brake firmly and block the wheels to prevent any vehicle

8.1 Pry out the release tab with a screwdriver

movement. As an added precaution, have an assistant in the driver's seat applying the brake pedal.

4   Grab the inner cable a few inches behind where it attaches to the throttle linkage and pull the cable forward. It should easily slide through the cable housing with no binding or jerky operation **(see illustration)**.

5   Release the cable and it should return to its original location with the cable stop against the cable terminal.

6   If the TV cable does not operate as above, the cause is a defective or misadjusted cable or damaged components at either end of the cable.

Adjustment

7   The engine should not be running during this adjustment.

1976 through 1980 models

Refer to illustration 9.8

8   Disengage the snap-lock and check that

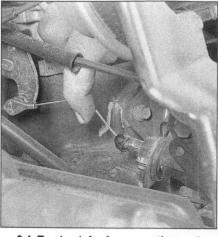

9.4 To check for free operation, pull forward on the throttle valve (TV) inner cable, feeling for smooth operation through the full range of travel - the cable should retract evenly and rapidly when released

the cable is free to slide through the lock **(see illustration)**.

9   Move the carburetor throttle lever to the fully open position.

10   Push the snap-lock down until it is flush and then gently release the throttle lever.

1981 and later models

Refer to illustration 9.11

11   Depress the cable release tab and move the slider back through the fitting away from the throttle linkage until the slider stops against the fitting **(see illustration)**.

12   Release the tab.

13   Turn the throttle lever to the wide open throttle (WOT) position, which will automatically adjust the cable. Release the throttle lever. **Caution:** *Don't use excessive force at the throttle lever to adjust the TV cable. If*

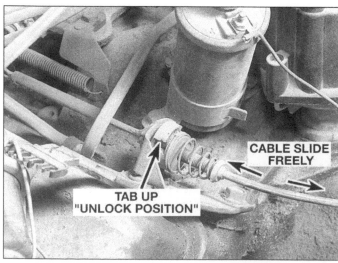

9.8 With the release tab in the "unlocked position" make sure the cable can move in and out freely

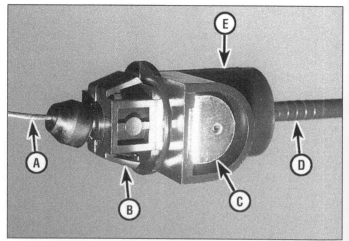

9.11 On 1981 models, depress the throttle valve (TV) cable release tab and pull the slider back until it rests on its stop - release the tab and open the throttle completely

A	TV cable	C	Release tab	E Slider
B	Locking lugs	D	Cable casing	

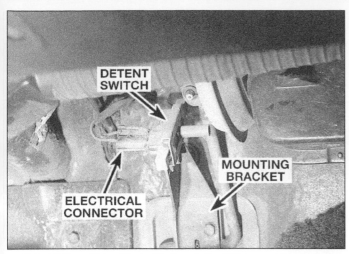

10.1 1972 and later Turbo Hydra-Matic 400 downshift (detent) switch details

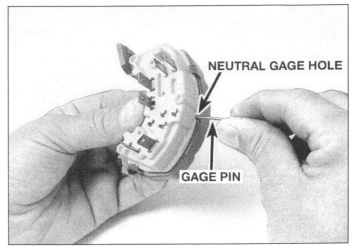

11.6a On earlier models, insert a pin into the gage hole to adjust the neutral start switch

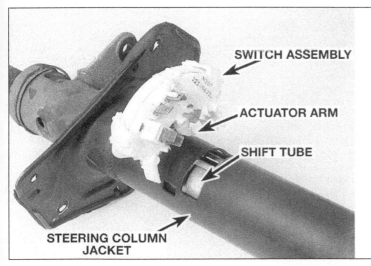

11.6b Later model neutral start switches are adjusted by moving the assembly housing to the Low gear position and then shifting into Park

great effort is required to adjust the cable, disconnect the cable at the transmission end and check for free operation. If it's still difficult, replace the cable. If it's now free, suspect a bent TV link in the transmission or a problem with a throttle lever.

14   After adjustment check for proper operation as described in Steps 3 through 6 above.

10   Turbo Hydra-Matic 400 downshift (detent) switch - adjustment

1970 and 1971 models

1   The detent switch on these models is located in the engine compartment by the throttle lever. Pull the detent switch driver rearwards until the hole in the switch body aligns with the hole in the driver. Insert a pin (0.092 in diameter) into the holes to a depth of 0.10 in to hold the driver in position.

2   Loosen the switch mounting bolt.

3   Depress the accelerator fully and then move the switch forward until the driver con-tacts the accelerator lever.

4   Tighten the switch bolt and extract the alignment pin.

1972 and later models

Refer to illustration 10.1

1   The switch is mounted on the pedal bracket (see illustration).

2   The switch is set by pushing the plunger as far forward as possible. At the first full depression of the accelerator pedal, the switch is automatically adjusted.

11   Neutral start switch - check and adjustment

Refer to illustrations 11.6a and 11.6b

1   When the switch is operating properly, the engine should crank over with the selector lever in Park or Neutral only. Also, the backup lights should come on when the lever is in Reverse.

2   If a new switch is being installed, set the shift lever against the "Neutral" gate by rotat-ing the lower lever on the shift tube in a coun-terclockwise direction as viewed from the driver's seat.

3   Locate the switch actuating tang in the shifter tube slot and then tighten the securing screws.

4   Connect the wiring harness and switch on the ignition and check that the starter motor will actuate.

5   If the switch operates correctly, move the shift lever out of Neutral, which will cause the alignment pin (installed during production of the switch) to shear.

6   If an old switch is being installed or readjusted, use a pin (0.093 to 0.097-inch diameter) to align the hole in the switch with the actuating tang. Insert the pin to a depth of 1/4-inch. Remove the pin before moving the shift lever out of Neutral (see illustration). On later models the switch automatically ratch-ets to the proper adjustment when the shift lever is moved to Park. To readjust a later model switch, move the switch assembly housing to the Low gear position and then shift into Park (see illustration).

12   Automatic transmission - removal and installation

Refer to illustrations 12.5, 12.6 and 12.21

Removal

1   Disconnect the negative cable at the battery. Place the cable out of the way so it cannot accidentally come in contact with the negative terminal of the battery, as this would once again allow power into the electrical system of the vehicle.

2   Raise the vehicle and support it securely.

3   Drain the transmission fluid (see Chap-ter 1).

4   Remove the torque converter cover.

5   Mark the relation of the flywheel and the

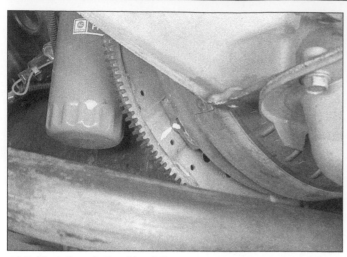

12.5 Mark the relationship of the torque converter to the flywheel from the vacuum modulator

12.6 Remove the torque converter-to-flywheel retaining bolts (here, a flywheel wrench is being used to prevent the flywheel from turning, but a wrench on the crankshaft vibration damper will also work)

torque converter so they can be reinstalled in the same position **(see illustration)**.

6   Remove the torque converter-to-flywheel nuts/bolts **(see illustration)**. Turn the crankshaft bolt for access to each nut in turn.

7   Remove the starter motor (see Chapter 5).

8   Remove the driveshaft (see Chapter 8).

9   Disconnect the speedometer cable.

10   Disconnect the electrical connectors from the transmission.

11   On models so equipped, disconnect the vacuum hose(s).

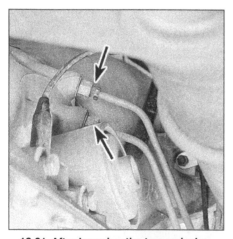

12.21 After lowering the transmission, use a flare-nut wrench to disconnect the cooler lines (arrows) and plug them to prevent fluid loss. It may be necessary to use a back-up wrench on the transmission-side fittings to prevent twisting the lines

12   Remove any exhaust component which will interfere with transmission removal (see Chapter 4).

13   Disconnect the TV linkage rod or cable.

14   Disconnect the shift linkage.

15   Support the engine using a jack and a block of wood under the oil pan to spread the load.

16   Support the transmission with a jack - preferably a jack made for this purpose. Safety chains will help steady the transmission on the jack.

17   Remove the rear mount-to-crossmember attaching bolts and the crossmember-to-frame attaching bolts.

18   Remove the two engine rear support-to-transmission extension housing attaching bolts.

19   Raise the transmission sufficiently to allow removal of the crossmember.

20   Remove the bolts securing the transmission to the engine.

21   Lower the transmission slightly and disconnect and plug the transmission cooler lines **(see illustration)**.

22   Remove the transmission fluid filler tube.

23   Move the transmission to the rear to disengage it from the engine block dowel pins and make sure the torque converter is detached from the flywheel. Secure the torque converter to the transmission so that it will not fall out during removal. Lower the transmission from the vehicle.

Installation

24   Make sure prior to installation that the torque converter hub is securely engaged in the pump.

25   With the transmission secured to the jack, raise the transmission into position, making sure to keep it level so the torque converter does not slide forward. Connect the transmission cooler lines.

26   Turn the torque converter to line up the torque converter and flywheel bolt holes. The white paint mark on the torque converter and the stud made during Step 5 must line up.

27   Move the transmission carefully forward until the dowel pins are engaged and the torque converter is engaged.

28   Install the transmission housing-to-engine bolts and nuts. Tighten the bolts and nuts to the specified torque.

29   Install the torque converter-to-flywheel nuts. Tighten the nuts to the specified torque.

30   Install the transmission mount crossmember and through-bolts. Tighten the bolts and nuts securely.

31   Remove the jacks supporting the transmission and the engine.

32   Install the fluid filler tube.

33   Install the starter.

34   Connect the vacuum hose(s) (if equipped).

35   Connect the shift and TV linkage.

36   Plug in the transmission electrical connectors.

37   Install the torque converter cover.

38   Connect the driveshaft.

39   Connect the speedometer.

40   Adjust the shift linkage.

41   Install any exhaust system components which were removed.

42   Lower the vehicle.

43   Fill the transmission with the specified fluid (see Chapter 1), run the vehicle and check for fluid leaks.

Chapter 8
Clutch and drivetrain

Contents

Specifications

Clutch

Clutch type	Single dry plate, diaphragm spring
Clutch pedal free-play and adjustment	see Chapter 1

Rear axle

Axleshaft end-play (B and O type)...	0.001 to 0.022 in

Torque specifications

	ft-lb
Clutch pressure plate bolts..	35
Transmission case to clutch bellhousing bolts..............................	55
Clutch bellhousing to engine bolts	30
Universal joint strap bolts	15
Universal joint U-bolt nuts	15
Universal joint flange bolts (double cardan type)	65
Differential carrier cover bolts..	20 to 25
Oil filler plug	20
Differential pinion lock	20

1   General information

The information in this Chapter deals with the components from the rear of the engine to the rear wheels, except for the transmission, which is dealt with in the previous Chapter. For the purposes of this Chapter, these components are grouped into three categories; clutch, driveshaft and rear axle. Separate Sections within this Chapter offer general descriptions and checking procedures for components in each of the three groups.

Since nearly all the procedures covered in this Chapter involve working under the vehicle, make sure it's securely supported on sturdy jackstands or on a hoist where the vehicle can be easily raised and lowered.

2   Clutch - description and check

Refer to illustration 2.1

1   All vehicles with a manual transmission use a single dry plate, diaphragm spring type clutch **(see illustration)**. The clutch disc has a splined hub that allows it to slide along the splines of the transmission input shaft. The clutch and pressure plate are held in contact by spring pressure exerted by the diaphragm in the pressure plate.

2   The mechanical release system includes the clutch pedal, the clutch linkage which actuates the clutch release lever and the release bearing.

3   When pressure is applied to the clutch pedal to release the clutch, mechanical pressure is exerted against the outer end of the clutch release lever. As the lever pivots, the shaft fingers push against the release bearing. The bearing pushes against the fingers of the diaphragm spring of the pressure plate assembly, which in turn releases the clutch plate.

4   Terminology can be a problem when discussing the clutch components because common names are in some cases different from those used by the manufacturer. For example, the driven plate is also called the clutch plate or disc, and the clutch release bearing is sometimes called a throwout bearing.

5   Other than to replace components with obvious damage, some preliminary checks should be performed to diagnose clutch problems.

a)   *To check "clutch spin down time," run the engine at normal idle speed with the transmission in Neutral (clutch pedal up - engaged). Disengage the clutch (pedal down), wait several seconds and shift the transmission into Reverse. No grinding noise should be heard. A grinding noise would most likely indicate a problem in the pressure plate or the clutch disc.*

b)   *To check for complete clutch release, run the engine (with the parking brake*

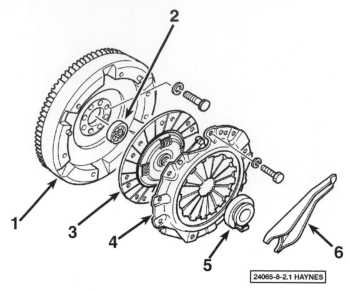

2.1 Exploded view of the clutch components

1   *Flywheel*
2   *Pilot bearing or bushing*
3   *Clutch disc*
4   *Clutch cover (pressure plate)*

5   *Clutch release bearing (throwout bearing)*
6   *Clutch release lever*

applied to prevent movement) and hold the clutch pedal approximately 1/2-inch from the floor. Shift the transmission between 1st gear and Reverse several times. If the shift is rough, component failure is indicated.

c)   *Visually inspect the pivot bushing at the top of the clutch pedal to make sure there is no binding or excessive play.*

d)   *On vehicles with mechanical release systems, a clutch pedal that is difficult to operate is most likely caused by faulty linkage. Check it for worn bushings and bent or distorted parts.*

e)   *Crawl under the vehicle and make sure the clutch release lever is solidly mounted on the ball stud.*

3   Clutch components - removal, inspection and installation

Warning: *Dust produced by clutch wear and deposited on clutch components contains asbestos, which is hazardous to your health. DO NOT blow it out with compressed air and DO NOT inhale it. DO NOT use gasoline or petroleum-based solvents to remove the dust. Brake system cleaner should be used to flush the dust into a drain pan. After the clutch components are wiped clean with a rag, dispose of the contaminated rags and cleaner in a covered, marked container.*

Removal

Refer to illustration 3.7

1   Access to the clutch components is normally accomplished by removing the transmission, leaving the engine in the vehicle. If,

of course, the engine is being removed for major overhaul, then check the clutch for wear and replace worn components as necessary. However, the relatively low cost of the clutch components compared to the time and trouble spent gaining access to them warrants their replacement anytime the engine or transmission is removed, unless they are new or in near perfect condition. The following procedures are based on the assumption the engine will stay in place.

2   Referring to Chapter 7 Part A, remove the transmission from the vehicle. Support the engine while the transmission is out. Preferably, an engine hoist should be used to support it from above. However, if a jack is used underneath the engine, make sure a piece of wood is positioned between the jack and oil pan to spread the load. **Caution:** *The pickup for the oil pump is very close to the bottom of the oil pan. If the pan is bent or distorted in any way, engine oil starvation could occur.*

3   Remove the return spring and the clutch release lever pushrod.

4   Remove the clutch housing-to-engine bolts and then detach the housing. It may have to be gently pried off the alignment dowels with a screwdriver or pry bar.

5   The clutch fork and release bearing can remain attached to the housing for the time being.

6   To support the clutch disc during removal, install a clutch alignment tool through the clutch disc hub.

7   Carefully inspect the flywheel and pressure plate for indexing marks. The marks are usually an X, an O or a white letter. If they cannot be found, scribe marks yourself so the pressure plate and the flywheel will be in

3.7 After removal of the transmission, this will be the view of the clutch components

| 1 | *Pressure plate* | 2 | *Flywheel* |

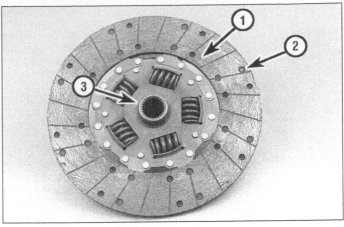

3.12 The clutch plate

1 **Lining** - *this will wear down in use*
2 **Rivets** - *these secure the lining and will damage the flywheel or pressure plate if allowed to contact the surfaces*
3 **Markings** - *"Flywheel side" or something similar*

the same alignment during installation **(see illustration)**.

8   Turning each bolt only 1/4-turn at a time, loosen the pressure plate-to-flywheel bolts. Work in a criss-cross pattern until all spring pressure is relieved. Then hold the pressure plate securely and completely remove the bolts, followed by the pressure plate and clutch disc.

Inspection

Refer to illustrations 3.12 and 3.14

9   Ordinarily, when a problem occurs in the clutch, it can be attributed to wear of the clutch driven plate assembly (clutch disc). However, all components should be inspected at this time.

10   Inspect the flywheel for cracks, heat checking, grooves and other obvious defects. If the imperfections are slight, a machine shop can machine the surface flat and smooth, which is highly recommended regardless of the surface appearance. Refer to Chapter 2 for the flywheel removal and installation procedure.

11   Inspect the pilot bearing (see Section 5).

12   Inspect the lining on the clutch disc. There should be at least 1/16 inch of lining above the rivet heads. Check for loose rivets, distortion, cracks, broken springs and other obvious damage **(see illustration)**. As mentioned above, ordinarily the clutch disc is routinely replaced, so if in doubt about the condition, replace it with a new one.

13   The release bearing should also be replaced along with the clutch disc (see Section 4).

14   Check the machined surfaces and the diaphragm spring fingers of the pressure plate **(see illustration)**. If the surface is grooved or otherwise damaged, replace the pressure plate. Also check for obvious damage, distortion, cracking, etc. Light glazing can be removed with medium grit emery cloth. If a new pressure plate is required, new and factory-rebuilt units are available.

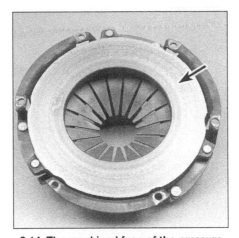

3.14 The machined face of the pressure plate must be inspected for score marks and other damage

Installation

Refer to illustration 3.16

15   Before installation, clean the flywheel and pressure plate machined surfaces with lacquer thinner or acetone. It's important that no oil or grease is on these surfaces or the lining of the clutch disc. Handle the parts only with clean hands.

16   Position the clutch disc and pressure plate against the flywheel with the clutch held in place with an alignment tool **(see illustration)**. Make sure it's installed properly (most replacement clutch plates will be marked "flywheel side" or something similar - if not marked, install the clutch disc with the damper springs toward the transmission).

17   Tighten the pressure plate-to-flywheel bolts only finger tight, working around the pressure plate.

18   Center the clutch disc by ensuring the alignment tool extends through the splined hub and into the pilot bearing in the

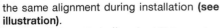

3.16 A clutch alignment tool can be purchased at most auto parts stores and eliminates all guesswork when centering the clutch plate in the pressure plate

crankshaft. Wiggle the tool up, down or side-to-side as needed to bottom the tool in the pilot bearing. Tighten the pressure plate-to-flywheel bolts a little at a time, working in a criss-cross pattern to prevent distorting the cover. After all of the bolts are snug, tighten them to the specified torque. Remove the alignment tool.

19   Using high-temperature grease, lubricate the inner groove of the release bearing (refer to Section 4). Also place grease on the release lever contact areas and the transmission input shaft bearing retainer.

20   Install the clutch release bearing as described in Section 4.

21   Install the clutch housing and tighten the bolts to the specified torque.

22   Install the transmission, slave cylinder and all components removed previously. Tighten all fasteners to the proper torque specifications.

23   Refer to Chapter 1 for clutch pedal free play check and adjustment information.

4.7 When installing the release bearing, make sure the fingers and the tabs fit into the bearing recess

4   Clutch release bearing and lever - removal, inspection and installation

Refer to illustration 4.7
Warning: *Dust produced by clutch wear and deposited on clutch components may contain asbestos, which is hazardous to your health. DO NOT blow it out with compressed air and DO NOT inhale it. DO NOT use gasoline or petroleum-based solvents to remove the dust. Brake system cleaner should be used to flush the dust into a drain pan. After the clutch components are wiped clean with a rag, dispose of the contaminated rags and cleaner in a covered, marked container.*

Removal

1   Disconnect the negative cable from the battery.
2   Remove the transmission (see Chapter 7A).
3   Remove the clutch housing (see Section 3).
4   Remove the clutch release lever from the ball stud, then remove the bearing from the lever.

Inspection

5   Hold the center of the bearing and rotate the outer portion while applying pressure. If the bearing doesn't turn smoothly or if it's noisy, replace it with a new one. Considering the difficulty in getting to this bearing, it is normally routinely replaced when removed. Wipe the bearing with a clean rag and inspect it for damage, wear and cracks. Don't immerse the bearing in solvent - it's sealed for life and to do so would ruin it.

Installation

6   Lightly lubricate the clutch lever crown and spring retention crown where they contact the bearing with high-temperature grease. Fill the inner groove of the bearing with the same grease .

7   Attach the release bearing to the clutch lever so that both fork tabs fit into the bearing recess **(see illustration)**.
8   Lubricate the clutch release lever ball socket with high-temperature grease and push the lever onto the ball stud until it's firmly seated.
9   Apply a light coat of high-temperature grease to the face of the release bearing, where it contacts the pressure plate diaphragm fingers.
10   Install the clutch housing and tighten the bolts to the specified torque.
11   Prior to installing the transmission, apply a light coat of grease to the transmission front bearing retainer.
12   The remainder of installation is the reverse of the removal procedure. Tighten all bolts to the specified torque.

5   Pilot bearing - inspection and replacement

Refer to illustration 5.9
1   The clutch pilot bearing is a needle roller type bearing which is pressed into the rear of the crankshaft. It is greased at the factory and does not require additional lubrication. Its primary purpose is to support the front of the transmission input shaft. The pilot bearing should be inspected whenever the clutch components are removed from the engine. Due to its inaccessibility, if you are in doubt as to its condition, replace it with a new one.
Note: *If the engine has been removed from the vehicle, disregard the following steps which do not apply.*
2   Remove the transmission (refer to Chapter 7A).
3   Remove the clutch components (see Section 3).
4   Inspect for any excessive wear, scoring, lack of grease, dryness or obvious damage. If any of these conditions are noted, the bearing should be replaced. A flashlight will be helpful to direct light into the recess.
5   Removal can be accomplished with a special puller and slide hammer, but an alternative method also works very well.
6   Find a solid steel bar which is slightly smaller in diameter than the bearing. Alternatives to a solid bar would be a wood dowel or a socket with a bolt fixed in place to make it solid.
7   Check the bar for fit - it should just slip into the bearing with very little clearance.
8   Pack the bearing and the area behind it (in the crankshaft recess) with heavy grease. Pack it tightly to eliminate as much air as possible.
9   Insert the bar into the bearing bore and strike the bar sharply with a hammer, which will force the grease to the back side of the bearing and push it out **(see illustration)**. Remove the bearing and clean all grease from the crankshaft recess.
10   To install the new bearing, lightly lubricate the outside surface with lithium-based

5.9 Pack the recess behind the pilot bearing with heavy grease and force it out hydraulically with a steel rod slightly smaller than the bore in the bearing - when the hammer strikes the rod, the bearing will pop out of the crankshaft

grease, then drive it into the recess with a soft-face hammer.
11   Install the clutch components, transmission and all other components removed previously, tightening all fasteners properly.

6   Clutch pedal - removal and installation

1   Disconnect the negative cable from the battery.
2   Disconnect and remove the starter safety switch (see Section 7).
3   Disconnect the clutch and brake pedal pushrods and pedal return springs.
4   Remove the pedal pivot nut or spring clip and slide the pedal to the left to remove it.
5   Wipe clean all parts; however, don't use cleaning solvent on the bushings. Replace all worn parts with new ones.
6   Installation is the reverse of removal. Check that the starter safety switch allows the vehicle to be started only with the clutch pedal fully depressed.

7   Clutch safety switch - removal and installation

1   Disconnect the cable from the negative terminal of the battery.
2   Unplug the electrical connector from the switch.
3   On 1970 models remove the retainer from the pins on the clutch pedal, then remove the switch retaining screw and remove the switch.
4   On 1971 through 1976 models, remove the mounting screws and compress the switch actuating shaft retainer, then detach the switch from the mounting bracket and clutch pedal.
5   To install the switch, reverse the

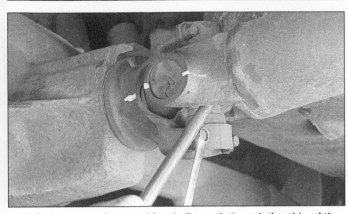

9.3 Before removing the driveshaft, mark the relationship of the driveshaft yoke to the differential flange - to prevent the driveshaft from turning when loosening the strap bolts, insert a screwdriver through the yoke

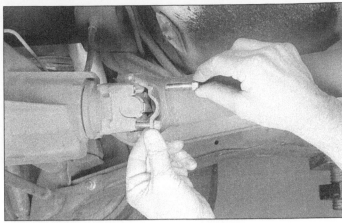

9.4 Remove the bolts and straps retaining the driveshaft to the differential yoke

removal procedure. No adjustment is necessary.
6   Check for proper operation.

8   Driveshaft and universal joints - description and check

1   The driveshaft is a tube running between the transmission and the differential. Universal joints are located at either end of the driveshaft and permit power to be transmitted to the rear wheels at varying angles.
2   The driveshaft features a splined yoke at the front, which slips into the extension housing of the transmission. This arrangement allows the driveshaft to slide back and forth within the transmission as the vehicle is in operation.
3   An oil seal is used to prevent leakage of fluid at this point and to keep dirt and contaminants from entering the transmission. If leakage is evident at the front of the driveshaft, replace the oil seal, referring to the procedures in Chapter 7.
4   The driveshaft assembly requires very little service. Most factory equipped universal joints are lubricated for life and must be replaced if problems develop. The driveshaft must be removed from the vehicle for this procedure.
5   Since the driveshaft is a balanced unit, it's important that no undercoating, mud, etc., be allowed to stay on it. When the vehicle is raised for service it's a good idea to clean the driveshaft and inspect it for any obvious damage. Also check that the small weights used to originally balance the driveshaft are in place and securely attached. Whenever the driveshaft is removed it's important that it be reinstalled in the same relative position to preserve this balance.
6   Problems with the driveshaft are usually indicated by a noise or vibration while driving the vehicle. A road test should verify if the problem is the driveshaft or another vehicle component:
a)   On an open road, free of traffic, drive the

vehicle and note the engine speed (rpm) at which the problem is most evident.
b)   With this noted, drive the vehicle again, this time manually keeping the transmission in 1st, then 2nd, then 3rd gear ranges and running the engine up to the engine speed noted.
c)   If the noise or vibration occurs at the same engine speed regardless of which gear the transmission is in, the driveshaft is not at fault because the speed of the driveshaft varies in each gear.
d)   If the noise or vibration decreased or was eliminated, visually inspect the driveshaft for damage, material on the shaft that would effect balance, missing weights and damaged universal joints. Another possibility for this condition would be tires that are out of balance.
7   To check for worn universal joints:
a)   On an open road, free of traffic, drive the vehicle slowly until the transmission is in High gear. Let off on the accelerator, allowing the vehicle to coast, then accelerate. A clunking or knocking noise will indicate worn universal joints.
b)   Drive the vehicle at a speed of about 10 to 15 mph and then place the transmission in Neutral, allowing the vehicle to coast. Listen for abnormal driveline noises.
c)   Raise the vehicle and support it securely on jackstands. With the transmission in Neutral, manually turn the driveshaft, watching the universal joints for excessive play.

9   Driveshaft - removal and installation

Refer to illustrations 9.3 and 9.4

Removal
1   Disconnect the negative cable from the battery.
2   Raise the vehicle and support it securely on jackstands. Place the transmission in

Neutral with the parking brake off.
3   Using a sharp scribe, white paint or a hammer and punch, place marks on the driveshaft and the differential flange in line with each other (see illustration). This is to make sure the driveshaft is reinstalled in the same position to preserve the balance.
4   Disconnect the rear universal joint by unscrewing and removing the nuts from the U-bolts or strap retaining bolts (see illustration), or by removing the flange bolts. Turn the driveshaft (or tires) as necessary to bring the bolts into the most accessible position.
5   Tape the bearing caps to the spider to prevent the caps from coming off during removal.
6   Lower the rear of the driveshaft and then slide the front out of the transmission.
7   To prevent loss of fluid and protect against contamination while the driveshaft is out, wrap a plastic bag over the transmission housing and hold it in place with a rubber band.

Installation
8   Remove the plastic bag on the transmission and wipe the area clean. Inspect the oil seal carefully. Procedures for replacement of this seal can be found in Chapter 7.
9   Slide the front of the driveshaft into the transmission.
10   Raise the rear of the driveshaft into position, checking to be sure that the marks are in alignment. If not, turn the rear wheels to match the pinion flange and the driveshaft.
11   Remove the tape securing the bearing caps and install the straps and bolts. Tighten the bolts to the torque listed in this Chapter's Specifications.

10   Universal joints - replacement

Note: *A press or large vise will be required for this procedure. It may be advisable to take the driveshaft to a local dealer or machine shop where the universal joints can be replaced for you, normally at a reasonable charge.*

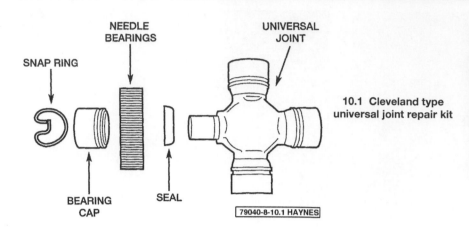

SNAP RING

NEEDLE
BEARINGS

UNIVERSAL
JOINT

BEARING
CAP

SEAL

10.1 Cleveland type universal joint repair kit

79040-8-10.1 HAYNES

10.2 To press the universal joint out of the driveshaft, set it up in a vise with the small socket (on the right) pushing the joint and bearing cap into the large socket

Cleveland type joint

Refer to illustrations 10.1 and 10.2

1   Clean away all dirt from the ends of the bearings on the yokes so that the snap-rings can be removed using a pair of snap-ring pliers. If the snap-rings are very tight, tap the end of the bearing cup (inside the snap-ring) to relieve the pressure **(see illustration)**.

2   Support the trunnion yoke on a short piece of tube or the open end of a socket, then use a suitably sized socket to press out the cross (trunnion) by means of a vise **(see illustration)**.

3   Press the trunnion through as far as possible, then grip the bearing cup in the jaws of a vise to fully remove it. Repeat the procedure for the remaining cups.

4   On some models, the slip yoke at the transmission end has a vent hole. When dismantling, ensure that this vent hole is not blocked.

5   A universal joint repair kit will contain a new trunnion, seals, bearings, cups and snap-rings. Some replacement universal joints are equipped with a grease fitting. Be sure it is offset in the proper direction (toward the driveshaft).

6   Commence reassembly by packing each of the reservoirs at the trunnion ends with lubricant.

7   Make sure that the dust seals are correctly located on the trunnion, so that the cavities in the seals are nearer the trunnion.

8   Using a vise, press one bearing cup into the yoke so that it enters not more than 1/4-inch.

9   Using a thick grease, stick each of the needle rollers inside the cup.

10   Insert the trunnion into the partially fitted bearing cup, taking care not to displace the needle rollers.

11   Stick the needle bearings into the opposite cup and then, holding the trunnion in correct alignment, press both cups fully home in the jaws of the vise.

12   Install the new snap-rings.

13   Repeat the operations on the other two bearing cups. After the caps and snap rings have been installed, strike the yoke firmly with a hammer to center the U-joint in the yoke **(see illustration 10.18)**.

14   In extreme cases of wear or neglect, it is possible that the bearing cup housings in the yoke will have worn so much that the cups are a loose fit in the yokes. In such cases, replace the complete driveshaft assembly.

15   Always check the wear in the sliding sleeve splines and replace the sleeve if worn.

Saginaw type joint

Refer to illustrations 10.16 and 10.18

16   Where a Saginaw joint is to be disassembled, the procedure given in the previous Steps 1 through 12 for pressing out the bearing cup is applicable. If the joint has been previously repaired it will be necessary to remove the snap-rings inboard of the yokes; if this is to be the first time that servicing has been carried out, there are no snap-rings to remove, but the pressing operation in the vise will shear the plastic molding material **(see illustration)**.

17   Having removed the cross (trunnion), remove the remains of the plastic material from the yoke. Use a small punch to remove the material from the injection holes.

18   Reassembly is similar to that given for the Cleveland type joint except that the snap-rings are installed inside the yoke. After the caps and snap rings have been installed, strike the yoke firmly with a hammer to center the joint in the yoke **(see illustration)**.

Double cardan type constant velocity joint

Refer to illustrations 10.19 and 10.21

19   An inspection kit containing two bearing cups and two retainers is available to permit the joint to be dismantled to the stage where

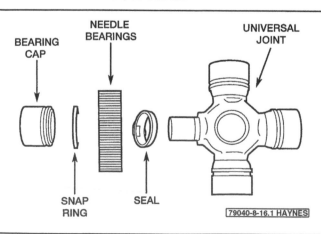

BEARING
CAP

NEEDLE
BEARINGS

UNIVERSAL
JOINT

10.16 Saginaw type universal joint repair kit

SNAP
RING

SEAL

79040-8-16.1 HAYNES

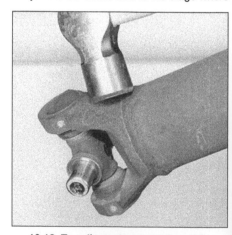

10.18 To relieve stress produced by pressing the bearing caps into the yokes, strike the yoke in the area shown

**10.19 CV joint alignment marks made
before disassembly**

**10.21 Place the companion flange in a vise
as shown and remove the seal and
centering ball**

11 Rear axle - description and check

Refer to illustration 11.3

Description

1   The rear axle assembly is a hypoid, semi-floating type (the centerline of the pinion gear is below the centerline of the ring gear). The differential carrier is a casting with a pressed steel cover, and the axle tubes are made of steel, pressed and welded into the carrier.

2   An optional locking rear axle is also available. This differential allows for normal differential operation until one wheel loses traction. The unit utilizes multi-disc clutch packs and a speed sensitive engagement mechanism which locks both axleshafts together, applying equal rotational power to both wheels.

3   In order to undertake certain operations, particularly replacement of the axleshafts, it's important to know the axle identification number. It's located on the front face of the right side axle tube. The third letter of the code identifies the manufacturer of the axle.

the joint can be inspected. Before any dismantling is started, mark the flange yoke and coupling yoke so that they can be reassembled in the same relative position **(see illustration)**.

20   Dismantle the joint by removing the bearing cups in a similar way to that described in the preceding Section, according to type.

21   Disengage the flange yoke and trunnion from the centering ball. Pry the seal from the ball socket and remove the washers, springs and three ball seats **(see illustration)**.

22   Clean the ball seat insert bushing and inspect for wear. If evident, the flange yoke and trunnion assembly must be replaced.

23   Clean the seal, ball seats, spring and washers and inspect for wear. If excessive wear is evident or parts are broken, a replacement service kit must be used.

24   Remove all plastic material from the groove of the coupling yoke (if this type of joint is used).

25   Inspect the centering ball; if damaged, it must be replaced.

26   Withdraw the centering ball from the stud using a suitable extractor. Provided that the ball is not to be re-used, it will not matter if it is damaged.

27   Press a new ball onto the stud until it seats firmly on the stud shoulder. It is extremely important that no damage to the ball occurs during this stage, and suitable protection must be given to it.

28   Using the grease provided in the repair kit, lubricate all the parts and insert them into the ball seat cavity in the following order: spring, washer (small OD), three ball seats (largest opening outwards to receive the ball), washer (large OD) and the seal.

29   Lubricate the seal lips and press it (lip inwards) into the cavity. Fill the cavity with the grease provided.

30   Install the flange yoke to the centering ball, ensuring that the alignment marks are correctly positioned.

31   Install the trunnion caps as described previously for the Cleveland or Saginaw types.

This is important, as axle design varies slightly among manufacturers. Manufacturer's code letter B indicates that the wheel bearings are pressed onto the axleshafts, whereas all other axles have bearings pressed into the axle tubes, and the axleshafts are retained with C-locks **(see illustration)**.

Check

4   Many times a fault is suspected in the rear axle area when, in fact, the problem lies elsewhere. For this reason, a thorough check should be performed before assuming a rear axle problem.

5   The following noises are those commonly associated with rear axle diagnosis procedures:

a)   *Road noise is often mistaken for mechanical faults. Driving the vehicle on different surfaces will show whether the road surface is the cause of the noise. Road noise will remain the same if the vehicle is under power or coasting.*

b)   *Tire noise is sometimes mistaken for mechanical problems. Tires which are worn or low on pressure are particularly susceptible to emitting vibrations and noises. Tire noise will remain about the same during varying driving situations, where rear axle noise will change during coasting, acceleration, etc.*

c)   *Engine and transmission noise can be deceiving because it will travel along the driveline. To isolate engine and transmission noises, make a note of the engine speed at which the noise is most pronounced. Stop the vehicle and place the transmission in Neutral and run the engine to the same speed. If the noise is the same, the rear axle is not at fault.*

6   Overhaul and general repair of the rear axle is beyond the scope of the home mechanic due to the many special tools and critical measurements required. Thus, the procedures listed here will involve axleshaft removal and installation, axleshaft oil seal replacement, axleshaft bearing replacement and removal of the entire unit for repair or replacement.

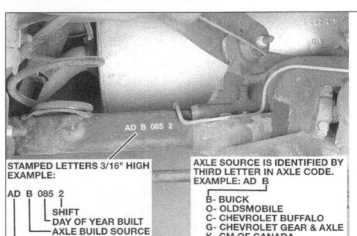

**11.3 Location of
the rear axle
identification
number - early
models have only
a three-letter code**

STAMPED LETTERS 3/16" HIGH
EXAMPLE:

AD B 085 2

SHIFT
DAY OF YEAR BUILT
AXLE BUILD SOURCE
AXLE RATIO

AXLE SOURCE IS IDENTIFIED BY
THIRD LETTER IN AXLE CODE.
EXAMPLE: AD B

B- BUICK
O- OLDSMOBILE
C- CHEVROLET BUFFALO
G- CHEVROLET GEAR & AXLE
K- GM OF CANADA
P- PONTIAC

12.3a Remove the pinion shaft lock bolt . . .

12.3b . . . then carefully remove the pinion shaft from the differential carrier (don't turn the wheels or the carrier after the shaft has been removed, or the spider gears may fall out)

12.4 Push the axle flange in, then remove the C-lock from the inner end of the axleshaft

12 Axleshaft - removal and installation (except B and O type axles)

Refer to illustrations 12.3a, 12.3b, 12.4 and 12.5

1   Raise the rear of the vehicle, support it securely and remove the wheel and brake drum (refer to Chapter 9).
2   Remove the cover from the differential

12.5 Carefully pull the axleshaft from the housing to avoid damaging the seal

carrier and allow the oil to drain into a container.
3   Remove the lock bolt from the differential pinion shaft. Remove the pinion shaft **(see illustrations)**.
4   Push the outer (flanged) end of the axleshaft in and remove the C-lock from the inner end of the shaft **(see illustration)**.
5   Withdraw the axleshaft, taking care not to damage the oil seal in the end of the axle housing as the splined end of the axleshaft passes through it **(see illustration)**.
6   Installation is the reverse of removal. Tighten the lock bolt to the specified torque.
7   Always use a new cover gasket and tighten the cover bolts to the specified torque.
8   Refill the axle with the correct quantity and grade of lubricant (see Chapter 1).

13 Axleshaft oil seal - replacement (except B and O type axles)

Refer to illustrations 13.2 and 13.3

1   Remove the axleshaft as described in

the preceding Section.
2   Pry the old oil seal out of the end of the axle housing, using a large screwdriver or the inner end of the axleshaft itself as a lever **(see illustration)**.
3   Using a large socket as a seal driver, tap the seal into position so that the lips are facing in and the metal face is visible from the end of the axle housing **(see illustration)**. When correctly installed, the face of the oil seal should be flush with the end of the axle housing. Lubricate the lips of the seal with gear oil.
4   Installation of the axleshaft is described in the preceding Section.

14 Axleshaft bearing - replacement (except B and O type axles)

Refer to illustrations 14.3 and 14.4

1   Remove the axleshaft (refer to Section 12) and the oil seal (refer to Section 13).
2   A bearing puller will be required, or a tool which will engage behind the bearing will have to be fabricated.
3   Attach a slide hammer and pull the

13.2 The axleshaft oil seal can sometimes be pried out with the end of the axle

13.3 A large socket can be used to install the new seal squarely

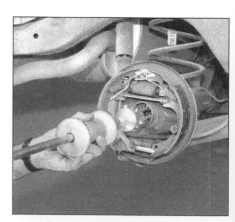

14.3 A slide hammer with a special bearing puller attachment is required to pull the axleshaft bearing from the axle housing

14.4 A special bearing driver, available at auto parts stores, is needed to install the axleshaft bearing without damaging it

15.3 Using a slide hammer to remove a B type axleshaft

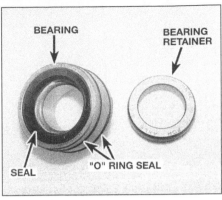

15.9 Typical rear axle bearing and retainer

16.3 Check the torque required to rotate the pinion with an in-lb torque wrench

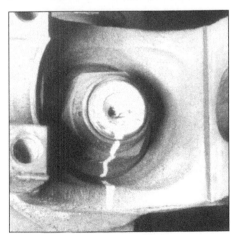

16.4 Mark the relative positions of the pinion, nut and flange before removing the nut

bearing from the axle housing **(see illustration)**.
4   Clean out the bearing recess and drive in the new bearing using a special tool available at most auto parts stores **(see illustration)**. **Caution:** *Failure to use this tool could result in bearing damage. Lubricate the new bearing with gear lubricant. Make sure that the bearing is tapped into the full depth of its recess and that the numbers on the bearing are visible from the outer end of the housing.*
5   Discard the old oil seal and install a new one, then install the axleshaft.

15 Axleshaft - removal, overhaul and installation (B and O type axles)

Refer to illustrations 15.3 and 15.9
1   The usual reason for axleshaft removal on this type of axle is that the shaft end play has become excessive. To check this, remove the wheel and brake drum (see Chapter 9) and attach a dial gauge with its stylus against the axleshaft end flange. If the shaft is then moved in and out by hand, the end play should not exceed 0.022-inch. If the end play is excessive, carry out the following operations, as the shaft bearing is probably worn.
2   Remove the brake drum and brake shoe assembly (see Chapter 9). Unscrew and remove the bolts that attach the axleshaft retainer plate to the brake backing plate.
3   Attach a slide hammer to the wheel mounting studs and withdraw the axleshaft. Do not attempt to pull the axleshaft by hand from the housing, as you will only succeed in pulling the vehicle off the support stands **(see illustration)**.
4   As the axleshaft is removed, it is possible that the bearing will become separated into three parts. This does not indicate that the bearing is unserviceable. If this happens, remove the two sections left behind from the axle tube.
5   Since the home mechanic does not usu-

ally have access to a heavy-duty hydraulic press, it is generally most convenient to take the axle assembly to a machine shop and have the old bearing pressed off and the new one pressed on. However, if you choose to do the job yourself, place the axleshaft in a vise so that the bearing retainer ring rests on the edges of the jaws.
6   Using a hammer and a sharp chisel, nick the retainer in two places. This will have the effect of spreading the retainer so that it will slide off the shaft. This may take some persistence to loosen the retainer adequately for removal. Do not damage the shaft in the process and never attempt to cut the retainer away with a torch, as the temper of the shaft will be ruined.
7   Using a heavy-duty press, press the bearing from the axleshaft.
8   Remove and discard the oil seal.
9   When installing the new bearing **(see illustration)**, make sure that the retainer plate and the seal are installed to the shaft first. Press on the bearing and the retaining ring tight up against it.
10   Before installing the axleshaft assembly, smear wheel bearing grease onto the bearing end and in the bearing recess in the axle housing tube.
11   Apply rear axle oil to the axleshaft splines.
12   Hold the axleshaft horizontal and insert it into the axle housing. Feel when the shaft splines have picked up those in the differential side gears and then push the shaft fully into position, using a soft faced hammer on the end flange as necessary.
13   Bolt the retainer plate to the brake backplate, install the brake drum and wheel and lower the vehicle to the ground.

16 Pinion oil seal - replacement

Refer to illustrations 16.3, 16.4, 16.6 and 16.10
1   Raise the rear of the vehicle and support it securely on jackstands placed under the axle. Block the front wheels to keep the vehi-

cle from rolling off the stands.
2   Disconnect the driveshaft and fasten it out of the way.
3   Use an inch-pound torque wrench to check the torque required to rotate the pinion. Record it for use later **(see illustration)**.
4   Scribe or punch alignment marks on the pinion shaft, nut and flange **(see illustration)**.

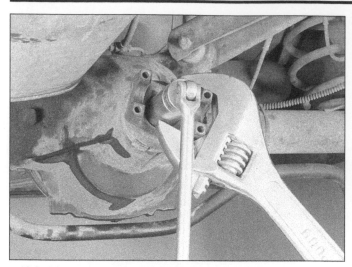

16.6 A large wrench or slip-joint pliers can be used to hold the companion flange while the pinion nut is loosened

16.10 Use a large socket or piece of pipe to tap the pinion oil seal into place

5   Count the number of threads visible between the end of the nut and the end of the pinion shaft and record it for use later.

6   Keep the companion flange from moving while the self-locking pinion nut is loosened **(see illustration).**

7   Remove the pinion nut.

8   Withdraw the companion flange. It may be necessary to use a two or three-jaw puller engaged behind the flange to draw it out. Do not attempt to pry behind the flange or hammer on the end of the pinion shaft.

9   Pry out the old seal and discard it.

10   Lubricate the lips of the new seal with high-temperature grease and tap it evenly into position with a seal installation tool or a large socket. Make sure it enters the housing squarely and is tapped in to its full depth **(see illustration).**

11   Align the mating marks made before disassembly and install the companion flange. If necessary, tighten the pinion nut to draw the flange into place. Do not try to hammer the flange into position.

12   Apply non-hardening sealant to the ends of the splines visible in the center of the flange so oil will be sealed in.

13   Install the washer (if equipped) and pinion nut. Tighten the nut carefully until the original number of threads are exposed.

14   Measure the torque required to rotate the pinion and tighten the nut in small increments until it matches the figure recorded in Step 5. In order to compensate for the drag of the new oil seal, the nut should be tightened more until the rotational torque of the pinion slightly exceeds what was recorded earlier, but not by more than 5 in-lbs.

15   Connect the driveshaft and lower the vehicle.

17   Rear axle assembly - removal and installation

1   Raise the rear of the vehicle and support it securely on jackstands placed under the body frame rails.

2   Remove the rear wheels and brake drums. See Chapter 9 for information if difficulty is experienced in removing the brake drums.

3   Disconnect the driveshaft from the differential (see Section 9), then hang the back end of the driveshaft up, out of the way, using a wire hung from underneath the floorpan.

4   Disconnect the axle vent hose from the top of the axle housing.

5   Disconnect the brake hose from the junction block on top of the axle housing and plug the hose to prevent leaks.

6   Disconnect the rear parking brake cables from the equalizer (see Chapter 9).

7   Position an adjustable floorjack under the differential housing and just take up the weight. Do not raise the jack sufficiently to take the weight of the vehicle from the frame stands.

8   Disconnect the lower shock absorber mountings (see Chapter 10).

9   Refer to Chapter 10 and disconnect the upper and lower control arms from the axle housing, then lower the axle assembly and remove the coil springs.

10   Carefully lower the axle assembly, then pull the floorjack to the rear and remove the axle assembly from under the vehicle.

11   Installation is a reversal of removal, but tighten all suspension bolts and nuts to the specified torque (see Chapter 10). Final-tighten the control arm through-bolts only after the vehicle has been lowered to the ground after reassembly (see Chapter 10).

12   Bleed the brakes (see Chapter 9).

18   Positraction (limited slip) type axles- description. check and precautions

1   This type of axle incorporates a special type of differential unit. Essentially the device provides more driving force to the wheel with traction when one wheel begins to spin.

2   Two types of assembly are used, one having clutch plates and the other having cones. Some makes of Positraction are not repairable and in the event of a fault occurring or wear developing, the complete assembly must be replaced.

3   An on-vehicle check can be carried out in the following way. If equipped with manual transmission, shift into 'Neutral'.

4   Raise the rear of the vehicle until the wheels are off the ground. Remove one wheel.

5   Using a suitable adapter connect a torque wrench to the axleshaft flange. Alternatively, a spring balance can be used.

6   Have an assistant hold the wheel still in position on the opposite hub quite firmly to prevent it from rotating, then measure the torque required to start the axleshaft (opposite to the one to which the torque wrench is attached) moving. A minimum of 40 ft-lb should be required for an axle which has seen considerable service and 70 ft-lb for a new, or nearly new assembly.

7   Always use fluid lubricant for topping up a Positraction type axle.

8   Never run the engine when one rear wheel is off the ground as the vehicle may obtain traction through the remaining wheel and jump off the jacks or support stands.

Chapter 9   Brakes

Contents

Specifications

Brake fluid type .... See Chapter 1

Disc brakes

Minimum brake pad thickness	See Chapter 1
Rotor discard thickness...	Refer to minimum thickness casting on rotor
Lateral runout	0.004 inch maximum
Rotor thickness variation (parallelism).....................................	0.0005 inch
Caliper-to-knuckle clearance	0.005 to 0.012 inch

Drum brakes

Minimum brake lining thickness	See Chapter 1
Drum discard thickness..	Refer to minimum thickness casting on drum
Out-of-round...	0.006 inch maximum
Taper...	0.003 inch maximum

Torque specifications

Ft-lbs (unless otherwise indicated)

Master cylinder mounting nuts	24
Power booster mounting nuts	24
Bleeder valves	110 in-lbs
Brake shoe anchor pin..	120 in-lbs
Wheel cylinder to flange plate	110 in-lbs
Caliper mounting bolt	35
Flexible hose to caliper...	22
Shield to steering knuckle	140 in-lbs
Brake pedal pivot bolt	30

Component location

Typical rear drum brake assembly

1	Return spring
2	Return spring
3	Hold-down spring
4	Primary brake shoe

5	Adjuster screw assembly
6	Actuator lever
7	Lever return spring

8	Actuator spring (not all models)
9	Hold-down spring
10	Secondary brake shoe

1   General information

The vehicles covered by this manual are equipped with hydraulically operated front and rear brake systems. The front brakes are disc or drum type and the rear brakes are drum type. Both the front and rear brakes are self adjusting. The front disc brakes automatically compensate for pad wear, while the drum brakes incorporate an adjustment mechanism which is activated as the brakes are applied when the vehicle is driven in Reverse.

Hydraulic system

The hydraulic system consists of two separate circuits. The master cylinder has separate reservoirs for the two circuits and in the event of a leak or failure in one hydraulic circuit, the other circuit will remain operative. A visual warning of circuit failure or air in the system is given by a warning light activated by displacement of the piston in the pressure differential switch portion of the combination valve from its normal "in balance" position.

Combination valve

A combination valve, located in the engine compartment below the master cylinder, consists of three sections providing the following functions. The metering section limits pressure to the front brakes until a predetermined front input pressure is reached and until the rear brakes are activated. There is no restriction at inlet pressures below three psi, allowing pressure equalization during non-braking periods. The proportioning section proportions outlet pressure to the rear brakes after a predetermined rear input pressure has been reached, preventing early rear wheel lock-up under heavy brake loads. The valve is also designed to assure full pressure to one brake system should the other system fail. The pressure differential warning switch incorporated into the combination valve is designed to continuously compare the front and rear brake pressure from the master cylinder and energize the dash warning light in the event of either a front or rear brake system failure. The design of the switch and valve are such that the switch will stay in the "warning"

Component location

Typical front disc brake assembly

1	Dust cap	3	Outer brake pad	5	Inner brake pad	7	Brake hose
2	Brake rotor	4	Caliper mounting bolts	6	Bleed screw	8	Caliper

position once a failure has occurred. The only way to turn the light off is to repair the cause of the failure and apply a brake pedal force of 450 psi.

Power brake booster

The power brake booster, utilizing engine manifold vacuum and atmospheric pressure to provide assistance to the hydraulically operated brakes, is mounted on the firewall in the engine compartment.

Parking brake

The parking brake operates the rear brakes only, through cable actuation. It's activated by a pedal mounted on the left side kick panel.

Service

After completing any operation involving disassembly of any part of the brake system, always test drive the vehicle to check for proper braking performance before resuming normal driving. When testing the brakes, perform the tests on a clean, dry flat surface. Conditions other than these can lead to inaccurate test results.

Test the brakes at various speeds with both light and heavy pedal pressure. The vehicle should stop evenly without pulling to one side or the other. Avoid locking the brakes, because this slides the tires and diminishes braking efficiency and control of the vehicle.

Tires, vehicle load and front-end alignment are factors which also affect braking performance.

2 Disc brake pads - replacement

Refer to illustrations 2.5 and 2.6a through 2.6g

Warning: *Disc brake pads must be replaced on both front wheels at the same time - never replace the pads on only one wheel. Also, the dust created by the brake system may contain asbestos, which is harmful to your health. Never blow it out with compressed air and don't inhale any of it. An approved filtering mask should be worn when working on the brakes. Do not, under any circumstances, use petroleum-based solvents to clean brake parts. Use brake cleaner or denatured alcohol only!*
Note: *When servicing the disc brakes, use only high quality, nationally recognized name-brand pads.*

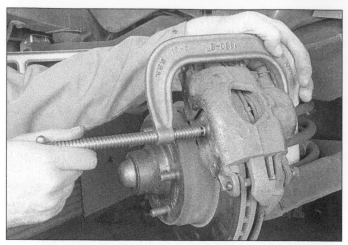

2.5 Using a large C-clamp, push the piston back into the caliper bore - note that one end of the clamp is on the flat area near the brake hose fitting and the other end (screw end) is pressing against the outer brake pad

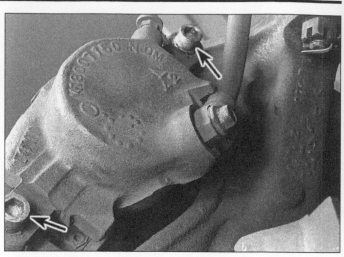

2.6a Remove the two caliper-to-steering knuckle mounting bolts (arrows) (this will require the use of an Allen head or Torx head socket wrench)

2.6b Slide the caliper up and off the rotor

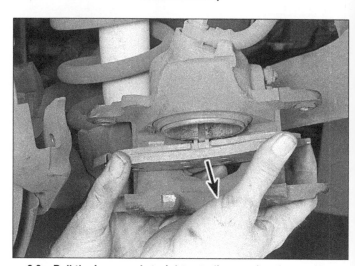

2.6c Pull the inner pad straight out, disengaging the retainer spring from the caliper piston

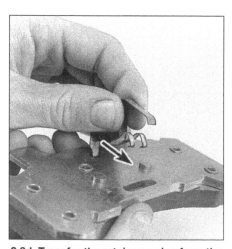

2.6d Transfer the retainer spring from the old inner pad to the new one - hook the end of the spring in the hole at the top of the pad, then insert the two prongs of the spring into the slot on the pad backing plate

1   Remove the cover from the brake fluid reservoir.

2   Loosen the wheel lug nuts, raise the front of the vehicle and support it securely on jackstands.

2.6e Lubricate the caliper mounting ears with multipurpose grease

3   Remove the front wheels. Work on one brake assembly at a time, using the assembled brake for reference if necessary.

4   Inspect the brake disc carefully as outlined in Section 4. If machining is necessary, follow the information in that Section to remove the disc, at which time the pads can be removed from the calipers as well.

5   Push the piston back into the bore to provide room for the new brake pads. A C-clamp can be used to accomplish this **(see illustration)**. As the piston is depressed to the bottom of the caliper bore, the fluid in the master cylinder will rise. Make sure it doesn't overflow. If necessary, siphon off some of the fluid.

6   Follow the accompanying illustrations, beginning with **illustration 2.6a**, for the actual pad replacement procedure. Be sure to stay in order and read the caption under each illustration.

7   When reinstalling the caliper, be sure to tighten the mounting bolts to the specified torque. After the job has been completed,

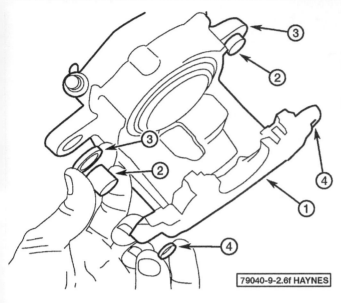

2.6f Push the mounting bolt sleeves out of the bores, remove the old bushings and install the new ones supplied with the brake pads

1   Caliper
2   Sleeves
3   Bushings
4   Bushings

2.6g Slide the caliper assembly over the rotor, install the mounting bolts, then insert a screwdriver between the rotor and outer brake pad, pry up, then strike the pad ears with a hammer to eliminate all play between the pad and caliper

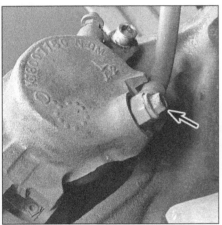

3.4 It's easier to remove the brake hose inlet fitting bolt (arrow) before removing the caliper mounting bolts

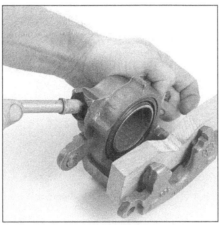

3.8 With the caliper padded to catch the piston, use compressed air to force the piston out of the bore - make sure your hands or fingers are not between the piston and caliper!

3.9 When prying the dust boot out of the caliper, be very careful not to scratch the bore surface

firmly depress the brake pedal a few times to bring the pads into contact with the disc.

3   Disc brake caliper - removal, overhaul and installation

Warning: *Dust created by the brake system may contain asbestos, which is harmful to your health. Never blow it out with compressed air and don't inhale any of it. An approved filtering mask should be worn when working on the brakes. Do not, under any circumstances, use petroleum-based solvents to clean brake parts. Use brake cleaner or denatured alcohol only!*

Note: *If an overhaul is indicated (usually because of fluid leakage) explore all options before beginning the job. New and factory rebuilt calipers are available on an exchange basis, which makes this job quite easy. If it's decided to rebuild the calipers, make sure a rebuild kit is available before proceeding. Always rebuild the calipers in pairs - never rebuild just one of them.*

Removal

Refer to illustration 3.4

1   Remove the cover from the brake fluid reservoir, siphon off two-thirds of the fluid into a container and discard it.

2   Loosen the wheel lug nuts, raise the front of the vehicle and support it securely on jackstands. Remove the front wheels.

3   Bottom the piston in the caliper bore **(see illustration 2.5)**.

4   **Note:** *Do not remove the brake hose from the caliper if you are only removing the caliper.* Remove the brake hose inlet fitting bolt and detach the hose **(see illustration)**. On models with a hydraulic fitting (tube nut), use a flare nut wrench. Have a rag handy to catch spilled fluid and wrap a plastic bag tightly around the end of the hose to prevent fluid loss and contamination.

5   Remove the two mounting bolts and detach the caliper from the vehicle (refer to Section 2 if necessary).

Overhaul

Refer to illustrations 3.8, 3.9, 3.10, 3.11, 3.15, 3.16, 3.17 and 3.18

6   Refer to Section 2 and remove the brake pads from the caliper.

7   Clean the exterior of the caliper with brake cleaner or denatured alcohol. Never use gasoline, kerosene or petroleum-based cleaning solvents. Place the caliper on a clean workbench.

8   Position a wooden block or several shop rags in the caliper as a cushion, then use compressed air to remove the piston from the caliper **(see illustration)**. Use only enough air pressure to ease the piston out of the bore. If the piston is blown out, even with the cushion in place, it may be damaged. **Warning:** *Never place your fingers in front of the piston in an attempt to catch or protect it when applying compressed air, as serious injury could occur.*

9   Carefully pry the dust boot out of the caliper bore **(see illustration)**.

10   Using a wood or plastic tool, remove the

3.10 Remove the piston seal with a wooden or plastic tool to avoid scratching the bore and seal groove

piston seal from the groove in the caliper bore **(see illustration)**. Metal tools may cause bore damage.

11   Remove the caliper bleeder screw, then remove and discard the sleeves and bushings from the caliper ears. Discard all rubber parts **(see illustration)**.

12   Clean the remaining parts with brake system cleaner or denatured alcohol, then blow them dry with compressed air.

13   Carefully examine the piston for nicks and burrs and loss of plating. If surface defects are present, the parts must be replaced.

14   Check the caliper bore in a similar way. Light polishing with crocus cloth is permissible to remove light corrosion and stains. Discard the mounting bolts if they're corroded or damaged.

15   When assembling, lubricate the piston bores and seal with clean brake fluid. Position the seal in the caliper bore groove **(see illustration)**.

16   Lubricate the piston with clean brake fluid, then install a new boot in the piston groove with the fold toward the open end of the piston **(see illustration)**.

17   Insert the piston squarely into the

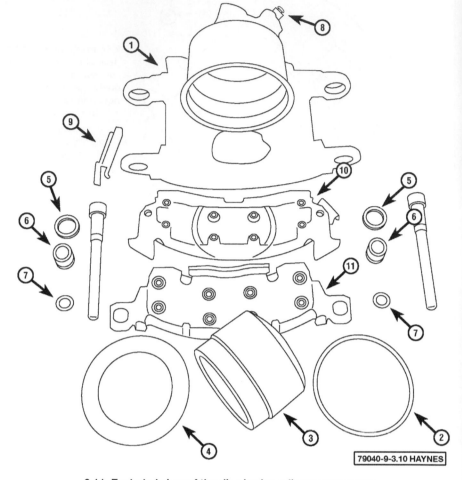

3.11 Exploded view of the disc brake caliper components

1	Caliper housing	5	Rubber bushing	9	Clip
2	Piston seal	6	Sleeve	10	Inner brake pad
3	Piston	7	Rubber bushing	11	Outer brake pad
4	Boot	8	Bleeder Screw		

caliper bore, then apply force to bottom it **(see illustration)**.

18   Position the dust boot in the caliper counterbore, then use a drift to drive it into

position **(see illustration)**. Make sure the boot is recessed evenly below the caliper face.

19   Install the bleeder screw.

3.15 Position the seal in the caliper bore groove, making sure it isn't twisted

3.16 Install the new dust boot in the piston groove (note that the folds are at the open end of the piston)

3.17 Install the piston squarely in the caliper bore then bottom it by pushing down evenly

3.18 Seat the boot in the counterbore (a seal driver is being used in this photo, but a drift punch will work if care is exercised)

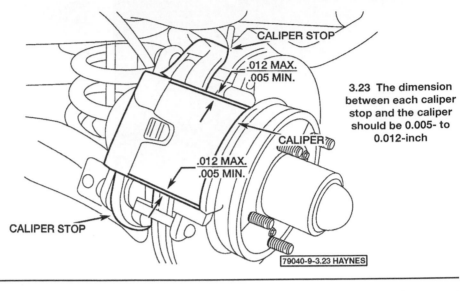

3.23 The dimension between each caliper stop and the caliper should be 0.005- to 0.012-inch

CALIPER STOP

.012 MAX.
.005 MIN.

CALIPER

.012 MAX.
.005 MIN.

CALIPER STOP

79040-9-3.23 HAYNES

20   Install new bushings in the mounting bolt holes and fill the area between the bushings with the silicone grease supplied in the rebuild kit. Push the sleeves into the mounting bolt holes.

Installation

Refer to illustration 3.23

21   Inspect the mounting bolts for excessive corrosion.
22   Place the caliper in position over the rotor and mounting bracket, install the bolts and tighten them to the specified torque.
23   Check to make sure the total clearance between the caliper and the bracket stops is between 0.005- and 0.012-inch **(see illustration)**.
24   Install the brake hose and inlet fitting bolt, using new copper washers, then tighten the bolt to the specified torque.
25   If the line was disconnected, be sure to bleed the brakes (Section 10).
26   Install the wheels and lower the vehicle.
27   After the job has been completed, firmly depress the brake pedal a few times to bring

the pads into contact with the disc.
28   Check brake operation before driving the vehicle in traffic.

4   Brake disc - inspection, removal and installation

Inspection

Refer to illustrations 4.3, 4.4a, 4.4b and 4.5

1   Loosen the wheel lug nuts, raise the vehicle and support it securely on jackstands. Remove the wheel.
2   Remove the brake caliper as outlined in Section 3. Remove caliper mounting bracket on models so equipped. It's not necessary to disconnect the brake hose. After removing the caliper bolts, suspend the caliper out of the way with a piece of wire. Don't let the caliper hang by the hose and don't stretch or twist the hose.
3   Visually check the disc surface for score marks and other damage. Light scratches

and shallow grooves are normal after use and may not always be detrimental to brake operation, but deep score marks - over 0.015-inch (0.38 mm) - require disc removal and refinishing by an automotive machine shop. Be sure to check both sides of the disc **(see illustration)**. If pulsating has been noticed during application of the brakes, suspect disc runout. Be sure to check the wheel bearings to make sure they're properly adjusted.
4   To check disc runout, place a dial indicator at a point about 1/2-inch from the outer edge of the disc **(see illustration)**. Set the indicator to zero and turn the disc. The indicator reading should not exceed the specified allowable runout limit. If it does, the disc should be refinished by an automotive machine shop. **Note:** *Professionals recommend resurfacing of brake discs regardless of the dial indicator reading (to produce a smooth, flat surface that will eliminate brake pedal pulsations and other undesirable symptoms related to questionable discs). At the very least, if you elect not to have the discs*

4.3 Check the rotor for deep grooves and score marks (be sure to inspect both sides of the rotor)

4.4a Check rotor runout with a dial indicator - if the reading exceeds the maximum allowable runout, the rotor will have to be resurfaced or replaced

4.4b Using a swirling motion, remove the glaze from the rotor with medium-grit emery cloth

4.5 A micrometer is used to measure rotor thickness - this can be done on the vehicle (as shown) or on the bench (the minimum thickness is cast into the inside of the rotor)

5.4a Drum brake components

1 Hold down pins
2 Backing plate
3 Parking brake lever (rear only)
4 Secondary shoe
5 Shoe guide
6 Parking brake strut (rear only)
7 Actuator lever
8 Actuator link
9 Return spring
10 Return spring
11 Hold down spring
12 Lever pivot
13 Lever return spring
14 Strut spring (rear only)
15 Adjusting screw assembly
16 Adjusting screw spring
17 Primary shoe

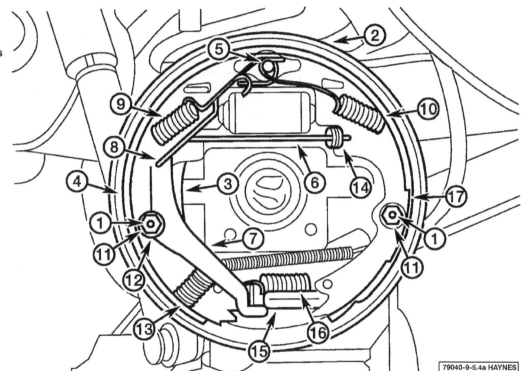

79040-9-5.4a HAYNES

resurfaced, deglaze them with medium-grit emery cloth (use a swirling motion to ensure a nondirectional finish) **(see illustration)**.

5   The disc must not be machined to a thickness less than the specified minimum refinish thickness. The minimum wear (or discard) thickness is cast into the inside of the disc. The disc thickness can be checked with a micrometer **(see illustration)**.

Removal

6   Refer to Chapter 1, Front wheel bearing check, repack and adjustment for the hub/disc removal procedure. Remove the two lug nuts that were installed to hold the disc in place and pull the disc from the hub.

Installation

7   Install the disc and hub assembly and adjust the wheel bearing (Chapter 1).
8   Install the caliper and brake pad assembly over the disc and position it on the steering knuckle (refer to Section 3 for the caliper installation procedure, if necessary). Tighten the caliper bolts to the specified torque.
9   Install the wheel, then lower the vehicle to the ground. Depress the brake pedal a few times to bring the brake pads into contact with the disc. Bleeding of the system will not be necessary unless the brake hose was disconnected from the caliper. Check the operation of the brakes carefully before placing the vehicle into normal service.

5   Drum brake shoes (front and rear) - replacement

Refer to illustrations 5.4a through 5.4v
Warning: *Drum brake shoes must be replaced on both wheels at the same time - never replace the shoes on only one wheel. Also, the dust created by the brake system may contain asbestos, which is harmful to your health. Never blow it out with compressed air and don't inhale any of it. An approved filtering mask should be worn when working on the brakes. Do not, under any circumstances, use petroleum-based solvents to clean brake parts. Use brake cleaner or denatured alcohol only!*

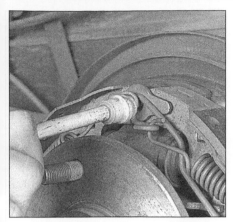

5.4b Remove the shoe return springs - the spring tool shown here is available at most auto parts stores and makes this job much easier and safer

5.4c Pull the bottom of the actuator lever toward the secondary brake shoe, compressing the lever return spring - the actuator link can now be removed

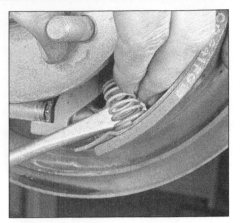

5.4d Pry the actuator lever spring out with a large screwdriver

1   Loosen the wheel lug nuts, raise the rear of the vehicle and support it securely on jack-stands. Block the front wheels to keep the vehicle from rolling.
2   Release the parking brake.
3   Remove the wheel. **Note:** *All four rear brake shoes must be replaced at the same time, but to avoid mixing up parts, work on only one brake assembly at a time.*
4   Follow the accompanying photos (**illustrations 5.4a through 5.4v**) for the inspection and replacement of the brake shoes. Be sure to stay in order and read the caption under each illustration. **Note:** *If the brake drum cannot be easily pulled off the axle and shoe assembly, make sure that the parking brake is completely released, then apply some penetrating oil at the hub-to-drum joint. Allow the oil to soak in and try to pull the drum off. If the drum still cannot be pulled off, the brake shoes will have to be retracted. This is accomplished by first removing the plug from the backing plate. With the plug removed, pull the lever off the adjusting star wheel with one small screwdriver while turning the adjusting wheel with another small screwdriver, moving the shoes away from the drum. The drum should now come off.*

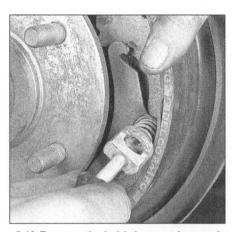

5.4e Slide the parking brake strut out from between the axle flange and primary shoe (rear drum brakes only)

5.4f Remove the hold-down springs and pins - the hold-down spring tool shown here is available at most auto parts stores

Caution: *Whenever the brake shoes are replaced, the retractor and hold-down springs should also be replaced. Due to the continuous heating/cooling cycle that the springs are subjected to, they lose their ten-*

sion over a period of time and may allow the shoes to drag on the drum and wear at a much faster rate than normal. When replacing the rear brake shoes, use only high quality nationally recognized brand-name parts.

5.4g Remove the actuator lever and pivot - be careful not to let the pivot fall out of the lever. On some models, the lever is two pieces, attached with a spring - you don't normally need to disassemble the lever

5.4h Spread the top of the shoes apart and slide the assembly around the axle

5.4i Unhook the parking brake lever from the secondary shoe (rear drum brake only)

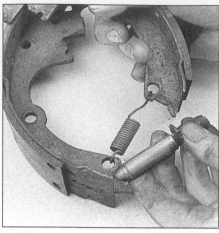

5.4j Spread the bottom of the shoes apart and remove the adjusting screw assembly

5.4k Clean the adjusting screw with solvent, dry it off and lubricate the threads and end with multi-purpose grease, then reinstall the adjusting screw assembly between the new brake shoes

5.4l Lubricate the shoe contact points on the backing plate with high temperature brake grease

5.4m Insert the parking brake lever into the opening in the secondary brake shoe (rear drum brake only)

5.4n Spread the shoes apart and slide them into position on the backing plate

5.4o Install the hold-down pin and spring through the backing plate and primary shoe

5.4p Insert the lever pivot into the actuator lever, place the lever over the secondary shoe hold-down pin and install the hold-down spring

5.4q Guide the parking brake strut behind the axle flange and engage the rear end of it in the slot on the parking brake lever - spread the shoes enough to allow the other end to seat against the primary shoe (rear drum brake only)

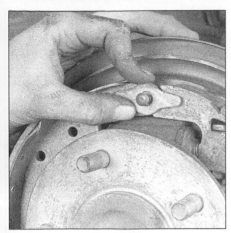

5.4r Place the shoe guide over the anchor pin

5.4s Hook the lower end of the actuator link to the actuator lever, then loop the top end over the anchor pin

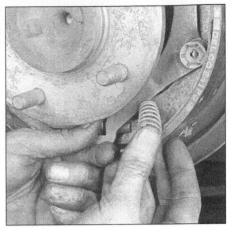

5.4t Install the lever return spring over the tab on the actuator lever, then push the spring up onto the brake shoe

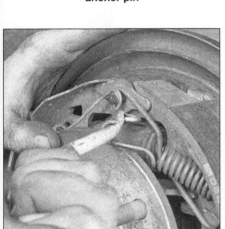

5.4u Install the primary and secondary shoe return springs

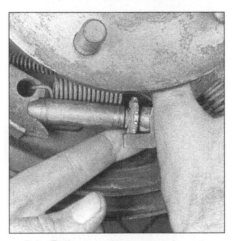

5.4v Pull out on the actuator lever to disengage it from the adjusting screw wheel, turn the wheel to adjust the shoes in or out as necessary - the brake drum should slide over the shoes and turn with a very slight amount of drag

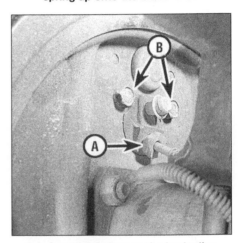

6.4 Completely loosen the brake line fitting (A) then remove the two wheel cylinder mounting bolts (B)

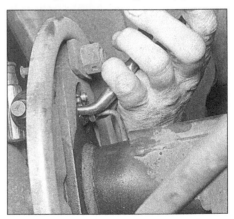

6.5a Using a pair of needle-nose pliers, squeeze the retaining clip together to release it from the wheel cylinder

5   Before reinstalling the drum it should be checked for cracks, score marks, deep scratches and hard spots, which will appear as small discolored areas. If the hard spots cannot be removed with fine emery cloth or if any of the other conditions listed above exist, the drum must be taken to an automotive machine shop to have it resurfaced. **Note:** *Professionals recommend resurfacing the drums whenever a brake job is done. Resurfacing will eliminate the possibility of out-of-round drums. If the drums are worn so much that they can't be resurfaced without exceeding the maximum allowable diameter (stamped into the drum), then new ones will be required. At the very least, if you elect not to have the drums resurfaced, remove the glazing from the surface with medium-grit emery cloth using a swirling motion.*
6   Install the brake drum on the axle flange.
7   Mount the wheel, install the lug nuts, then lower the vehicle.
8   Make a number of forward and reverse stops to adjust the brakes until satisfactory pedal action is obtained.
9   Check brake operation before driving the vehicle in traffic.

6   Wheel cylinder - removal, overhaul and installation

Note: *If an overhaul is indicated (usually because of fluid leakage or sticky operation) explore all options before beginning the job. New wheel cylinders are available, which makes this job quite easy. If it's decided to rebuild the wheel cylinder, make sure that a rebuild kit is available before proceeding. Never overhaul only one wheel cylinder - always rebuild both of them at the same time.*

Removal

Refer to illustrations 6.4, 6.5a and 6.5b
1   Raise the rear of the vehicle and support it securely on jackstands. Block the front wheels to keep the vehicle from rolling.
2   Remove the brake shoe assembly (Section 5).
3   Remove all dirt and foreign material from around the wheel cylinder.

4   Unscrew the brake line fitting **(see illustration)**. Don't pull the brake line away from the wheel cylinder.
5   Remove the wheel cylinder mounting bolts. On models using retaining clips, use curved needle-nose pliers, to remove the

6.5b Insert two awls in between the wheel cylinder and the retaining clip tangs and pull back to release the wheel cylinder (cylinder removed for clarity)

wheel cylinder retaining clip **(see illustration)**, from the rear of the wheel cylinder. If curved needle-nose pliers are not available, insert two awls into the access slots between the cylinder pilot and the retainer locking tabs and bend both tabs away at the same time **(see illustration)**.

6   Detach the wheel cylinder from the brake backing plate and place it on a clean workbench. Immediately plug the brake line to prevent fluid loss and contamination. **Note:** *If the brake shoe linings are contaminated with brake fluid, install new brake shoes.*

Overhaul

Refer to illustration 6.7

7   Remove the bleeder screw, cups, pistons, boots and spring assembly from the wheel cylinder body **(see illustration)**.

8   Clean the wheel cylinder with brake fluid, denatured alcohol or brake system cleaner. **Warning:** *Do not, under any circumstances, use petroleum based solvents to clean brake parts!*

9   Use compressed air to remove excess fluid from the wheel cylinder and to blow out the passages.

10   Check the cylinder bore for corrosion and score marks. Crocus cloth can be used to remove light corrosion and stains, but the cylinder must be replaced with a new one if the defects cannot be removed easily, or if the bore is scored.

11   Lubricate the new cups with brake fluid.

12   Assemble the wheel cylinder components. Make sure the cup lips face in **(see illustration)**.

Installation

13   Place the wheel cylinder in position and install the bolts.

14   Connect the brake line and tighten the fitting. Install the brake shoe assembly.

15   Bleed the brakes (Section 10).

16   Check brake operation before driving the vehicle in traffic.

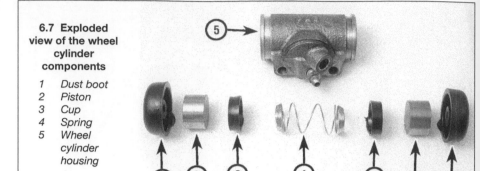

6.7 Exploded view of the wheel cylinder components

1   *Dust boot*
2   *Piston*
3   *Cup*
4   *Spring*
5   *Wheel cylinder housing*

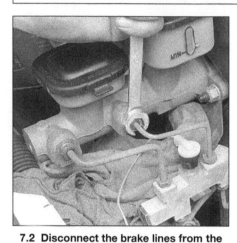

7.2 Disconnect the brake lines from the master cylinder - a flare-nut wrench should be used

7   Master cylinder - removal, overhaul and installation

Removal

Refer to illustrations 7.2 and 7.6

Warning: *Failure to fully depressurize the powermaster unit before performing service operations could result in personal injury and damage to painted surfaces. The use of rubber hoses other than those furnished specifically for the powermaster may lead to functional problems requiring major overhaul. To depressurize the powermaster unit, make sure the ignition switch is Off, and apply and release the brake pedal a minimum of ten times, using approximately 50 pounds of force on the pedal.*

Note: *Before deciding to overhaul the master cylinder, investigate the availability and cost of a new or factory rebuilt unit and also the availability of a rebuild kit.*

1   Place rags under the brake line fittings and prepare caps or plastic bags to cover the ends of the lines once they are disconnected. **Caution:** *Brake fluid will damage paint. Cover all body parts and be careful not to spill fluid during this procedure.*

2   Loosen the tube nuts at the ends of the brake lines where they enter the master cylinder. To prevent rounding off the flats on these

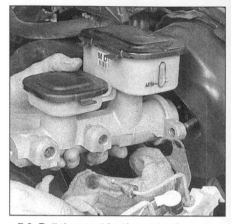

7.6 Pull the combination valve forward, being careful not to bend or kink the lines, then slide the master cylinder off the mounting studs

nuts, a flare-nut wrench, which wraps around the nut, should be used **(see illustration)**.

3   Pull the brake lines away from the master cylinder slightly and plug the ends to prevent contamination.

4   On manual and powermaster brakes, disconnect the pushrod at the brake pedal inside the car.

5   On the powermaster unit, disconnect the electrical connectors at the front and rear of the unit.

6   Remove the two master cylinder mounting nuts. If there is a bracket retaining the combination valve, move it forward slightly, taking care not to bend the hydraulic lines running to the combination valve, and remove the master cylinder from the vehicle **(see illustration)**.

7   Remove the reservoir covers and reservoir diaphragms, then discard any fluid remaining in the reservoir.

Overhaul

Late model master cylinder (aluminum body)

Refer to illustrations 7.9a, 7.9b, 7.10, 7.11, 7.16, 7.17a, 7.17b, 7.17c, 7.17d and 7.18

8   Mount the master cylinder in a vise. Be sure to line the vise jaws with blocks of wood to prevent damage to the cylinder body.

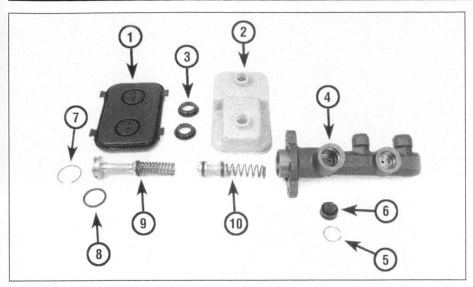

7.9a Exploded view of a later model master cylinder assembly

1	Reservoir cover and diaphragm	6	Quick take-up valve
2	Reservoir	7	Lock ring
3	Reservoir grommets	8	Primary piston O-ring
4	Master cylinder body	9	Primary piston assembly
5	Quick take-up valve circlip	10	Secondary piston assembly

7.9b Push the primary piston in and remove the lock ring

7.10 Pull the primary piston and spring assembly out of the bore

7.11 To remove the secondary piston, tap the cylinder against a block of wood

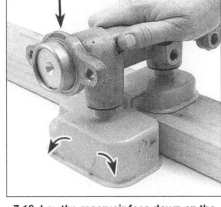

7.16 Lay the reservoir face down on the bench, with the secondary reservoir propped up on a block of wood - push the master cylinder straight down over the reservoir tubes using a rocking motion

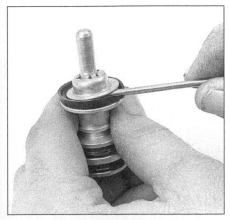

7.17a Pry the secondary piston spring retainer off with a small screwdriver, then remove the seal

7.17b Remove the secondary seals from the piston (some only have one seal)

9   Remove the primary piston lock ring by depressing the piston and prying the ring out with a screwdriver **(see illustrations)**.

10   Remove the primary piston assembly from the cylinder bore **(see illustration)**.

11   Remove the secondary piston assembly from the cylinder bore. It may be necessary to remove the master cylinder from the vise and invert it, carefully tapping it against a block of wood to expel the piston **(see illustration)**.

12   Pry the reservoir from the cylinder body with a screwdriver. Remove the grommets.

13   Do not attempt to remove the quick take-up valve from the master cylinder body - it's not serviceable.

14   Inspect the cylinder bore for corrosion and damage. If any corrosion or damage is found, replace the master cylinder body with a new one, as abrasives cannot be used on the bore.

15   Lubricate the new reservoir grommets with clean brake fluid and press them into the master cylinder body. Make sure they're properly seated.

16   Lay the reservoir on a hard surface and press the master cylinder body onto the reservoir, using a rocking motion **(see illustration)**.

17   Remove the old seals from the secondary piston assembly and install the new secondary seals with the lips facing away

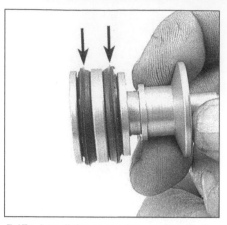

7.17c Install the secondary seals with the lips facing away from each other (on the single seal design, the seal lip should face away from the center of the piston)

7.17d Install a new primary seal on the secondary piston with the seal lip facing in the direction shown

7.18 Install a new spring retainer over the end of the secondary piston and push it into place with a socket

from each other **(see illustrations)**. The lip on the primary seal must face in **(see illustration)**.

18 Attach the spring retainer to the secondary piston assembly **(see illustration)**.

19 Lubricate the cylinder bore with clean brake fluid and install the spring and secondary piston assembly.

20 Install the primary piston assembly in the cylinder bore, depress it and install the lock ring.

21 Inspect the reservoir cover and diaphragm for cracks and deformation. Replace any damaged parts with new ones and attach the diaphragm to the cover.

22 **Note:** *Whenever the master cylinder is removed, the complete hydraulic system must be bled. The time required to bleed the system can be reduced if the master cylinder is filled with fluid and bench bled (refer to Steps 40 through 44) before the master cylinder is installed on the vehicle.*

Early model master cylinder (cast iron body)

Refer to illustrations 7.25, 7.26a, 7.26b, 7.29a, 7.29b, 7.30, 7.32a, 7.32b and 7.33

23 Drain all fluid from the master cylinder and place the unit in a vise. Use wood blocks to cushion the jaws of the vise.

24 On manual brakes remove the pushrod retaining ring.

25 Remove the secondary stop bolt from the bottom of the front fluid reservoir (Delco Moraine) or from the base of the master cylinder body (Bendix) **(see illustration)**.

26 Remove the retaining ring from the groove and take out the primary piston assembly **(see illustration)**. Following the primary piston out of the bore will be the secondary piston, spring and retainer. A piece of bent stiff wire can be used to draw these assemblies out of the cylinder bore **(see illustration)**.

27 Examine the inside surface of the master cylinder and the secondary piston. If there is evidence of scoring or "bright' wear areas,

7.25 Removing secondary stop bolt

7.26b Drawing out piston assembly with wire

the entire master cylinder should be replaced with a new one.

28 If the components are in good condition, wash them with brake system cleaner. Discard all the rubber components and the primary piston. Purchase a rebuild kit which will contain all the necessary parts for the overhaul.

7.26a Removing retaining ring

7.29a Removing line seats

29 Inspect the line seats which are located in the master cylinder body where the lines connect. If they appear damaged they should be replaced with new ones, which come in the overhaul kit. They are forced out of the body by threading a screw into the tube and then prying outwards **(see illustration)**. The new ones are forced into place using a spare brake line nut **(see illustration)**. All parts nec-

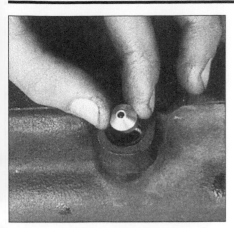

7.29b Installing new line seat (note the cone-shaped end is installed facing up)

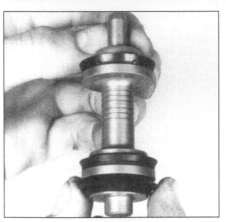

7.30 Seals correctly installed on piston

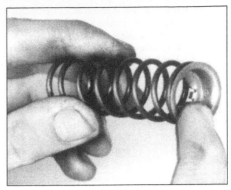

7.32a Assembling retainer and spring

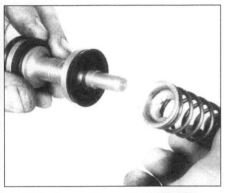

7.32b Assembling spring and piston

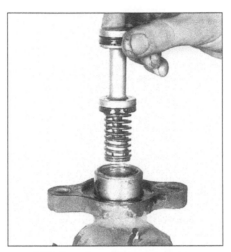

7.33 Pushing piston into master cylinder bore

essary for this should be included in the rebuild kit.

30   Place the new secondary seals in the grooves of the secondary piston **(see illustration)**.

31   Assemble the primary seals and seal protector over the end of the secondary piston.

32   Lubricate the cylinder bore and secondary piston with clean brake fluid. Insert the spring retainer into the spring then place the retainer and spring over the end of the secondary piston **(see illustrations)**. The retainer should locate inside the primary seal lips.

33   With the master cylinder vertical, push the secondary piston into the bore to seat its spring **(see illustration)**.

34   Coat the seals of the primary piston with brake fluid and fit it into the cylinder bore. Hold it down while the retaining ring is installed in the cylinder groove.

35   Continue to hold the piston down while the stop screw is installed.

36   Install the reservoir diaphragm into the reservoir cover plate, making sure it is fully collapsed inside the recess lid.

37   **Note:** *Whenever the master cylinder is removed, the complete hydraulic system must be bled. The time required to bleed the system can be reduced if the master cylinder is filled with fluid and bench bled (refer to Steps 40 through 44) before the master cylinder is installed on the vehicle.*

Powermaster brake system

38   Because of the special tools, equipment and expertise required to disassemble, overhaul and reassemble this assembly, it is recommended that it be left to a dealer service department or a repair shop. This unit can be removed and installed by a competent do-it-yourselfer.

39   **Note:** *Whenever the master cylinder is removed, the complete hydraulic system must be bled. The time required to bleed the system can be reduced if the master cylinder is filled with fluid and bench bled (refer to Steps 40 through 44) before the master cylinder is installed on the vehicle.*

Bench bleeding master cylinder

Note: *On powermaster systems, only the master cylinder portion can be bench bled. See Steps 45 through 50 for powermaster bleed and fill.*

40   Insert threaded plugs of the correct size into the cylinder outlet holes and fill the reservoirs with brake fluid. The master cylinder should be supported in such a manner that brake fluid will not spill during the bench bleeding procedure.

41   Loosen one plug at a time and push the piston assembly into the bore to force air from the master cylinder. To prevent air from being drawn back into the cylinder, the appropriate plug must be replaced before allowing the piston to return to its original position.

42   Stroke the piston three or four times for each outlet to ensure that all air has been expelled.

43   Since high pressure is not involved in the bench bleeding procedure, an alternative to the removal and replacement of the plugs with each stroke of the piston assembly is available. Before pushing in on the piston assembly, remove one of the plugs completely. Before releasing the piston, however, instead of replacing the plug, simply put your finger tightly over the hole to keep air from being drawn back into the master cylinder. Wait several seconds for the brake fluid to be drawn from the reservoir to the piston bore, then repeat the procedure. When you push down on the piston it will force your finger off the hole, allowing the air inside to be expelled. When only brake fluid is being ejected from the hole, replace the plug and go on to the other port.

44   Refill the master cylinder reservoirs and install the diaphragm and cover assembly. **Note:** *The reservoirs should only be filled to the top of the reservoir divider to prevent overflowing when the cover is installed.*

Powermaster bleed and fill

Note: *Bench bleed the master cylinder portion of powermaster before installing the unit on the vehicle.*

45   Fill both sides of the reservoir to the Full marks on the inside of the reservoir. Use only clean, new brake fluid meeting DOT specifications shown on the reservoir cover.

46   Turn the ignition On. With the pump running, the brake fluid level in the booster side of the reservoir should decrease as brake fluid is moved to the accumulator. If the booster side of the reservoir begins to run dry, add brake fluid to just cover the reservoir pump or until the pump stops. **Note:** *The pump must shut off within 20 seconds. Turn ignition Off after 20 seconds have elapsed. Check for leaks or flow back into the reservoir from the booster return port.*

47   Install the reservoir cover assembly to the reservoir.

48   Check that the ignition is Off and apply and release the brake pedal 10 times. Remove the reservoir cover and adjust the booster fluid level to the Full mark.
49   Turn the ignition On. The pump will run and refill the accumulator. Make sure that the pump does not run longer than 20 seconds and that the fluid level remains above the pump sump port in the reservoir.
50   Install the reservoir cover. With the ignition On, apply and release the brake pedal on and off 10 to 15 times, to cycle the pump and remove air from the booster section. Do not allow the pump to run more than 20 seconds for each cycle. Recheck the high and low reservoir fluid levels per Steps 45 and 46. If fluid levels do not stabilize at the high and low levels, if the pump runs more than 20 seconds, or if the pump cycles without brake applications, have your car towed to the nearest dealer or repair shop. Do not drive with powermaster brakes not working properly.

Installation

51   Carefully install the master cylinder by reversing the removal steps, then bleed the brakes (refer to Section 10).

---

8   Combination valve - check and replacement

Check

1   Disconnect the wire from the pressure differential switch. **Note:** When unplugging the connector, squeeze the side lock releases, moving the inside tabs away from the switch, then pull up. Pliers may be used as an aid if necessary.
2   Using a jumper wire, connect the switch wire to a good ground, such as the engine block.
3   Turn the ignition key to the On position. The warning light in the instrument panel should light.
4   If the warning light does not light, either the bulb is burned out or the electrical circuit is defective. Replace the bulb (refer to Chapter 12) or repair the electrical circuit as necessary.
5   When the warning light functions correctly, turn the ignition switch off, disconnect the jumper wire and reconnect the wire to the switch terminal.
6   Make sure the master cylinder reservoirs are full, then attach a bleeder hose to one of the rear wheel bleeder valves and immerse the other end of the hose in a container partially filled with clean brake fluid.
7   Turn the ignition switch On.
8   Open the bleeder valve while a helper applies moderate pressure to the brake pedal. The brake warning light on the instrument panel should light.
9   Close the bleeder valve before the helper releases the brake pedal.
10   Reapply the brake pedal with moderate

to heavy pressure. The brake warning light should go out.
11   Attach the bleeder hose to one of the front brake bleeder valves and repeat Steps 8 through 10. The warning light should react in the same manner as in Steps 8 and 10.
12   Turn the ignition switch Off.
13   If the warning light did not come on in Steps 8 and 11, but does light when a jumper is connected to ground, the warning light switch portion of the combination valve is defective and the combination valve must be replaced with a new one since the components of the combination valve are not individually serviceable.

Replacement

14   Place a container under the combination valve and protect all painted surfaces with newspapers or rags.
15   Disconnect the hydraulic lines at the combination valve, then plug the lines to prevent further loss of fluid and to protect the lines from contamination.
16   Disconnect the electrical connector from the pressure differential switch.
17   Remove the bolt holding the valve to the mounting bracket and remove the valve from the vehicle.
18   Installation is the reverse of the removal procedure.
19   Bleed the entire brake system.

---

9   Brake hoses and lines - inspection and replacement

Inspection

1   About every six months, with the vehicle raised and supported securely on jackstands, the rubber hoses which connect the steel brake lines with the front and rear brake assemblies should be inspected for cracks, chafing of the outer cover, leaks, blisters and other damage. These are important and vulnerable parts of the brake system and inspection should be complete. A light and mirror will be helpful for a thorough check. If a hose exhibits any of the above conditions, replace it with a new one.

Replacement

Front brake hose

Refer to illustration 9.2

2   Using a back-up wrench, disconnect the brake line from the hose fitting, being careful not to bend the frame bracket or brake line (see illustration).
3   Use a pair of pliers to remove the U-clip from the female fitting at the bracket, then detach the hose from the bracket.
4   Unscrew the brake hose from the caliper. At the caliper end of the hose, remove the bolt from the fitting block, then remove the hose and the copper washers on either side of the fitting block.
5   To install the hose, first thread it into the caliper, tightening it securely. When installing

the hose, always use new copper washers on either side of the fitting block and lubricate all bolt threads with clean brake fluid before installation.
6   With the fitting engaged with the caliper locating ledge, attach the hose to the caliper, tightening the fitting bolt to the specified torque.
7   Without twisting the hose, install the female fitting in the hose bracket. It will fit the bracket in only one position.
8   Install the U-clip retaining the female fitting to the frame bracket.
9   Using a back-up wrench, attach the brake line to the hose fitting.
10   When the brake hose installation is complete, there should be no kinks in the hose. Make sure the hose doesn't contact any part of the suspension. Check this by turning the wheels to the extreme left and right positions. If the hose makes contact, remove it and correct the installation as necessary. Bleed the system (Section 10).

Rear brake hose

11   Using a back-up wrench, disconnect the hose at the frame bracket, being careful not to bend the bracket or steel lines.
12   Remove the U-clip with a pair of pliers and separate the female fitting from the bracket.
13   Disconnect the two hydraulic lines at the junction block, then unbolt and remove the hose.
14   Bolt the junction block to the axle housing and connect the lines, tightening them securely. Without twisting the hose, install the female end of the hose in the frame bracket.
15   Install the U-clips retaining the female end to the bracket.
16   Using a back-up wrench, attach the steel line fittings to the female fittings. Again, be careful not to bend the bracket or steel line.
17   Make sure the hose installation did not loosen the frame bracket. Tighten the bracket if necessary.
18   Fill the master cylinder reservoir and bleed the system (refer to Section 10).

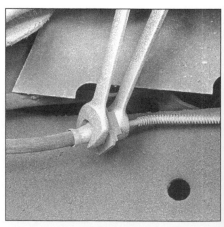

9.2 Place a wrench on the hose fitting to prevent it from turning and disconnect the line with a flare-nut wrench

10.8 When bleeding the brakes, a hose is connected to the bleeder valve and then submerged in brake fluid - air will be seen as bubbles in the container and the hose

11.3 To adjust the parking brake cable, turn the adjusting nut on the equalizer while preventing the cable from turning with a pair of locking pliers clamped to the end of the adjuster rod (later model cable)

Metal brake lines

19   When replacing brake lines be sure to use the correct parts. Don't use copper tubing for any brake system components. Purchase steel brake lines from a dealer or auto parts store.

20   Prefabricated brake line, with the tube ends already flared and fittings installed, is available at auto parts stores and dealers. These lines are also bent to the proper shapes.

21   If prefabricated lines are not available, obtain the recommended steel tubing and fittings to match the line to be replaced. Determine the correct length by measuring the old brake line (a piece of string can usually be used for this) and cut the new tubing to length, allowing about 1/4 -inch extra for flaring the ends.

22   Install the fitting over the cut tubing and flare the ends of the line with a flaring tool. A double-flare is the only acceptable type for automotive brake system applications.

23   If necessary, carefully bend the line to the proper shape. A tube bender is recommended for this **Warning:** *Do not crimp or damage the line.*

24   When installing the new line make sure it's securely supported in the brackets and has plenty of clearance between moving or hot components.

25   After installation, check the master cylinder fluid level and add fluid as necessary. Bleed the brake system as outlined in the next Section and test the brakes carefully before driving the vehicle in traffic.

10   Brake system bleeding

Refer to illustration 10.8

Warning: *Wear eye protection when bleeding the brake system. If the fluid comes in contact with your eyes, immediately rinse them with water and seek medical attention.*

Note: *Bleeding the hydraulic system is necessary to remove any air that manages to find its way into the system when it's been opened during removal and installation of a hose, line, caliper or master cylinder.*

1   It will probably be necessary to bleed the system at all four wheels if air has entered the system due to low fluid level, or if the brake lines have been disconnected at the master cylinder.

2   If a brake line was disconnected only at a wheel, then only that caliper or wheel cylinder must be bled.

3   If a brake line is disconnected at a fitting located between the master cylinder and any of the calipers or wheel cylinders, that part of the system served by the disconnected line must be bled.

4   Remove any residual vacuum from the brake power booster by applying the brake several times with the engine off.

5   Remove the master cylinder reservoir cover and fill the reservoir with brake fluid. Reinstall the cover. **Note:** *Check the fluid level often during the bleeding operation and add fluid as necessary to prevent the fluid level from falling low enough to allow air bubbles into the master cylinder.*

6   Have an assistant on hand, as well as a supply of new brake fluid, a clear container partially filled with clean brake fluid, a length of 3/16-inch plastic, rubber or vinyl tubing to fit over the bleeder valve and a wrench to open and close the bleeder valve.

7   Beginning at the right rear wheel, loosen the bleeder valve slightly, then tighten it to a point where it is snug but can still be loosened quickly and easily.

8   Place one end of the tubing over the bleeder valve and submerge the other end in brake fluid in the container **(see illustration)**.

9   Have the assistant pump the brakes slowly a few times to build pressure in the system, then hold the pedal firmly depressed.

10   While the pedal is held depressed, open the bleeder valve just enough to allow a flow of fluid to leave the valve. Watch for air bub-

bles to exit the submerged end of the tube. When the fluid flow slows after a couple of seconds, close the valve and have your assistant release the pedal.

11   Repeat Steps 9 and 10 until no more air is seen leaving the tube, then tighten the bleeder valve and proceed to the left rear wheel, the right front wheel and the left front wheel, in that order, and perform the same procedure. Be sure to check the fluid in the master cylinder reservoir frequently.

12   Never use old brake fluid. It contains moisture which will deteriorate the brake system components.

13   Refill the master cylinder with fluid at the end of the operation.

14   Check the operation of the brakes. The pedal should feel solid when depressed, with no sponginess. If necessary, repeat the entire process. **Warning:** *Do not operate the vehicle if you are in doubt about the effectiveness of the brake system.*

11   Parking brake - adjustment

Refer to illustration 11.3

1   The adjustment of the parking brake cable may be necessary whenever the rear brake cables have been disconnected or the parking brake cables have stretched due to age and stress.

2   Depress the parking brake pedal exactly three ratchet clicks and then raise the car for access underneath, supporting it securely on jackstands.

3   Tighten the adjusting nut until the left rear tire can just barely be turned in a rearward motion **(see illustration)**. The tire should be completely locked from moving in a forward rotation.

4   Carefully release the parking brake pedal and check that the tire is able to rotate freely in either direction. It is important that there is no drag on the rear brakes with the pedal released.

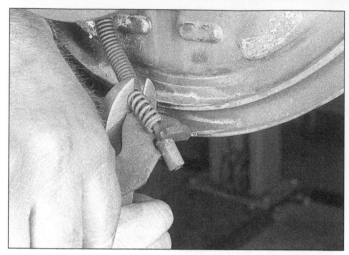

12.4 To disconnect the cable end from the parking brake lever, pull back on the return spring and maneuver the cable out of the slot in the lever

12.5 Depress the retention tangs (arrows) to free the cable and housing from the backing plate

12   Parking brake cables - replacement

Rear cables

Refer to illustrations 12.4 and 12.5

1   Loosen the wheel lug nuts, raise the rear of the vehicle and support it securely on jackstands. Remove the wheel(s).
2   Loosen the equalizer nut to slacken the cables, then disconnect the cable to be replaced from the equalizer.
3   Remove the brake drum from the axle flange. Refer to Section 5 if any difficulty is encountered.
4   Remove the brake shoe assembly far enough to disconnect the cable end from the parking brake lever **(see illustration)**.
5   Depress the tangs on the cable housing retainer and push the housing and cable through the backing plate **(see illustration)**.
6   To install the cable, reverse the removal procedure and adjust the cable as described in the preceding Section.

Front cable

Refer to illustration 12.8

7   Raise the vehicle and support it securely on jackstands.
8   Loosen the equalizer assembly and remove the cable adjusting nut to provide slack in the cable **(see illustration)**.
9   Disconnect the front cable from the cable joiner bracket.
10   Disconnect the cable from the pedal assembly.
11   Free the cable from the routing clips and push the cable and grommet through the firewall.
12   To install the cable, reverse the removal procedure and adjust the cable as described in the preceding Section.

13   Parking brake pedal - removal and installation

1   Disconnect the battery ground cable and the parking brake warning switch wire.
2   Remove the clip and ball from the clevis (if necessary, the equalizer nut can be loosened) **(see illustration 12.8)**. Depress the retention tangs and remove the parking brake cable from the pedal assembly.
3   Remove the pedal rear mounting bolt and the nuts from the mounting studs at the front of the dash panel (under the hood).
4   Remove the pedal assembly.
5   Installation is the reverse of the removal procedure.

14   Brake pedal - removal and installation

1970 through 1973

1   On manual transmission models, dis-

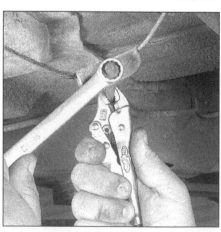

12.8 Use a pair of pliers to hold the front cable end and loosen the equalizer adjusting nut

connect the clutch pedal return spring and disconnect the clutch pushrod from the pedal arm.
2   If a power booster is not fitted, disconnect the brake pedal return spring and then disconnect the master cylinder pushrod from the pedal arm. Remove the retainer from the right-hand side of the pedal pivot shaft. Slide the clutch pedal to the left and remove it from the support braces. Remove the brake pedal and nylon bushings.
3   If a power booster is fitted, loosen the booster mounting nuts enough to allow the pushrod to slide off the pedal pin. Remove the clip retainer from the pedal pin and remove the pushrod. Remove the pedal pivot shaft nut and slide the pivot bolt out of the mounting bracket. The brake pedal and bushings can now be removed. Replace any worn bushings and lubricate upon reassembly, which is the reversal of the removal and dismantling sequence.
4   On vehicles without a power booster, adjust the brake pedal by releasing the pushrod locknut and turning the pushrod in or out to give between 1/16- and 1/4-inch of free movement. Tighten the locknut securely.
5   Adjust the brake light switch as described in Section 16.
6   Check the clutch pedal free play as described in Chapter 1.

1974 and later

7   Disconnect the clutch pedal return spring (if equipped).
8   Remove the clip retainer from the pushrod pin.
9   Unscrew the nut from the end of the pedal pivot shaft and withdraw the shaft far enough to be able to remove the pedal spacer and bushings.
10   Replace and/or lubricate the bushings as necessary.
11   Installation is the reverse of removal.

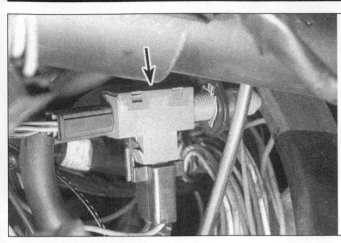

16.1 Typical brake light switch with cruise control

15 Power brake booster - inspection, removal and installation

1   The power brake booster unit requires no special maintenance apart from periodic inspection of the vacuum hose and the case.
2   Dismantling of the power unit requires special tools and is not ordinarily done by the home mechanic. If a problem develops, install a new or factory rebuilt unit.
3   Remove the nuts attaching the master cylinder to the booster and carefully pull the master cylinder forward until it clears the mounting studs. Don't bend or kink the brake lines.
4   Disconnect the vacuum hose where it attaches to the power brake booster.
5   From the passenger compartment, disconnect the power brake pushrod from the top of the brake pedal.

6   Also from this location, remove the nuts attaching the booster to the firewall.
7   Carefully lift the booster unit away from the firewall and out of the engine compartment.
8   To install, place the booster in position and tighten the nuts to the Specifications. Connect the brake pedal.
9   Install the master cylinder and vacuum hose.
10   Carefully test the operation of the brakes before placing the vehicle in normal service.

16 Brake light switch - check, adjustment and replacement

Refer to illustration 16.1

1   The brake light switch **(see illustration)** is located on a flange or bracket protruding

from the brake pedal support.
2   With the brake pedal in the fully released position, the plunger of the switch should be fully pressed in. When the pedal is depressed, the plunger releases and sends electrical current to the brake lights at the rear of the car.
3   To check the brake light switch, simply note whether the brake lights come on when the pedal is depressed and go off when the pedal is released. If they don't, adjust the switch. To do this, depress the brake pedal about 1/2-inch, push the switch forward until the body (not the plunger) of the switch contacts the brake pedal, then release the pedal. Don't pull up on the pedal; it will push the switch into the proper position by itself.
4   If the lights still don't come on, either the switch is not getting voltage, the switch itself is defective, or the circuit between the switch and the lights is defective. There is always the remote possibility that all of the brake light bulbs are burned out, but this is not very likely.
5   Use a voltmeter or test light to verify that there's voltage present at one side of the switch connector. If no voltage is present, troubleshoot the circuit from the switch to the fuse box. If there is voltage present, check for voltage on the other terminal when the brake pedal is depressed. If no voltage is present, replace the switch. If there is voltage present, troubleshoot the circuit from the switch to the brake lights (see the *Wiring diagrams* at the end of Chapter 12).
6   To replace a faulty switch, simply unplug the electrical connectors (one for the switch circuit, one for cruise control) and pull the switch out of its bracket. Installation is the reverse of removal. Be sure to adjust the switch as described in Step 2.

Notes

Chapter 10
Steering and suspension systems

Contents

Specifications

Torque specifications

Front suspension   **Ft-lbs** (unless otherwise indicated)

Upper control arm pivot shaft-to-frame nuts	
1970	50
1971 through 1975	55
1976 and 1977	75
1978 through 1984	45
1985 and later	48
Upper control arm pivot shaft-to-bushing bolts	
1970 through 1973	45
Upper control arm pivot shaft-to-bushing nuts	
1974	68
1975 through 1977	75
1978 and later	85
Upper balljoint-to-control arm bolts	
Replacement type only	10
Upper balljoint stud-to-steering knuckle nut	
1970 through 1973	50
1974 through 1979	60
1980 through 1984	65
1985 through 1988	52
Lower balljoint stud-to-steering knuckle nut	
1970	80
1971	90
1972 through 1973	80
1974 through 1979	83
1980 through 1984	90
1985 and later	70

Torque specifications   **Ft-lbs** (unless otherwise indicated)

Front suspension (continued)

Lower control arm bushing-to-frame bolt and nut
    1970 through 1971 80
    1972 through 1973 85
    1974 through 1975 95
    1976 through 1977 100
    1978 and later 65
Front shock absorber upper stud nut .... 100 in-lbs
Front shock absorber-to-lower control arm bolts 20
Front stabilizer bar link nut .... 156 in-lbs
Front stabilizer bar bracket-to-frame bolts 24

Rear suspension

Control arm bolt and nuts
    Upper 80
    Lower 80
Rear shock absorber-to-lower mount 65
Rear shock absorber-to-upper mount .... 144 in-lbs
Rear stabilizer bar-to-lower control arm nut
    1970 through 1978 55
    1979 and later 37

Steering

Steering wheel-to-steering shaft nut
    1970 through 1978 30
    1979 35
    1980 and later 30
Outer tie-rod end-to-steering knuckle nut
    1970 through 1979 35
    1980 and 1981 40
    1982 and later 30
Inner tie-rod end-to-intermediate rod nut
    1970 and 1978 35
    1979 and later 40
Tie-rod adjuster sleeve clamp nut
    1970 .... 132 in-lbs
    1971 through 1978 22
    1979 and later 15
Steering gear-to-frame bolts
    1970 through 1986 70
    1987 and later 80
Pitman arm-to-steering gear shaft nut
    1970 140
    1971 and later 185
Pitman arm-to-intermediate rod nut
    1970 and 1971 50
    1972 and later 45
Idler arm-to-intermediate rod nut
    1970 through 1973 50
    1974 through 1979 35
    1980 and later 40
Idler arm-to-frame nuts
    1970 through 1973 35
    1974 through 1979 40
    1980 and later 60
Intermediate shaft pinch bolts
    1970 through 1978 30
    1979 and later
        Upper 46
        Lower 30
Wheel lug nuts
    1970 65
    1971 through 1975 70
    1976 and later 80

Component location

1.1 Front suspension and steering components

1	Stabilizer bar	4	Idler arm	7	Inner tie-rod	10	Lower control arm
2	Steering gear	5	Outer tie-rod	8	Intermediate rod	11	Lower balljoint
3	Pitman arm	6	Adjuster sleeve	9	Shock absorber		

1   General information

Refer to illustration 1.1

Warning: *Whenever any suspension or steering fasteners are loosened or removed they must be inspected and, if necessary, replaced with new ones of the same part number or of original equipment quality and design. Cotter pins, used extensively on steering components, should never be reused - always replace them with new ones. Torque specifications must be followed for proper reassembly and component retention. Never attempt to heat, straighten or weld any steering or suspension component. Always replace bent or* damaged parts with new ones.

Each front wheel is connected to the frame by upper and lower control arms (A-frames), through a steering knuckle and upper and lower balljoints. A coil spring is installed between the lower control arm and the frame, and a telescopic shock absorber is positioned inside the coil spring. A stabilizer bar connected to the frame rails and to the lower control arms on each side aids in controlling body roll **(see illustration)**.

The rear suspension consists of a solid axle located by upper and lower control arms, with coil springs supplying the suspension. All models utilize telescopic shock absorbers installed between the spring mounts or seats and the frame rails.

The steering system consists of the steering wheel and column, an articulated intermediate shaft, a recirculating ball steering gear and steering linkage. Power steering is standard on all models.

2   Front stabilizer bar - removal and installation

Refer to illustrations 2.2 and 2.3

Removal

1   Raise the vehicle and support it securely on jackstands. Apply the parking brake.

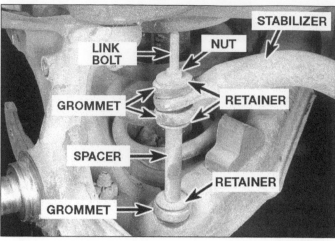

2.2 Typical front stabilizer bar-to-lower control arm linkage

2.3 Remove the brackets and rubber bushings on each side which attach the stabilizer bar to the frame

2   Remove the stabilizer bar link nuts, noting how the washers and bushings are positioned **(see illustration)**. Clamp a pair of locking pliers to the stabilizer bar link to prevent it from turning.

3   Remove the stabilizer bar bracket bolts and detach the bar from the vehicle **(see illustration)**.

4   Pull the brackets off the stabilizer bar and inspect the bushings for cracks, hardening and other signs of deterioration. If the bushings are damaged, cut them from the bar.

Installation

5   Position the stabilizer bar bushings on the bar with the slits facing the front of the vehicle. **Note:** *The offset in the bar must face down.*

6   Push the brackets over the bushings and raise the bar up to the frame. Install the bracket bolts but don't tighten them completely at this time.

7   Install the stabilizer bar link nuts, washers and rubber bushings and tighten the nuts securely.

8   Tighten the bracket bolts to the torque listed in this Chapter's Specifications.

3   Front shock absorbers - removal and installation

Refer to illustration 3.2

Removal

1   Loosen the wheel lug nuts, raise the vehicle and support it securely on jackstands. Apply the parking brake. Remove the wheel.

2   Remove the upper shock absorber stem nut **(see illustration)**. Use an open end wrench to keep the stem from turning. If the nut won't loosen because of rust, squirt some penetrating oil on the stem threads and allow it to soak in for awhile. It may be necessary to keep the stem from turning with a pair of locking pliers, since the flats provided for a wrench are quite small.

3   Remove the two lower shock mount bolts and pull the shock absorber out through the bottom of the lower control arm. Remove the washers and the rubber grommets from the top of the shock absorber.

Installation

4   Extend the new shock absorber as far as possible. Position a new washer and rubber grommet on the stem and guide the

shock up through the coil spring and into the upper mount.

5   Install the upper rubber grommet and washer and wiggle the stem back-and-forth to ensure that the grommets are centered in the mount. Tighten the stem nut securely.

6   Install the lower mounting bolts and tighten them securely.

4   Balljoints - replacement

Refer to illustrations 4.4, 4.5 and 4.8

Upper balljoint

1   Loosen the wheel lug nuts, raise the vehicle and support it securely on jackstands. Apply the parking brake. Remove the wheel.

2   Place a jack or a jackstand under the lower control arm. **Warning:** *The jack or jackstand must remain under the control arm during removal and installation of the balljoint to hold the spring and control arm in position.*

3   Remove the cotter pin from the balljoint stud and back off the nut two turns.

4   Separate the balljoint from the steering knuckle with a special tool, available at most auto parts stores, to press the balljoint out of the steering knuckle). An equivalent tool can

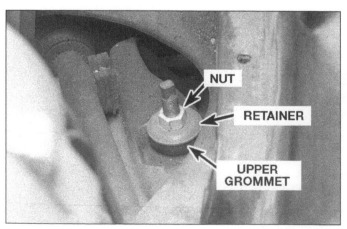

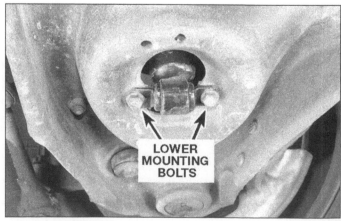

3.2 Typical front shock absorber mounting details

4.4 A special tool is used to push the balljoint out of the steering knuckle, but an alternative tool can be fabricated from a large bolt, nut, washer and socket

be fabricated from a large bolt, nut, washer and socket **(see illustration)**. Countersink the center of the bolt head with a large drill bit to prevent the tool from slipping off the balljoint stud. Install the tool as shown in the illustration, hold the bolt head with a wrench and tighten the nut against the washer until the balljoint pops out. Notice that the balljoint nut hasn't been completely removed.

5   Using a 1/8-inch drill bit, drill a 1/4-inch deep hole in the center of each rivet head **(see illustration)**.

6   Using a 1/2-inch diameter drill bit, drill off the rivet heads.

7   Use a punch to knock the rivet shanks out, then remove the balljoint from the control arm.

8   Position the new balljoint on the control arm and install the bolts and nuts supplied in the kit **(see illustration)**. Be sure to tighten the nuts to the torque specified on the balljoint kit instruction sheet.

9   Insert the balljoint stud into the steering knuckle and install the nut, tightening it to the specified torque.

10   Install a new cotter pin, tightening the nut slightly if necessary to align a slot in the

nut with the hole in the balljoint stud.

11   Install the balljoint grease fitting and fill the joint with grease.

12   Install the wheel, tightening the lug nuts to the specified torque.

13   Drive the vehicle to an alignment shop to have the front end alignment checked and, if necessary, adjusted.

Lower balljoint

14   The lower balljoint is a press fit in the lower control arm and requires special tools to remove and replace it. Refer to Section 6, remove the lower control arm and take it to an automotive machine shop to have the old balljoint pressed out and the new balljoint pressed in.

---

5   Upper control arm - removal and installation

Refer to illustration 5.4

Removal

1   Loosen the wheel lug nuts, raise the front of the vehicle and support it securely on jackstands. Apply the parking brake. Remove the wheel.

2   Support the lower control arm with a jack or jackstand. The support point must be as close to the balljoint as possible to give maximum leverage on the lower control arm.

3   Disconnect the upper balljoint from the steering knuckle (refer to Section 4). **Note: *DO NOT use a "pickle fork" type balljoint separator - it may damage the balljoint seals.***

4   Remove the control arm-to-frame nuts and bolts, recording the position of any alignment shims. They must be reinstalled in the same location to maintain wheel alignment **(see illustration)**.

5   Detach the control arm from the vehicle. **Note:** *The control arm bushings are pressed into place and require special tools for removal and installation. If the bushings must be replaced, take the control arm to a dealer service department or an automotive machine shop to have the old bushings pressed out and the new ones pressed in.*

4.5 Drill pilot holes into the heads of the balljoint rivets with a 1/8-inch bit, then use a 1/2-inch bit to cut the rivet heads off - be careful not to enlarge the holes in the control arm

Installation

6   Position the control arm on the frame and install the bolts and nuts. Install any alignment shims that were removed. Tighten the nuts to the specified torque.

7   Insert the balljoint stud into the steering knuckle and tighten the nut to the specified torque. Install a new cotter pin, tightening the nut slightly, if necessary, to align a slot in the nut with the hole in the balljoint stud.

8   Install the wheel and lug nuts and lower the vehicle. Tighten the lug nuts to the specified torque.

9   Drive the vehicle to an alignment shop to have the front end alignment checked and, if necessary, adjusted.

---

6   Lower control arm - removal and installation

Removal

1   Loosen the wheel lug nuts, raise the vehicle and support it securely on jackstands. Apply the parking brake. Remove the wheel.

4.8 Install the replacement balljoint in the upper control arm

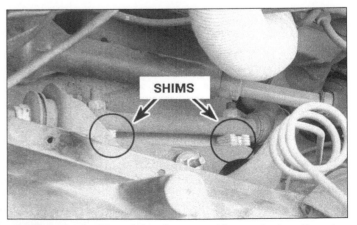

5.4 Note the position of the alignment shims and return them to their original positions

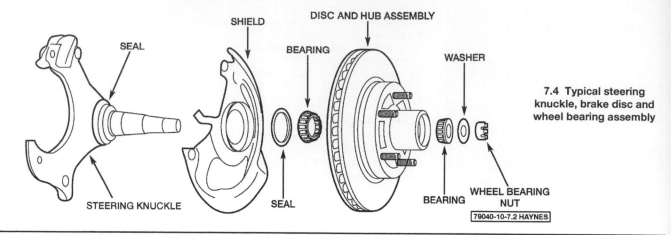

SHIELD

DISC AND HUB ASSEMBLY

SEAL

BEARING

WASHER

7.4 Typical steering knuckle, brake disc and wheel bearing assembly

STEERING KNUCKLE

SEAL

BEARING

WHEEL BEARING NUT

79040-10-7.2 HAYNES

2   Remove the shock absorber (see Section 3).
3   Disconnect the stabilizer bar from the lower control arm (see Section 2).
4   Remove the coil spring as described in Section 8.
5   Remove the cotter pin and back off the lower control arm balljoint stud nut two turns. Separate the balljoint from the steering knuckle (see Section 4).
6   Remove the control arm from the vehicle. **Note:** *The control arm bushings are pressed into place and require special tools for removal and installation. If the bushings must be replaced, take the control arm to a dealer service department or an automotive machine shop to have the old bushings pressed out and the new ones pressed in.*

Installation

7   Insert the balljoint stud into the steering knuckle, tighten the nut to the specified torque and install a new cotter pin. If necessary, tighten the nut slightly to align a slot in the nut with the hole in the balljoint stud.
8   Install the coil spring (see Section 8) and the lower control arm pivot bolts and nuts, but don't completely tighten them at this time.
9   Connect the stabilizer bar to the lower control arm.
10   Position the floor jack under the lower control arm balljoint and raise the arm to simulate normal ride height. Tighten the lower control arm pivot bolt nuts to the specified torque.
11   Install the wheel and lug nuts, lower the vehicle and tighten the lug nuts to the specified torque.
12   Drive the vehicle to an alignment shop to have the front end alignment checked and, if necessary, adjusted.

7   Steering knuckle - removal and installation

Refer to illustration 7.4

Removal

1   Loosen the wheel lug nuts, raise the vehicle and support it securely on jackstands

placed under the frame. Apply the parking brake. Remove the wheel.
2   Remove the brake caliper and suspend it with a piece of wire (see Chapter 9). Do not let it hang by the brake hose!
3   Remove the brake rotor and hub assembly (see Chapter 1).
4   Remove the splash shield from the steering knuckle **(see illustration)**.
5   Separate the tie-rod end from the steering arm (see Section 18).
6   If the steering knuckle must be replaced, remove the dust seal from the spindle by prying it off with a screwdriver. If it's damaged, replace it with a new one.
7   Position a floor jack under the lower control arm and raise it slightly to take the spring pressure off the suspension stop. The jack must remain in this position throughout the entire procedure.
8   Remove the cotter pins from the upper and lower balljoint studs and back off the nuts two turns each.
9   Break the balljoints loose from the steering knuckle with a balljoint separator (see Section 4). **Note:** *A pickle fork type balljoint separator may damage the balljoint seals.*
10   Remove the nuts from the balljoint studs, separate the control arms from the steering knuckle and remove the knuckle from the vehicle.

Installation

11   Place the knuckle between the upper and lower control arms and insert the balljoint studs into the knuckle, beginning with the lower balljoint. Install the nuts and tighten them to the specified torque. Install new cotter pins, tightening the nuts slightly to align the slots in the nuts with the holes in the balljoint studs, if necessary.
12   Install the splash shield.
13   Connect the tie-rod end to the steering arm and tighten the nut to the specified torque. Be sure to use a new cotter pin.
14   Install the brake rotor and adjust the wheel bearings following the procedure outlined in Chapter 1.
15   Install the brake caliper.

16   Install the wheel and lug nuts. Lower the vehicle to the ground and tighten the nuts to the specified torque.

8   Front coil spring - removal and installation

Removal

Refer to illustrations 8.5 and 8.7

1   Loosen the front wheel lug nuts, raise the vehicle and place it securely on jackstands. Remove the wheel.
2   Remove the shock absorber (see Section 3).
3   Remove the stabilizer bar link bolt (see Section 2).
4   Disconnect the outer tie-rod end from the steering knuckle (see Section 18).
5   Install a suitable internal type spring compressor in accordance with the tool manufacturer's instructions **(see illustration)**. Compress the spring enough to relieve all pressure from the spring seats (but don't compress it any more than necessary, or it could be ruined). When you can wiggle the

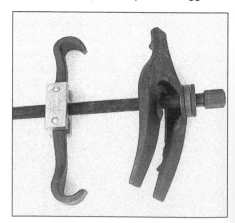

8.5 A typical aftermarket internal type spring compressor: The hooked arms grip the upper coils of the spring, the plate is inserted between the lower coils, and when the nut on the threaded rod is turned, the spring is compressed

8.7 Lower control arm pivot bolt and nut locations

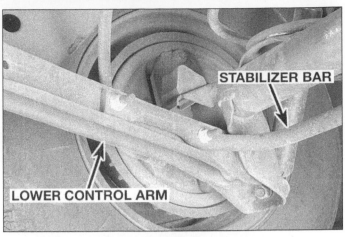

9.2 The rear stabilizer bar bolts to the lower control arm on each side

spring, it's compressed enough. (You can buy a suitable spring compressor at most auto parts stores or rent one from a tool rental yard.)

6   Support the lower control arm with a floor jack.

7   Remove the control arm pivot bolts and nuts **(see illustration)**.

8   Pull the lower control arm down and to the rear, then guide the compressed coil spring out.

9   If the coil spring is being replaced, carefully unscrew the spring compressor and transfer it to the new spring.

Installation

10   Inspect the upper and lower spring insulators. If either insulator is cracked or excessively worn, replace it. Inspect the coil spring for chips in the corrosion protection coating. If the coating has been chipped or damaged, replace the spring.

11   If the coil spring is being replaced, install the spring compressor and compress the spring.

12   With the lower spring insulator in place, position the spring on the lower control arm with the flat end of the spring facing up and

the tapered end facing down. Make sure the tapered end seats on the lower control arm with the lower end of the spring seated in the lowest part of the spring seat. The end of the spring must cover all or part of one of the drain holes in the lower control arm, but the other hole must not be covered.

13   Put the floor jack under the lower control arm and raise the arm into position in the frame. Install the control arm pivot bolts and nuts. Tighten the nuts until they're snug but don't torque them yet.

14   Remove the spring compressor.

15   Reattach the outer tie-rod end to the steering knuckle (see Section 18).

16   Reattach the stabilizer bar link to the lower control arm (see Section 2).

17   Install the shock absorber (see Section 3).

18   Position the floor jack under the lower control arm balljoint and raise the arm to simulate normal ride height. Tighten the lower control arm pivot bolt nuts to the torque listed in this Chapter's Specifications.

19   Install the wheel, remove the jackstands and lower the vehicle. Tighten the wheel lug nuts to the torque listed in the Chapter 1 Specifications.

9   Rear stabilizer bar - removal and installation

Refer to illustration 9.2

1   Raise the rear of the vehicle and support it securely on jackstands.

2   Remove the two nuts and bolts on each side securing the stabilizer bar to the lower control arms **(see illustration)**.

3   Remove the stabilizer bar and, if present, the shims installed between the bar and lower control arm on each side.

4   Installation is the reverse of the removal procedure, being sure to replace the shims (if any were removed) and tightening the nuts and bolts to the specified torque.

10   Rear shock absorbers - removal and installation

Refer to illustration 10.3

1   Raise the rear of the vehicle and support it securely on jackstands.

2   Support the rear axle with a floor jack placed under the differential.

3   On models where the lower end of the

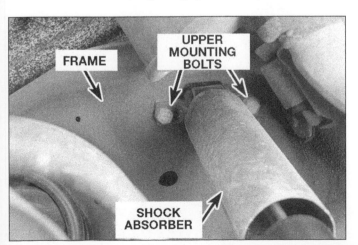

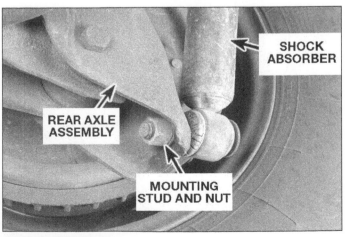

10.3 Typical rear shock absorber mounting details

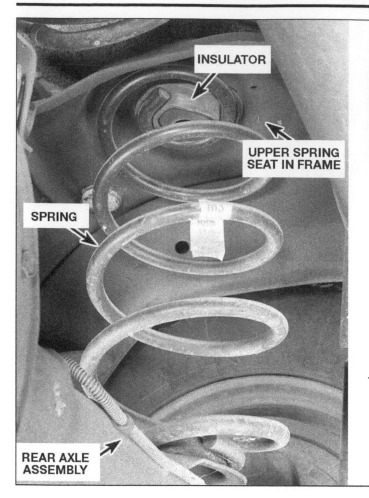

11.6 Typical rear
coil spring and
insulator details

12.3 Typical rear upper control arm mounting details

shock absorber is retained to the axle by a stud, remove the nut and washer **(see illustration)**.

4   On models where the lower end of the shock absorber is retained to the axle by a through-bolt, remove the nut, lockwasher and bolt.

5   Remove the two upper bolts securing the shock absorber to the frame and remove the shock absorber from the vehicle.

6   Installation is the reverse of the removal procedure.

11 Rear coil springs - removal and installation

Refer to illustration 11.6

1   Raise the rear of the vehicle and support it securely on jackstands under the frame rails.

2   Remove the rear wheels.

3   Place a floor jack under the center of the differential housing and raise the axle until the coil springs just start to compress. **Warning:** *When the shock absorbers are disconnected, the floor jack will be the sole support for the rear axle assembly, so be sure it is positioned correctly.*

4   Disconnect the lower ends of both shock absorbers from the rear axle.

5   Remove the clip attaching the brake line to the frame crossmember.

6   Carefully lower the jack just enough to remove the spring and rubber insulator (on top of the spring) **(see illustration)**. **Note:** *When removing the spring make note of the position of the coil ends, top and bottom. The spring must be reinstalled with the coil ends in the same position.*

7   Installation is the reverse of the removal procedure.

12 Rear control arms - removal and installation

Refer to illustrations 12.3 and 12.9
Warning: *Remove and install one control arm*

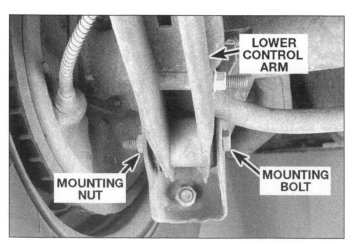

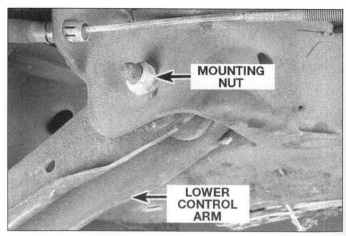

12.9 Typical rear lower control arm mounting details

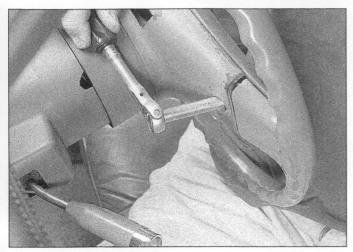

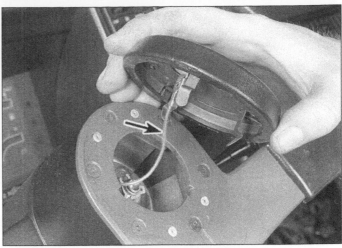

13.2a To detach the horn pad from most steering wheels, remove the two retaining screws located on the backside of steering wheel

13.2b On other steering wheels, the horn pad is removed by gripping it firmly and pulling it from the steering wheel - be sure to detach the horn wire (arrow)

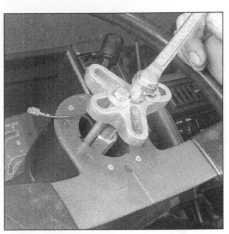

13.3 Snap-ring pliers are used to remove the safety clip from the steering shaft (not all models have this clip)

13.4 Check to be sure that there are alignment marks on the steering shaft and steering wheel (arrow) - if they aren't there or don't line up, scribe or paint new marks

13.5 Remove the steering wheel from the shaft with a puller - do not hammer on the shaft!

at a time to prevent the axle from rocking to the front or rear.

1   Raise the vehicle and support it securely with jackstands placed under the frame rails.

Upper arm

2   Support the rear axle with a floor jack placed under the differential.

3   Remove the pivot bolt holding the rear of the control arm to the ear on the axle housing **(see illustration)** and pry the control arm upwards until it clears the ear.

4   Remove the control arm pivot bolt at the frame crossmember.

5   Remove the control arm.

6   Installation is the reverse of the removal procedure.

7   Raise the rear axle to simulate normal ride height, then tighten the fasteners to the torque listed in this Chapter's Specifications.

Lower control arm

8   Support the differential with a floor jack.

9   Remove the pivot bolt holding the rear of the control arm to the bracket on the axle

housing **(see illustration)**.

10   Remove the control arm pivot bolt at the frame rail.

11   Remove the control arm.

12   Installation is the reverse of the removal procedure.

13   Lower the vehicle to the ground and tighten the control arm bolts to the specified torque.

13 Steering wheel - removal and installation

Refer to illustrations 13.2a, 13.2b, 13.3, 13.4 and 13.5

1   Disconnect the cable from the negative terminal of the battery.

2   Remove the horn pad from the steering wheel and disconnect the wire to the horn switch **(see illustrations)**.

3   Remove the safety clip from the steering shaft, if equipped **(see illustration)**.

4   Remove the steering wheel retaining nut then mark the relationship of the steering shaft to the hub (if marks don't already exist or don't line up) to simplify installation and ensure steering wheel alignment **(see illustration)**.

5   Use a puller to disconnect the steering wheel from the shaft **(see illustration)**.

6   To install the wheel, align the mark on the steering wheel hub with the mark on the shaft and slip the wheel onto the shaft. Install the nut and tighten it to the specified torque. Install the safety clip, if equipped.

7   Connect the horn wire and install the horn pad.

8   Connect the negative battery cable.

14 Intermediate shaft - removal and installation

Refer to illustrations 14.2 and 14.3

1   Turn the front wheels to the straight ahead position.

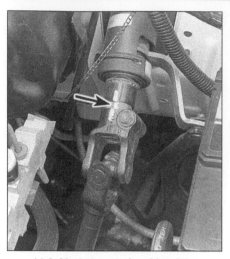

14.2 Mark the relationship of the intermediate shaft to the steering shaft (arrow) and the steering gear input shaft

15.4 Paint alignment marks on the Pitman arm and the steering gear output shaft, then remove the nut and washer

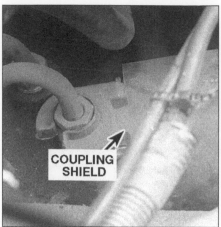

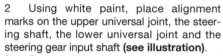

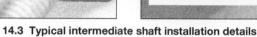

14.3 Typical intermediate shaft installation details

2   Using white paint, place alignment marks on the upper universal joint, the steering shaft, the lower universal joint and the steering gear input shaft **(see illustration)**.
3   Remove the upper and lower universal joint pinch bolts **(see illustration)**. Some models require the steering gear to be lowered for shaft removal.
4   Pry the intermediate shaft off of the steering gear shaft and the steering column shaft with a large screwdriver, then remove the shaft from the vehicle.
5   Installation is the reverse of the removal procedure. Be sure to align the marks and tighten the pinch bolts to the specified torque.

15   Steering gear - removal and installation

Refer to illustrations 15.4, 15.5 and 15.6

Removal

1   Raise the front of the vehicle and sup-

port it securely on jackstands. Apply the parking brake.
2   Place a drain pan under the steering gear (power steering only). Remove the power steering pressure and return lines and cap the ends to prevent excessive fluid loss and contamination.
3   Mark the relationship of the lower intermediate shaft universal joint to the steering gear input shaft. Remove the lower intermediate shaft pinch bolt.
4   Mark the relationship of the Pitman arm to the Pitman shaft so it can be installed in the same position **(see illustration)**. Remove the nut and washer.
5   Remove the Pitman arm from the shaft with a two-jaw puller **(see illustration)**.
6   Support the steering gear and remove the steering gear-to-frame mounting bolts **(see illustration)**. Lower the unit, separate the intermediate shaft from the steering gear input shaft and remove the steering gear from the vehicle.

Installation

7   Raise the steering gear into position and

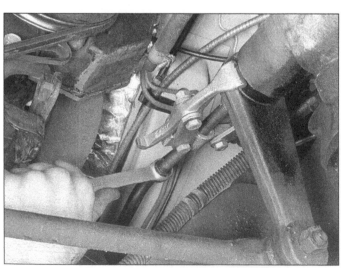

15.5 Use a puller to remove the Pitman arm from the steering gear output shaft

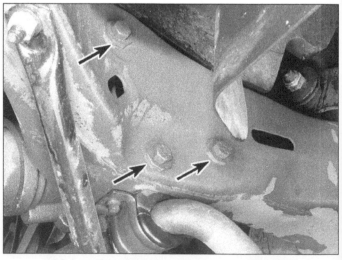

15.6 Location of the steering gear mounting bolts

16.3 Remove the power steering pump mounting bolts and nuts (arrows)

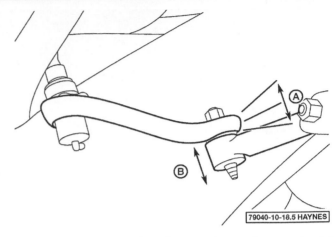

79040-10-18.5 HAYNES

18.5 To check for excessive idler arm play, apply approximately 25 lbs. of force up and down (B) on the idler arm - if the total movement (A) is greater than 1/4-inch, replace the idler arm

connect the intermediate shaft, aligning the marks.

8   Install the mounting bolts and washers and tighten them to the specified torque.

9   Slide the Pitman arm onto the Pitman shaft, ensuring that the marks are aligned. Install the washer and nut and tighten the nut to the specified torque.

10   Install the lower intermediate shaft pinch bolt and tighten it to the specified torque.

11   Connect the power steering pressure and return hoses to the steering gear and fill the power steering pump reservoir with the recommended fluid (see Chapter 1).

12   Lower the vehicle and bleed the steering system as outlined in Section 17.

16   Power steering pump - removal and installation

Refer to illustration 16.3

1   Disconnect the hydraulic lines from either the pump or the steering gear and support them in a raised position to keep the fluid from draining.

2   Loosen the pump mounting bolts and pivot the pump inwards (towards the engine) until the drivebelt can be removed.

3   Remove the pump mounting bolts and braces and remove the pump **(see illustration)**.

4   Using a special power steering pump pulley puller (available at most automotive parts stores) pull the pulley off of the power steering pump and remove the bolts retaining the pump to the mounting bracket.

5   Reinstall the pump by reversing the removal procedure. Before installing the drivebelt, prime the pump by filling the reservoir with fluid, then turn the pulley in the opposite direction to normal rotation until no more air bubbles are observed in the reservoir.

6   Install the drivebelt (see Chapter 1) and bleed the system as described in Section 17.

17   Power steering system - bleeding

1   Following any operation in which the power steering fluid lines have been disconnected, the power steering system must be bled to remove all air and obtain proper steering performance.

2   With the front wheels in the straight ahead position, check the power steering fluid level and, if low, add fluid until it reaches the Cold mark on the dipstick.

3   Start the engine and allow it to run at fast idle. Recheck the fluid level and add more if necessary to reach the Cold mark on the dipstick.

4   Bleed the system by turning the wheels from side-to-side, without hitting the stops. This will work the air out of the system. Keep the reservoir full of fluid as this is done.

5   When the air is worked out of the system, return the wheels to the straight ahead position and leave the vehicle running for several more minutes before shutting it off.

6   Road test the vehicle to be sure the steering system is functioning normally and noise free.

7   Recheck the fluid level to be sure it is up to the Hot mark on the dipstick while the engine is at normal operating temperature. Add fluid if necessary (see Chapter 1).

18   Steering linkage - inspection, removal and installation

Warning: *Whenever any of the suspension or steering fasteners are loosened or removed they must be inspected and if necessary, replaced with new ones of the same part number or of original equipment quality and*

design. Torque specifications must be followed for proper reassembly and component retention. Never attempt to heat, straighten or weld any suspension or steering component. Instead, replace any bent or damaged part with a new one.

Caution: *DO NOT use a "pickle fork" type balljoint separator - it may damage the balljoint seals.*

Inspection

Refer to illustration 18.5

1   The steering linkage connects the steering gear to the front wheels and keeps the wheels in proper relation to each other. The linkage consists of the Pitman arm, fastened to the steering gear shaft, which moves the intermediate rod back-and-forth. The intermediate rod is supported on the other end by a frame-mounted idler arm. The back-and-forth motion of the intermediate rod is transmitted to the steering knuckles through a pair of tie-rod assemblies. Each tie-rod is made up of an inner and outer tie-rod end, a threaded adjuster tube and two clamps.

2   Set the wheels in the straight ahead position and lock the steering wheel.

3   Raise one side of the vehicle until the tire is approximately 1-inch off the ground.

4   Mount a dial indicator with the needle resting on the outside edge of the wheel. Grasp the front and rear of the tire and using light pressure, wiggle the wheel back-and-forth and note the dial indicator reading. The gauge reading should be less than 0.108-inch. If the play in the steering system is more than specified, inspect each steering linkage pivot point and ball stud for looseness and replace parts if necessary.

5   Raise the vehicle and support it on jackstands. Push up, then pull down on the end of the idler arm, exerting a force of approximately 25 pounds each way. Measure the total distance the end of the arm travels **(see illustration)**. If the play is greater than 1/4-

18.9 Use a puller to press the tie-rod end out of the steering knuckle

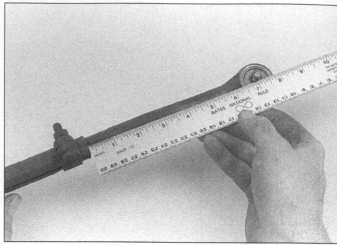

18.11 Measure the distance from the adjuster tube to the ball stud centerline so the new tie-rod end can be set to the same length

inch, re-place the idler arm.

6   Check for torn ball stud boots, frozen joints and bent or damaged linkage components.

Removal and installation

Refer to illustrations 18.9, 18.11, 18.13 and 18.15

Tie-rod

7   Loosen the wheel lug nuts, raise the vehicle and support it securely on jackstands. Apply the parking brake. Remove the wheel.

8   Remove the cotter pin and loosen, but do not remove, the castellated nut from the ball stud.

9   Using a two jaw puller, separate the tie-rod end from the steering knuckle **(see illustration)**. Remove the castellated nut and pull the tie-rod end from the knuckle.

10   Remove the nut securing the inner tie-rod end to the intermediate rod. Separate the

inner tie-rod end from the intermediate rod in the same manner as in Step 9.

11   If the inner or outer tie-rod end must be replaced, measure the distance from the end

of the adjuster tube to the center of the ball stud and record it **(see illustration)**. Loosen the adjuster tube clamp and unscrew the tie-rod end.

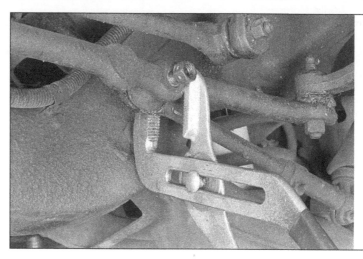

18.13 It may be necessary to force the ball stud into the tapered hole to keep it from turning while the nut is tightened

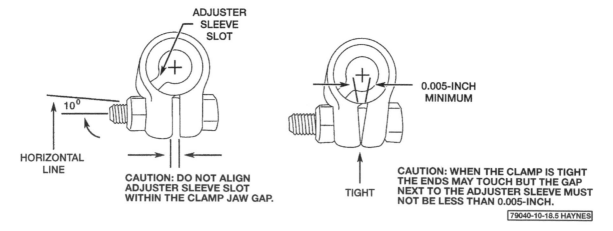

ADJUSTER SLEEVE SLOT

10°

HORIZONTAL LINE

CAUTION: DO NOT ALIGN ADJUSTER SLEEVE SLOT WITHIN THE CLAMP JAW GAP.

0.005-INCH MINIMUM

TIGHT

CAUTION: WHEN THE CLAMP IS TIGHT THE ENDS MAY TOUCH BUT THE GAP NEXT TO THE ADJUSTER SLEEVE MUST NOT BE LESS THAN 0.005-INCH.

79040-10-18.5 HAYNES

18.15 Tie-rod adjuster tube and clamp position details

12   Lubricate the threaded portion of the tie-rod end with chassis grease. Screw the new tie-rod end into the adjuster tube and adjust the distance from the tube to the ball stud to the previously measured dimension. The number of threads showing on the inner and outer tie-rod ends should be equal within three threads. Don't tighten the clamp yet.

13   To install the tie-rod, insert the inner tie-rod end ball stud into the intermediate rod until it's seated. Install the nut and tighten it to the specified torque. If the ball stud spins when attempting to tighten the nut, force it into the tapered hole with a large pair of pliers **(see illustration)**.

14   Connect the outer tie-rod end to the steering knuckle and install the castellated nut. Tighten the nut to the specified torque and install a new cotter pin. If necessary, tighten the nut slightly to align a slot in the nut with the hole in the ball stud.

15   Tighten the clamp nuts. The center of the bolt should be nearly horizontal and the adjuster tube slot must not line up with the gap in the clamps **(see illustration)**.

16   Install the wheel and lug nuts, lower the vehicle and tighten the lug nuts to the specified torque. Drive the vehicle to an alignment shop to have the front end alignment checked and, if necessary, adjusted.

Idler arm

17   Raise the vehicle and support it securely on jackstands. Apply the parking brake.

18   Loosen but do not remove the idler arm-to-intermediate rod nut.

19   Separate the idler arm from the intermediate rod with a two jaw puller. Remove the nut.

20   Remove the idler arm-to-frame bolts.

21   To install the idler arm, position it on the frame and install the bolts, tightening them to the specified torque.

22   Insert the idler arm ball stud into the intermediate rod and install the nut. Tighten the nut to the specified torque. If the ball stud spins when attempting to tighten the nut, force it into the tapered hole with a large pair of pliers.

Intermediate rod

23   Raise the vehicle and support it securely on jackstands. Apply the parking brake.

24   Separate the two inner tie-rod ends from the intermediate rod.

25   Separate the intermediate rod from the Pitman arm.

26   Separate the idler arm from the intermediate rod.

27   Installation is the reverse of the removal procedure. If the ball studs spin when attempting to tighten the nuts, force them into the tapered holes with a large pair of pliers. Be sure to tighten all of the nuts to the specified torque.

Pitman arm

28   Refer to Section 16 of this Chapter for the Pitman arm removal procedure.

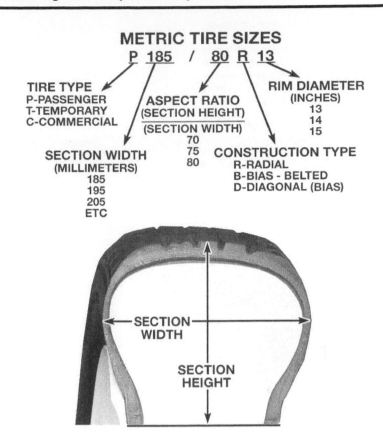

19.1 Metric tire size code

19   Wheels and tires - general information

Refer to illustration 19.1

All vehicles covered by this manual are equipped with metric-sized fiberglass or steel belted radial tires **(see illustration)**. Use of other size or type of tires may affect the ride and handling of the vehicle. Don't mix different types of tires, such as radials and bias belted, on the same vehicle as handling may be seriously affected. It's recommended that tires be replaced in pairs on the same axle, but if only one tire is being replaced, be sure it's the same size, structure and tread design as the other.

Because tire pressure has a substantial effect on handling and wear, the pressure on all tires should be checked at least once a month or before any extended trips (see Chapter 1).

Wheels must be replaced if they are bent, dented, leak air, have elongated bolt holes, are heavily rusted, out of vertical symmetry or if the lug nuts won't stay tight. Wheel repairs that use welding or peening are not recommended.

Tire and wheel balance is important in the overall handling, braking and performance of the vehicle. Unbalanced wheels can adversely affect handling and ride characteristics as well as tire life. Whenever a tire is installed on a wheel, the tire and wheel should be balanced by a shop with the proper equipment.

20   Front end alignment - general information

Refer to illustration 20.1

A front end alignment refers to the adjustments made to the front wheels so they are in proper angular relationship to the suspension and the ground. Front wheels that are out of proper alignment not only affect steering control, but also increase tire wear. The front end adjustments normally required are camber, caster and toe-in **(see illustration)**.

Getting the proper front wheel alignment is a very exacting process, one in which complicated and expensive machines are necessary to perform the job properly. Because of this, you should have a technician with the proper equipment perform these tasks. We will, however, use this space to give you a basic idea of what is involved with front end alignment so you can better understand the process and deal intelligently with the shop that does the work.

Toe-in is the turning in of the front wheels. The purpose of a toe specification is to ensure parallel rolling of the front wheels. In a vehicle with zero toe-in, the distance between the front edges of the wheels will be the same as the distance between the rear edges of the wheels. The actual amount of toe-in is normally only a fraction of an inch. Toe-in adjustment is controlled by the length of the tie-rods. Incorrect toe-in will cause the tires to wear improperly by making them scrub against the road surface.

Camber is the tilting of the front wheels from the vertical when viewed from the front of the vehicle. When the wheels tilt out at the top, the camber is said to be positive (+). When the wheels tilt in at the top the camber is negative (-). The amount of tilt is measured in degrees from the vertical and this measurement is called the camber angle. This angle affects the amount of tire tread which contacts the road and compensates for changes in the suspension geometry when the vehicle is cornering or traveling over an undulating surface. Camber is adjusted by adding or subtracting shims at the upper control arm pivot mounting points.

Caster is the tilting of the top of the front steering axis from the vertical. A tilt toward the rear is positive caster and a tilt toward the front is negative caster. Caster is adjusted by moving shims from one end of the upper control arm mount to the other.

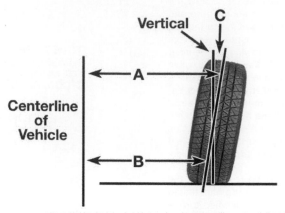

CAMBER ANGLE (FRONT VIEW)

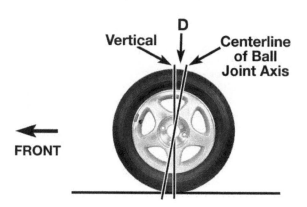

CASTER ANGLE (SIDE VIEW)

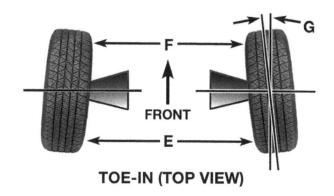

TOE-IN (TOP VIEW)

20.1 Typical front end alignment details

A minus B =C (degrees camber)   E minus F = toe-in (measured in inches)
D = degrees caster   G = toe-in (expressed in degrees)

Chapter 11 Body

Contents

1   General information

These models have a separate frame and body. Certain components are particularly vulnerable to accident damage and can be unbolted and repaired or replaced. Among these parts are the body moldings, bumpers, the hood and trunk lids and all glass.

Only general body maintenance practices and body panel repair procedures within the scope of the do-it-yourselfer are included in this Chapter.

2   Body - maintenance

1   The condition of your vehicle's body is very important, because the resale value depends a great deal on it. It's much more difficult to repair a neglected or damaged body than it is to repair mechanical components. The hidden areas of the body, such as the wheel wells, the frame and the engine compartment, are equally important, although they don't require as frequent attention as the rest of the body.

2   Once a year, or every 12,000 miles, it's a good idea to have the underside of the body steam cleaned. All traces of dirt and oil will be removed and the area can then be inspected carefully for rust, damaged brake lines, frayed electrical wires, damaged cables and other problems. The front suspension components should be greased after completion of this job.

3   At the same time, clean the engine and the engine compartment with a steam cleaner or water soluble degreaser.

4   The wheel wells should be given close attention, since undercoating can peel away and stones and dirt thrown up by the tires can cause the paint to chip and flake, allowing rust to set in. If rust is found, clean down to the bare metal and apply an anti-rust paint.

5   The body should be washed about once a week. Wet the vehicle thoroughly to soften the dirt, then wash it down with a soft sponge and plenty of clean soapy water. If the surplus dirt is not washed off very carefully, it can wear down the paint.

6   Spots of tar or asphalt thrown up from the road should be removed with a cloth soaked in solvent.

7   Once every six months, wax the body and chrome trim. If a chrome cleaner is used to remove rust from any of the vehicle's plated parts, remember that the cleaner also removes part of the chrome, so use it sparingly.

3   Vinyl trim - maintenance

Don't clean vinyl trim with detergents, caustic soap or petroleum-based cleaners. Plain soap and water works just fine, with a soft brush to clean dirt that may be ingrained. Wash the vinyl as frequently as the rest of the vehicle.

After cleaning, application of a high quality rubber and vinyl protectant will help prevent oxidation and cracks. The protectant can also be applied to weatherstripping, vacuum lines and rubber hoses, which often fail as a result of chemical degradation, and to the tires.

4   Upholstery and carpets - maintenance

1   Every three months remove the carpets or mats and clean the interior of the vehicle (more frequently if necessary). Vacuum the upholstery and carpets to remove loose dirt and dust.

2   Leather upholstery requires special care. Stains should be removed with warm water and a very mild soap solution. Use a clean, damp cloth to remove the soap, then wipe again with a dry cloth. Never use alcohol, gasoline, nail polish remover or thinner to clean leather upholstery.

3   After cleaning, regularly treat leather upholstery with a leather wax. Never use car wax on leather upholstery.

4   In areas where the interior of the vehicle is subject to bright sunlight, cover leather seats with a sheet if the vehicle is to be left out for any length of time.

5   Use of an interior-type windshield sun deflector is also recommended.

5   Body repair - minor damage

See photo sequence

1   If the scratch is superficial and does not penetrate to the metal of the body, repair is very simple. Lightly rub the scratched area with a fine rubbing compound to remove loose paint and built up wax. Rinse the area with clean water.

2   Apply touch-up paint to the scratch, using a small brush. Continue to apply thin layers of paint until the surface of the paint in the scratch is level with the surrounding paint. Allow the new paint at least two weeks to harden, then blend it into the surrounding paint by rubbing with a very fine rubbing compound. Finally, apply a coat of wax to the scratch area.

3   If the scratch has penetrated the paint and exposed the metal of the body, causing the metal to rust, a different repair technique is required. Remove all loose rust from the bottom of the scratch with a pocket knife, then apply rust-inhibiting paint to prevent the formation of rust in the future. Using a rubber or nylon applicator, coat the scratched area with glaze-type filler. If required, the filler can be mixed with thinner to provide a very thin paste, which is ideal for filling narrow scratches. Before the glaze filler in the scratch hardens, wrap a piece of smooth cotton cloth around the tip of a finger. Dip the cloth in thinner and then quickly wipe it along the surface of the scratch. This will ensure that the surface of the filler is slightly hollow. The scratch can now be painted over as described earlier in this Section.

Repair of dents

4   When repairing dents, the first job is to pull the dent out until the affected area is as close as possible to its original shape. There is no point in trying to restore the original shape completely as the metal in the damaged area will have stretched on impact and cannot be restored to its original contours. It is better to bring the level of the dent up to a point which is about 1/8-inch below the level of the surrounding metal. In cases where the dent is very shallow, it is not worth trying to pull it out at all.

5   If the back side of the dent is accessible, it can be hammered out gently from behind using a soft-face hammer. While doing this, hold a block of wood firmly against the opposite side of the metal to absorb the hammer blows and prevent the metal from being stretched.

6   If the dent is in a section of the body which has double layers, or some other factor makes it inaccessible from behind, a different technique is required. Drill several small holes through the metal inside the damaged area, particularly in the deeper sections. Screw long, self-tapping screws into the holes just enough for them to get a good grip in the metal. Now the dent can be pulled out by pulling on the protruding heads of the screws with locking pliers.

7   The next stage of repair is the removal of paint from the damaged area and from an inch or so of the surrounding metal. This is easily done with a wire brush or sanding disk in a drill motor, although it can be done just as effectively by hand with sandpaper. To complete the preparation for filling, score the surface of the bare metal with a screwdriver or the tang of a file, or drill small holes in the affected area. This will provide a good grip for the filler material. To complete the repair, see the Section on filling and painting.

Repair of rust holes or gashes

8   Remove all paint from the affected area and from an inch or so of the surrounding metal using a sanding disk or wire brush mounted in a drill motor. If these are not available, a few sheets of sandpaper will do the job just as effectively.

9   With the paint removed, you will be able to determine the severity of the corrosion and decide whether to replace the whole panel, if possible, or repair the affected area. New body panels are not as expensive as most people think, and it is often quicker to install a new panel than to repair large areas of rust.

10   Remove all trim pieces from the affected area except those which will act as a guide to the original shape of the damaged body, such as headlight shells, etc. Using metal snips or a hacksaw blade, remove all loose metal and any other metal that is badly affected by rust. Hammer the edges of the hole inward to create a slight depression for the filler material.

11   Wire brush the affected area to remove the powdery rust from the surface of the metal. If the back of the rusted area is accessible, treat it with rust-inhibiting paint.

12   Before filling is done, block the hole in some way. This can be done with sheet metal riveted or screwed into place, or by stuffing the hole with wire mesh.

13   Once the hole is blocked off, the affected area can be filled and painted. See the following subsection on filling and painting.

Filling and painting

14   Many types of body fillers are available, but generally speaking, body repair kits which contain filler paste and a tube of resin hardener are best for this type of repair work. A wide, flexible plastic or nylon applicator will be necessary for imparting a smooth and contoured finish to the surface of the filler material. Mix up a small amount of filler on a clean piece of wood or cardboard (use the hardener sparingly). Follow the manufacturer's instructions on the package, otherwise the filler will set incorrectly.

15   Using the applicator, apply the filler paste to the prepared area. Draw the applicator across the surface of the filler to achieve the desired contour and to level the filler surface. As soon as a contour that approximates the original one is achieved, stop working the paste. If you continue, the paste will begin to stick to the applicator. Continue to add thin layers of paste at 20-minute intervals until the level of the filler is just above the surrounding metal.

16   Once the filler has hardened, the excess can be removed with a body file. From then on, progressively finer grades of sandpaper should be used, starting with a 180-grit paper and finishing with 600-grit wet-or-dry paper. Always wrap the sandpaper around a flat rubber or wooden block, otherwise the surface of the filler will not be completely flat. During the sanding of the filler surface, the wet-or-dry paper should be periodically rinsed in water. This will ensure that a very smooth finish is produced in the final stage.

17   At this point, the repair area should be surrounded by a ring of bare metal, which in turn should be encircled by the finely feathered edge of good paint. Rinse the repair area with clean water until all of the dust produced by the sanding operation is gone.

18   Spray the entire area with a light coat of primer. This will reveal any imperfections in the surface of the filler. Repair the imperfections with fresh filler paste or glaze filler and once more smooth the surface with sandpaper. Repeat this spray-and-repair procedure until you are satisfied that the surface of the filler and the feathered edge of the paint are perfect. Rinse the area with clean water and allow it to dry completely.

19   The repair area is now ready for painting. Spray painting must be carried out in a warm, dry, windless and dust-free atmosphere. These conditions can be created if you have access to a large indoor work area, but if you are forced to work in the open, you will have to pick the day very carefully. If you are working indoors, dousing the floor in the work area with water will help settle the dust which would otherwise be in the air. If the

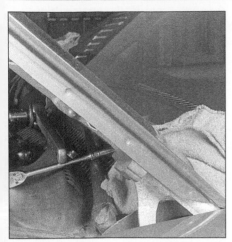

8.1 Pad the back corners of the hood with cloths so the windshield won't be damaged if the hood accidentally swings rearward

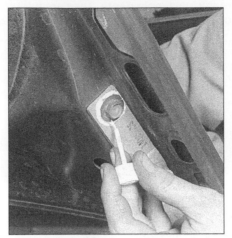

8.2 Use white paint or a scribe to mark the hood bolt locations

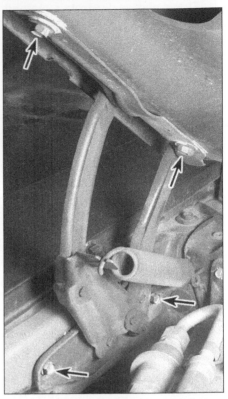

8.4 Typical hood hinge mounting bolt locations (arrows)

repair area is confined to one body panel, mask off the surrounding panels. This will help minimize the effects of a slight mismatch in paint color. Trim pieces such as chrome strips, door handles, etc., will also need to be masked off or removed. Use masking tape and several thickness of newspaper for the masking operations.

20   Before spraying, shake the paint can thoroughly, then spray a test area until the spray painting technique is mastered. Cover the repair area with a thick coat of primer. The thickness should be built up using several thin layers of primer rather than one thick one. Using 600-grit wet-or-dry sandpaper, rub down the surface of the primer until it is very smooth. While doing this, the work area should be thoroughly rinsed with water and the wet-or-dry sandpaper periodically rinsed as well. Allow the primer to dry before spraying additional coats.

21   Spray on the top coat, again building up the thickness by using several thin layers of paint. Begin spraying in the center of the repair area and then, using a circular motion, work out until the whole repair area and about two inches of the surrounding original paint is covered. Remove all masking material 10 to 15 minutes after spraying on the final coat of paint. Allow the new paint at least two weeks to harden, then use a very fine rubbing compound to blend the edges of the new paint into the existing paint. Finally, apply a coat of wax.

6   Body repair - major damage

1   Major damage must be repaired by an auto body shop specifically equipped to perform unibody repairs. These shops have the specialized equipment required to do the job properly.

2   If the damage is extensive, the body must be checked for proper alignment or the

vehicle's handling characteristics may be adversely affected and other components may wear at an accelerated rate.

3   Due to the fact that all of the major body components (hood, fenders, etc.) are separate and replaceable units, any seriously damaged components should be replaced rather than repaired. Sometimes the components can be found in a wrecking yard that specializes in used vehicle components, often at considerable savings over the cost of new parts.

7   Hinges and locks - maintenance

Once every 3000 miles, or every three months, the hinges and latch assemblies on the doors, hood and trunk should be given a few drops of light oil or lock lubricant. The door latch strikers should also be lubricated with a thin coat of grease to reduce wear and ensure free movement. Lubricate the door and trunk locks with spray-on graphite lubricant.

8   Hood - removal, installation and adjustment

Refer to illustrations 8.1, 8.2 and 8.4
Note: *The hood is heavy and somewhat awkward to remove and install - at least two people should perform this procedure.*

Removal and installation

1   Use blankets or pads to cover the cowl area of the body and the fenders **(see illustration)**. This will protect the body and paint as the hood is lifted off.

2   Scribe or paint alignment marks around the bolt heads to insure proper alignment during installation **(see illustration)**.

3   Disconnect any cables or wire harnesses which will interfere with removal.

4   Have an assistant support the weight of the hood. Remove the hinge-to-hood nuts or

bolts **(see illustration)**.

5   Lift off the hood.

6   Installation is the reverse of removal.

Adjustment

7   Fore-and-aft and side-to-side adjustment of the hood is done by moving the hood in relation to the hinge plate after loosening the bolts or nuts.

8   Scribe a line around the entire hinge plate so you can judge the amount of movement **(see illustration 8.2)**.

9   Loosen the bolts or nuts and move the hood into correct alignment. Move it only a little at a time. Tighten the hinge bolts or nuts and carefully lower the hood to check the alignment.

10   If necessary after installation, the entire hood lock assembly can be adjusted up and down as well as from side to side on the radiator support so the hood closes securely and is flush with the fenders. To do this, scribe a line around the hood lock mounting bolts to provide a reference point (Section 12). Then loosen the bolts and reposition the latch assembly as necessary. Following adjustment, retighten the mounting bolts.

11   Finally, adjust the hood bumpers on the radiator support so the hood, when closed, is flush with the fenders.

12   The hood latch assembly, as well as the hinges, should be periodically lubricated with white lithium-base grease to prevent sticking and wear.

These photos illustrate a method of repairing simple dents. They are intended to supplement *Body repair - minor damage* in this Chapter and should not be used as the sole instructions for body repair on these vehicles.

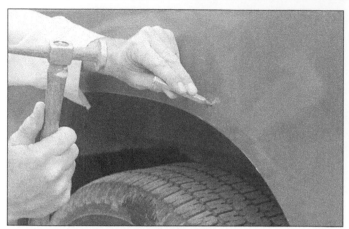

1 If you can't access the backside of the body panel to hammer out the dent, pull it out with a slide-hammer-type dent puller. In the deepest portion of the dent or along the crease line, drill or punch hole(s) at least one inch apart . . .

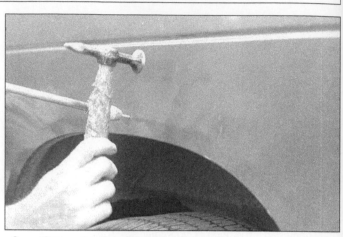

2 . . . then screw the slide-hammer into the hole and operate it. Tap with a hammer near the edge of the dent to help 'pop' the metal back to its original shape. When you're finished, the dent area should be close to its original contour and about 1/8-inch below the surface of the surrounding metal

3 Using coarse-grit sandpaper, remove the paint down to the bare metal. Hand sanding works fine, but the disc sander shown here makes the job faster. Use finer (about 320-grit) sandpaper to feather-edge the paint at least one inch around the dent area

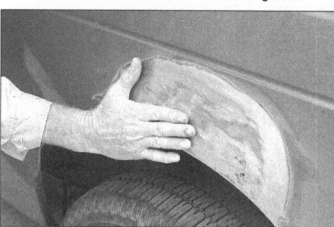

4 When the paint is removed, touch will probably be more helpful than sight for telling if the metal is straight. Hammer down the high spots or raise the low spots as necessary. Clean the repair area with wax/silicone remover

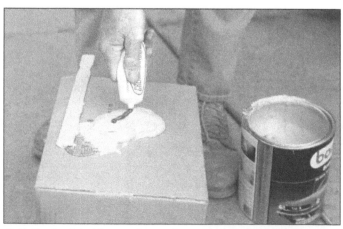

5 Following label instructions, mix up a batch of plastic filler and hardener. The ratio of filler to hardener is critical, and, if you mix it incorrectly, it will either not cure properly or cure too quickly (you won't have time to file and sand it into shape)

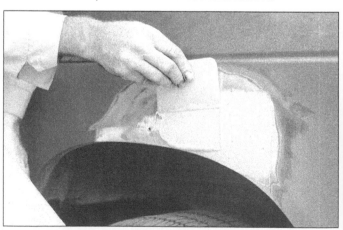

6 Working quickly so the filler doesn't harden, use a plastic applicator to press the body filler firmly into the metal, assuring it bonds completely. Work the filler until it matches the original contour and is slightly above the surrounding metal

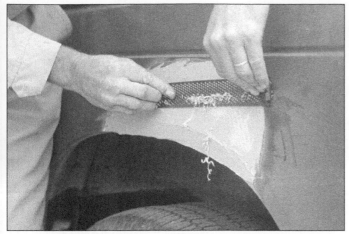

7 Let the filler harden until you can just dent it with your fingernail. Use a body file or Surform tool (shown here) to rough-shape the filler

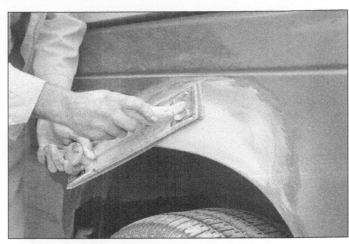

8 Use coarse-grit sandpaper and a sanding board or block to work the filler down until it's smooth and even. Work down to finer grits of sandpaper - always using a board or block - ending up with 360 or 400 grit

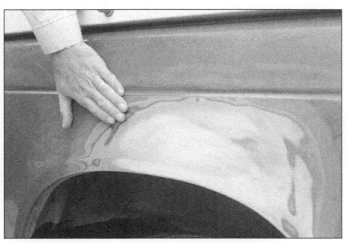

9 You shouldn't be able to feel any ridge at the transition from the filler to the bare metal or from the bare metal to the old paint. As soon as the repair is flat and uniform, remove the dust and mask off the adjacent panels or trim pieces

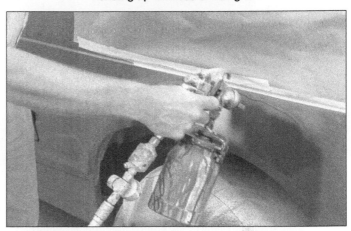

10 Apply several layers of primer to the area. Don't spray the primer on too heavy, so it sags or runs, and make sure each coat is dry before you spray on the next one. A professional-type spray gun is being used here, but aerosol spray primer is available inexpensively from auto parts stores

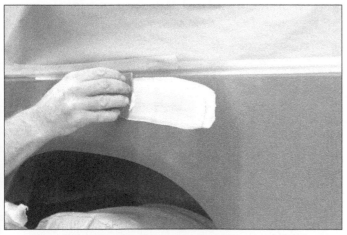

11 The primer will help reveal imperfections or scratches. Fill these with glazing compound. Follow the label instructions and sand it with 360 or 400-grit sandpaper until it's smooth. Repeat the glazing, sanding and respraying until the primer reveals a perfectly smooth surface

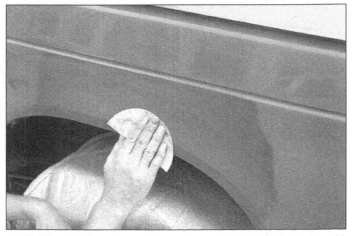

12 Finish sand the primer with very fine sandpaper (400 or 600-grit) to remove the primer overspray. Clean the area with water and allow it to dry. Use a tack rag to remove any dust, then apply the finish coat. Don't attempt to rub out or wax the repair area until the paint has dried completely (at least two weeks)

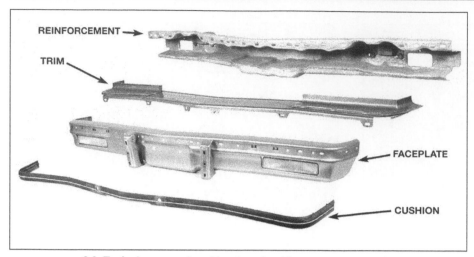

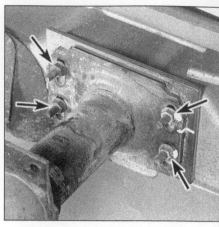

9.9a Mounting bolts for the energy absorber used on 1973 and later model bumpers (front shown)

9.6 Typical energy-absorbing type front bumper components

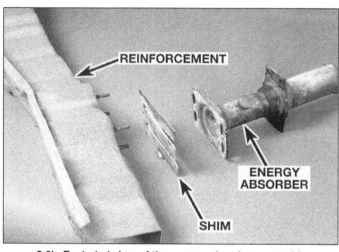

9.9b Exploded view of the energy absorber assembly

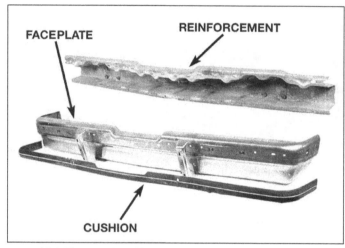

10.4 Energy-absorbing type rear bumper components

9   Bumpers (front) - removal and installation

1970 through 1972 models

1   Raise the hood and disconnect the battery.
2   Remove the radiator grille.
3   Remove the parking lights from their mounts and place to one side; also remove the side marker lights.
4   Remove the valance panel (if equipped).
5   Unbolt the bumper bracket from the frame and remove the bumper.

1973 and later models

Refer to illustrations 9.6, 9.9a and 9.9b
Note: *Vehicles manufactured from 1973 on are fitted with energy absorbing devices between the bumpers and vehicle frame to meet Federal Safety Standards under impact load. These are filled with hydraulic fluid and gas under high pressure and should not be interfered with or subjected to heat.*
6   Unscrew and remove the nuts which

hold the reinforcement to the energy absorber **(see illustration)**.
7   Withdraw the bumper.
8   If the reinforcement must be removed, detach the license plate bracket and other reinforcement brackets.
9   If the energy absorber must be removed, unbolt it from the frame and slide it from the vehicle **(see illustrations)**.
10   Installation of all types is a reversal of removal, but move the position of the bumper up and down or from side to side in order to equalize the bumper-to-body clearance before tightening the mounting bolts and nuts.

10   Bumpers (rear) - removal and installation

1970 through 1972 models

1   Disconnect the license plate light.
2   Remove the valance panel, if so equipped.
3   Unscrew and remove all bumper mount-

ing bolts and lower the bumper to the floor.
4   Installation is the reverse of removal.

1973 and later models

Refer to illustration 10.4
4   On these bumpers with energy absorbers, the removal operations are similar to those described in Steps 6 to 9 of the preceding Section **(see illustration)**.

11   Front fender and wheelhouse - removal and installation

Note: *The fender and wheel house can be heavy and awkward to remove and install. We recommend you have an assistant available.*
1   Disconnect the battery ground cable. If the right-hand fender and wheelhouse are to be removed, remove the battery also.
2   Raise the vehicle with a jack placed under the suspension or frame, then remove the wheel. Remove the front bumper.
3   Remove the hood and hinges.
4   Disconnect all electrical wiring, clips and

12.1 Typical hood latch mounting bolt locations (arrows)

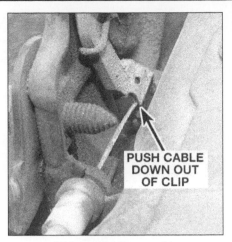

12.2 Hood release cable installation details

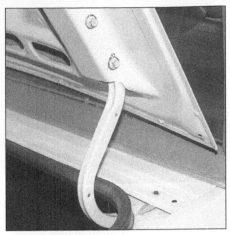

13.3 Scribe or paint around the trunk lid attaching bolts before loosening them

other items attached to the assembly.

5   Disconnect the filler panel, header panel, valance panel and brace.

6   Remove the headlight and bezel (see Chapter 12).

7   Remove the headlight retainer support screws where applicable.

8   Remove the fender and wheelhouse assembly attachment screws. Note the number of shims fitted to assist in installation.

9   Remove the side-marker light connections as the assembly is being carefully lifted away from the vehicle.

10   Installation is the reverse of removal except that the fender and wheelhouse should be guided in at the bottom adjacent to the door first of all. Adjust the assembly before finally tightening the screws using the original shims, then add or remove shims as necessary.

12   Hood latch mechanism - removal and installation

Refer to illustrations 12.1 and 12.2

1   Scribe a line around the latchplate to aid alignment when installing, then remove the screws retaining it to the radiator support **(see illustration)**.

2   Disconnect the hood release cable **(see illustration)**.

3   Installation is the reverse of the removal procedure. Adjust the hood latch bolts so that the hood engages securely when closed and the hood bumpers are slightly compressed.

13   Trunk lid - removal, installation and adjustment

Refer to illustration 13.3

1   Open the trunk lid and cover the edges of the trunk compartment with pads or cloths to protect the painted surfaces when the lid is removed.

2   Disconnect any cables or wire harness connectors attached to the trunk lid that would interfere with removal.

3   Scribe or paint alignment marks around the hinge bolt mounting flanges **(see illustration)**.

4   While an assistant supports the lid, remove the hinge bolts from both sides and lift it off.

5   Installation is the reverse of removal. **Note:** *When reinstalling the trunk lid, align the hinge bolt flanges with the marks made during removal.*

6   After installation, close the lid and see if it's in proper alignment with the surrounding panels. Fore-and-aft and side-to-side adjustments of the lid are controlled by the position of the hinge bolts in the slots. To adjust it, loosen the hinge bolts, reposition the lid and retighten the bolts.

7   The height of the lid in relation to the surrounding body panels when closed can be adjusted by loosening the lock striker bolts, repositioning the striker and retightening the bolts.

14   Trunk lid lock cylinder - removal and installation

Early models

1   Open the trunk and remove the lock cylinder retainer screw(s).

2   Pull the retainer clip down or away from the lock cylinder and remove the cylinder from the body.

3   Installation is the reverse of the removal procedure.

Later models

Refer to illustration 14.4

4   The procedure for later models is basically as described above, but in some instances the retainer is secured with stud nuts or rivets **(see illustration)**.

5   On models with swivel-type emblems, it

14.4 This rivet must be removed before the lock cylinder retainer can be detached

will be necessary to drill out the rivets and remove the emblem so the lock cylinder can be withdrawn.

6   When drilling out rivets, use a 1/8-inch drill, taking care not to enlarge the rivet hole. A 1/8 x 5/16-inch pop rivet is suitable for a replacement when installing.

15   Trunk lid lock and striker - removal and installation

1   Remove the lid lock cylinder (Section 14).

2   The lid lock is retained by bolts which can readily be removed with a wrench.

3   When installing, ensure that the lid lock is correctly aligned before finally tightening the bolts.

4   The striker is retained with bolts or screws. Before removing a striker, scribe around the adjacent panel to facilitate installation in the original position. If necessary, adjustment can be made by repositioning. **Note:** *On some models the striker is welded to the rear panel of the trunk compartment and can not be removed or adjusted.*

16 Trunk lid torque rods - adjustment and removal

Refer to illustration 16.2

1   Torque rods are used to control the amount of effort needed to operate the trunk lid. These can be adjusted, if necessary, by moving the rods to different notches.

2   To move the rod, push a length of 1/2-inch internal diameter pipe over its end and use the pipe as a lever **(see illustration)**.

3   Removal is carried out in a similar way, but relieve the torque rod tension gently, then slide out the rod.

17 Windshield and fixed glass - replacement

Replacement of the windshield and fixed glass requires the use of special fast-setting adhesive/caulk materials and some specialized tools and techniques. These operations should be left to a dealer service department or a shop specializing in glass work.

18 Rear quarter trim panel - removal and installation

1   Remove the rear seat cushion and back.

2   Remove the rear quarter upper trim.

3   Remove the window crank (if equipped) and the trim panel securing screws.

4   Remove the door sill plate and molding and slide the trim panel forward to remove it.

5   Installation is the reverse of removal.

19 Rear quarter window glass and regulator- removal and installation

Rear quarter window glass

1   Push in the trim panel and use a hooked piece of wire or screwdriver to remove the

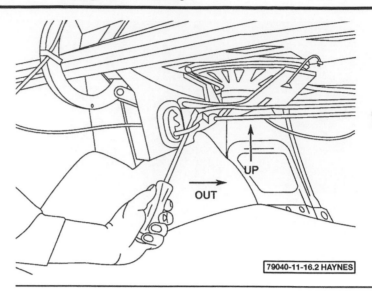

16.2 Using a piece of pipe or large screwdriver, pry the torque rod end out of the notch

`79040-11-16.2 HAYNES`

retaining clip from the window crank.

2   Lower the glass halfway down. Remove the window handle and trim panel (see Section 18).

3   Remove the rear stop bolt and lower bolt.

4   Remove the upper support bolts and lower guide bolts.

5   Disengage the regulator lift arm from the sash channel cam and remove the entire assembly through the access hole.

6   Installation is the reverse of removal. When installing the winder handle, place the clip into the handle and then push the handle onto the shaft.

Rear quarter window regulator

7   Remove the glass/guide assembly as described in Step 4.

8   If power-operated windows are fitted, disconnect the wiring from the motor.

9   Remove the quarter window rear up stop (manual regulator), also the down stop (power-operated).

10 Unscrew and remove the regulator mounting bolts and withdraw the regulator through the door opening.

20 Door trim panel - removal and installation

Refer to illustrations 20.3a, 20.3b, 20.3c, 20.3d, 20.3e, 20.5, 20.6, 20.7, 20.8a, 20.8b and 20.9

1   Disconnect the negative cable from the battery.

2   Unscrew and remove the door lock knob (if equipped).

3   Remove all the trim panel screws **(see illustrations)**. Some screws are visible, but others are hidden behind covers. Check the armrest and door pull areas carefully, since they usually have screws. On earlier models, be sure to check beneath the armrest assembly - there are usually two or more screws here, deeply recessed in the armrest.

4   There is usually a trim plate around the inside door latch handle that is secured by one or more screws. Remove the screw(s), lift the handle away from the door and carefully work the trim plate over the handle.

5   On models with manual windows, remove the window crank handle **(see illustration)**.

20.3a Some door panel screws are visible, such as these at the mirror trim plate

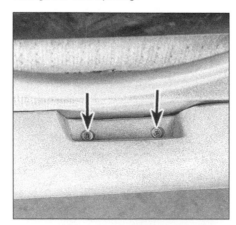

20.3b Other screws will require a little searching, such as these within the armrest opening (arrows)

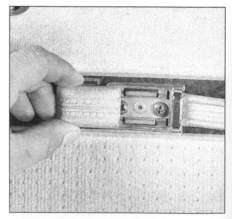

20.3c Still other trim panel screws are hidden beneath pry-off covers, such as these at the ends of the door pull strap

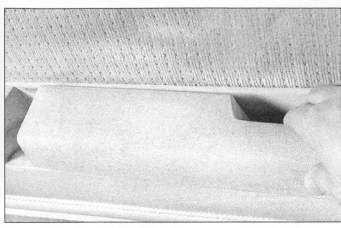

20.3d On models with an armrest assembly like this, remove the cushion . . .

20.3e . . . for access to the screws (arrows)

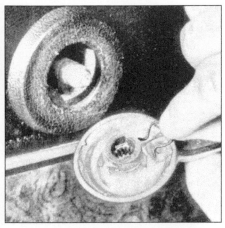

20.5 To remove the window crank handle, use a special tool (available at auto parts stores) or a hooked tool to pull out this clip

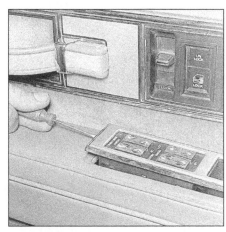

20.6 Remove any screws on the face of the switch panel, then pry it up gently

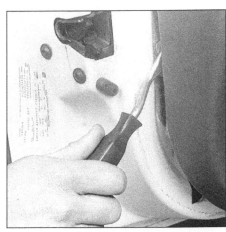

20.7 Carefully pry loose the door panel clips around the edge of the panel - use a special tool like this, if possible

6   On power window models, remove any screws on the switch panel, then pry it up gently **(see illustration)**. Some early models have screws, but most later models simply clip into place. After the panel is lifted up, disconnect the electrical connectors from the underside of the panel.

7   Using a special clip removal tool (available at auto parts stores), carefully work around the perimeter of the door panel to find all the trim panel clips and pry them gently loose from the door **(see illustration)**. If a special tool is not available, you can sometimes use a putty knife without damaging the door panel, but be careful.

8   Carefully pull the panel away from the door, lift it up and move it away from the door. If any resistance is felt, you've missed a screw or fastener somewhere, so inspect the panel again. Note that some panels are in two pieces, so remove them both **(see illustrations)**.

9   After pulling the door panel loose from

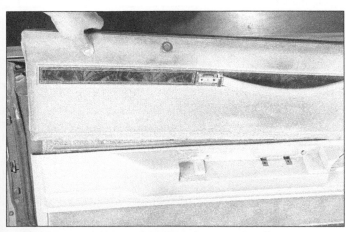

20.8a Many models have two-piece door panels - first remove the upper panel . . .

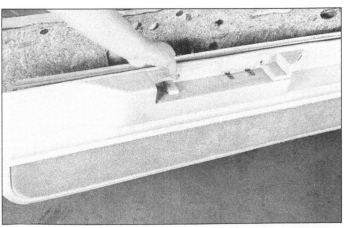

20.8b . . . then the lower panel

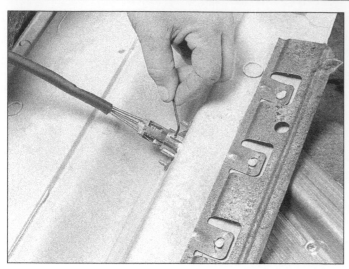

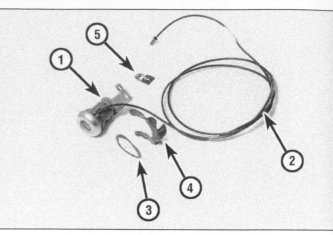

21.1 Typical door lock cylinder assembly

1	Lock cylinder	3	Body sealing gasket
2	Fiber optic cable (if equipped)	4	Lock cylinder retaining clip
		5	Lock rod retaining clip

20.9 The mirror remote control cable is usually attached with a small, hex-headed set screw

the door there will usually be electrical connectors to unplug. On some models you will also have to disconnect the remote mirror cable before the panel can be lifted away **(see illustration)**.

10   For access to the inner door, carefully peel back the plastic watershield.

11   Prior to installation of the door panel, be sure to reinstall any clips in the panel which may have come out during the removal procedure and remain in the door itself.

12   Plug in the wire harness connectors and place the panel in position in the door. Press the door panel into place until the clips are seated and install the armrest/door pulls. Install the manual window crank or power window switch assembly. When installing the manual window crank, put the clip into position on the crank, then firmly press or tap the handle into place.

21 Door lock cylinder and latch - removal and installation

Front door

Lock cylinder

Refer to illustration 21.1

1   To remove the lock cylinder, remove the trim panel and the water deflector. Raise the window, then use a screwdriver to slide the lock cylinder retaining clip out of engagement **(see illustration)**. Disengage the lock rod, disconnect the fiber optic cable (if equipped),

then pull the cylinder out from the outside. Installation is the reverse of the removal procedure.

Latch assembly

Refer to illustration 21.4

2   To remove the latch assembly, raise the window, remove the trim panel and water deflector.

3   Working through the large access hole, disengage the remote control-to-latch connecting rod (spring clip). It may be necessary to disengage the inside locking rod from the latch, which can be achieved by sliding the plastic retaining sleeves toward each other.

4   Remove the three screws securing the latch to the rear of the door **(see illustration)**. Remove the latch assembly from the door. On some models the inside locking rod must be removed from the latch after removal of the latch assembly.

5   Installation is the reverse of removal.

22 Door handle - removal and installation

Outside handle

Refer to illustrations 22.3a and 22.3b

1   Different designs of outside door handles are used according to vehicle model and date of production, but removal is similar for all types.

2   Raise the window completely, remove the door trim and peel away the upper corner of the water deflector in order to gain access to the outside handle mounting nuts or bolts.

3   Unscrew the nuts or bolts, disconnect the control rod and remove the handle and gaskets **(see illustrations)**.

4   Installation is the reverse of removal.

Inside handle

Refer to illustration 22.6

5   Remove the door panel and watershield

21.4 Door latch mounting screws (arrows)

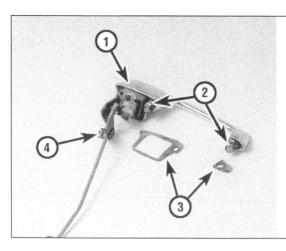

22.3a Typical outside door handle assembly (push-button type)

1   Door handle
2   Mounting bolts
3   Body sealing gaskets
4   Spring clip

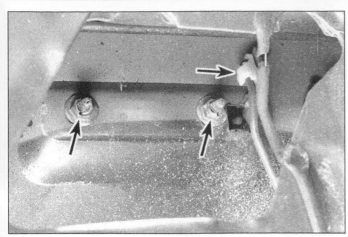

22.3b Liftbar-type door handle mounting nuts and rod retaining clip (arrows)

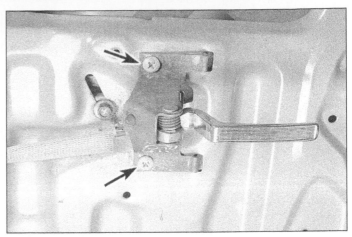

22.6 Remove the screws (arrows), separate the handle from the door and disconnect the operating rod

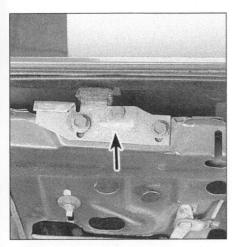

23.2a Typical window stabilizer guide assembly (arrow)

23.2b Typical rear travel stop

23 Front door glass - removal and installation

Refer to illustrations 23.2a, 23.2b, 23.3 and 23.4

1   Remove the door trim panel and water deflector.

2   Remove the weatherstrip clips, the travel stops and the stabilizer guide assembly **(see illustrations)**.

3   Mark the position of the vertical guide bolts **(see illustration)** and remove them, disengage the guide from the roller and rest the guide in the bottom of the door.

4   Set the window to the half-raised position and remove the lower sash channel nuts **(see illustration)**. Now raise the window glass completely and remove the window roller nuts.

5   Tilt the top of the glass until the rear roller is clear of the inner panel and then lift the glass from the door.

6   Installation is the reverse of removal, but adjust the channels, guides and stops as necessary to give smooth operation before finally tightening their mounting nuts.

(see Section 20). Note that on some earlier models the door handle screws can be accessed without removing the trim panel; remove the handle trim plate to see if you can access the screws.

6   Remove the screws securing the handle assembly to the door **(see illustration)**.

7   Lift the handle assembly away from the door and disengage the operating rod from the handle.

8   Installation is the reverse of removal.

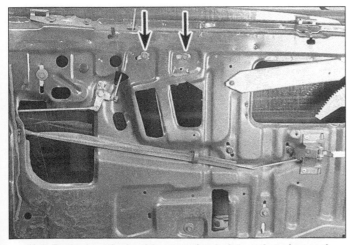

23.3 Typical vertical guide mounting bolt locations (arrows)

23.4 The window glass channel mounting nuts (arrow) can be removed through the inner panel access holes

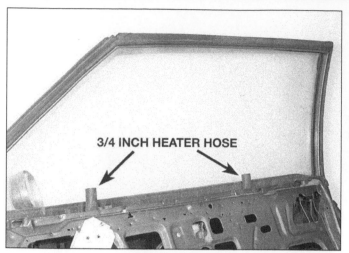

24.2 Use two pieces of 3/4-inch heater hose, wedged between the glass and the door frame, to hold the glass in the raised position

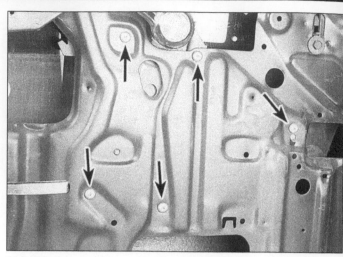

24.3 Window regulator mounting rivets (arrows) - Late model with power windows shown

24 Front door window regulator - removal and installation

Refer to illustrations 24.2 and 24.3

1   Remove the door trim panel and water deflector.

2   Raise the window completely and secure in this position by wedging two short sections of 3/4 inch heater hose between the glass and the door frame **(see illustration)**.

3   Mark the position of the cam attaching bolts and remove the regulator bolts or rivets **(see illustration)**.

4   With electrically operated regulators, disconnect the wiring harness. **Warning:** *It is imperative that the regulator sector gear on electrically operated regulators be locked into position before removing the motor from the regulator. The control arms are under pres-sure and can cause serious injury if the motor is removed without performing the operation described in Chapter 12, Section 26, Steps 12 through 18.*

5   Slide the front regulator upper arm from the sash, then slide the rear lift arm forward from the sash cam.

6   Slide the regulator to the rear and withdraw it through the lower rear access hole.

7   Installation is the reverse of removal.

25 Door - removal and installation

1   Remove the door trim panel. Disconnect any wire harness connectors and push them through the door opening so they won't interfere with door removal.

2   Place a jack or jackstand under the door or have an assistant on hand to support it when the hinge bolts are removed. **Note:** *If a jack or jackstand is used, place a rag between it and the door to protect the door's painted surfaces.*

3   Scribe around the door hinges.

4   Remove the hinge-to-door bolts or drive out the pins and carefully lift off the door.

5   Installation is the reverse of removal.

6   Following installation of the door, check the alignment and adjust it if necessary as follows:

a)   *Up-and-down and forward-and-backward adjustments are made by loosening the hinge-to-body bolts and moving the door as necessary.*

b)   *The door lock striker can also be adjusted both up-and-down and sideways to provide positive engagement with the lock mechanism. This is done by loosening the mounting bolts and moving the striker as necessary.*

Chapter 12
Chassis electrical system

Contents

1   General information

The electrical system is a 12-volt, negative ground type. Power for the lights and all electrical accessories is supplied by a lead/acid-type battery which is charged by the alternator.

This Chapter covers repair and service procedures for the various electrical components not associated with the engine. Information on the battery, alternator, distributor and starter motor can be found in Chapter 5.

It should be noted that when portions of the electrical system are serviced, the negative battery cable should be disconnected from the battery to prevent electrical shorts and/or fires.

2   Electrical troubleshooting - general information

A typical electrical circuit consists of an electrical component, any switches, relays, motors, fuses, fusible links or circuit breakers related to that component and the wiring and connectors that link the component to both the battery and the chassis. To help you pinpoint an electrical circuit problem, wiring diagrams are included at the end of this book.

Before tackling any troublesome electrical circuit, first study the appropriate wiring diagrams to get a complete understanding of what makes up that individual circuit. Trouble spots, for instance, can often be narrowed down by noting if other components related to the circuit are operating properly. If several components or circuits fail at one time, chances are the problem is in a fuse or ground connection, because several circuits are often routed through the same fuse and ground connections.

Electrical problems usually stem from simple causes, such as loose or corroded connections, a blown fuse, a melted fusible link or a bad relay. Visually inspect the condition of all fuses, wires and connections in a problem circuit before troubleshooting it.

If testing instruments are going to be utilized, use the diagrams to plan ahead of time where you will make the necessary connections in order to accurately pinpoint the trouble spot.

The basic tools needed for electrical troubleshooting include a circuit tester or voltmeter (a 12-volt bulb with a set of test leads can also be used), a continuity tester, which includes a bulb, battery and set of test leads, and a jumper wire, preferably with a circuit breaker incorporated, which can be used to bypass electrical components. Before attempting to locate a problem with test instruments, use the wiring diagram(s) to decide where to make the connections.

Voltage checks

Voltage checks should be performed if a circuit is not functioning properly. Connect one lead of a circuit tester to either the negative battery terminal or a known good ground. Connect the other lead to a connector in the circuit being tested, preferably nearest to the battery or fuse. If the bulb of the tester lights, voltage is present, which means that the part of the circuit between the connector and the battery is problem free. Continue checking the rest of the circuit in the same fashion. When you reach a point at which no voltage is present, the problem lies between that point and the last test point with voltage. Most of the time the problem can be traced to a loose connection. **Note:** *Keep in mind that some circuits receive voltage only when the ignition key is in the Accessory or Run position.*

Finding a short

One method of finding shorts in a circuit is to remove the fuse and connect a test light or voltmeter in its place to the fuse terminals. There should be no voltage present in the circuit. Move the wiring harness from side to side while watching the test light. If the bulb goes on, there is a short to ground somewhere in that area, probably where the insulation has rubbed through. The same test can be performed on each component in the circuit, even a switch.

Ground check

Perform a ground test to check whether a component is properly grounded. Disconnect the battery and connect one lead of a self-powered test light, known as a continuity tester, to a known good ground. Connect the other lead to the wire or ground connection being tested. If the bulb goes on, the ground is good. If the bulb does not go on, the ground is not good.

Continuity check

A continuity check is done to determine if there are any breaks in a circuit - if it is passing electricity properly. With the circuit off (no power in the circuit), a self-powered continuity tester can be used to check the circuit. Connect the test leads to both ends of the circuit (or to the "power" end and a good ground), and if the test light comes on, the circuit is passing current properly. If the light doesn't come on, there is a break somewhere in the circuit. The same procedure can be used to test a switch, by connecting the continuity tester to the switch terminals. With the switch turned On, the test light should come on.

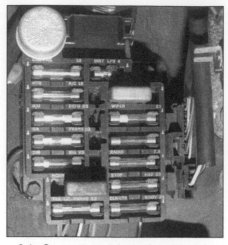

3.1a On most models the fuse block is located under the dash to the left of the driver (early style fuse block, with glass fuses shown)

3.1b Later models use miniaturized fuses in the fuse block

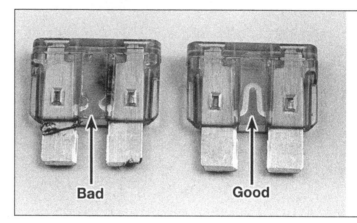

3.4 When a fuse blows, the element between the terminals melts - the fuse on the left is blown, the fuse on the right is good

Finding an open circuit

When diagnosing for possible open circuits, it is often difficult to locate them by sight because oxidation or terminal misalignment are hidden by the connectors. Merely wiggling a connector on a sensor or in the wiring harness may correct the open circuit condition. Remember this when an open circuit is indicated when troubleshooting a circuit. Intermittent problems may also be caused by oxidized or loose connections.

Electrical troubleshooting is simple if you keep in mind that all electrical circuits are basically electricity running from the battery, through the wires, switches, relays, fuses and fusible links to each electrical component (light bulb, motor, etc.) and to ground, from which it is passed back to the battery. Any electrical problem is an interruption in the flow of electricity to and from the battery.

---

3   Fuses - general information

Refer to illustrations 3.1a, 3.1b and 3.4

The electrical circuits of the vehicle are protected by a combination of fuses, circuit breakers and fusible links. The fuse block is located under the instrument panel on the left side of the dashboard **(see illustrations)**. On some later models, the fuse block is incorporated into the wiring harness adjacent to the steering column.

Each of the fuses is designed to protect a specific circuit, and the various circuits are identified on the fuse panel itself.

A blown fuse on earlier models can be readily identified by inspecting the element inside the glass tube. If this metal element is broken, the fuse is inoperable and must be replaced with a new one.

On later models, miniaturized fuses are employed in the fuse block. These compact fuses, with blade terminal design, allow fingertip removal and replacement. If an electrical component fails, always check the fuse first. The easiest way to check fuses is with a test light. Check for power at the exposed terminal tips of each fuse. If power is present on one side of the fuse but not the other, the fuse is blown. A blown fuse can also be confirmed by visually inspecting it **(see illustration)**.

Be sure to replace blown fuses with the correct type. Fuses of different ratings are physically interchangeable, but only fuses of the proper rating should be used. Replacing a fuse with one of a higher or lower value than specified is not recommended. Each electrical circuit needs a specific amount of protection. The amperage value of each fuse is molded into the fuse body.

If the replacement fuse immediately fails, don't replace it again until the cause of the problem is isolated and corrected. In most cases, the cause will be a short circuit in the wiring caused by a broken or deteriorated wire.

4 Fusible links - general information

Some circuits are protected by fusible links. The links are used in circuits which are not ordinarily fused, such as the ignition circuit.

Although the fusible links appear to be of a heavier gauge than the wire they are protecting, the appearance is due to the thick insulation. All fusible links are four wire gauges smaller than the wire they are designed to protect.

Fusible links cannot be repaired, but a new link of the same size wire can be put in its place. The procedure is as follows:

a) *Disconnect the negative cable from the battery.*
b) *Disconnect the fusible link from the wiring harness.*
c) *Cut the damaged fusible link out of the wiring just behind the connector.*
d) *Strip the insulation back approximately 1/2-inch.*
e) *Position the connector on the new fusible link and crimp it into place.*
f) *Use rosin core solder at each end of the new link to obtain a good solder joint.*
g) *Use plenty of electrical tape around the soldered joint. No wires should be exposed.*
h) *Connect the battery ground cable. Test the circuit for proper operation.*

5 Circuit breakers - general information

Circuit breakers protect components such as power windows, power door locks and headlights. Some circuit breakers are located in the fuse box.

On some models the circuit breaker resets itself automatically, so an electrical overload in a circuit breaker protected system will cause the circuit to fail momentarily, then come back on. If the circuit does not come back on, check it immediately. Once the condition is corrected, the circuit breaker will resume its normal function. Some circuit breakers must be reset manually.

6 Relays - general information

Several electrical accessories in the vehicle use relays to transmit the electrical signal to the component. If the relay is defective, that component will not operate properly.

The various relays are grouped together in several locations. If a faulty relay is suspected, it can be removed and tested by a dealer service department or a repair shop. Defective relays must be replaced as a unit.

7 Turn signal and hazard flashers - check and replacement

Turn signal flasher

1   The turn signal flasher, a small canister-shaped unit located in the fuse block or in the wiring harness under the dash, flashes the turn signals.
2   When the flasher unit is functioning properly, an audible click can be heard during its operation. If the turn signals fail on one side or the other and the flasher unit does not make its characteristic clicking sound, a faulty turn signal bulb is indicated.

3   If both turn signals fail to blink, the problem may be due to a blown fuse, a faulty flasher unit, a broken switch or a loose or open connection. If a quick check of the fuse box indicates that the turn signal fuse has blown, check the wiring for a short before installing a new fuse.
4   To replace the flasher, simply pull it out of the fuse block or wiring harness.
5   Make sure that the replacement unit is identical to the original. Compare the old one to the new one before installing it.
6   Installation is the reverse of removal.

Hazard flasher

7   The hazard flasher, a small canister-shaped unit located in the fuse block or the wiring harness, flashes all four turn signals simultaneously when activated.
8   The hazard flasher is checked in a fashion similar to the turn signal flasher (see Steps 2 and 3).
9   To replace the hazard flasher, pull it from the back of fuse block.
10   Make sure the replacement unit is identical to the one it replaces. Compare the old one to the new one before installing it.
11   Installation is the reverse of removal.

8 Headlight - removal and installation

Refer to illustrations 8.2, 8.4 and 8.5

1   Whenever replacing a headlight, be careful not to turn the spring-loaded adjusting screws of the headlight, as this will alter the aim.
2   Remove the headlight bezel screws and remove the decorative bezel **(see illustration)**.
3   Use a cotter pin removal tool or similar device to unhook the spring from the retaining ring.
4   Remove the two screws which secure the retaining ring and withdraw the ring **(see illustration)**. Support the light as this is done.

8.2 Use a Phillips-head screwdriver to remove the headlight bezel screws (arrow)

8.4 Remove the retaining screws (don't confuse them with the adjustment screws) and lift the headlight ring off (round headlight shown, rectangular headlight similar)

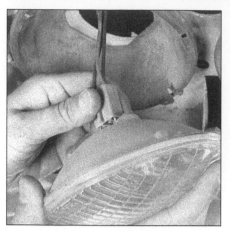

8.5 Pull the headlight forward and unplug the connector

9.1 Typical headlight adjusting screw locations

5   Pull the sealed beam unit outward slightly and disconnect the electrical connector from the rear of the light **(see illustration)**. Remove the light from the vehicle.

6   Position the new unit close enough to connect the electrical connector. Make sure that the numbers molded into the lens are at the top.

7   Install the retaining ring with its mounting screws and spring.

8   Install the decorative bezel and check for proper operation. If the adjusting screws were not altered, the new headlight will not need to have its aim adjusted.

9   Headlights - adjustment

Refer to illustrations 9.1 and 9.3
Note: *The headlights must be aimed correctly. If adjusted incorrectly they could blind the driver of an oncoming vehicle and cause a serious accident or seriously reduce your ability to see the road. The headlights should be checked for proper aim every 12 months and any time a new headlight is installed or front end body work is performed. It should be emphasized that the following procedure is only an interim step which will provide temporary adjustment until the headlights can be adjusted by a properly equipped shop.*

1   Headlights have two spring loaded adjusting screws, one on the top controlling up-and-down movement and one on the side controlling left-and-right movement **(see illustration)**.

2   There are several methods of adjusting the headlights. The simplest method requires a blank wall 25 feet in front of the vehicle and a level floor.

3   Position masking tape vertically on the wall in reference to the vehicle centerline and the centerlines of both headlights **(see illustration)**.

4   Position a horizontal tape line in reference to the centerline of all the headlights. **Note:** *It may be easier to position the tape on the wall with the vehicle parked only a few inches away.*

5   Adjustment should be made with the vehicle sitting level, the gas tank half-full and no unusually heavy load in the vehicle.

6   Starting with the low beam adjustment, position the high intensity zone so it is two inches below the horizontal line and two inches to the side of the headlight vertical line, away from oncoming traffic. Adjustment is made by turning the top adjusting screw clockwise to raise the beam and counterclockwise to lower the beam. The adjusting screw on the side should be used in the same manner to move the beam left or right.

7   With the high beams on, the high intensity zone should be vertically centered with the exact center just below the horizontal line. **Note:** *It may not be possible to position the headlight aim exactly for both*

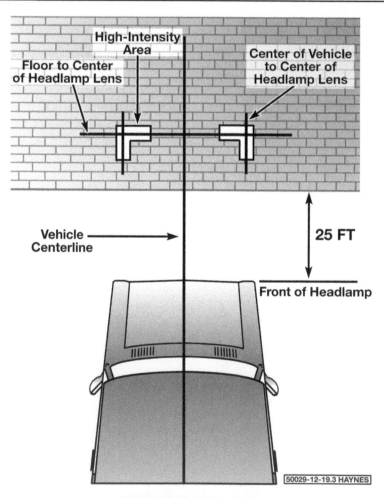

9.3 Headlight aiming details

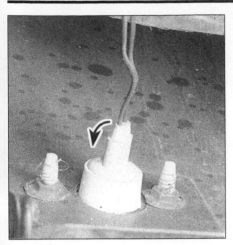

10.4 Twist the bulb holder 1/4-turn counterclockwise and pull it out of the lens assembly

11.2 Twist the bulb holder counterclockwise to remove it from the housing

high and low beams. If a compromise must be made, keep in mind that the low beams are the most used and have the greatest effect on driver safety.

8   Have the headlights adjusted by a dealer service department or service station at the earliest opportunity.

10 Bulb replacement - front end

Parking light

1970 through 1973 models

1   Remove the lens screws, take the lens off and withdraw the bulb from its holder.

1974 and later models

2   Twist the socket out of the rear of the housing.
3   Pull the bulb straight out of the holder.

Front side marker light

Refer to illustration 10.4

4   On most models, twist the bulb holder at the rear of the light 1/4-turn to release it **(see illustration)**. The bulb can then be withdrawn from its socket.
5   If the bulb holder cannot be accessed from the rear, the side marker light lens can be taken off after removing two screws.

11 Bulb replacement - rear end

Refer to illustration 11.2

Stop, tail, turn signal and back-up lights

1970 through 1972 models

1   Access to all bulbs is obtained by removing the lens screws and the lens.

1973 and later models

2   The bulb holders can be accessed from

the backs of the housings by working within the trunk **(see illustration)**.
3   Twist the bulb holder at the rear of the light 1/4-turn to release it.

License plate light

1970 through 1973 models

4   Access to the bulb is obtained after removing the lens securing screws and taking off the lens.

1974 and later models

5   Remove the light mounting screws and lower the light until the bulb holder can be twisted from its socket in the light body.

Rear marker light

6   Twist the bulb holder from its socket after reaching into the trunk or up under the rear fender, according to design.

12 Bulb replacement - interior

Center console lights

1   Pry up the switch assembly from the console and remove the bulb from its socket.
2   The courtesy light bulb is accessible after removing the lens screws and taking off the lens.

Automatic floor shift quadrant lights

3   Remove the quadrant trim plate from the console and withdraw the light socket.

Interior (roof) light

4   Pinch the sides of the plastic lens together and remove it. It may be necessary to use a small screwdriver to pry off the lens.
5   The festoon-type bulb can now be carefully pried from between the spring contacts. Pry only on the metal ends - not on the glass.

13 Bulb replacement - instrument cluster

1970 through 1972 models

1   To replace a bulb in the instrument panel cluster, extract the 6 screws which attach the panel pad and remove the pad.
2   By reaching over the top of the panel. the bulb holders can be reached. Use a mirror if necessary.
3   To replace an indicator bulb. pry off the 3 screw covers, remove the screws and bezel.

1973 and later models

4   To reach the instrument panel bulbs, the instrument panel bezel must first be withdrawn. To do this, disconnect the battery ground cable.
5   Remove the clock and radio control knobs.
6   Remove the instrument panel bezel retaining screws.
7   Withdraw the bezel far enough to be able to disconnect the rear defogger switch `if so equipped. Remove the bezel.
8   It may be necessary to remove certain instruments to gain access to all of the wedge-base bulbs.

14 Headlight switch - removal and installation

Early models

Refer to illustration 14.3

1   Disconnect the battery ground cable.
2   Pull the headlight control knob to the On position.
3   Reach up under the instrument panel and depress the switch shaft retainer button **(see illustration)**. While holding the button, pull out the knob and shaft assembly.
4   Remove the bezel nut and switch from

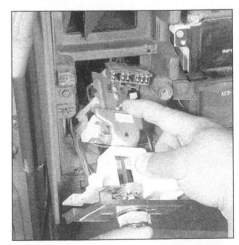

14.3 Reach up under the dash to depress the button and remove the headlight knob and shaft (switch shown removed for clarity)

16.3 Use a screwdriver to pry up the cover plate

16.4 A special tool is required to depress the steering shaft lock plate so the retaining ring can be removed

the instrument panel and then disconnect the wiring plug.

Later models

5   Removal is the same as for earlier models except that the instrument panel pad is fixed, and access to the switch is gained by removing the instrument bezel. The switch is then removed by removing the screws and pulling it out of the instrument panel.

---

15 Windshield wiper/washer switch - removal and installation

1   Disconnect the battery ground cable.

1970 through 1985 models

2   On 1970 through 1973 models the wiper/washer switch is accessed from the rear of the instrument panel. On these models, remove the screws securing the instrument cluster and pull the cluster outward slightly to gain better access to the switch (see Section 19). Disconnect the electrical connector from the rear of the switch. Remove the screws securing the switch to the instrument panel and remove the switch.

3   On 1974 through 1985 models the wiper/washer switch is accessed from the front of the instrument panel. On these models, remove the instrument cluster bezel (see Section 19). Remove the screws securing the switch to the instrument panel. Withdraw the switch from the instrument panel and disconnect the electrical connector from the rear of the switch.

1986 and later models

4   The wiper/washer switch is located on the left-hand side of the steering column, under the turn signal switch.
5   To remove the wiper/washer switch, first remove the steering wheel (Chapter 10) and the turn signal switch (Section 16).
6   Remove the upper mounting screw on the ignition and dimmer switch. This releases the dimmer switch and actuator rod assembly. Be careful not to move the ignition switch. If this does happen, it will require adjustment before reassembly.
7   The wiper/washer switch and pivot assembly can now be removed from the column housing after disconnecting the wire harness.
8   During installation, place the wiper/washer switch and pivot assembly into the housing and feed the connector down through the bowl and shroud assembly.
9   Install the switch as outlined in Section 16.
10   Install the pinched end of the dimmer switch actuator rod into the dimmer switch.

Feed the other end of the rod through the hose in the shroud and into the wiper/washer switch and pivot assembly drive. Do not tighten the mounting screw yet. Depress the dimmer switch slightly and use a 3/32-inch drill bit to locate the switch correctly. Push the switch up to remove the play between the ignition and dimmer switches and the actuator rod. Tighten the switch screw securely. Remove the drill bit and check the dimmer switch function by operating the lever.

---

16 Turn signal switch - removal and installation

Refer to illustrations 16.3, 16.4, 16.5 and 16.8
1   Disconnect the negative battery cable and remove the steering wheel (Chapter 10).
2   Remove the steering column trim cover located at the base of the dashboard.
3   At the end of the steering column, late models have a plastic cover plate which should be pried out of the column, using a screwdriver in the slots provided **(see illustration)**.
4   The lock plate will now have to be removed from the steering column. This is held in place with a snap-ring which fits into a groove in the steering shaft. The lock plate must be depressed to relieve pressure on the snap-ring. A special U-shaped tool which fits on the shaft should be used to depress the lock plate as the snap-ring is removed from its groove **(see illustration)**.

16.5 Lift off the canceling cam and spring

16.8 Remove the turn signal switch assembly mounting screws

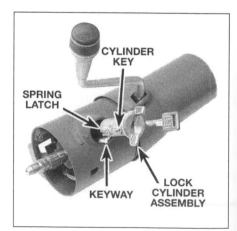

17.3 On 1970 through 1978 models, the lock cylinder is retained by a spring latch

CYLINDER KEY

SPRING LATCH

KEYWAY

LOCK CYLINDER ASSEMBLY

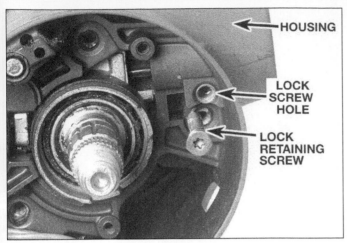

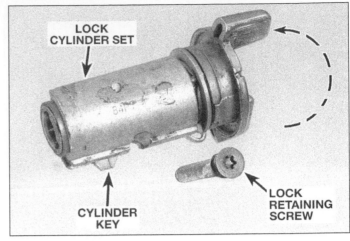

17.12 On 1979 and later models, the lock cylinder is held in place by a screw

5   Slide the canceling cam, upper bearing preload spring and thrust washer off the end of the shaft **(see illustration)**.
6   Remove the turn signal lever attaching screw and withdraw the turn signal lever from the side of the column.
7   Push in the hazard warning knob and unscrew the knob from the threaded shaft. On later models, you'll need to remove a screw.
8   Remove the three turn signal assembly mounting screws **(see illustration)**.
9   Disconnect the long, thin connector for the turn signal switch, located at the base of the steering column.
10   Pull the switch wiring connector out of the bracket on the steering column jacket. Tape the connector terminals to prevent damage. Feed the wiring connector up through the column support bracket and pull the switch, wiring harness and connectors out the top of the steering column.
11   Installation is the reverse of removal; however, make sure the wiring harness is in the protector as it is pulled into position. Before installing the thrust washer, upper bearing preload spring and canceling cam, make sure the switch is in the neutral position and the warning knob is pulled out. Always use a new snap-ring on the shaft for the lock plate.

---

17   Ignition lock cylinder - removal and installation

1970 through 1978 models

Refer to illustration 17.3

1   The lock cylinder is located on the upper right-hand side of the steering column. On these models, the lock cylinder should be removed only in the Run position, otherwise damage to the warning buzzer switch may occur.
2   Remove the steering wheel (Chapter 10) and turn signal switch (Section 16). **Note:** *The turn signal switch need not be fully removed*

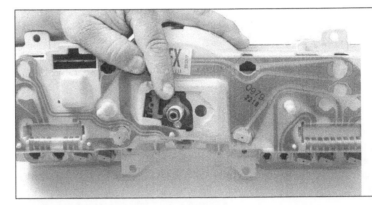

18.1 Reach behind the cluster and depress the clip to release the speedometer cable

provided that it is pushed to the rear far enough for it to be slipped over the end of the shaft. Do not pull the harness out of the column.
3   Insert a thin bladed screwdriver into the slot in the turn signal switch housing. Break the housing flash loose and at the same time depress the spring latch at the lower end of the lock cylinder. Holding the latch depressed, withdraw the lock cylinder from the housing **(see illustration)**.
4   To install the new lock cylinder/sleeve assembly, hold the sleeve and rotate the lock clockwise against the stop.
5   Insert the cylinder/sleeve assembly into the housing so the key on the cylinder sleeve is aligned with the housing keyway.
6   Insert a 0.070-inch diameter drill between the lock bezel and the housing and then rotate the cylinder counterclockwise, maintaining pressure on the cylinder until the drive section mates with the sector.
7   Press in the lock cylinder until the snap-ring engages in the grooves and secures the cylinder in the housing. Remove the drill and check the lock action.
8   Install the turn signal switch and the steering wheel.

1979 and later models

Refer to illustration 17.12

9   The lock cylinder should be removed in the Run position only.

10   Remove the steering wheel (Chapter 10) and turn signal switch (Section 16). It is not necessary to completely remove the switch. Pull it up and over the end of the steering shaft. Do not pull the wiring harness out of the column.
11   Remove the ignition key warning switch.
12   Using a magnetized screwdriver, remove the lock retaining screw. Do not allow this screw to drop down into the column, as this will require a complete disassembly of the steering column to retrieve the screw **(see illustration)**.
13   Pull the lock cylinder out of the side of the steering column.
14   To install, rotate the lock cylinder set and align the cylinder key with the keyway in the steering column housing.
15   Push the lock all the way in and install the retaining screw.
16   Install the remaining components referring to the appropriate Sections.

---

18   Speedometer cable - replacement

Refer to illustrations 18.1 and 18.2

1   The instrument cluster must first be removed as described in Section 19. Depress the retaining clip on the back of the cluster to release the speedometer cable **(see illustration)**.

18.2 Use pliers to pull the speedometer cable out of the housing

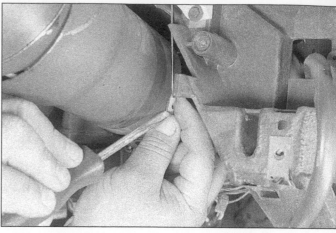

19.24 Use a screwdriver to detach the shift indicator cable from the steering column (later models)

2   Once the end of the speedometer drive cable is exposed, grip the inner cable with pliers and draw it out of the housing **(see illustration)**.

3   Installation is the reverse of removal, but on 1970 through 1974 models, lubricate the lower three-quarters of the inner cable before inserting it. On later models, lubricate the entire cable. Use special speedometer cable lubricant, not oil, for this purpose.

4   Insert the cable into the housing using a twisting movement until the lower end is felt to engage with the pinion gear at the transmission.

19   Instrument cluster and bezel - removal and installation

1970 through 1972 models

1   Disconnect the battery ground cable.

2   Lower the upper end of the steering column and support it .

3   Disconnect the parking brake hand release rod attachment.

4   Disconnect the speedometer cable.

5   Remove the instrument panel pad (16 screws).

6   Disconnect the radio speaker support bracket from the instrument panel.

7   If air conditioning is installed, remove the center outlet (3 screws) and the control assembly (4 screws).

8   Disconnect the speaker from the radio, also the antenna and power leads. Pull off the radio knobs, washers and bezels.

9   Unbolt the radio support braces and withdraw the radio from under the instrument panel.

10   Remove the 7 instrument panel bolts and withdraw the instrument panel. The help of an assistant will be required for this job.

11   Extract the three telltale snap covers and remove the three telltale housing mounting screws and then remove the housing.

12   Remove the clock setting knob.

13   Separate the instrument cluster from the

19.25 Typical instrument cluster screw locations (arrows)

carrier by extracting the 8 securing screws.

14   Remove the 3 instrument cluster lens screws and withdraw the lens.

15   The individual instruments are now accessible.

16   Installation is the reverse of removal.

1973 through 1981 models

Bezel

17   Disconnect the negative cable from the battery.

18   Remove the clock setting knob if equipped. Remove the radio control knobs if the model your working on incorporates the radio with the instrument cluster bezel.

19   Remove the screws and take off the lower trim panel, located directly underneath the instrument panel. On most models, removing this panel will expose the lower instrument bezel retaining screws.

20   Remove the headlight switch knob and shaft assembly, referring to Section 14, Step 3. Many earlier models have a bezel nut around the switch assembly that must be removed before the bezel will come off. Usually, the nut has two recesses that a small set of needle-nose pliers will fit into, allowing you to loosen the nut.

21   Remove the bezel retaining screws or nuts. On most models, there are upper and lower screws only. Some earlier models are

attached with screws or nuts from the back - you'll have to work up under the dash to access these.

22   Carefully lift the bezel away from the dashboard and disconnect the rear defogger switch if so equipped and remove the bezel. If *any* resistance is encountered, find the cause - there's probably a screw you missed somewhere. In some cases, clips hold part of the bezel in place, so you'll need to pry gently at these spots with a padded screwdriver.

Cluster

Refer to illustrations 19.24 and 19.25

23   Remove the bezel, as described in Steps 17 through 22.

24   Remove the shift indicator (early models) or detach the indicator cable on later models **(see illustration)**.

25   Remove the cluster retaining screws or nuts **(see illustration)**. Pull the cluster out and disconnect the speedometer cable **(see illustration 18.1)**. On some later models it may be necessary to disconnect the cable at the transmission to provide sufficient slack for the cluster to be pulled out for access to the connectors.

26   Unplug any electrical connectors that would interfere with removal.

27   Detach the cluster from the instrument panel.

28   Installation is the reverse of removal.

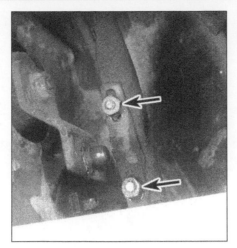
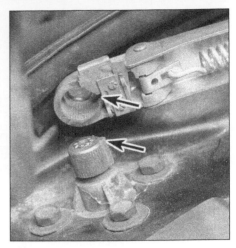

21.9 Typical windshield wiper motor installation details

20   Windshield wiper arm - removal and installation

1   Make sure the wiper arms are in the self-parked position, the motor having been switched off in the low speed mode.
2   Note carefully the position of the wiper arm in relation to the windshield lower reveal molding. Use tape on the windshield to mark the exact location of the wiper arm on the glass.
3   Using a hooked tool or a small screwdriver, pull aside the small spring tang which holds the wiper arm to the splined transmission shaft and at the same time pull the arm from the shaft.
4   Installation is the reverse of removal, but do not push the arm fully home on the shaft until the alignment of the arm has been checked. If necessary, the arm can be pulled off again and turned through one or two serrations of the shaft to correct the alignment without the necessity of pulling aside the spring tang.
5   Finally, press the arm fully home on its shaft and then wet the windshield glass and operate the motor on low speed to ensure the arc of travel is correct.

21   Windshield wiper motor - removal and installation

1970 models

1   Make sure that the wiper motor is in 'Park' position.
2   Disconnect the wiring and washer hoses from the wiper motor.
3   Remove the plastic access cover and then loosen the nut which retains the crank arm to the motor assembly.
4   Remove the 3 motor mounting screws and withdraw the motor.

1971 and later models

Refer to illustration 21.9
5   Raise the hood and remove the cowl screen.
6   Reaching through the opening, loosen the drive link-to-crank arm nuts.
7   Remove the transmission drive link from the motor crank arm.
8   Disconnect the wiring and the washer hoses from the wiper motor.
9   Remove the three motor mounting screws and withdraw the motor, guiding the crank arm through the hole in the firewall

22.8 Remove the hex nuts (arrows) located behind the radio control knobs

(see illustration).
10   Installation is the reverse of removal, but before connecting the drive link, check that the motor is in Park.

22   Radio - removal and installation

Removal

Refer to illustrations 22.8, 22.11 and 22.12
1   Disconnect the negative terminal of the battery.

1970 and 1971 models

2   Remove the ashtray and the ashtray housing.
3   Remove the radio knobs, controls, washers, trim plate and nuts.
4   Remove the hoses from the center air distribution duct.
5   Disconnect all the radio leads and then extract the screw from the radio rear mounting bracket and remove the radio.

1973 through 1984 models

6   If the vehicle is equipped with air conditioning, remove the lap cooler duct.
7   Turn the radio control knobs until the slot in the base of the knob is visible, depress the metal retainer with a small screwdriver and pull off the knobs.
8   Using a socket, remove the control shaft nuts and their washers **(see illustration)**.
9   Working under the instrument panel, remove the nut from the left-hand side rear bracket which secures the radio to the instrument panel.
10   Push the radio towards the front of the vehicle so that the control shafts can clear the instrument panel. Pull the radio down from the instrument panel until the wiring can be reached and disconnected. Remove the radio.

1985 and later models

11   On later models, remove the radio trim

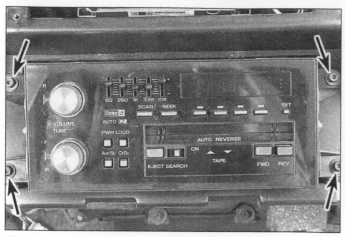

22.11 Remove the radio mounting screws (1985 and later models)

22.12 Pull the radio out from the dash and unplug the antenna and all electrical connectors (arrows)

23.4a When measuring the voltage at the rear window defogger grid, wrap a piece of aluminum foil around the probe of the voltmeter and press the foil against the wire with your finger

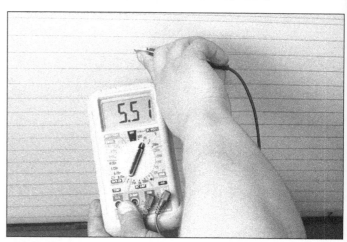

23.4b To determine if a heating element has broken, check the voltage at the center of each element - if the voltage is 6-volts, the element is unbroken

cover. Remove the retaining screws **(see illustration)**.

12   Pull the radio from the instrument panel towards the rear of the vehicle until the wiring can be reached and disconnected **(see illustration)**. Remove the radio.

Installation

All models

13   Installation is a reversal of removal but observe the following precautions. Never switch on power to the receiver until the speaker leads, the antenna lead and the heat sink are all plugged into it.

---

23   Electric grid-type rear defogger - testing and repair

Refer to illustrations 23.4a, 23.4b and 23.11

1   This option consists of a rear window with a number of horizontal elements that are baked into the glass surface during the glass forming operation.

2   Small breaks in the element system can

be successfully repaired without removing the rear window.

3   To test the grids for proper operation, start the engine and turn on the system.

4   Place the negative lead against the negative (ground) bus bar. Check the voltage at the center of each heating element **(see illustrations)**. If the voltage is 6-volts, the element is okay (there is no break). If the voltage is 10-volts or more, the element is broken somewhere between the mid-point and ground. If the voltage is 0-volts the element is broken between the mid-point and the positive side.

5   To find the break, slide the probe toward the positive side. The point at which the voltmeter deflects from zero to several volts is the point at which the heating element is broken. **Note:** *If the heating element is not broken, the voltmeter will indicate 12-volts at the positive side and gradually decrease to 0-volts as you slide the positive probe toward the ground side.*

6   To repair a break in a grid line it is recommended that a repair kit specifically for this purpose be purchased from a your local dealer. Included in the repair kit will be a

decal, a container of silver plastic and hardener, a mixing stick and instructions.

7   To repair a break, first turn off the system and allow it to de-energize for a few minutes.

8   Lightly buff the grid line area with fine steel wool and then clean the area thoroughly with alcohol.

9   Use the decal supplied in the repair kit, or use electrician's tape above and below the area to be repaired. The space between the pieces of tape should be the same as existing grid lines. This can be checked from outside the car. Press the tape tightly against the glass to prevent seepage.

10   Mix the hardener and silver plastic thoroughly.

11   Using the brush from the repair kit, apply the silver plastic mixture between the pieces of tape, overlapping the damaged area slightly on either end **(see illustration)**.

12   Carefully remove the decal or tape and apply a constant stream of hot air directly to the repaired area. A heat gun set at 500 degrees to 700 degrees F is recommended. Hold the gun about one inch from the glass

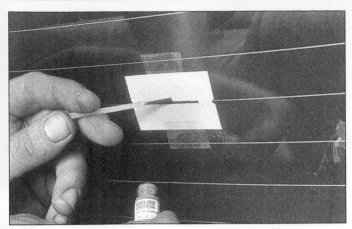

23.11 To use a defogger repair kit, apply masking tape to the inside of the window at the damaged area, then brush on the special conductive coating

25.4 Typical power door lock solenoid installation details

for one to two minutes.

13   If the new grid line appears off color, tincture of iodine can be used to clean the repair and bring it back to the proper color. This mixture should not remain on the repair for more than 30 seconds.

14   Although the defogger is now fully operational, the repaired area should not be disturbed for at least 24 hours.

24   Cruise control system - description and check

The cruise control system maintains vehicle speed with a vacuum actuated servo motor, located in the engine compartment, which is connected to the throttle linkage by a cable. The system consists of the servo motor, clutch switch, brake switch, control switches, a relay and associated vacuum hoses.

Diagnosis can usually be limited to simple checks of the wiring and vacuum connections for minor faults which can be easily repaired. These include:

a) *Inspecting the cruise control actuating switches for broken wires and loose connections.*

b) *Checking the cruise control fuse.*

c) *Checking the hoses in the engine compartment for tight connections, cracks and obvious vacuum leaks. The cruise control system is operated by vacuum, so it's critical that all vacuum switches, hoses and connections are secure.*

25   Power door lock system - description and solenoid replacement

Refer to illustration 25.4

1   This optional system incorporates a solenoid actuator inside each door. The solenoid is electrically operated from a con-

trol switch on the instrument panel and operates the lock through a linkage. Each actuator has an internal circuit breaker which may require one to three minutes to reset.

2   To remove the solenoid, raise the door window and remove the door panel trim pad as described in Chapter 11.

3   After prying away the water deflector, the solenoid can be seen through the large access hole. The solenoid can be mounted to either the rear door lock pillar or the inner metal door panel.

4   Early models use attaching screws through the door panel and into the solenoid bracket. Later models use rivets to secure the solenoid to the pillar **(see illustration)**. These must be drilled out using a 1/4-inch drill bit.

5   Once the securing devices are removed, disconnect the wiring harness at the solenoid and the actuating link, held in place with a metal clip. Remove the solenoid from the door cavity.

6   To install, place the solenoid in position and connect the electrical connector and actuating link. If rivets were drilled out, new aluminum rivets (1/4 x 0.500-inch size) can be used upon reassembly. Optionally, 1/4 - 20 screws and U-nuts can be used.

7   Check the operation of the door locks before installing the water deflector and trim panel.

26   Power window system - description and motor replacement

1   This system incorporates an electric motor and an independent control switch for each of the door windows. The driver's door has a master control switch permitting operation of all windows.

2   The electric motor which powers the window regulator is a reversible-direction motor and operates on 12 volts. It features an internal circuit breaker for protection. The motor is secured to the regulator with bolts.

3   The electrical motor can be removed from the regulator with the remainder of the window system intact only if the door glass is intact and attached to the regulator. If the door glass is broken or removed from the door, the motor must be separated after the regulator is removed from inside the door.

Glass intact and attached

4   Raise the window and remove the door trim panel and water deflector as described in Chapter 11.

5   Reach inside the door access cavity and disconnect the wiring harness at the motor.

6   It is imperative at this point that the window glass be taped or blocked in the up position. This will prevent the glass from falling into the door and possibly causing injury or damage.

7   Since the bolts used to secure the motor to the regulator are inaccessible, it is necessary to drill three large access holes in the metal door panel. The position of these holes is critical. Use the full-size templates included with the new motor. 1978 and later models have locating dimples stamped into the inner door skin to eliminate the need for templates. The template should be positioned on the door with tape after properly aligning it with the regulator attaching rivets (late models) or bolts (early models).

8   Use a centerpunch to dimple the panel at the center of the template access holes and then cut the 3/4-inch holes with a hole saw.

9   Reach in through the access hole and support the motor as the attaching bolts are removed. Remove the motor through the access hole, being careful that the window glass is firmly supported in the up position.

10   Before installation, the motor drive gear and regulator sector teeth should be lubricated.

11   Upon positioning the motor, make sure the drive gear properly engages with the regulator sector teeth. Install remaining components in the reverse order of removal. Waterproof tape can be used to seal the three access holes drilled in the metal inner panel.

Glass broken or not attached

12　Remove the window regulator as described in Chapter 11. Make sure the wiring harness to the motor is disconnected first.

13　It is imperative that the regulator sector gear be locked into position before removing the motor from the regulator. The control arms are under pressure and can cause serious injury if the motor is removed without performing the following operation.

14　Drill a hole through the regulator sector gear and backplate. Install a bolt and nut to lock the gear in position. Do not drill closer than 1/2-inch to the edge of the sector gear or backplate.

15　Remove the three motor attaching bolts and remove the motor assembly from the regulator.

16　Prior to installation, the motor drive gear and regulator sector teeth should be lubricated. The lubricant should be cold weather approved to at least -20 degrees F.

17　When installing the motor to the regulator, make sure the sector gear teeth and drive gear teeth properly mesh.

18　Once the motor attaching bolts are tightened, the locking nut and bolt can be removed. Install the regulator as described in Chapter 11. Don't forget to connect the motor wiring.

27　Ignition switch - removal and installation

1　As a precaution against theft of the vehicle, the ignition switch is located along the side of the steering column and remotely controlled by a rod and a rack assembly from the ignition lock cylinder.

1970 through 1976 models

Refer to illustration 27.3

2　To remove the ignition switch, the steer-

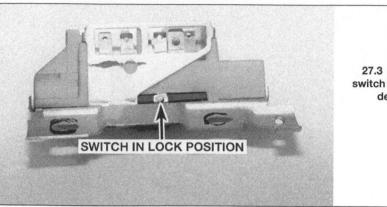

SWITCH IN LOCK POSITION

27.3 Ignition switch assembly details

ing column must either be removed or lowered and well supported. There is no need to remove the steering wheel.

3　Before removing the switch set it to the 'LOCK' position **(see illustration)**. If the lock cylinder and actuating rod have already been removed, the 'LOCK' position of the switch can be determined by inserting a screwdriver in the actuating rod slot and then moving the switch slide up until a definite stop is felt then moving it down one detent.

4　Installation is a reversal of removal but again the switch must be in the 'LOCK' position and the original or identical type securing screws must be used. Longer or thicker screws could cause the collapsible design of the steering column to become inoperative.

1977 and later models

5　On these models, the switch should be set to the 'OFF-UNLOCKED' position before removal. If the lock cylinder has already been removed, the switch actuating rod should be pulled up until a definite stop is felt then pushed down two detents.

6　Before installing the switch, set it to the 'OFF-UNLOCK' position and set the gearshift lever in neutral. Setting the switch should be

carried out in the following way. Move the switch slide two positions to the right from 'ACCESSORY' to 'OFF/UNLOCK.' Fit the actuator rod into the slide hole and assemble to the steering column using two screws. These screws must be of the original type and only tighten the lower one to 35 in-lb torque.

28　Wiring diagrams - general information

Since it isn't possible to include all wiring diagrams for every year covered by this manual, the following diagrams are those that are typical and most commonly needed.

Prior to troubleshooting any circuits, check the fuse and circuit breakers (if equipped) to make sure they're in good condition. Make sure the battery is properly charged and check the cable connections (Chapter 1).

When checking a circuit, make sure that all connectors are clean, with no broken or loose terminals. When unplugging a connector, do not pull on the wires. Pull only on the connector housings themselves.

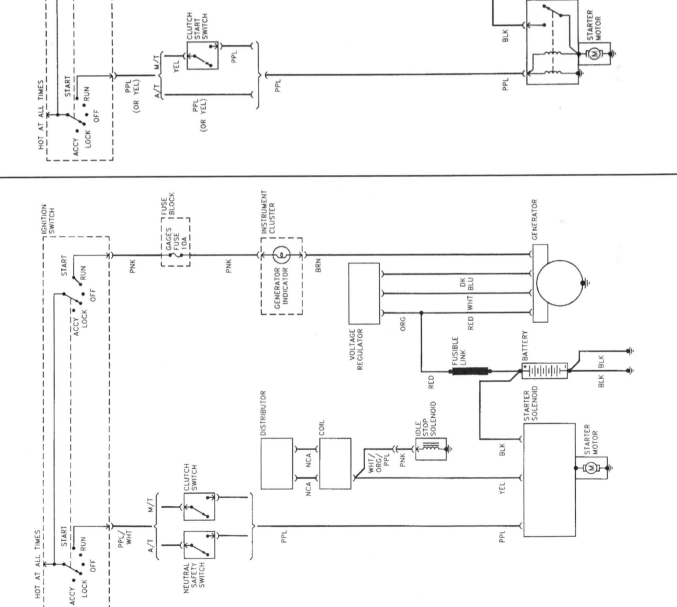

Typical charging, starting and ignition system (1978 and 1979 models)

Typical charging, starting and ignition system (1977 and earlier models)

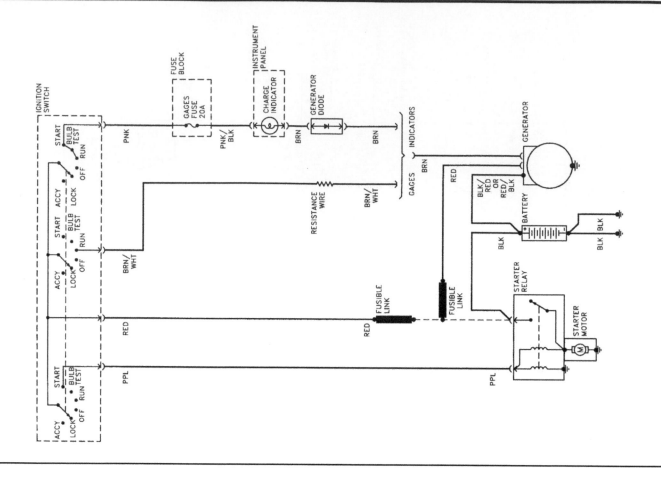

Typical charging and starting (1986 and later models)

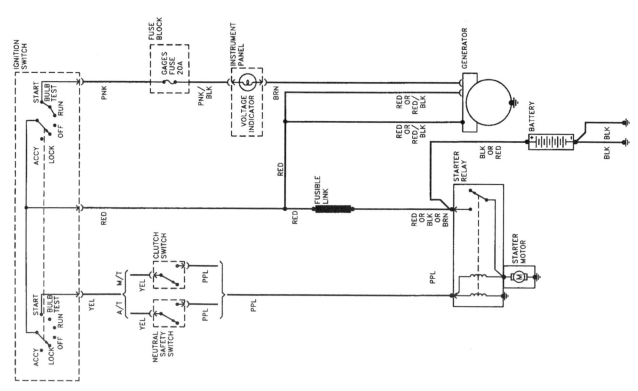

Typical charging and starting (1980 through 1985 models)

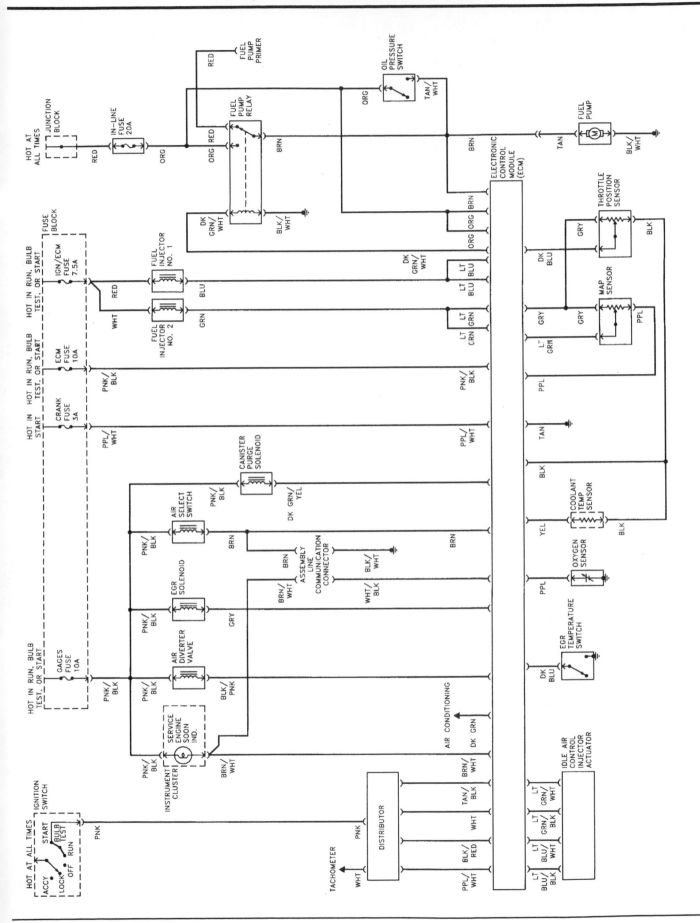

Typical electronic engine control system for V6 engines

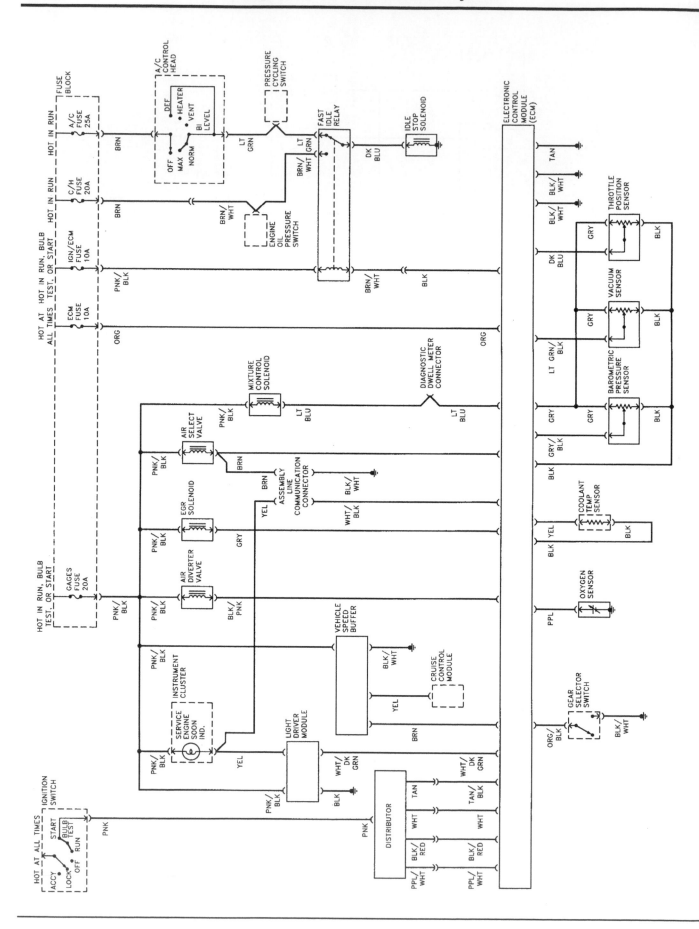

Typical electronic engine control system for V8 engines (1980 through 1987 models)

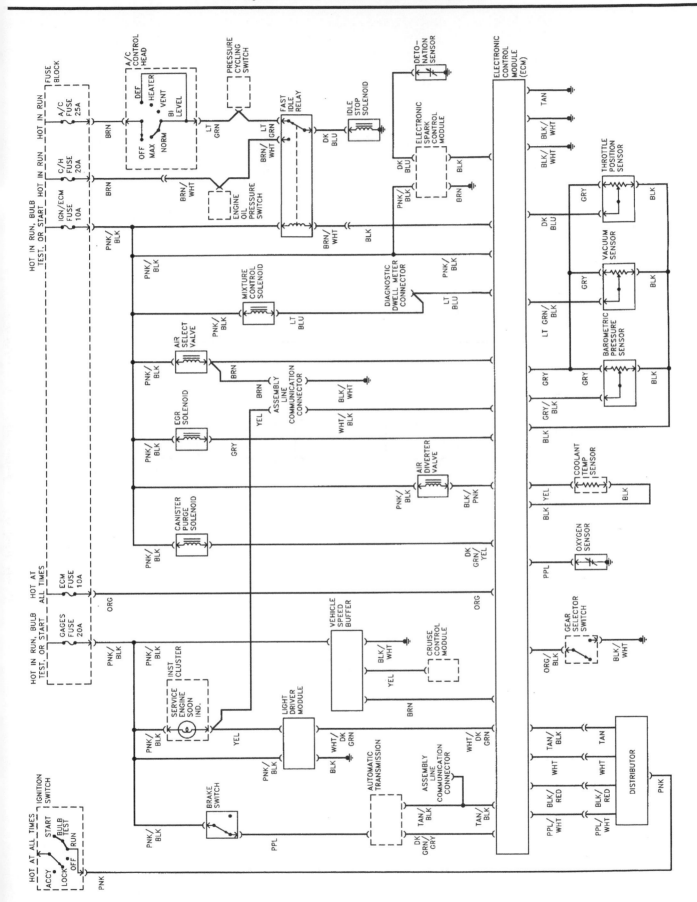

Typical electronic engine control system for V8 engines (1988 models)

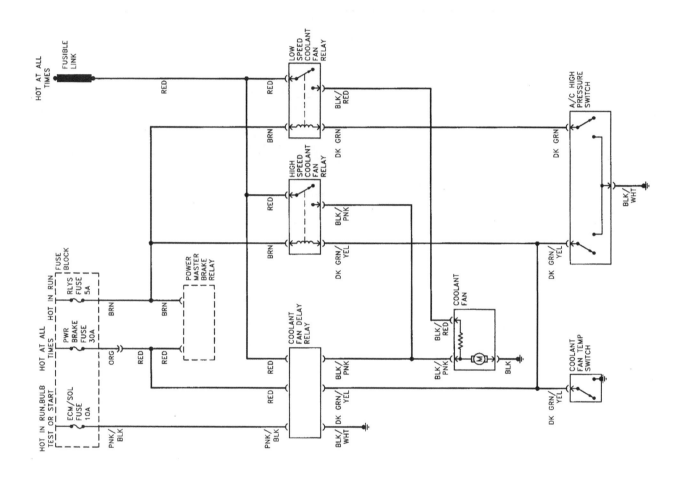

Typical engine cooling system (turbocharged engines)

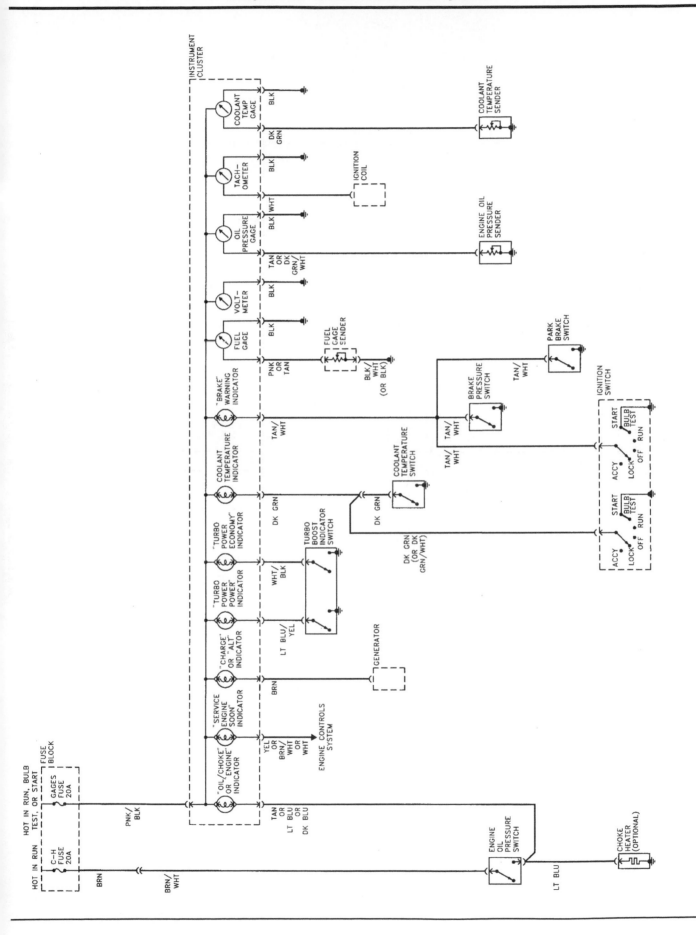

Typical instrument panel gauges and warning light system

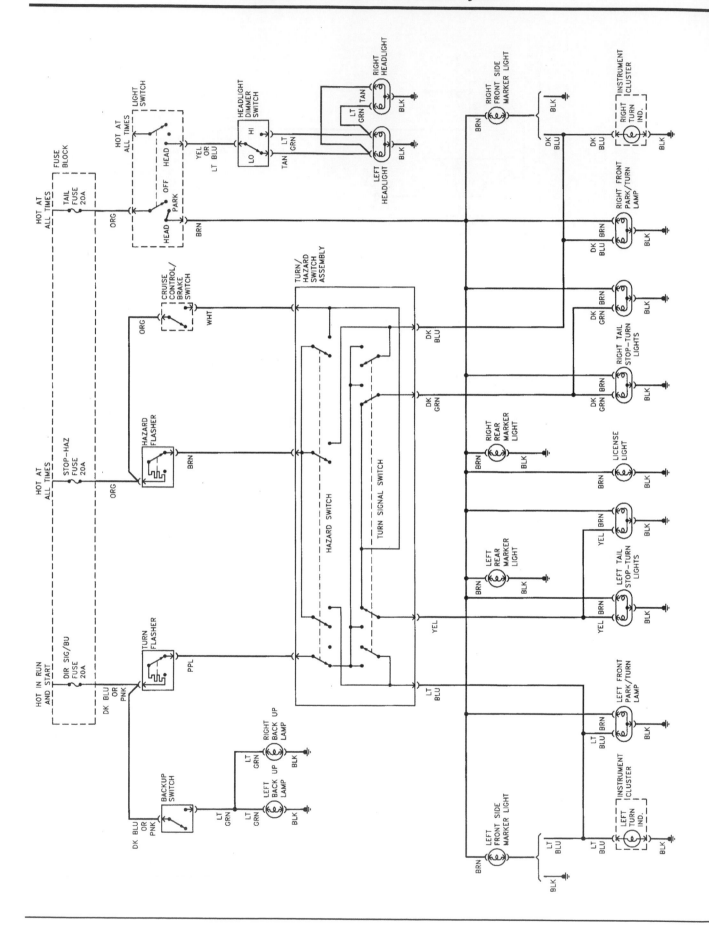

Typical exterior lighting system (1979 and earlier models)

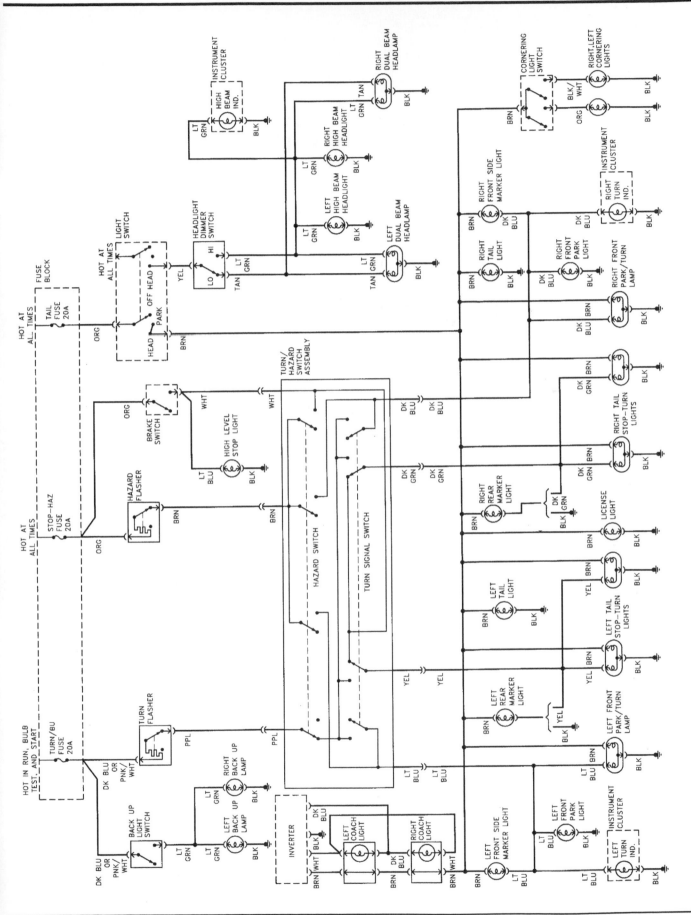

Typical exterior lighting system (1980 and later models)

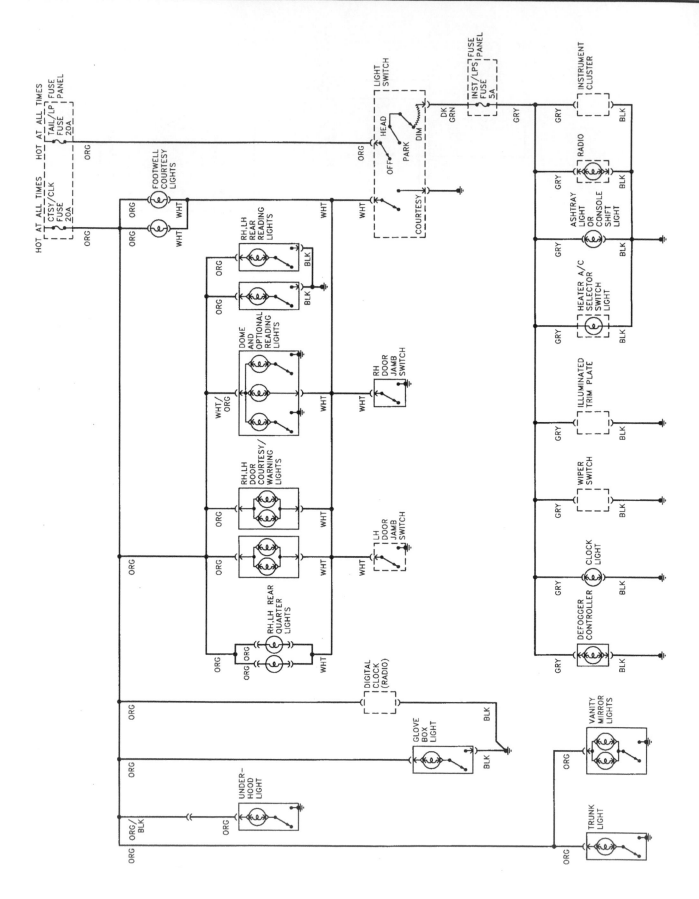

Typical interior lighting system (1985 and earlier models)

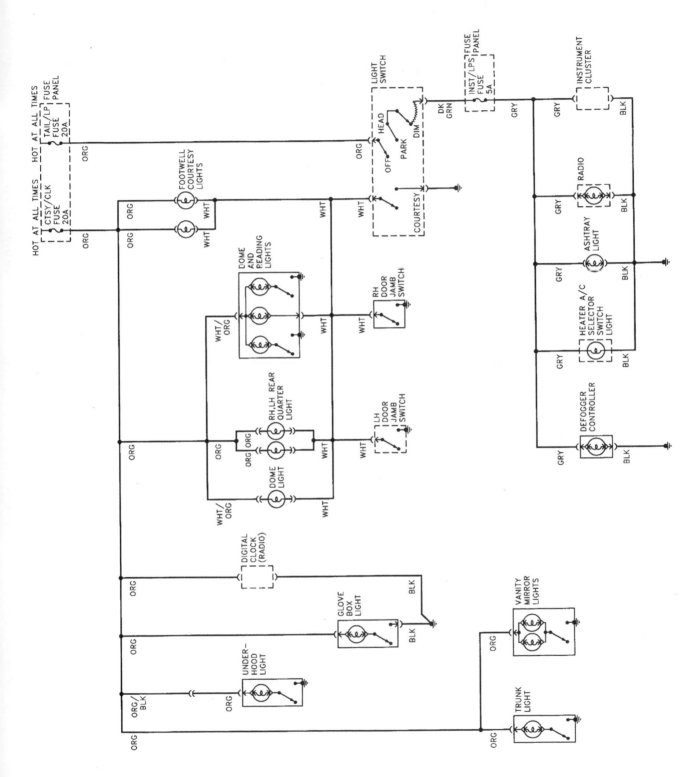

Typical interior lighting system (1986 and later models)

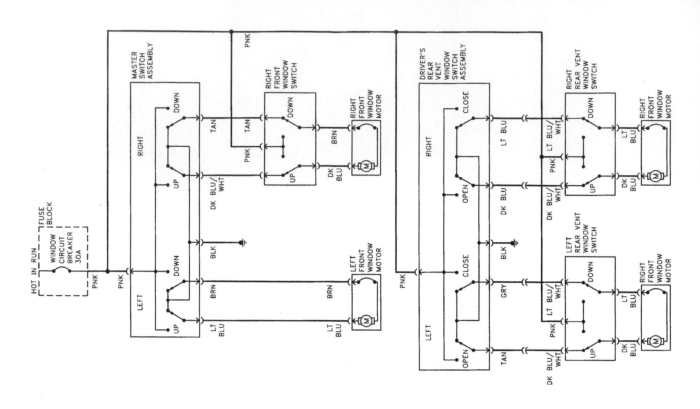

Typical power window system (1984 and later models)

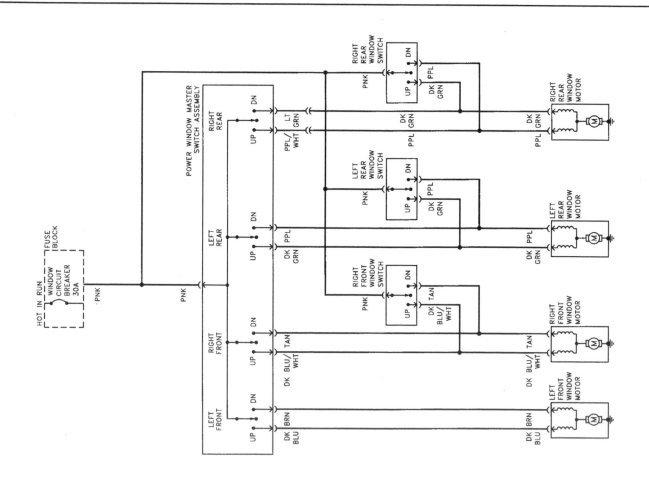

Typical power window system (1983 and earlier models)

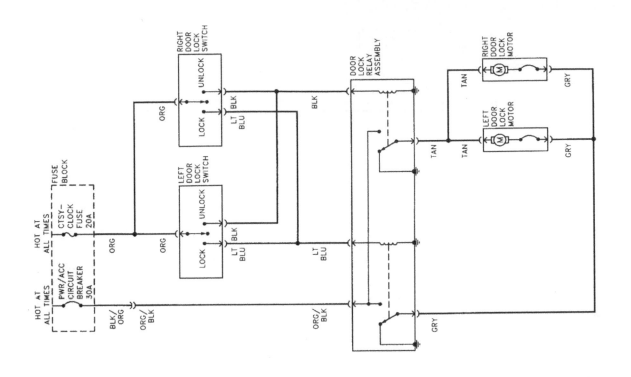

Typical power door lock system (1986 and later models)

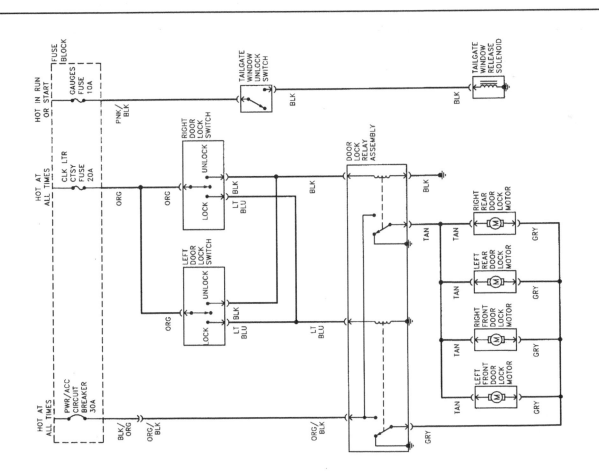

Typical power door lock system (1985 and earlier models)

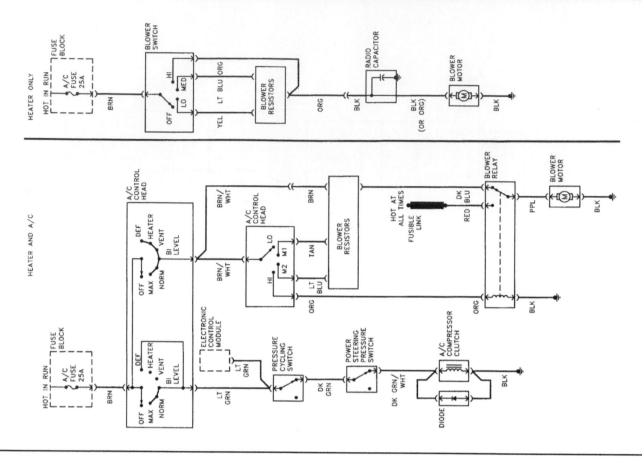

Typical heating and air conditioning system (1986 and later models)

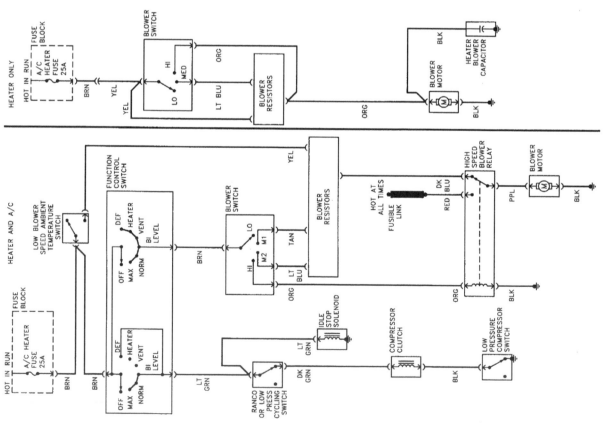

Typical heating and air conditioning system (1985 and earlier models)

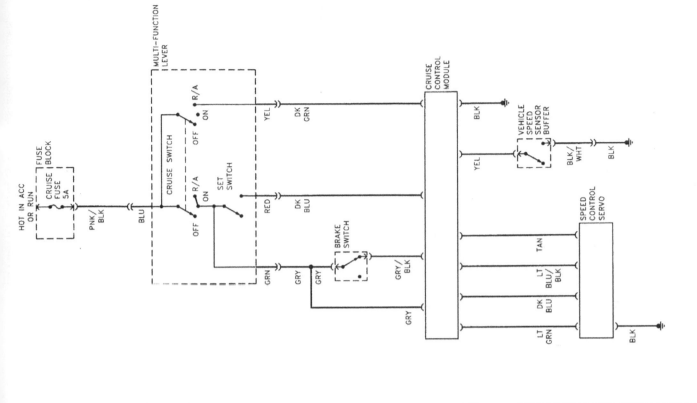

Typical cruise control system (1985 and later models)

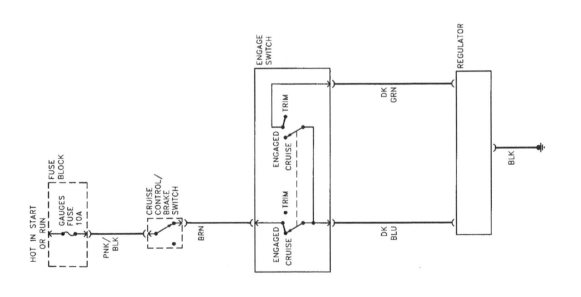

Typical cruise control system (1984 and earlier models)

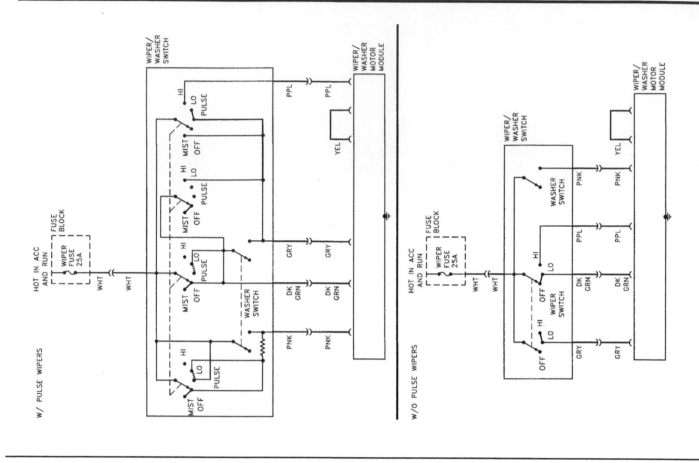

Typical windshield wiper and washer system (1985 and later models)

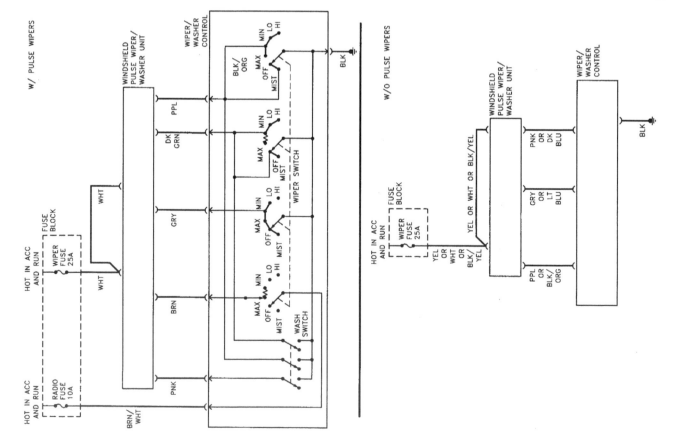

Typical windshield wiper and washer system (1984 and earlier models)

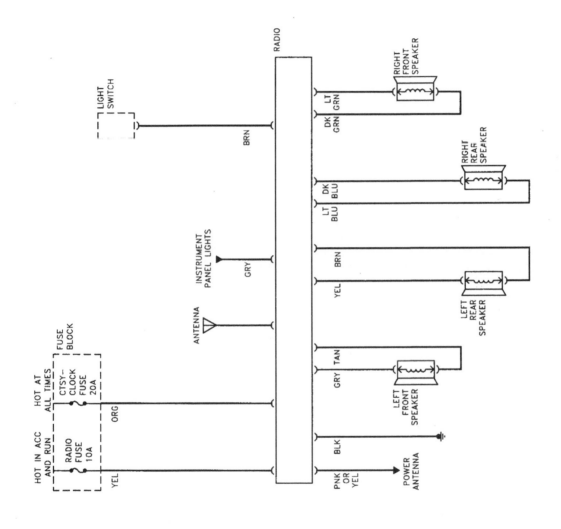

Typical audio system

Notes

Index

Haynes Automotive Manuals

NOTE: If you do not see a listing for your vehicle, please visit our website haynes.com for the latest product information.

ACURA
12020	Integra '86 thru '89 & Legend '86 thru '90
12021	Integra '90 thru '93 & Legend '91 thru '95
	Integra '94 thru '00 - see HONDA Civic (42025)
	MDX '01 thru '07 - see HONDA Pilot (42037)
12050	Acura TL all models '99 thru '08

AMC
	Jeep CJ - see JEEP (50020)
14020	Concord/Hornet/Gremlin/Spirit '70 thru '83
14025	(Renault) Alliance & Encore '83 thru '87

AUDI
15020	4000 all models '80 thru '87
15025	5000 all models '77 thru '83
15026	5000 all models '84 thru '88
	Audi A4 '96 thru '01 - see VW Passat (96023)
15030	Audi A4 '02 thru '08

AUSTIN
	Healey Sprite - see MG Midget (66015)

BMW
18020	3/5 Series '82 thru '92
18021	3 Series including Z3 models '92 thru '98
18022	3-Series incl. Z4 models '99 thru '05
18023	3-Series '06 thru '10
18025	320i all 4 cyl models '75 thru '83
18050	1500 thru 2002 except Turbo '59 thru '77

BUICK
19010	Buick Century '97 thru '05
	Century (front-wheel drive) - see GM (38005)
19020	Buick, Oldsmobile & Pontiac Full-size (Front wheel drive) '85 thru '05
19025	Buick, Oldsmobile & Pontiac Full-size (Rear wheel drive) '70 thru '90
19030	Mid-size Regal & Century '74 thru '87
	Regal - see GENERAL MOTORS (38010)
	Skyhawk - see GM (38015)
	Skylark - see GM (38020, 38025)
	Somerset - see GENERAL MOTORS (38025)

CADILLAC
21015	CTS & CTS-V '03 thru '12
21030	Cadillac Rear Wheel Drive '70 thru '93
	Cimarron, Eldorado & Seville - see GM (38015, 38030, 38031)

CHEVROLET
10305	Chevrolet Engine Overhaul Manual
24010	Astro & GMC Safari Mini-vans '85 thru '05
24015	Camaro V8 all models '70 thru '81
24016	Camaro all models '82 thru '92
	Cavalier - see GM (38015)
	Celebrity - see GM (38005)
24017	Camaro & Firebird '93 thru '02
24020	Chevelle, Malibu, El Camino '69 thru '87
24024	Chevette & Pontiac T1000 '76 thru '87
	Citation - see GENERAL MOTORS (38020)
24027	Colorado & GMC Canyon '04 thru '10
24032	Corsica/Beretta all models '87 thru '96
24040	Corvette all V8 models '68 thru '82
24041	Corvette all models '84 thru '96
24045	Full-size Sedans Caprice, Impala, Biscayne, Bel Air & Wagons '69 thru '90
24046	Impala SS & Caprice and Buick Roadmaster '91 thru '96
	Impala '00 thru '05 - see LUMINA (24048)
24047	Impala & Monte Carlo all models '06 thru '11
	Lumina '90 thru '94 - see GM (38010)
24048	Lumina & Monte Carlo '95 thru '05
	Lumina APV - see GM (38035)
24050	Luv Pick-up all 2WD & 4WD '72 thru '82
	Malibu - see GM (38026)
24055	Monte Carlo all models '70 thru '88
	Monte Carlo '95 thru '01 - see LUMINA
24059	Nova all V8 models '69 thru '79
24060	Nova/Geo Prizm '85 thru '92
24064	Pick-ups '67 thru '87 - Chevrolet & GMC
24065	Pick-ups '88 thru '98 - Chevrolet & GMC
24066	Pick-ups '99 thru '06 - Chevrolet & GMC
24067	Chevy Silverado & GMC Sierra '07 thru '12
24070	S-10 & GMC S-15 Pick-ups '82 thru '93
24071	S-10, Sonoma & Jimmy '94 thru '04
24072	Chevrolet TrailBlazer, GMC Envoy & Oldsmobile Bravada '02 thru '09
24075	Sprint '85 thru '88, Geo Metro '89 thru '01
24080	Vans - Chevrolet & GMC '68 thru '96
24081	Full-size Vans '96 thru '10

CHRYSLER
10310	Chrysler Engine Overhaul Manual
25015	Chrysler Cirrus, Dodge Stratus, Plymouth Breeze, '95 thru '00
25020	Full-size Front-Wheel Drive '88 thru '93
	K-Cars - see DODGE Aries (30008)
	Laser - see DODGE Daytona (30030)
25025	Chrysler LHS, Concorde & New Yorker, Dodge Intrepid, Eagle Vision, '93 thru '97
25026	Chrysler LHS, Concorde, 300M, Dodge Intrepid '98 thru '04
25027	Chrysler 300, Dodge Charger & Magnum '05 thru '09
25030	Chrysler/Plym. Mid-size '82 thru '95
	Rear-wheel Drive - see DODGE (30050)
25035	PT Cruiser all models '01 thru '10
25040	Chrysler Sebring '95 thru '06, Dodge Stratus '01 thru '06, Dodge Avenger '95 thru '00

DATSUN
28005	200SX all models '80 thru '83
28007	B-210 all models '73 thru '78
28009	210 all models '78 thru '82
28012	240Z, 260Z & 280Z Coupe '70 thru '78
28014	280ZX Coupe & 2+2 '79 thru '83
	300ZX - see NISSAN (72010)
28018	510 & PL521 Pick-up '68 thru '73
28020	510 all models '78 thru '81
28022	620 Series Pick-up all models '73 thru '79
	720 Series Pick-up - see NISSAN (72030)
28025	810/Maxima all gas models '77 thru '84

DODGE
	400 & 600 - see CHRYSLER (25030)
30008	Aries & Plymouth Reliant '81 thru '89
30010	Caravan & Ply. Voyager '84 thru '95
30011	Caravan & Ply. Voyager '96 thru '02
30012	Challenger/Plymouth Saporro '78 thru '83
	Challenger '67-'76 - see DART (30025)
30013	Caravan, Chrysler Voyager, Town & Country '03 thru '07
30016	Colt/Plymouth Champ '78 thru '87
30020	Dakota Pick-ups all models '87 thru '96
30021	Durango '98 & '99, Dakota '97 thru '99
30022	Durango '00 thru '03, Dakota '00 thru '04
30023	Durango '04 thru '09, Dakota '05 thru '11
30025	Dart, Challenger/Plymouth Barracuda & Valiant 6 cyl models '67 thru '76
30030	Daytona & Chrysler Laser '84 thru '89
	Intrepid - see Chrysler (25025, 25026)
30034	Dodge & Plymouth Neon '95 thru '99
30035	Omni & Plymouth Horizon '78 thru '90
30036	Dodge and Plymouth Neon '00 thru '05
30040	Pick-ups all full-size models '74 thru '93
30041	Pick-ups all full-size models '94 thru '01
30042	Pick-ups full-size models '02 thru '08
30045	Ram 50/D50 Pick-ups & Raider and Plymouth Arrow Pick-ups '79 thru '93
30050	Dodge/Ply./Chrysler RWD '71 thru '89
30055	Shadow/Plymouth Sundance '87 thru '94
30060	Spirit & Plymouth Acclaim '89 thru '95
30065	Vans - Dodge & Plymouth '71 thru '03

EAGLE
	Talon - see MITSUBISHI (68030, 68031)
	Vision - see CHRYSLER (25025)

FIAT
34010	124 Sport Coupe & Spider '68 thru '78
34025	X1/9 all models '74 thru '80

FORD
10320	Ford Engine Overhaul Manual
10355	Ford Automatic Transmission Overhaul
11500	Mustang '64-1/2 thru '70 Restoration Guide
36004	Aerostar Mini-vans '86 thru '97
	Aspire - see FORD Festiva (36030)
36006	Contour/Mercury Mystique '95 thru '00
36008	Courier Pick-up all models '72 thru '82
36012	Crown Victoria & Mercury Grand Marquis '88 thru '10
36016	Escort/Mercury Lynx '81 thru '90
36020	Escort/Mercury Tracer '91 thru '02
	Expedition - see FORD Pick-up (36059)
36022	Escape & Mazda Tribute '01 thru '11
36024	Explorer & Mazda Navajo '91 thru '01
36025	Explorer/Mercury Mountaineer '02 thru '10
36028	Fairmont & Mercury Zephyr '78 thru '83
36030	Festiva & Aspire '88 thru '97
36032	Fiesta all models '77 thru '80
36034	Focus all models '00 thru '11
36036	Ford & Mercury Full-size '75 thru '87
36044	Ford & Mercury Mid-size '75 thru '86
36045	Ford Fusion & Mercury Milan '06 thru '10
36048	Mustang V8 all models '64-1/2 thru '73
36049	Mustang II 4 cyl, V6 & V8 '74 thru '78
36050	Mustang & Mercury Capri '79 thru '93
36051	Mustang all models '94 thru '04
36052	Mustang '05 thru '10
36054	Pick-ups and Bronco '73 thru '79
36058	Pick-ups and Bronco '80 thru '96
36059	Pick-ups & Expedition '97 thru '09
36060	Super Duty Pick-up, Excursion '99 thru '10
36061	F-150 Pick-ups '04 thru '10
36062	Pinto & Mercury Bobcat '75 thru '80
36066	Probe all models '89 thru '92
	Probe '93 thru '97 - see MAZDA 626 (61042)
36070	Ranger/Bronco II gas models '83 thru '92
36071	Ford Ranger '93 thru '10 & Mazda Pick-ups '94 thru '09
36074	Taurus & Mercury Sable '86 thru '95
36075	Taurus & Mercury Sable '96 thru '01
36078	Tempo & Mercury Topaz '84 thru '94
36082	Thunderbird/Mercury Cougar '83 thru '88
36086	Thunderbird/Mercury Cougar '89 thru '97
36090	Vans all V8 Econoline models '69 thru '91
36094	Vans full size '92 thru '10
36097	Windstar Mini-van '95 thru '07

GENERAL MOTORS
10360	GM Automatic Transmission Overhaul
38005	Buick Century, Chevrolet Celebrity, Olds Cutlass Ciera & Pontiac 6000 '82 thru '96
38010	Buick Regal, Chevrolet Lumina, Oldsmobile Cutlass Supreme & Pontiac Grand Prix front wheel drive '88 thru '07
38015	Buick Skyhawk, Cadillac Cimarron, Chevrolet Cavalier, Oldsmobile Firenza Pontiac J-2000 & Sunbird '82 thru '94
38016	Chevrolet Cavalier/Pontiac Sunfire '95 thru '05
38017	Chevrolet Cobalt & Pontiac G5 '05 thru '11
38020	Buick Skylark, Chevrolet Citation, Olds Omega, Pontiac Phoenix '80 thru '85
38025	Buick Skylark & Somerset, Olds Achieva, Calais & Pontiac Grand Am '85 thru '98
38026	Chevrolet Malibu, Olds Alero & Cutlass, Pontiac Grand Am '97 thru '03
38027	Chevrolet Malibu '04 thru '10
38030	Cadillac Eldorado & Oldsmobile Toronado '71 thru '85, Seville '80 thru '85, Buick Riviera '79 thru '85
38031	Cadillac DeVille & Seville '86 thru '91, DeVille & Buick Riviera '86 thru '93, Fleetwood & Olds Toronado '86 thru '92
38032	DeVille '94 thru '05, Seville '92 thru '04 Cadillac DTS '06 thru '10
38035	Chevrolet Lumina APV, Olds Silhouette & Pontiac Trans Sport '90 thru '96
38036	Chevrolet Venture, Olds Silhouette, Pontiac Trans Sport & Montana '97 thru '05
	GM Full-size RWD - see BUICK (19025)
38040	Chevrolet Equinox '05 thru '09 Pontiac Torrent '06 thru '09
38070	Chevrolet HHR '06 thru '11

GEO
	Metro - see CHEVROLET Sprint (24075)
	Prizm - see CHEVROLET (24060) or TOYOTA (92036)
40030	Storm all models '90 thru '93
	Tracker - see SUZUKI Samurai (90010)

GMC
	Vans & Pick-ups - see CHEVROLET

HONDA
42010	Accord CVCC all models '76 thru '83
42011	Accord all models '84 thru '89
42012	Accord all models '90 thru '93
42013	Accord all models '94 thru '97
42014	Accord all models '98 thru '02
42015	Accord '03 thru '07
42020	Civic 1200 all models '73 thru '79
42021	Civic 1300 & 1500 CVCC '80 thru '83
42022	Civic 1500 CVCC all models '75 thru '79
42023	Civic all models '84 thru '91
42024	Civic & del Sol '92 thru '95
42025	Civic '96 thru '00, CR-V '97 thru '01, Acura Integra '94 thru '00
42026	Civic '01 thru '10, CR-V '02 thru '09
42035	Odyssey all models '99 thru '10
	Passport - see ISUZU Rodeo (47017)

HYUNDAI
43010	Elantra all models '96 thru '10
43015	Excel & Accent all models '86 thru '09
43050	Santa Fe all models '01 thru '06
43055	Sonata all models '99 thru '08

INFINITI
	G35 '03 thru '08 - see NISSAN 350Z (72011)

ISUZU
	Hombre - see CHEVROLET S-10 (24071)
47017	Rodeo, Amigo & Honda Passport '89 thru '02
47020	Trooper '84 thru '91, Pick-up '81 thru '93

JAGUAR
49010	XJ6 all 6 cyl models '68 thru '86
49011	XJ6 all models '88 thru '94
49015	XJ12 & XJS all 12 cyl models '72 thru '85

JEEP
50010	Cherokee, Comanche & Wagoneer Limited all models '84 thru '01
50020	CJ all models '49 thru '86
50025	Grand Cherokee all models '93 thru '04
50026	Grand Cherokee '05 thru '09
50029	Grand Wagoneer & Pick-up '72 thru '91
50030	Wrangler all models '87 thru '11
50035	Liberty '02 thru '07

KIA
54050	Optima '01 thru '10
54070	Sephia '94 thru '01, Spectra '00 thru '09, Sportage '05 thru '10

LEXUS
	ES 300/330 - see TOYOTA Camry (92007) (92008)
	RX 330 - see TOYOTA Highlander (92095)

LINCOLN
	Navigator - see FORD Pick-up (36059)
59010	Rear Wheel Drive all models '70 thru '10

MAZDA
61010	GLC (rear wheel drive) '77 thru '83
61011	GLC (front wheel drive) '81 thru '85
61012	Mazda3 '04 thru '11
61015	323 & Protegé '90 thru '03
61016	MX-5 Miata '90 thru '09
61020	MPV all models '89 thru '98
	Navajo - see FORD Explorer (36024)
61030	Pick-ups '72 thru '93
	Pick-ups '94 on - see Ford (36071)
61035	RX-7 all models '79 thru '85
61036	RX-7 all models '86 thru '91
61040	626 (rear wheel drive) '79 thru '82
61041	626 & MX-6 (front wheel drive) '83 thru '92
61042	626 '93 thru '01, & MX-6/Ford Probe '93 thru '02
61043	Mazda6 '03 thru '11

MERCEDES-BENZ
63012	123 Series Diesel '76 thru '85
63015	190 Series 4-cyl gas models, '84 thru '88
63020	230, 250 & 280 6 cyl sohc '68 thru '72
63025	280 123 Series gas models '77 thru '81
63030	350 & 450 all models '71 thru '80
63040	C-Class: C230/C240/C280/C320/C350 '01 thru '07

MERCURY
64200	Villager & Nissan Quest '93 thru '01
	All other titles, see FORD listing.

MG
66010	MGB Roadster & GT Coupe '62 thru '80
66015	MG Midget & Austin Healey Sprite Roadster '58 thru '80

MINI
67020	Mini '02 thru '11

MITSUBISHI
68020	Cordia, Tredia, Galant, Precis & Mirage '83 thru '93
68030	Eclipse, Eagle Talon & Plymouth Laser '90 thru '94
68031	Eclipse '95 thru '05, Eagle Talon '95 thru '98
68035	Galant '94 thru '10
68040	Pick-up '83 thru '96, Montero '83 thru '93

NISSAN
72010	300ZX all models incl. Turbo '84 thru '89
72011	350Z & Infiniti G35 all models '03 thru '08
72015	Altima all models '93 thru '06
72016	Altima '07 thru '10
72020	Maxima all models '85 thru '92
72021	Maxima all models '93 thru '01
72025	Murano '03 thru '10
72030	Pick-ups '80 thru '97, Pathfinder '87 thru '95
72031	Frontier Pick-up, Xterra, Pathfinder '96 thru '04
72032	Frontier & Xterra '05 thru '11
72040	Pulsar all models '83 thru '86
72050	Sentra all models '82 thru '94
72051	Sentra & 200SX all models '95 thru '06
72060	Stanza all models '82 thru '90
72070	Titan pick-ups '04 thru '10, Armada '05 thru '10

OLDSMOBILE
73015	Cutlass '74 thru '88
	For other OLDSMOBILE titles, see BUICK, CHEVROLET or GM listings.

PLYMOUTH
	For PLYMOUTH titles, see DODGE.

PONTIAC
79008	Fiero all models '84 thru '88
79018	Firebird V8 models except Turbo '70 thru '81
79019	Firebird all models '82 thru '92
79025	G6 all models '05 thru '09
79040	Mid-size Rear-wheel Drive '70 thru '87
	Vibe '03 thru '11 - see TOYOTA Matrix (92060)
	For other PONTIAC titles, see BUICK, CHEVROLET or GM listings.

PORSCHE
80020	911 Coupe & Targa models '65 thru '89
80025	914 all 4 cyl models '69 thru '76
80030	924 all models incl. Turbo '76 thru '82
80035	944 all models incl. Turbo '83 thru '89

RENAULT
	Alliance, Encore - see AMC (14020)

SAAB
84010	900 including Turbo '79 thru '88

SATURN
87010	Saturn all S-series models '91 thru '02
87011	Saturn Ion '03 thru '07
87020	Saturn all L-series models '00 thu '04
87040	Saturn VUE '02 thru '07

SUBARU
89002	1100, 1300, 1400 & 1600 '71 thru '79
89003	1600 & 1800 2WD & 4WD '80 thru '94
89100	Legacy models '90 thru '99
89101	Legacy & Forester '00 thru '06

SUZUKI
90010	Samurai/Sidekick/Geo Tracker '86 thru '01

TOYOTA
92005	Camry all models '83 thru '91
92006	Camry all models '92 thru '96
92007	Camry/Avalon/Solara/Lexus ES 300 '97 thru '01
92008	Toyota Camry, Avalon and Solara & Lexus ES 300/330 all models '02 thru '06
92009	Camry '07 thru '11
92015	Celica Rear Wheel Drive '71 thru '85
92020	Celica Front Wheel Drive '86 thru '99
92025	Celica Supra all models '79 thru '92
92030	Corolla all models '75 thru '79
92032	Corolla rear wheel drive models '80 thru '87
92035	Corolla front wheel drive models '84 thru '92
92036	Corolla & Geo Prizm '93 thru '02
92037	Corolla models '03 thru '11
92040	Corolla Tercel all models '80 thru '82
92045	Corona all models '74 thru '82
92050	Cressida all models '78 thru '82
92055	Land Cruiser FJ40/43/45/55 '68 thru '82
92056	Land Cruiser FJ60/62/80/FZJ80 '80 thru '96
92060	Matrix & Pontiac Vibe '03 thru '11
92065	MR2 all models '85 thru '87
92070	Pick-up all models '69 thru '78
92075	Pick-up all models '79 thru '95
92076	Tacoma, 4Runner & T100 '93 thru '04
92077	Tacoma all models '05 thru '09
92078	Tundra '00 thru '06, Sequoia '01 thru '07
92079	4Runner all models '03 thru '09
92080	Previa all models '91 thru '95
92081	Prius '01 thru '08
92082	RAV4 all models '96 thru '10
92085	Tercel all models '87 thru '94
92090	Sienna all models '98 thru '09
92095	Highlander & Lexus RX-330 '99 thru '07

TRIUMPH
94007	Spitfire all models '62 thru '81
94010	TR7 all models '75 thru '81

VW
96008	Beetle & Karmann Ghia '54 thru '79
96009	New Beetle '98 thru '11
96016	Rabbit, Jetta, Scirocco, & Pick-up gas models '75 thru '92 & Convertible '80 thru '92
96017	Golf, GTI & Jetta '93 thru '98, Cabrio '95 thru '02
96018	Golf, GTI & Jetta '98 thru '05
96019	Jetta, Rabbit, GTI & Golf '05 thru '11
96020	Rabbit, Jetta, Pick-up diesel '77 thru '84
96023	Passat '98 thru '05, Audi A4 '96 thru '01
96030	Transporter 1600 all models '68 thru '79
96035	Transporter 1700, 1800, 2000 '72 thru '79
96040	Type 3 1500 & 1600 '63 thru '73
96045	Vanagon air-cooled models '80 thru '83

VOLVO
97010	120, 130 Series & 1800 Sports '61 thru '73
97015	140 Series all models '66 thru '74
97020	240 Series all models '76 thru '93
97040	740 & 760 Series all models '82 thru '88

TECHBOOK MANUALS
10205	Automotive Computer Codes
10206	OBD-II & Electronic Engine Management
10210	Automotive Emissions Control Manual
10215	Fuel Injection Manual, '78 thru '85
10220	Fuel Injection Manual, '86 thru '99
10225	Holley Carburetor Manual
10230	Rochester Carburetor Manual
10240	Weber/Zenith/Stromberg/SU Carburetor
10305	Chevrolet Engine Overhaul Manual
10310	Chrysler Engine Overhaul Manual
10320	Ford Engine Overhaul Manual
10330	GM and Ford Diesel Engine Repair
10333	Engine Performance Manual
10340	Small Engine Repair Manual
10345	Suspension, Steering & Driveline
10355	Ford Automatic Transmission Overhaul
10360	GM Automatic Transmission Overhaul
10405	Automotive Body Repair & Painting
10410	Automotive Brake Manual
10415	Automotive Detailing Manual
10420	Automotive Electrical Manual
10425	Automotive Heating & Air Conditioning
10430	Automotive Reference Dictionary
10435	Automotive Tools Manual
10440	Used Car Buying Guide
10445	Welding Manual
10450	ATV Basics
10452	Scooters 50cc to 250cc

SPANISH MANUALS
98903	Reparación de Carrocería & Pintura
98904	Manual de Carburador Modelos Holley & Rochester
98905	Códigos Automotrices de la Computadora
98906	OBD-II & Sistemas de Control Electrónico del Motor
98910	Frenos Automotriz
98913	Electricidad Automotriz
98915	Inyección de Combustible '86 al '99
99040	Chevrolet & GMC Camionetas '67 al '87
99041	Chevrolet & GMC Camionetas '88 al '98
99042	Chevrolet Camionetas Cerradas '68 al '95
99043	Chevrolet/GMC Camionetas '94 al '04
99048	Chevrolet/GMC Camionetas '99 al '06
99055	Dodge Caravan/Ply. Voyager '84 al '95
99075	Ford Camionetas y Bronco '80 al '94
99076	Ford F-150 '97 al '09
99077	Ford Camionetas Cerradas '69 al '91
99088	Ford Modelos de Tamaño Mediano '75 al '86
99089	Ford Camionetas Ranger '93 al '10
99091	Ford Taurus & Mercury Sable '86 al '95
99095	GM Modelos de Tamaño Grande '70 al '90
99100	GM Modelos de Tamaño Mediano '70 al '88
99106	Jeep Cherokee, Wagoneer & Comanche '84 al '00
99110	Nissan Camionetas & Pathfinder '80 al '96
99118	Nissan Sentra '82 al '94
99125	Toyota Camionetas y 4-Runner '79 al '95

Over 100 Haynes motorcycle manuals also available

5/22

Haynes North America, Inc., 859 Lawrence Drive, Newbury Park, CA 91320 • (805) 498-6703 • www.haynes.com